McFarlin Library
WITHDRAWN

World Cattle III

Cattle of North America

UNIVERSITY OF OKLAHOMA PRESS : NORMAN

World Cattle III

Cattle of North America

by John E. Rouse

By John E. Rouse

World Cattle, I: *Cattle of Europe, South America, Australia, and New Zealand* (Norman, 1970)
World Cattle, II: *Cattle of Africa and Asia* (Norman, 1970)
World Cattle, III: *Cattle of North America* (Norman, 1973)

Library of Congress Cataloging in Publication Data

Rouse, John E.
    Cattle of North America.

    (His World cattle, 3)
    Bibliography: p:
    1. Cattle—America. I. Title.
SF196.A43R68   636.2'0097   72–861

Copyright 1973 by the University of Oklahoma Press, Publishing Division of the University. Composed and printed at Norman, Oklahoma, U.S.A., by the University of Oklahoma Press. First edition.

Preface

Cattle of North America is the third volume of WORLD CATTLE and completes the task begun in 1962 to present in one work an illustrated conspectus of the many kinds of cattle in the world and the ways in which they are husbanded. The same eyewitness approach was followed for this volume as was employed in gathering the material for the first two volumes. This involved three years of travel throughout the countries of North America and the surrounding areas. Representative cattle in all these countries were seen, and the methods of handling them were discussed with ranchers, farmers, dairymen, packing-plant operators, breed associations, and livestock specialists in universities and experiment stations.

The illustrations are all from photographs taken by either my wife or me except those of historical nature where the source is shown. The effort was to obtain pictures that are fair representations of the cattle common to a country and to avoid specially fitted show types. Meaningless duplication of illustrations of the same breed in different countries has been avoided.

The small nations of the Caribbean area were reached by air. Travel through all the other countries, from Panama through Canada, was by automobile, logging over 75,000 miles. This required several trips across the United States to visit every state in the Union at least once, and twice across Canada to see the cattle in every province.

Obviously an author writes of those matters that are of interest to him, hoping that his subject will appeal to his readers. My concern with cattle dates from the closing years of the past century when I was sent to summer on a central Illinois farm. My indoctrination to rural life was by way of a small Jersey cow that I milked twice a day. I can now see that she was a distinctive member of a small herd of what will later be described as Native American cattle. There were fourteen cows in all. My Jersey and two nearly all red cows, which my older cousins referred to as Durhams, were the only animals which, as I look back on them, showed any particular breed characteristics. There was a large blue roan which nearly filled a three-gallon milk pail at every milking, and a red-and-white spotted cow with strawberry-roan forequarters that was also an exceptional milker. The rest of the herd, quite nondescript, I now realize were fair representatives of Native cattle.

My familiarity with these cows lasted throughout the grade school years and into the first term in high school. In the final year of my association with them an Angus bull calf was purchased as a future sire. The advantages of a polled breed and of the Angus in particular were well advertised in the *Breeders Gazette*, and, I am sure, had much to do with the choice of this bull. My early farming days, however, were terminated before his progeny were calved.

In retrospect, this herd holds an absorbing interest for me as I write. Here were representatives of two of the first breeds to be established in the United States, a Jersey and two Shorthorns, and the other eleven were the now-forgotten Native cattle. Further, here upgrading to the Angus breed was being initiated, which regretfully I failed to see consummated.

Some thirty years after my boyhood farm days, I managed to acquire a farm with a small dairy herd on the edge of the Chicago milkshed. This was at the time when the controversy over the brucellosis vaccination program reached fever pitch. I followed the advocates of the no-vaccination policy and ended up losing my 40-cow herd. That catastrophe led to a 250-head steer-feeding operation. The concluding step was taken twenty years ago when I traded the Corn Belt farm for a Wyoming mountain ranch and began to develop an Angus herd.

Such varied ties to the industry over three-quarters of a century have intensified my interest in cattle and also have provided a background for the observations that are recorded here on man's major source of milk and meat and, until quite recently, draft power.

This volume covers 19 countries and brings to 105 the total number in which I have seen the cattle and talked with the people who cared for them. Cuba was the last country to be visited. Just as the completed manuscript went to the publisher, a unique opportunity to visit this country presented itself. Since Cuba has been off limits to the United

Preface

States citizen for over a decade, happenings there hold an absorbing interest for him today. A major cattle-improvement program was discovered under way there which has therefore been presented in some detail.

My wife, Roma, accompanied me on all these travels. She typed the story of the cattle in each country as we went along. Her criticism added greatly to the final composition. As with the first two volumes, this is also as much her work as mine. My sister, Mary Jane Rouse, spent many a long day in the final arrangement of the text and illustrations.

I am indebted to many other people who have helped me in many ways in putting the book together. They cover a wide range of human activities. A widow in the mountains of Guatemala left her churning to drive in her herd of ten cows so that I might photograph them; the chairman of the animal science department of a major land-grant college gave a day of his time and a tour of his experiment station. With such a broad spectrum of personal assistance the most meaningful acknowledgment I can make is to say that the book is actually the product of the many people who have helped in countless ways in accumulating and compiling the material. In more specific acknowledgment I express my thanks to the following:

Charles R. Koch, the widely known livestock editor of Oxford, Ohio, edited the entire manuscript for Volume III and added immeasurably to its substance and accuracy. In a few concepts we have not seen eye to eye, and he should in no way be held responsible for matters of opinion which have crept into the text.

Dr. Ralph W. Phillips, Director, International Organizations Staff of the USDA, offered a word of encouragement from time to time, made helpful comments, and furnished me with copies of his wide range of writings on many kinds of cattle.

Four door openers paved the way for my travels through the various areas of North America. Such introductions added immeasurably to the co-operation that was accorded me in the many places that I visited. For them I am indebted to:

W. D. Farr, Past President of the American National Cattlemen's Association, who directed me to many cattlemen in all parts of the United States and also reviewed some parts of the book.

Horace J. Davis, of the Foreign Agricultural Service of the USDA, who arranged with the agricultural attachés and economic officers of the United States embassies in all the countries visited to lend me a helping hand.

Cattle of North America

G. L. Locking, of the Livestock Division, Production and Marketing Branch, Canada Department of Agriculture, who planned an itinerary throughout Canada and put me in touch with the Livestock Division representative in each province.

Dr. Everett J. Warwick, Agricultural Research Service, USDA, who arranged visits for me with specialists in various branches of the Agricultural Research Service and also helped with pertinent comment in many fields of animal husbandry.

The various chapters in the section on the United States were reviewed by an authority in the particular field covered. The sections on foreign countries were reviewed by an authority on cattle husbandry in the general area. To the following who served in this regard I express my appreciation for their timely comments and suggestions and reiterate that any expression involving matters of opinion are mine alone.

Caribbean

James Armstrong, Dominican Republic
Steve Bennett, Trinidad and Tobago
R. V. Billingsley, College Station, Texas
H. G. Byron, Surinam
Raowl De Paz, Martinique
C. Gaztambide, Puerto Rico
Gustave Menager, Haiti
G. E. Redshaw, Jamaica
R. A. Wilkins, Guyana

Central America

Byron K. Montgomery, San José, Costa Rica

Mexico

Manuel Ibarguengoitia, Zacatecas   William L. Rodman, Mexico City

Canada

Ralph K. Bennett, Ottawa

United States

G. E. Blake, Washington, D.C.
J. S. Brinks, Fort Collins, Colorado
H. A. Hancock, Laramie, Wyoming
J. W. Oxley, Fort Collins, Colorado
L. S. Pope, College Station, Texas
Lew Sullivan, Greeley, Colorado
John C. Todd, Greeley, Colorado
Richard S. Willham, Ames, Iowa

At Large

John P. Maule, Edinburgh, Scotland

JOHN E. ROUSE

Saratoga, Wyoming
July 15, 1972

Introduction

THE SAME FORMAT has been followed in the preparation of *The Cattle of North America* as was employed in the first two volumes of WORLD CATTLE. This embodies a thumbnail sketch of each country, followed by a presentation of the breeds found in the country, marketing procedures, abattoirs or packing plants, cattle-disease problems, governmental policies involving the cattle industry, and the outlook for cattle.

The countries covered in this volume are the island nations of the Caribbean area, in which section the Guianas, located in the northeast corner of South America, have been included; Middle America, which comprises all the countries from Panama to Mexico; and Canada and the United States.

Collection of the data for the compilation of the three volumes of WORLD CATTLE began in the fall of 1962 and was completed in the spring of 1972. This decade witnessed major changes in cattle husbandry and the activities allied to it in Canada and the United States.

"Exotic" or "new" breeds were designations that were not used in 1962. Today, nearly any discussion involving beef cattle sooner or later includes mention of the European breeds which for the past few years have been imported to Canada and thence to the United States and which are now starting to come direct to the United States.

Crossbreeding has been known and employed in varying degrees ever since purebred cattle were first imported to the United States. Intermixing different breeds has been a common practice, particularly in the South, for more than half a century. But during the past few years crossbreeding has become a byword across the land largely because of the impact of the new breeds.

The immediate goal in writing WORLD CATTLE was to provide cattlemen in the United States with a ready reference to the various kinds of cattle and the ways they are handled in the rest of the world. Particular emphasis was placed on the unusual, such as little-known breeds and unique husbandry practices.

The history of cattle raising in the United States involves a complex pattern of local cattle populations in the various areas that eventually became a part of the nation. Only a hundred years ago "Native cattle" were recognized as a definite type in the United States by authorities who wrote at that time. A reference to Native cattle today brings forth the comment that there were *no* cattle native to North America. The mid-nineteenth-century cattle were the melting-pot product of the various breeds brought to America in previous years—a native cattle which have since been bred out and forgotten. The Piedmont of the Carolinas was the first open-range cattle country in the United States with the equivalent in all but name of later-day ranches and cowboys. Some features of such bovine history are presented here.

Throughout the world the people who are involved in various aspects of the cattle and dairy industries have an intense interest in what the United States is doing with its herds. The director of an experiment station in Senegal asks what breeds of cattle do best in the southern United States, or an Israeli professor wants details on the United States beef-grading system.

In order to provide a reference to the wide spectrum of cattle husbandry in North America—the new breeds and crossbreeding, the historical aspects, and a detailed account of the many phases of the cattle industry as they exist today—the sections on the United States and Canada were expanded to give a broader account of activities than was done for other countries.

The United States has more types and breeds of cattle, and more of their derivatives, than any other area of the world. Canada is a close second but lacks the Zebu, which is an important element in the cattle population of the southern states. These two countries are now involved in the most widespread diversification of breeds that has ever been attempted. Only the future can determine the outcome of this intermingling of many breeds. The elements from which this movement is evolving are presented here.

Contents

	Volume III	
Preface		*page* v
Introduction		ix
NORTH AMERICA		3
Part I:	THE CARIBBEAN	7
	Barbados	17
	Cuba	23
	Dominican Republic	55
	Haiti	71
	Jamaica	81
	Martinique	99
	Trinidad and Tobago	105
	The Guianas	
	French Guiana	117
	Guyana	121
	Surinam	135
Part II:	MIDDLE AMERICA	147
	Central America	149
	Costa Rica	155
	El Salvador	169
	Guatemala	181

Cattle of North America

	Honduras	193
	Nicaragua	207
	Panama	219
	Mexico	229
Part III:	CANADA	259
	Cattle Breeds	266
	Management Practices	307
	Marketing	325
	Packing Plants	332
	Cattle Diseases	333
	Government and Cattle	335
	Outlook for Cattle	338
Part IV:	UNITED STATES	343
	Background of the Cattle Industry in the United States	348
	Sources of Cattle	349
	Early Cattle Populations	358
	Cattle Breeds	378
	Management Practices	485
	Marketing	578
	Packing Plants	602
	Cattle Diseases	609
	Government and Cattle	618
	Outlook for Cattle	620

Appendix	
Human and Cattle Populations of North America	631
United States Cattle Population	632
Breed Associations in Canada	634
Breed Associations in the United States	635
Bibliography	637
Index	643

Figures and maps

The countries of the Caribbean	8
Sixteenth-century routes of the Spanish galleons through the Caribbean	12
Cuba	24
Middle America	146
Mexico	230
Canada	260
Land acquisitions which formed the United States	345
The six major cattle-raising regions of the United States	508
Precipitation map of the United States	510
Mean temperature charts of the United States, January and July	511
Physiographic map of the United States	512
Alaska	556
Hawaii	563

NORTH AMERICA

North America

THE CONTINENT OF NORTH AMERICA includes all of the land mass north of the Colombian-Panamanian border, which lies at latitude 7 degrees north, and all of the surrounding islands. Point Barrow in Alaska at latitude 71 degrees north is the northern-most extremity. The southern part of the continent, from the United States–Mexican border to Colombia, is commonly referred to as Middle America and includes Mexico, the countries of Central America, and Panama. The Caribbean Islands, the West Indies, considered a part of Middle America by some authorities, are discussed in a separate section in this book.

The basic physiographic features are the Canadian Shield in the northeast and the Appalachian Mountains in the east; the Central Lowlands which extend from the Gulf of Mexico to the Arctic Ocean; and the Western Cordilleras, which are nearly continuous from Panama through Alaska.

The early aborigines migrated from Asia to Alaska when these areas were connected by now submerged lands or ice masses. These people were the ancestors of the Indian population which spread southward and eastward across the continent. None of the Indian tribes had domesticated livestock. The only domesticated animal prior to the advent of the European was the dog.

The first Europeans to reach North America were the Norsemen around the year A.D. 1000. Bjarni Herjolfsson is credited by some sources as reaching the continent in A.D. 986. These early settlers had disappeared before the arrival of Columbus in the West Indies in 1492.

Cattle of North America

Following the landings by Columbus on the islands of the Caribbean, the Spaniards proceeded rapidly with exploration and settlement in both North and South America. During the sixteenth century, strongholds were established in Cuba and other Caribbean Islands, in Mexico, throughout Central America, and in Florida. The other European nations made no major effort to obtain a foothold in the New World until the opening of the seventeenth century.

The French founded Port Royal in Nova Scotia in 1605; the British settled Jamestown in 1607; and the Dutch were on Manhattan Island in 1609. A Swedish colony was soon established in Delaware. Nothing substantial had come of earlier efforts of the European powers to settle along the Atlantic coast. Throughout the seventeenth century the French and British extended their claims, and the Dutch and Swedish influence disappeared.

At the opening of the eighteenth century the continent was in the hands of the Spaniards, the French, and the British. Spain held claim to Mexico, the Gulf of Mexico coast from Florida to the Mississippi River, and the vast region of what is now the southwestern United States. The French and British were both in Canada. The British controlled the Atlantic seaboard, and the French claimed most of the Mississippi River drainage. Spain, Britain, France, and the Netherlands all were represented in the Caribbean Islands, which in this period were considered the most desirable region in North America.

In 1743 the Russians entered Alaska from Siberia and founded substantial settlements to handle their fur trade. They did not progress far from the coastal areas and sold their claims to the United States in 1867.

During the nineteenth century the fortunes of war and the wave of independence which spread across the Western Hemisphere resulted in the political entities which now encompass the North American continent. With the exception of a minor British holding in Central America, all of the North American mainland countries are now autonomous.

PART ONE: **The Caribbean**

The West Indies and the Guianas

THE CARIBBEAN ISLANDS, known in Columbus' time as "The Indies," extend in a wide arc which starts with Cuba, 125 miles off the coast of Yucatán, and reaches eastward for 1,500 miles. It then curves southward for 800 miles and ends at the island of Trinidad just off the northeastern shore of South America. Three islands of the Greater Antilles—Cuba, Puerto Rico, and Hispaniola (now divided into Haiti and the Dominican Republic)—separate the Caribbean Sea from the Gulf of Mexico in the northwest and from the Atlantic Ocean on the north. Jamaica, the other island of the Greater Antilles, lies 120 miles south of Cuba. The chain of smaller islands, the Lesser Antilles, extends from the Virgin Islands just off the coast of Puerto Rico to Trinidad. The Lesser Antilles divide the Caribbean Sea from the Atlantic Ocean to the east. The Caribbean Islands have been placed in a separate section because of the similarity in their agricultural and livestock development although they are sometimes included by geographers as part of Middle America.

The islands of the Greater Antilles are the tops lying above water line of a submerged mountain range which rises from an immense land mass on the ocean floor. If these mountains could be seen from that floor, they would equal the Himalayas in height and at some points would be higher. The Lesser Antilles are much more recent geologically and are of volcanic origin with the exception of Trinidad, which is a disconnected extension of South America, having been separated from that continent in geologically recent times. Some of the smaller islands are characterized by mountains up to 4,000 feet above sea level and by recent volcanic activity.

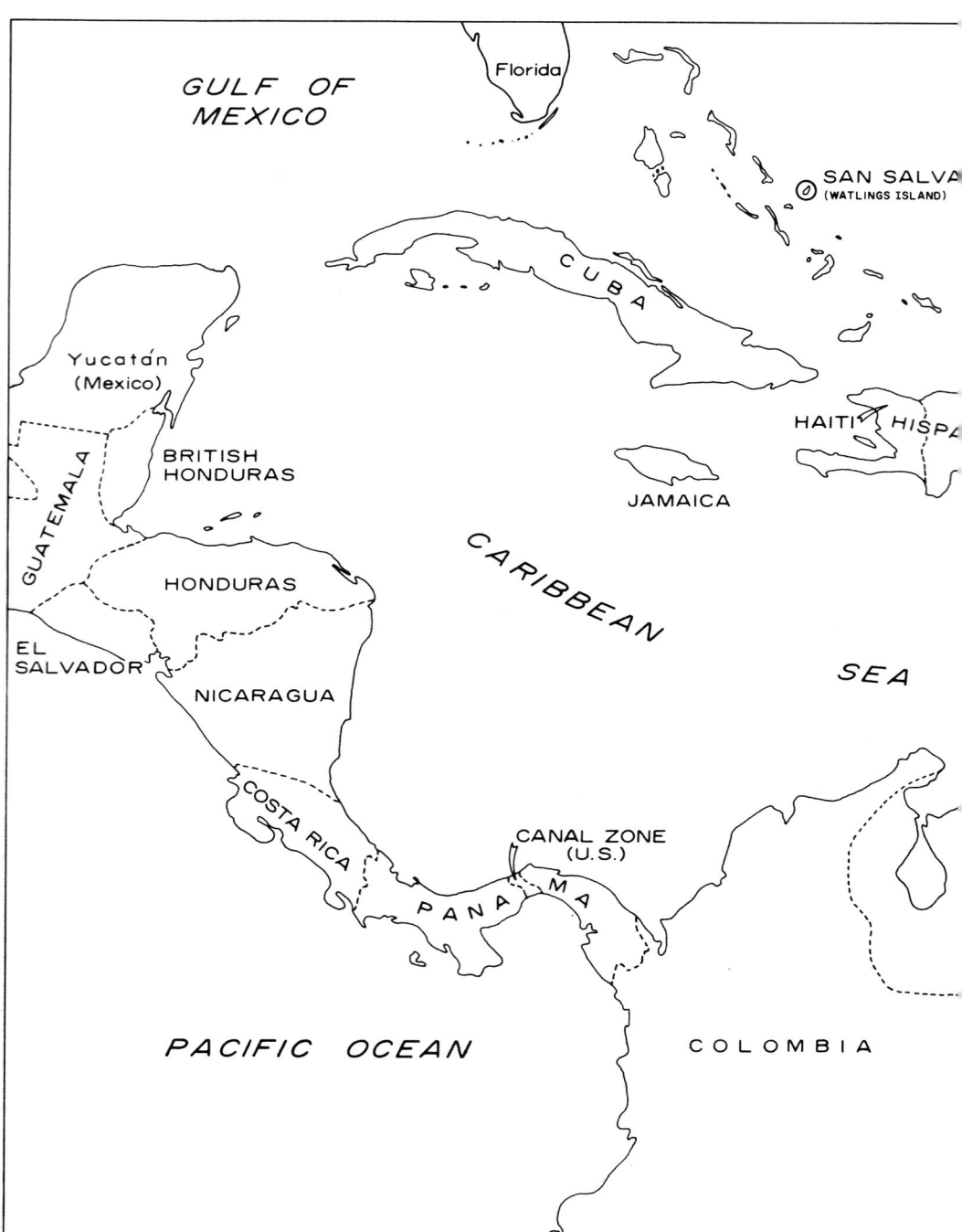

The countries of the Caribbean.

All the Caribbean islands lie within the tropics. With the exception of the inland parts of the large islands, however, they enjoy a warm and equable climate because of the surrounding water. Rainfall varies widely and is greatly influenced by the trade winds. In general, moisture is adequate for sustained plant growth and in some coastal regions is excessive. There are also some areas in the lee of the northeast trade winds where the annual rainfall is in the 20-inch range and droughts sometimes occur.

The Guianas—French Guiana, Guyana, and Surinam—lie in the northeast corner of South America, isolated from the rest of the continent by vast rain forests and mountains. Culturally, economically, and in the similarity of their cattle populations, these three countries are more closely allied to the island countries of the Caribbean than they are to the other countries of South America. Until the days of air transportation, practically the only communication between the Guianas and the rest of South America was by sea.

Until recent times the economy of this region has relied on sugar to the same extent as the West Indies. Some writers have described this corner of South America as "a sugar island surrounded by rain forest and the Atlantic Ocean," the "sugar island" comparison being with the islands of the West Indies.

French Guiana, Guyana, and Surinam are accordingly covered in a subsection following the Caribbean countries proper.

Spain's conquest of the New World expanded from the Greater Antilles to the other islands. After his first landfall on San Salvador (Watlings Island) and a few weeks' stay in other islands of the Bahamas, Columbus on his discovery voyage proceeded to the northern coast of Hispaniola. Cuba, Puerto Rico, and Jamaica were his next important landings, and, of these, Cuba developed into the main supply point for the exploration and conquest of Mexico, Panama, and South America. Although Columbus and his followers made landings on many other islands of the Lesser Antilles, these, with the exception of Trinidad, were largely ignored by the Spaniards.

Spain established two major trade routes for her galleons, which entered the Caribbean Sea at a point off the island of Guadeloupe. The southern route touched ports on the Spanish Main (the northern coast of South America) and the Isthmus of Panama; the northern route covered the ports of call on the southern shores of Puerto Rico, Hispaniola, and Cuba and extended on to Mexico. Annual expeditions followed these circuits. The fleet sailing the southern route would meet the northern fleet off the western tip of Cuba where they combined

The Caribbean

forces as a safety measure for the homeward voyage back to Spain.

The other European powers displayed little interest in the Caribbean area during most of the sixteenth century. At the beginning of the seventeenth century, however, Britain, France, and Holland cast a jealous eye on Spain's acquisitions in the New World and began to seek footholds in the Caribbean. One of their early protests to the Spanish dominion over both the lands and water of this part of the world took the form of the privateer. This was an armed ship, privately owned and manned but commissioned by a recognized government to harass Spanish shipping and settlements. At first, some sort of international excuse was contrived for the activities of these marauding vessels, but this formality was soon considered unnecessary. The loot from the Spanish galleons was proving to be rich indeed, and an atmosphere of lawlessness prevailed on the high seas. Then came the buccaneer. Like the privateer, the buccaneer at first had some semblance of authorization from a major power to attack the Spanish galleons or to pillage a town. Later he need only be reasonably circumspect in conduct. For example, if claiming British connections, he could attack only Spanish shipping. The final stage was the out-and-out pirate who fought and took whomever and whatever he could.

The privateers and buccaneers were of material assistance to the other European powers in gaining a position in the predominantly Spanish Caribbean area. The British occupied Barbados in 1624, and the French took over Martinique in 1635. Later, Spain resisted such intrusions to the limit of her capabilities, but Britain defeated the Spanish forces on Jamaica in 1697, and in the same year the French took Haiti. Spain managed to hold Cuba and Puerto Rico, her final footholds in the Western Hemisphere, until the Spanish-American War in 1898.

At the time Columbus reached the Caribbean, peaceful Arawak Indians practicing a primitive agriculture were living on Hispaniola, Cuba, and some of the other islands. On many of the Lesser Antilles were the Caribs, who had driven off the Arawaks who had preceded them. The Carib Indians were a war-making, hunter type, given to cannibalism where their enemies were concerned (the word cannibal derives from Carib). Both types of Indians had moved up to the islands from South America. The Caribs had actually arrived only shortly before Columbus. Attempts were made by the early Spaniards to train both types as slaves. Such efforts were usually unsuccessful, and the Indians soon died out, either the victims of the white man's diseases or killed outright.

As the islands developed, they followed an identical pattern of

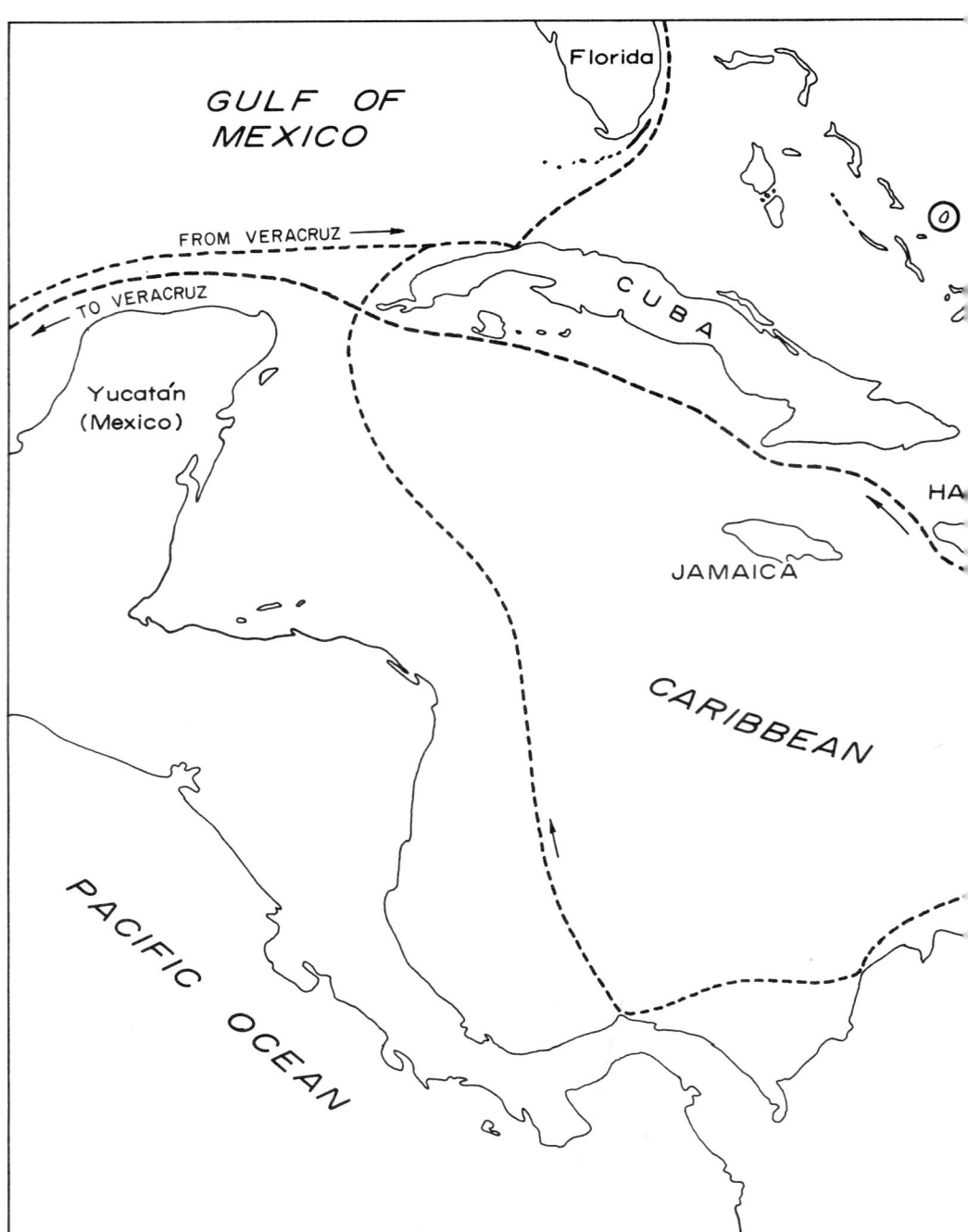

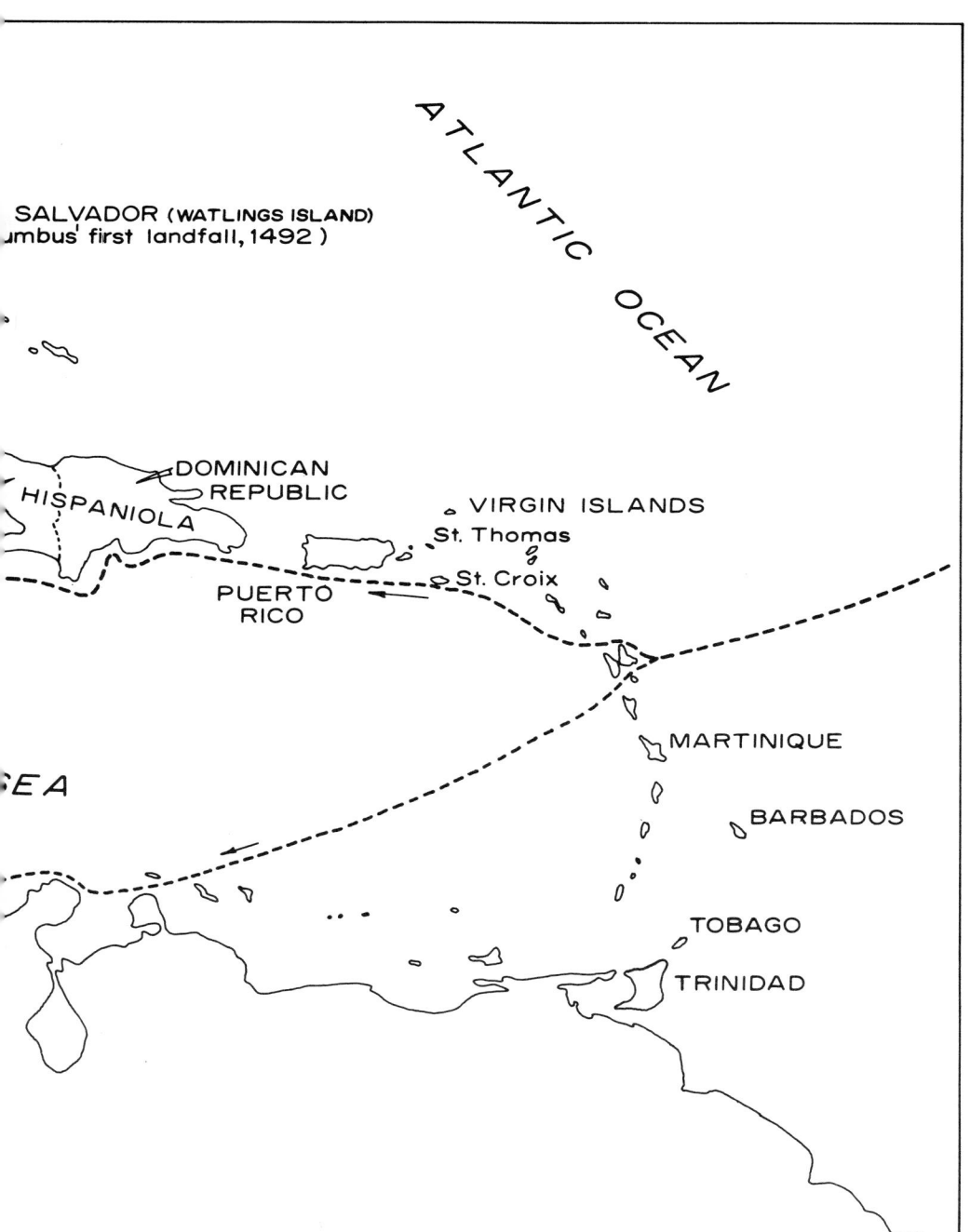

Sixteenth-century routes of Spanish galleons through the Caribbean.

plantation agriculture. In the early days there was some cultivation of spices and cocoa, but eventually the basic crop was sugar cane. African slaves were brought in for plantation labor. After the slaves were freed, considerable use was made of indentured East Indian labor on some of the British islands. In recent years the depressed world sugar market and labor problems with the cane workers have led to attempts to diversify agriculture. Such crops as cotton, tobacco, vegetables for export to European markets, and cattle, principally for beef, are all contributing in this endeavor.

Many of the political divisions in the West Indies were in a colonial or possession status until after the end of World War II. The major exceptions were Cuba, Haiti, and the Dominican Republic, which had previously acquired independence, and Puerto Rico, which had commonwealth status under the United States. The next major move toward political independence for the Caribbean area was Britain's organization of the West Indies Federation in 1958. This ostensibly was a voluntary association of ten colonies. The terminal objective was group independence for its members. Internal friction, taxation problems, and political jealousies led Jamaica to secede in 1961, Trinidad and Tobago the following year, and the Federation then fell apart. The other participants in the Federation had been Antigua, St. Kitts–Nevis–Anguilla, Montserrat, Barbados, Grenada, Dominica, St. Lucia, and St. Vincent. From this time on, movements for independence were almost entirely by individual local political units against their governments.

Most of the islands of any size had acquired independent status by the mid 1960's. Those that had not, such as Martinique and Guadeloupe, had settled for a high degree of self-government and recognition as an integral part of their national governments.

The early Spanish explorers left cattle on the islands throughout the Caribbean area. For fresh milk a vessel may have had a cow or two as part of the ship's stores, which, being no longer needed once the Indies had been reached, were unloaded at the first landfall. The Spanish king and queen were insistent on colonizing the new world—a viewpoint Columbus and his followers frequently did not share, being more interested in gold and silver—and sizable shipments of cattle were made to the Indies during the first part of the sixteenth century to stock early agricultural attempts. As population increased on the islands, there was some interisland shipping, which undoubtedly led to the movement of livestock between islands. On the larger islands, cattle which had wandered from their owners existed in a semiwild state and were hunted for their hides. This occurred particularly on Cuba, Hispaniola, and

The Caribbean

Trinidad, but was also the case on other islands. Such was the nature of the cattle population of the Caribbean area for over three centuries. Even the small islands, many of which the Spanish explorers bypassed after a landing or so, became populated with Spanish cattle. The first English, French, and Dutch settlers, finding the islands stocked with cattle, had no reason to ship in animals of their own until a much later day.

The first cattle on the Caribbean islands were the same as those which became established in northern South America, Central America, and Mexico. The Criollo cattle in all these regions were the descendants of those brought over from Spain in the early days of exploration and settlement.

During the eighteenth century sugar came to be the major factor in the economy of the West Indies. Development of large plantations required a better draft animal than the small Criollo; and about the middle of the nineteenth century Zebu cattle from India and later from Brazil began to be imported. This movement continued until well into the twentieth century when Zebu and Zebu-type cattle from the United States were brought to some of the islands. On the two larger islands under British influence, Jamaica and Trinidad, as well as on some of the smaller islands, representatives of several British breeds were brought in. There was also one known importation of African cattle—some of the native N'Dama breed from Senegal were shipped to St. Croix in the Virgin Islands when they were still a Danish possession. Water buffalo were taken to Trinidad early in the twentieth century and used for draft in the cane fields.

The principal effect of the nineteenth-century wave of cattle importations to the West Indies was the breeding out to a large extent of the Criollo (Creole in islands of French or English influence) descendants of the original Spanish cattle. Some of the Zebu cattle were bred straight, but the common practice was to cross them on the predominantly Criollo population. There was little planned crossing; any mixture of these two cattle types sufficed, gave a better draft animal than the straight Criollo, and was equally heat tolerant and resistant to tick-borne diseases. On some of the better-managed plantations, however, an upgrading program was followed. Zebu bulls were used continuously on the progeny of the Criollo cows until practically pure Zebu resulted. In recent years improvement in the over-all economy and political change has increased the demand for milk and beef, and this has led to a greater interest in animal husbandry. This, in turn, has stimulated new imports of cattle, mainly from the United States and Canada. The

Cattle of North America

Holstein-Friesian for milk and the American Brahman for beef have been most heavily represented in this movement, and both breeds are exerting a major influence on the cattle population of many of the islands.

In 1970 the cattle population of the Caribbean area varied widely in its composition. The Criollo of Spanish origin has been largely bred out except on Cuba. This usually has left a small, nondescript, Zebu-type animal which constitutes the major part of the cattle population. Cuba alone preserved the Spanish type in the pure, or nearly pure, Criollo which is seen there in sizable herds today. On some islands new breeds were developed which show real promise in the tropical environment— the Jamaica Hope, Jamaica Black, Jamaica Red, and Jamaica Brahman in Jamaica; the Senepol on St. Croix; the Romana Red in the Dominican Republic; and lastly, the buffalo for beef on Trinidad. An outstanding development of cattle is underway in Cuba for both beef and dairy animals. Some success is being achieved in maintaining northern dairy breeds in the tropical climate—particularly the Holstein-Friesian and, to a lesser extent, the Brown Swiss. The predominant nondescript Zebu population on most of the islands, which is the result of the breeding out of the Criollo, now will probably be bred out itself to more productive types. Animal husbandry, with particular emphasis on cattle, is receiving more attention than at any previous time.

The people of the Caribbean, with the exception of Cuba and Puerto Rico, were in a greater state of flux than their cattle in the opening years of the 1970's. The newly acquired independence of some governments and the high degree of autonomy of others had fallen into a pattern that was quite general. A layer of black government is in undisputed authority; next a veneer of social superiority and financial power cloaks the prerogatives of a small European element; beneath both of them the masses of the very poor struggle for a livelihood—the familiar heritage of all Latin America. This situation, coupled with some of the largest population densities in the world and birth rates among the highest, suggests a future with many unsolved problems. These human relationships will have to be resolved into a stabilized pattern that displays a reasonable degree of continuity before long-range programs for cattle of the new countries in the West Indies can be achieved. The communist government of Cuba is in complete control, and life proceeds in an orderly if quite spartan manner. Puerto Rico, where people are United States citizens, is covered under the section on the United States. Life on the island is now comparable in many respects to that in some of the southern states.

Barbados

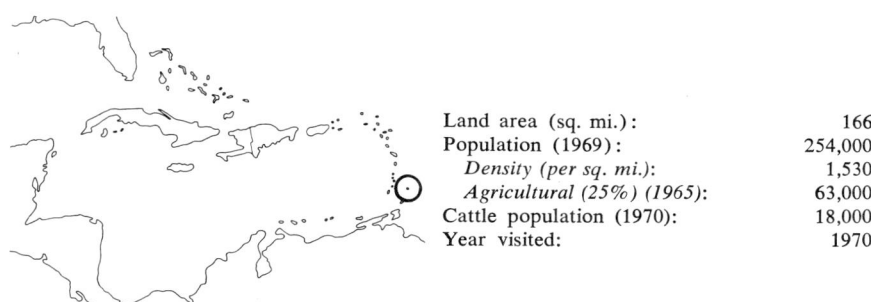

Land area (sq. mi.):	166
Population (1969):	254,000
Density (per sq. mi.):	1,530
Agricultural (25%) (1965):	63,000
Cattle population (1970):	18,000
Year visited:	1970

BARBADOS is a ham-shaped, low coral island lying in the Atlantic Ocean just below the 13th parallel. It is the most easterly of the West Indies. The climate is warm and pleasant because of the nearly constant northeast winds.

Although visited by the Spaniards, the island was ignored by them and was first claimed by the British in 1624. Active settlement started three years later. The Indians who had been on the island at one time had abandoned it for reasons of their own before the British came. Sugar plantations were the principal agricultural development and were the basis of the economy until recent years. When the wildfire of independence began to sweep the Caribbean area after World War II, Britain endeavored to pave the way for a practical unit of government and fostered the West Indies Federation composed of ten of her island colonies. Barbados initially joined this association in 1958 but unilaterally withdrew from it four years later. The political craving for a place in the sun led the small island, only nine-tenths the area of the city of New Orleans, to obtain independence in 1966 as a member of the Commonwealth of Nations.

Barbados escaped the influx of the indentured East Indians which occurred on some other Caribbean islands when slavery was abolished. The population is now 80 per cent Negro, possibly 15 per cent of mixed Negro and Caucasian extraction, and around 5 per cent Caucasian.

The basic sugar economy has been supplemented in recent years by a phenomenal growth in tourism, and the income derived therefrom is sufficient to maintain a favorable trade balance. There is a current trend to shift from producing sugar to growing bananas and vegetables for export. The difficulty in obtaining labor for work in the cane fields is the primary cause of this change in crop production. A few plantations are converting marginal cane land to pangolagrass pastures for raising cattle, particularly in areas where the terrain is too hilly to permit mechanization. Most of the island enjoys good rainfall, although on the northern tip precipitation averages only 35 inches annually, which is insufficient for growing cane.

CATTLE BREEDS

Spanish cattle were the first to arrive on Barbados. These were either brought in from other islands or were the semiwild progeny of cattle from the ships' stores released from Spanish vessels which touched the island before the British settled it. These Creole cattle were later bred to Zebu bulls which were brought in to develop a better draft animal. Occasionally an animal is seen, belonging to a plantation worker or small farmer, which is the straight Creole type. After World War II the draft ox disappeared rapidly. This was the typical Zebu-Creole cross as grown throughout the Caribbean Islands.

A general improvement in the economy of the island occurred after World War II and generated an increased demand for milk. Dairy cattle were introduced in sizable numbers, starting in the early 1950's. These were maintained both as small purebred herds to raise breeding stock and as bulls to use in upgrading the mixed Zebu-Creole cattle. The Holstein-Friesian, imported from Canada, constituted the large majority of this influx, although there were representatives of Brown Swiss, Guernsey, Jersey, and a few Friesian from England. The national dairy herd is now predominantly Holstein-Friesian, consisting of both purebred and grade stock.

A few plantations which are putting their marginal cane land into grass are raising beef cattle. The predominant practice has been to upgrade the mixed Zebu-Creole cows by continued use of American Brahman-type bulls. There is also some use of Santa Gertrudis bulls, and within the past few years the Charolais have attracted attention. Most of the few beef herds have a Charolais bull and a few crosses of that breed.

The government introduced a few Senepol cattle from St. Croix at the Central Experiment Station in 1961, but little was done with them and their traces have disappeared.

Cow showing strong Creole influence, owned by a farm worker.

MANAGEMENT PRACTICES

There are five large dairies on the island, each with over 100 head of cattle. The largest is a herd of 240 cows. These are recently established modern facilities, equipped with milking machines and milk parlors. Milk is cooled in refrigerated tanks, but transported to the pasteurizing plant in cans. Cows are brought in from pasture for milking in the early morning and kept during the day in paved yards provided with some type of shade. Following the afternoon's milking, they are returned to pasture. A supplemental ration consisting mostly of imported grain is fed during the day. Average milk production of 7,000 to 8,000 pounds per lactation is obtained. Considerable difficulty is experienced in getting a cow with calf, however, and the production is somewhat less than this figure per cow year. Heifer calves are raised to breeding age, but bull calves are slaughtered for veal under one week of age.

Many small owners milk from three to five cows and either sell their

Dairy-herd loafing shed at Hope Plantation.

production to the pasteurizing plant or privately. Their stock is usually tethered for grazing along the roadside or on such wasteland as is available.

The few plantation owners who have gone into raising beef cattle have fenced paddocks for rotational grazing. Bulls are run with the herd the year round, and calf crops as high as 65 per cent are obtained. Steers are marketed at four to five years of age, weighing around 900 pounds. Most slaughter cattle, however, are discarded dairy animals.

A government artificial breeding service was started in 1950 and is now up to 4,000 services a year. Most dairy cattle are bred in this manner. The frozen semen used is kept in liquid nitrogen. Some of the large dairies import semen direct from the United States and do their own breeding.

MARKETING

The price of cattle on the hoof early in 1970 was 17 to 18 cents a pound with little regard for quality. Milk was selling for five cents a pound to the pasteurizing plant.

Barbados

CATTLE DISEASES

Barbados is free of foot-and-mouth disease, tuberculosis, and blackleg. The general practice in the well-managed herds is to spray for ticks at intervals of two weeks to a month, depending on the tick population in the area. There is said to be very little trouble with internal parasites.

The major health difficulty in raising cattle is low fertility, particularly in the dairy herds. The Central Government Livestock Station in St. Michael, after thirty years effort, is now getting its herd bred at an average of 409-day intervals, but this is far above the island average. The cause of this breeding difficulty has not been determined. Dairy cattle in Barbados give the appearance of being in a thriftier condition than those on some other Caribbean islands where management levels are comparable. The fertility problem must have a deeper cause than nutrition unless it arises from a trace-mineral deficiency.

OUTLOOK FOR CATTLE

The future of the tiny island, with 1,500 inhabitants per square mile, is involved in the problem of population growth. Although the current rate of population increase is phenomenally low, slightly less than 1 per cent, it is still sufficient to create major problems in the future. Intensively developed for cattle raising, the island could become self-supporting in both milk and beef. But the land that would have to be diverted from cash crops to accomplish this end would leave little in the way of agricultural production to support the economy. Some further increase in the national cattle herd will probably develop as marginal cane land is converted to pangolagrass pastures. Such a movement, however, cannot be expected to extend to land that can be advantageously cropped for bananas or vegetables. All things considered, space for cattle on Barbados is limited.

Cuba

Land area, (sq. mi.):	44,218
Population:	8,600,000
Density (per sq. mi.):	196
Cattle population:	7,600,000
Year visited:	1972

THE REPUBLIC OF CUBA occupies the island of that name, the Isle of Pines thirty-five miles off its southwest coast, and some 1,600 small surrounding islands and keys. The total land area is nearly the same as that of Louisiana, and the population is more than twice the number in that state. The main island is the most westward of the Antilles and lies between the Gulf of Mexico and the Atlantic Ocean on the north and the Caribbean Sea on the south.

The land mass of the main island extends 746 miles from east to west and is 124 miles across at the widest point. Topographically, a broad anticline runs the length of the island with gentle slopes and wide, level areas well adapted to agriculture. The mountainous regions, which reach an elevation of 6,500 feet, cover one-fourth of the surface, but most of the remainder is suitable for cultivation. Originally, savannah, tropical grasslands with some brush and trees, covered one-fourth of the surface. Precipitation is seasonal, averaging 50 inches, with the rainy season from May to October. Rich, alluvial and limestone soils predominate in the lowlands. Sugar cane, first cultivated around 1550, has always been the money crop.

Cuba was the third landfall made by Columbus on his discovery voyage to the Western Hemisphere, and the bases established here were the steppingstones from which the conquistadors launched the conquest of Mexico. The sparse Indian population was enslaved by the Spaniards, but had practically disappeared by 1550 because of brutality and lack of resistance to European diseases.

For four centuries after the discovery by Columbus, Cuba and Puerto

Cuba.

Rico were the most prized possessions of Spain in the Caribbean and were the last vestige of her power in the New World. During the seventeenth century other European powers began to cast an envious eye on Cuba, and buccaneers and pirates took their toll of the island's riches. Spain's control was contested in varying degree into the eighteenth century. Britain captured Havana in 1762, but the island was returned to Spain the following year in exchange for the Floridas.

At the beginning of the nineteenth century the seeds of independence sprouted, and Spain maintained control with increasing difficulty. Organized rebellion began in 1868. The last phase of the thirty-year struggle of the island for independence from Spain began in 1895, and in 1898, following the Spanish-American War, Cuba attained independent status. Then for a period of seventeen months the island remained under United States military administration until the first Cuban government was established May 20, 1902.

The first half-century of Cuban independence was marked by periods of widespread unrest. There were two interventions by the United States to maintain order and protect American interests, one in 1906 and one in 1936, after which Batista, a former Cuban army sergeant, dominated the government. He became dictator in 1952 and widespread corruption followed. After the revolution led by Fidel Castro the revolutionary government took over in 1959. Cuba soon adopted the communistic ideology, diplomatic relations and trade with the United States terminated, and the country entered a period of privation. The economy, at first supported almost solely by the Soviet Union, suffered severely, but as other countries, particularly England, France, Italy, and Canada, established diplomatic and trading relations, economic conditions improved.

Agriculture, dominated by sugar, for two centuries has been the mainstay of the Cuban economy. Cattle have been an important element in the resources of the island from the days of the first Spanish colonizers, who prized them for their hides. Descendants of the original Spanish cattle were the principal means of draft power and in later years were used for milk and beef.

As is typical in communist countries, the economy revolves entirely around state-operated establishments. Only a portion of the agricultural sector is exempt from such strictures. Public utilities, stores, manufacturing facilities, services such as hotels and restaurants, are all operated by the state. But a sizable part of agricultural production, largely in grain, vegetables, and fruit, remains in private hands. Land ownership is limited to a maximum holding of 67.1 hectares (168 acres) per person. The production from the private farms is sold to the state agency concerned. Of the total agricultural land of 18,893,000 acres, something over 40 per cent is held by individuals, while the remaining 11,400,000 acres are in state farms which average over 20,000 acres each in size.

All agricultural workers on these state farms are on a 44-hour week, five 8-hour days and 4 hours on Saturday, and are paid a fixed salary. Necessary Sunday work is arranged by staggered shifts. A cowboy earns $138 per month.

The people are predominantly of Spanish extraction, but there is a sizable Negro element, the descendants of former slaves. The population is usually considered to be one-eighth Negro, nearly one-eighth mulatto, and three-fourths Caucasian, the last, however, sometimes carrying a minor degree of Negro influence.

The Cuba of today dates from "the Revolution," a term that is inescapable in any discussion of events in the country. January 1, 1959, was the day Castro took over. Differences in the ideologies of governments should not be permitted to cloud facts. The people of means of the old Cuba are gone; their property is now state property. The circumstances involved are now history, and the cycle in which they occurred is irreversible. The machinery of government, commerce, industry, and agriculture is in the hands of young, educated people who are employees of the state. They are not ashamed to be patriots. Both men and women are included in this management force of the country. They are making progress—and have tightened their belts along with the rest of the populace. The worker and his family have more than they had before the revolution—better housing, more and better food, better medical attention, better education. Castro is a revered leader to the vast majority of the people. These are facts which it serves no useful purpose to attempt to hide. Emphasis on the happenings which preceded the present status can in no way change the past.

The attitude of the Cuban people here touched on is a factor in the progress which is being made with their cattle.

CATTLE BREEDS

The first cattle in the Western Hemisphere were the Spanish landrace brought by Columbus on his later voyages to Cuba and Hispaniola. Cuba was the focal point from which cattle were introduced to Mexico early in the sixteenth century and from there to the southwestern United States.

Cattle had little value in Cuba except for their hides until the rapid development of sugar production during the eighteenth century. Then they found their place in the yoke, drawing carts and powering the early sugar mills. For almost 300 years the human influence on their development was minor and natural selection prevailed.

During the nineteenth century Zebu cattle were brought to Cuba and there was some mixing of the humped with the Spanish types. Later some of the British and European breeds were introduced. The bovine population, however, remained predominantly Criollo, descendants of the Spanish cattle, until the opening of the twentieth century.

Along with the development of the large cane plantations toward the end of the nineteenth century, large cattle operations were begun in the savannah country of Camagüey and Las Villas provinces.

In the final years of the war with Spain, which began in 1895, the cattle population was decimated. Writers of the time describe the deserted pastures in Camagüey Province and mention the fact that cattle had practically disappeared from the island. In 1899 the cattle population was estimated at only 371,000 head. Importations were encouraged after hostilities ceased, and reached a total of 925,000 for the three-year period 1900–1902. Breeding stock was admitted free, and there was only a nominal duty on cattle to be grown out on pasture for slaughter. Females could not be killed unless declared incapable of reproduction. About half of the cattle entering Cuba at the time came from Mexico. The United States was the next heaviest exporter and then Colombia, followed by several other countries of tropical America.

It is impossible to determine the effect of different breeds in rebuilding the Cuban population. In the immediate postwar period most of

Cuba

the cattle that were brought in probably were grown out for slaughter. The average price was $19 a head for Mexican cattle and $16 for those from the United States. Such low prices indicate that the imported stock was probably mostly young animals to be pastured and killed.

The important additions to Cuban breeding stock began during the period of reconstruction which followed the war with Spain and ended in 1919. Criollo cattle were obtained from Puerto Rico, Charolais from France, and Zebu breeds from Jamaica and Mexico. There is also evidence of Angus, Hereford, Holstein-Friesian, Jersey, Guernsey, and Red Poll in Cuba prior to World War I. Later the American Brahman was imported in considerable numbers and was followed by the Santa Gertrudis. For the most part, these introduced breeds were maintained pure on the large plantations although there was inadvertent crossing. The Zebu characteristics, however, had become a distinguishable feature of many Cuban cattle.

There is a blind spot in the story of cattle in Cuba that lies behind the agonies that followed the revolution. To attempt to explore this area would involve a quest into the past that is beyond the objective of this book.

There were beautiful herds of cattle in Cuba before the revolution and there are beautiful herds there now, which to a large extent trace back to the prerevolutionary days. The Cuban cattleman of an earlier day was justly proud of his stock, and husbandry practices were among the foremost in Latin America. The bovine population that is seen today in Cuba was not produced in one decade. The improvement by modern selection practices in both beef and dairy breeds that is now underway had an excellent base on which to build.

In 1948 the cattle population was estimated to be 90 per cent Criollo and mixed Criollo-Zebu types. In 1959, at the time of the revolution, the two outstanding introduced breeds were the Santa Gertrudis and Charolais. The recent importations of foreign cattle began in small numbers after 1962 and became quite sizable starting in 1966. Holstein-Friesian cows from Canada constituted the major part of these shipments, but other breeds were represented—for the most part by sires in the artificial insemination centers. There have also been continuing importations of semen from Canada.

The husbandry of cattle in Cuba today encompasses the development and improvement of specific breeds, as well as extensive crossbreed programs. This work is the province of specialized large state dairy farms and cattle ranches. The principal breeds now represented in the country are the following.

Cuban Criollo bull showing minor Zebu influence, five years old, 2,000 pounds. Rosa Fe Signet Artificial Insemination Center.

Cuban Criollo.—Cuban Criollo is the designation for the pure, or nearly pure, descendants of the Spanish cattle brought to the island in the closing years of the fifteenth century.

The Cuban Criollo is a solid tan color over most of the body with only minor white markings on the under side and occasionally on the legs. This coloring varies in shade from a dark reddish tan to a light fawn. Animals of other than tan color are not considered Criollo. The horns are thick at the base, rather wide and upswept on the cow, short and laterally extended on the bull. The dewlap is thin and quite pronounced on both cow and bull. The sheath shows considerable variation, quite prominent on some individuals and nearly nonexistent on others. A similar variation is noted in the umbilical fold on the cow which is fairly pendulous on some and almost unnoticeable on others. The tail stock on both bull and cow is often quite pronounced, thick at the base and tapering gradually. A large exposed vulva, similar but not as prominent as on the Zebu, is seen in many cows, but this condition is not universal. All these characteristics, appearing in varying degree,

Cuban Criollo bull showing marked Zebu influence, six years old, 1,800 pounds. Papi Lastre, Oriente Province.

are indicative of Zebu influence. However, there is no semblance of a hump on either male or female.

The Criollos are concentrated mainly in Oriente Province, and the large herds which were assembled by the present government for selection and breeding were chosen as the best representatives of the breed. This collection procedure could account for the differences noted in individual animals, for, since they came from different areas, the degree of Zebu influence may have varied.

A small-type Criollo was developed in the cultivated areas of Pinar del Río for use in cultivating between the rows of tobacco plants. In the cattle country of Las Villas and Camagüey provinces a heavy draft and a beef type was sought. There also had to be effective selection for milk to have produced the occasional 12,000-pound dairy cow that is seen today.

The Cuban Criollo stands well off the ground, and its conformation is more of a beef than a dairy type, showing good muscling. It is a large animal for a nonhumped breed native to the tropics. Cows weigh

Mature Cuban Criollo cow Bijirita, four and one-half years old, 1,300 pounds. Highest lactation 13,750 pounds, 4.5 per cent, 305 days. Bijirita Dairy Farm.

upward of 1,100 pounds, with individuals up to 1,300. A mature bull weighs 1,800 pounds, some as much as 2,000.

Size and uniform color indicate that the Criollo has undergone controlled selection for a considerable period. It is said that selection for a beef conformation started in 1940. Even prior to this time many breeders had selected for an improved type that went back for many generations. Their standards were the tan color, size, and conformation. Today, in areas where cattle are still used for draft, the Criollo is considered faster and more rugged than the Zebu.

In Middle America, the Caribbean countries, Venezuela, and Colombia the tan coloring is frequently seen, but many other color patterns are also common among the Criollo population. In the selected Criollo herds in Venezuela and Costa Rica, however, which were developed for high milk production, the nearly solid tan color, identical to that of the Cuban Criollo, is universal.

The Criollo in Cuba is now being selected by production records for a dual-purpose, milk-beef breed. The artificial insemination center at Bayamo in Oriente Province has 30 selected Criollo bulls whose progeny will be tested for milk and beef production. Selection is being made on the same basis in the 3,200-cow herd at the Papi Lastre state farm also in Oriente. Here the top-producing cow, on pasture and adequate supplement, gave 13,200 pounds of 4.5 per cent butterfat milk in 242 days when she was 11 years of age. She weighed 1,360 pounds. The average production of all cows under test in the herd is 6,160 pounds per 244-day lactation. This does not include the milk consumed by the calf. The Criollo cow has the reputation of milking well up to 17 years of age.

Among Cuban cattlemen some distinction is made between the "pink"-muzzled and the "black"-muzzled Criollo. It is claimed that animals with the black, or heavily pigmented, muzzle are descendants of those imported from Puerto Rico during the period after Cuba became independent and the cattle population was being rebuilt. There does not, however, appear to be any particular significance to color of muzzle as regards the commercial aspects of the breed.

The double-muscle characteristic has been noted in the Criollo, and some effort has been made to develop this trait in a beef type called the Tinima. The characteristic seems to be due to a recessive factor, since double-muscled offspring are produced from parents which do not display this abnormality. Calving difficulty is mentioned in connection with the Tinima strain.

Today the Criollos constitute only a small part of the total cattle population, generally estimated as around 1 per cent. The breed is unique in being the major example in the Western Hemisphere of the original Spanish cattle that has been maintained relatively pure down to the present day. Although the Cuban Criollo generally carries some degree of Zebu influence, it appears to be minor, since the hump has disappeared.

Cuban Zebu.—The term Zebu in Cuba is often used to designate any humped cattle which show a marked influence of the *Bos indicus*. Zebu is also used in the generic sense to refer to the various Zebu breeds, especially to the American Brahman. Cuban Zebu is used here to refer to the rather small humped animal of mixed breeding and nondescript appearance seen throughout the country.

Size varies considerably in the Cuban Zebu, the cow weighing 1,000 pounds or less. The hump is always present, though varying in size. The dewlap, sheath, and general conformation all show the Zebu influence.

American Brahman bull, mature weight 2,000 pounds. Santa Isabel Farm near Camagüey.

Few bulls are now being saved, and American Brahman bulls are used for upgrading when this can be arranged.

In the early 1900's, Zebu cattle were imported from Colombia, Venezuela, and Puerto Rico and probably other parts of tropical North America. Zebu and Zebu-cross bulls of both Hereford and Holstein-Friesian breeding were shown at the Agricultural Exposition in Havana in 1914. Photographs of the Zebu cattle indicate a preponderance of Guzerat and Nellore breeding. Importations of Zebu cattle were also made at about this time from Jamaica.

American Brahman.—During the period that the American Brahman breed was being developed in the United States there were importations of these Zebu cattle to Cuba; but the American Brahman as seen in Cuba today traces primarily to the numerous representatives of the breed that were brought in after the American Brahman Breeders As-

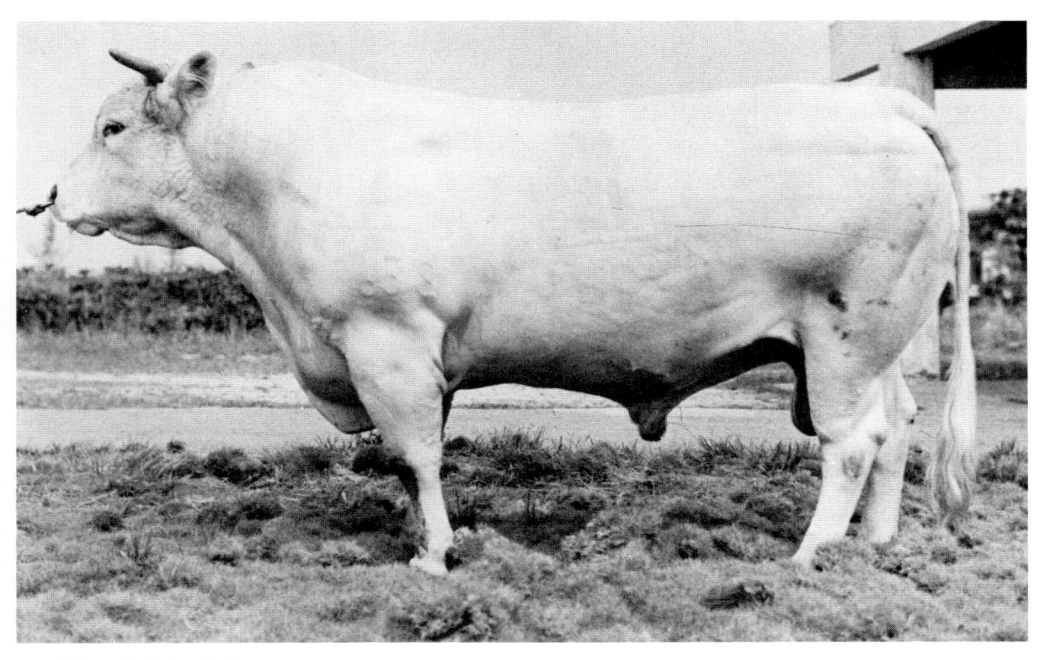

Charolais bull, five years old, 2,200 pounds. Rosa Fe Signet Artificial Insemination Center.

sociation was founded in 1924. Both pure and upgraded stock are referred to as "Brahman." The characteristics of the breed in Cuba conform to the description as given in the section "American Brahman" (in Part IV, "United States").

Large purebred herds are maintained for the production of breeding stock and of bulls for the artificial insemination centers. The Santa Isabel Farm near Camagüey in the province of that name has a beautiful group of sixty Brahman sires in its artificial insemination center.

Charolais.—Cuba was probably the site of the first introduction of Charolais cattle into the Western Hemisphere. The Marquis de la Real exhibited a number of Charolais cattle at the Agricultural Exposition in Havana in 1914. These had been raised on his ranch at Cayamas in Oriente Province and included a four-year-old cow which won first prize and a three-and-a-half-year-old bull that was awarded the second prize for the breed. This would place the Marquis' foundation stock as having arrived in Cuba no later than 1910. It is possible that representatives of the Charolais breed had reached the French island of Marti-

Charolais cow herd, average 1,600 pounds. Villa Matilde Genetic Center near Havana.

nique at about this same time or possibly even earlier. If so, no continuity of the breed was established on that island, the few bulls now seen there being recent imports. Cuba can rightfully claim to have been the first center of Charolais breeding in the Western World.

The Cuban Charolais are typical of this breed wherever seen, though somewhat smaller in size than in France and the United States. The cows are said to have little difficulty in calving, a factor that might be attributed to the smaller size. The mature cow is from 200 to 300 pounds lighter than in more northern countries, and the bull is around 400 pounds lighter. The tropical climate together with considerable inbreeding for many generations has probably caused this difference in size.

The Charolais is the major European beef breed now under development on the island. The Manuel Fajardo ranch at Jiguani in Oriente Province has a herd of 1,125 selected, purebred cows and an artificial insemination stud of performance-tested bulls. All were raised in Cuba.

Santa Gertrudis bulls in the production herd. Turiguanó, Camagüey.

The cows average 1,200 pounds and the bulls from 2,000 to 2,200. Artificial insemination is used, and a comprehensive selection program is followed based on the performance testing of young bulls and their progeny.

Santa Gertrudis.—The first Santa Gertrudis to arrive in Cuba was a bull purchased from the King Ranch in 1936 by Avaro Sanchez who at the time was managing his father's ranch, the Alvo, in Camagüey Province. The King Ranch brought in 931 heifers and 79 bulls in 1952 to form the foundation herd on their ranch in north Camagüey Province. The breed became popular in the cattle country and was adopted by a number of ranchers. After the revolution many of the King Ranch cattle were shipped to the Soviet Union, and the others were transferred to Turiguanó, known as "the island," a 15,000-acre ranch in the northwestern part of the province. This herd is being improved by breeding performance-tested bulls artificially to selected cow groups of 250 head each.

The Santa Gertrudis cattle in Cuba are typical of the breed as seen in the southern United States. The large herd at Turiguanó is maintained pure as are some other smaller herds throughout the country. There is, however, considerable experimental crossing of Santa Gertrudis bulls on cows of other breeds. At Turiguanó such crossing is with Shorthorn, Charolais, and Hereford females.

Santa Gertrudis cow herd in one of the detection corrals. Turiguanó, Camagüey.

Holstein-Friesian.—The Holstein-Friesian, generally referred to as Holstein, is the most important dairy breed. There has been extensive development of the breed by the importation of bulls and semen from Canada in the years following the revolution. Starting in 1966 there were several large shipments of purebred females. Herds of purebred and high-producing grade cows are now maintained on many state farms, and all artificial insemination centers have Holstein-Friesian sires. Modern selection practices are followed for milk production and type.

The tropical environment takes its toll of the northern-bred Holstein even under the best of care. Milk production declines, and hair coat tends to lose its bloom. Most of the national herd has not been in Cuba long enough to overcome such environmental effects, but with time and under the selection and management practices that are being followed, a better-adapted animal can be expected. Eventually a "Cuban Holstein," analogous to the Israeli Friesian, should evolve.

The Holstein in Cuba is otherwise identical to the breed in northern countries, and the purebred sires in the artificial insemination centers, both those raised in the country and the imported animals, are excellent examples of the breed.

Red-and-white Holstein-Friesians have been selected from Cuban-raised cattle, and some have also been imported from Canada. An exceptionally fine herd of 36 cows of this color pattern is maintained at the Pradera Roja dairy farm on the outskirts of Havana. Their average production is 12,500 pounds per lactation. There are a few red-and-white bulls and some black-and-white bulls carrying the red factor in the artificial insemination centers.

Other Breeds.—Many British breeds had been brought to Cuba in the

Progeny-proven Holstein-Friesian bull, six years old, 2,475 pounds. Rosa Fe Signet Artificial Insemination Center, Havana.

Red Holstein-Friesian cow herd. Pradera Roja Dairy Farm.

Guzerat bull, thirteen and a half years old, 1,800 pounds. Rosa Fe Signet Artificial Insemination Center.

years before the revolution. The Red Poll, in particular, was prominent in the herds of some breeders as it was in other tropical countries of the Western Hemisphere. Breeds that were held in limited numbers included the Hereford, Brown Swiss, and Jersey. As planned development progressed after the revolution, further importations were made to form the nucleus of experimental herds. These included Ayrshire, Guernsey, Jersey, Red Poll, Milking Shorthorn, and Devon from Britain; Shorthorn, South Devon, Angus (both black and red), Hereford (both horned and polled) from Canada; Jersey from New Zealand; and the Illawara Shorthorn from Australia. The Limousin is being used in crossbreeding experiments employing semen imported from Canada.

At various times in the past there were also introductions of several Zebu breeds, among which were the Indo-Brazil, Gyr, Guzerat, and Nellore. A few bulls of these breeds are maintained in some of the artificial insemination centers for use in experimental breeding projects.

Indo-Brazil bull, five years old, 2,150 pounds. National Artificial Insemination Headquarters.

At the large dairy complex under construction in eastern Havana Province, Agrupación Genética del Este Habana, it is planned to have individual purebred herds of Ayrshire, Brown Swiss, Guernsey, Jersey, and Red Poll, as well as Holstein cows, in the 288-cow units.

MANAGEMENT PRACTICES

During the years between Cuba's emergence as an independent country in 1898 and World War I cattle numbers increased rapidly. There were heavy importations of cattle both for growing out for slaughter and for breeding purposes. The total cattle population of less than 400,000 in 1899 increased to over 4 million in 1929. The number of cattle in 1971 was estimated at 7.6 million.

Organized cattle raising had developed by the 1950's but was confined mostly to the large ranches with representative cow herds of a few thousand head. Good management prevailed at some of these ranches,

Cattle of North America

where the calf crop was said to run from 80 to 90 per cent. Cattle were maintained on pasture the year round, and in the good cattle country of central Cuba stocking rates of 1 to 1½ acres a head were achieved. Steers went to market weighing around 1,000 pounds at two and one-half years of age. Blackleg and screwworm were the principal health hazards. The Cuban stockman had no superior anywhere in Latin America. More than half the cattle population, however, was comprised of the Zebu or Criollo ox team and milk cow in the hands of the small farmer.

Around the large cities there were modern dairies with the European dairy breeds, although many of the units supplying milk, particularly to the small towns, were substandard. Milk was generally in short supply, and price placed it beyond the reach of the poorer element in the population centers. In the country the Zebu or Criollo cow that raised her calf and was milked once a day supplied the milk.

In the years immediately following the revolution comprehensive plans were formulated for a nationwide improvement in beef and dairy cattle production. These involved consolidation of existing properties into large-scale units, provision for the wide use of artificial insemination, improved nutrition, and modern management practices.

With the exception of sugar production, which is the province of a separate ministry, Cuban agriculture is under the direction and management of the National Institute of Agrarian Reform, an entity of ministerial status. The private sector of agriculture is relatively unimportant in cattle, and management is elementary. While private farms include 40 per cent of the arable land, they account for only 20 per cent of the country's meat (including poultry and pork) and an insignificant amount of the milk going to the pasteurizing plants.

The administrative responsibility for operations involving cattle are divided between two agencies, one of which controls production while the other directs the activities of the "genetic centers."

The genetic centers are farms or ranches, often of considerable size, devoted to breed improvement. Such a center may include a dairy complex designed eventually to handle 10,000 milk cows or a ranch breeding several thousand cows. They hold a preferred position in the livestock economy of the country and enjoy first call on available concentrates. Practically all grain and protein supplements, other than fish meal, are imported and are in short supply.

Breeding.—Artificial insemination has been used extensively since 1966. The first artificial insemination center was established in 1962 at Cotarro, near Havana, where the National Artificial Insemination Headquarters is now located. Existing facilities on former plantation properties were first employed as bull studs. Modern facilities were built later, in such numbers, in fact, that they appear to be in excess of the country's needs. There are now a total of 17 artificial insemination centers including one under construction in Matanzas Province.

The Rosa Fe Signet Center at San José de las Lagos in Havana Province, opened in 1968, is the newest facility. It has capacity for 150 bulls, is excellently equipped, and is breeding 250,000 cows annually in the Havana area.

There are 1,600 bulls in the centers, a few of which are crossbreds. Dairy sires number 900, of which 100 are considered proven.

The pellet (referred to as "drops") method of semen preparation is used throughout the country, though glass ampules are used for special breeding in the genetic centers as insurance for sire identification. Liquid nitrogen is used for semen storage. The rectal method of impregnation is employed.

Technicians are given a two-year training course in the artificial insemination centers, the course including some time at "cowboying" and detecting cows in heat. Breeding service is free to private cow owners, but, as in the state herds, the bulls to be used are selected by the center.

In 1971 one million cows were bred artificially (700,000 dairy, 300,000 beef), and the number is increasing at the rate of 10 to 12 per cent annually. The average conception rate is said to be 65 per cent on first service based on palpation after 90 days. The concentration of semen per pellet or ampule is 50 to 60 million sperm, which leaves 20 to 25 million live sperm after freezing. This is double the number of live sperm usually considered adequate, a minimum of 10 million being standard practice in many United States artificial insemination centers.

In beef herds where natural service is used, bulls are usually run with the cows the year round, sometimes being alternated, one month with the cows and one month out. One bull to 25 cows is the common ratio. The nationwide calving rate is said to be 50 per cent. On some of the large state farms with good management 90 per cent is obtained.

Much effort is now devoted to crossbreeding in an effort to improve both milk and beef production. At the Bijirita Dairy Farm south of Havana, a ⅝ Holstein-Friesian, ⅜ Brahman cross is being developed as a new breed, paralleling an upgraded line starting with Brahman cows and using Holstein bulls continuously.

Comparison of these two lines is to be made for milk production and compatability to the Cuban environment. It is recognized that the

Cattle of North America

Holstein-Friesian–American Brahman bull, six years old, 2,475 pounds.

Holstein-Friesian–American Brahman cows. Average production 4,322 pounds, 244-day lactation. Bijirita Dairy Farm.

Holstein-Friesian–Criollo cow. Average production 5,600 pounds, 305-day lactation. Bijirita Dairy Farm.

upgraded Holstein cow will probably excel in milk production, but the object is to determine if the improved adaptability of the new breed will give it over-all superiority. In the same program, pure Criollo cows are also being upgraded to Holsteins.

At another nearby center a herd of pure Criollos is being bred with Limousin semen from Canada in an effort to develop a more productive beef animal; while at still another center, a Charolais herd is being crossed with Brahman, Shorthorn, Santa Gertrudis, Criollo, Holstein, and Brown Swiss bulls.

Nutrition.—Plans for development of the beef and dairy industries have recognized the importance of adequate nutrition and the difficulty in maintaining good pastures and providing balanced concentrates in the tropical environment.

The dairy farms that have recently been established have pastures of the more productive tropical grasses such as pangolagrass, guinea grass, and coastal Bermuda. Napiergrass and alfalfa are raised for green chop, although difficulty is encountered in growing alfalfa because of the

Loading silage, Turiguanó Ranch.

climate and a stand usually lasts only two or three years. Wide use is made of sprinkler irrigation during the dry season from November through April. Water is obtained from wells or from reservoirs built to collect surface water. On the well-managed, irrigated pastures a cow fed concentrate in proportion to her milk production can be maintained the year round on half an acre.

Dairy herds are on pasture except during the heat of the day when they are brought to loafing yards provided with sunshade. There is wide use of small, well-fenced paddocks of an acre or two for rotational grazing. In these the dry cows frequently follow those in milk.

The beef ranches use large pastures. While some have improved tropical grasses, much native forage is still used. Sprinkler irrigation is employed to some extent in the dry season, and rotational grazing is common practice. Stocking rates of one head an acre are obtained on well-developed land.

Considerable use is made of grass silage well laced with molasses. The newer dairy installations use bunkers with concrete walls. The beef ranches make large piles of silage from which a tractor-operated fork can load into wagons or trucks. There is little use of upright silos, and there are no pit silos.

The concentrate fed to dairy cows in milk is basically corn and soybean meal, the latter obtained mostly from China. High-producing cows are fed one pound of concentrate per liter of milk after the first five liters. Following is a typical dairy concentrate formula with a 20 per cent protein content:

	Per Cent
Corn	40
Ground oats	10
Wheat meal	20
Soybean meal	25
Bone meal	5
Total	100

Some dairy cows are fed molasses free choice when they are in the loafing sheds during the day. The intake is usually around 4 pounds a day, which reduces the concentrate requirement by about 3 pounds. Higher consumption of molasses, if forced by feeding less concentrate, has been found to result in decreased milk production.

In feed lots and for some progeny-testing of bulls molasses is fed extensively. Calves weighing 500 pounds conditioned to a molasses diet will consume 13 pounds a day of a molasses mixture containing 3 per cent urea and 0.5 per cent salt. This increases to 18 pounds by the time the animal is finished at 900 pounds. The ration also includes one pound a day of fish meal and a limited quantity of green chop (about nine pounds a day).

The conditioning procedure essential to these high intakes of molasses involves a gradual reduction in roughage and a compensating increase in molasses in the ration. Usually calves are fed molasses free choice until reaching a weight of around 500 pounds. This is either offered on pasture or with a full feed of green chop if the calves are confined. Increase in the intake of molasses is then induced by limiting the time on pasture or by reducing the amount of green chop. Normally after two weeks a molasses intake of around 13 pounds a day can be obtained, and the forage intake drops to 9 pounds of green chop containing 20 to 25 per cent dry matter. The molasses contains 77 per cent dry matter and is fortified with 3 per cent urea.

After a 14-day conditioning period, one pound a day of fish meal is fed separately. Water and salt are always present. Yearling calves on this ration, starting at 500 pounds, will reach 950 pounds in ten months, a gain of 1.5 pounds a day.

When a supplement is needed during the dry season, molasses is often fed free choice to dairy cows and to other cattle on pasture.

Feed Lots and Testing Stations.—Feed lots, which are not new to Cuba,

Feed lot, 2,000-head capacity, molasses–fish-meal ration. Turiguanó.

are receiving increased attention. At Turiguanó the *cebadero* (feed lot) has a capacity of 2,000 head and finishes around 5,000 head of 950-pound animals per year. To achieve this, considerable feeding is done by a restricted use of pasture during the first part of the period. Most cattle fed for slaughter are young bulls. No heifers are fed, as nearly all females are kept for replacements and to increase the cow herd.

There are a number of bull testing stations on the genetic center farms and one at the Institute of Animal Science near Havana. The usual practice in performance testing is to start calves on full feed at three to four months and at a weight of 350 pounds, thus canceling out as much as possible the effect of mothering ability and milk production of the dam. The calves are fed to a final weight of 450 kilograms (990 pounds) in about 300 days.

The Institute of Animal Science is performance-testing bull calves and progeny-testing sires that have proven superior on performance test.

Culled Charolais bull calves being fed on pasture for slaughter. Manuel Fajardo, Jiguani, Oriente Province.

Young Charolais bulls off performance test. Institute of Animal Science.

Cattle of North America

Progeny-bull testing station, 480-head capacity. Institute of Animal Science.

Only Charolais calves from the main Charolais breeding center at Manuel Fajardo in the Oriente are used. A preliminary cut is made here and then only the best-gaining calves, about 250 annually, are sent to the testing station. The culled calves are fed out on pasture for slaughter.

In a special test at the Institute of Animal Science, twelve progeny of superior performance-tested sires are group fed to a final weight of 990 pounds. Forty groups are fed at a time, and daily rate of gain, feed conversion, and carcass data emphasizing dressing percentage and the proportion of high- to low-price cuts are obtained. Bulls under progeny test are fed the molasses diet.

Farm Facilities.—Large dairy farm complexes called "plans" are being built around Havana in order to increase the milk supply for the community. Each complex will have a number of individual dairy units that are designed to handle from 72 to 288 cows. Such a unit consists of a milking facility, usually a milk parlor, although a few are designed for hand milking, and a steel storage tank; an adjacent loafing yard with sunshade; and a number of well-fenced one- to two-acre paddocks to provide rotational grazing. Sprinkler irrigation is provided for use during the dry season.

Some units have individual pens for young calves, group pens to handle older calves in age groups, and small rotational pastures for

Agrupación Genética del Este Habana Dairy Complex. 12,000 cows in 44 units—to be one-half completed in 1972. Milk parlor and sunshade of one unit in middle foreground.

raising young stock. From other units the calves are taken to a central location to be grown out. Convenient lanes are provided for the movement of cattle from pastures to loafing yards and milking facilities. Some units are provided with concrete bunker-type silos; others depend on green chop for feed to supplement pastures when necessary.

To provide for the workers in a group of units in a dairy complex, living facilities are built at a central location. Typically these consist of five four-story apartment buildings, each housing 48 families. The apartments have a living room, bath, and two or three bedrooms and are equipped with electrical appliances. Stores and recreational facilities are provided in an effort to make living in the country as attractive as in the city. By early 1972 a number of dairy units in three complexes were in operation and some of the apartment houses were occupied.

Typical dairy barn in a 288-cow unit. Agrupación Genética del Este Habana.

The older dairy farms have been developed by making use of such facilities as were available on plantations that were taken over at the time of the revolution. While not as modern as the new dairies, these have been equipped adequately for a good operation. Pastures have been fenced for rotational grazing, good facilities for hand or machine milking have been installed, and bulk storage has been provided.

The state beef farms have been built up on the old plantations and cattle ranches. The herds usually have several thousand mother cows. Additional fencing has been provided and is still under construction for better management of pastures.

Registration and Recording.—The Centro de Control Pecuarto (Cattle Control Center) in Havana handles the production testing of dairy cattle and the recording of purebred stock of both beef and dairy breeds.

The testing of dairy herds was centralized in 1970. Inspectors weigh the individual cow's milk production and take samples for butterfat content every thirty days. These data are put on IBM cards, and printouts are made on an accounting machine. Copies of these records are furnished the dairy farm. Currently, 15,000 cows are under the production test program, and the number is increasing rapidly. Data for sire evaluation are also recorded on IBM cards, and it was anticipated that a computer would soon be available for data processing.

Inspectors from the Control Center make a visual appraisal of all purebred cattle and approve each individual as to type before registration. Certificates of registration are then issued showing the pedigree. The procedure employed is patterned after that of the Canadian National Livestock Records. The breeds which are now being registered are Angus, Brahman (American Brahman), Charolais, Criollo, Guernsey, Hereford, Holstein-Friesian, Jersey, and Santa Gertrudis.

CATTLE DISEASES

Animal health is the domain of the National Institute of Veterinary Medicine in Havana. Considering the fact that it lies wholly within the tropics, Cuba is remarkably free of cattle diseases. Internal and external parasites are the worst plague to cattle, and heat stress affects all the northern breeds. The Criollo, Zebu, and cattle carrying a strong Zebu influence such as the Santa Gertrudis have a strong tolerance to these conditions. The Charolais, which has been maintained pure in Cuba for over half a century, has acclimatized to a considerable degree, but other European and British breeds are more susceptible to the tick-borne diseases and to heat stress.

The claim is made that Cuba has always been free of foot-and-mouth disease and is recognized as a foot-and-mouth-free country by the Food and Agriculture Organization (FAO) of the United Nations.

Cattle are imported only from countries that are free of the disease. Strict surveillance is maintained throughout the country by the Institute of Veterinary Medicine for possible outbreaks of the disease, and the comprehensive measures which would be taken on an outbreak can be gauged by the methods employed in dealing with the 1972 outbreak of African swine fever. When it was discovered, over 500,000 hogs were slaughtered in Havana Province, the site of the outbreak.

The incidence of brucellosis is said to be 0.2 per cent based on a total of 6,000,000 tests. Vaccination with Strain 19 was discontinued in 1962. Test and slaughter are now the methods of control, and two million cows are tested annually before calving.

The incidence of tuberculosis is 0.12 per cent based on the same number of tests as for brucellosis. Test and slaughter are also the means employed in controlling this disease.

There is some blackleg, and practically all calves are vaccinated against this disease when under one year of age. On some state farms two innoculations are given, one at three months of age and another at nine months. The vaccine is prepared in Cuba.

Vibriosis has not been identified. The high calving rate obtained on some of the state farms indicates that, if present, the incidence of this disease is not serious.

There is said to be no rabies in cattle.

Both pyroplasmosis and anaplasmosis are carried by ticks. Control is obtained by systematic dipping, and no attempt has been made at eradication of the tick population except for an experimental program on one small island. On the large state farms all cattle are dipped at intervals of one week to thirty days, dependent on the season and the tick population in the area. Asuntol, a preparation manufactured in West Germany, or Neocidol, a Swiss product, are used.

The screwworm is generally present and is combated by treating the navel of newborn calves with an iodine preparation. No attempt has been made at eradication.

Both stomach and intestinal worms are endemic and cause serious loss of condition in young calves unless they are treated. Three or four treatments a year are usually given, one of which is three months before cattle go to a feed lot.

Liver fluke is encountered in cattle on wet pastures but this is not a serious problem. Both affected animals and accumulations of stagnant water are treated as a means of control.

There is very little measles.

MARKETING

In the controlled economy of the communist state the only market transactions are those between private cattle owners and the state. All cattle for slaughter go direct to the slaughterhouses. Private owners are compensated for their animals in cash. The state farms are credited with the value of cattle killed, and the slaughterhouses are credited with the value of the beef going to retail stores or other consumers, such as the army. These transactions are on a value level that supports a retail price of 43 cents a pound for first-class beef—that of a quality suitable for broiling, frying, or roasting—and 38 cents for second-class beef, which is used for stews and the preparation of soup.

The dairy farms are credited with 14 cents per liter for whole milk going to the pasteurizing plants. Bottled milk is distributed to the stores and sold to the consumer for 20 cents per liter.

OUTLOOK FOR CATTLE

During the decade that followed World War II the cattle population of Cuba remained nearly constant, at a little over four million head, according to the FAO. A substantial increase was indicated by 1963, when the number of cattle as estimated by FAO was 5,750,000. The cattle population is now said to be 7,600,000. The number of cattle per 100 inhabitants increased from 75 in the 1947–51 period to 85 in 1966 and 89 in 1971. From 1956 to 1966 beef production increased by only 14 per cent, but milk produced increased 65 per cent. The per capita production of beef declined from 71 pounds in 1956 to 55 pounds in 1966, but milk production per capita increased from 265 pounds to 336 pounds in this period. Such statistics, however, are not in accord with the quantities of livestock products available to the populace. Meat, as bought in the shops, is rationed at 36 pounds annually per person, which fails to correspond to a production of 55 pounds of carcass beef even after allowing for the consumption in restaurants and the high ration allowed the military. Certain trends, however, are indicated. The increase in total cattle numbers reflects the effect of keeping al female calves in order to increase the cattle population, and this practice also could account for the slow increase in beef production. The increase in milk production seems to reflect the result of efforts to expand the dairy industry.

The progress of the Cuban agricultural economy is commonly evaluated on the volume of the annual sugar production. As the principal source of foreign exchange, sugar is of primary importance to the national economy. Variations in the annual production of sugar, however, do not indicate the level of other agricultural progress. This is particularly the case with cattle of both beef and dairy types. Since the revolution special emphasis has been given to the livestock sector of agriculture and especially to cattle. Progress in the production of both beef and milk has been made and appears certain to increase in the years ahead.

Management in all its phases follows modern concepts. The bull testing stations are providing better beef sires, and the progeny testing of the dairy bulls will improve the quality of the dairy sires. The attention that is given to pasture management and the use of supplemental feeds is increasing the productivity of both beef and dairy cattle. These programs are on a national scale and are utilizing modern practices in breed improvement to an extent that is not exceeded in any other Latin-American country.

Dominican Republic

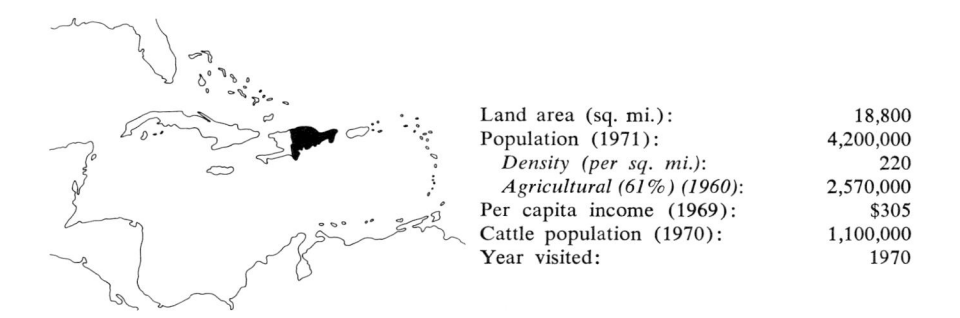

Land area (sq. mi.):	18,800
Population (1971):	4,200,000
Density (per sq. mi.):	220
Agricultural (61%) (1960):	2,570,000
Per capita income (1969):	$305
Cattle population (1970):	1,100,000
Year visited:	1970

THE DOMINICAN REPUBLIC, 18,800 square miles in area, occupies the eastern two-thirds of the island of Hispaniola which lies between Cuba and Puerto Rico in the north-central Caribbean Sea. An ill-defined border nearly 200 miles long separates the Dominican Republic from Haiti, which occupies the western third of the island. The land area of the republic is about one-third the size of Florida. The central part of the country is mountainous, and somewhat less than half the land is devoted to agriculture.

Columbus landed on Hispaniola in 1492 on his first voyage to the Western Hemisphere. He established his headquarters at what is now the city of Santo Domingo, and it remained the seat of his explorations in the Caribbean for a number of years. It was the first sizable settlement in the New World and the center of Spanish influence until the establishment in 1535 of the Viceroyalty of New Spain, as Mexico was then known. The French gained a foothold on the western side of the island during the seventeenth century.

In 1697 this area was ceded to France by Spain, and one hundred years later this French colony became independent as the Republic of Haiti. In 1822, Haiti conquered Santo Domingo, as the area of the present Dominican Republic was known at that time. The Haitians ruled the whole island for 22 years. After they were driven out of Santo Domingo, the Dominican Republic was established. The republic voluntarily returned to Spanish rule in 1861 for a period of four years, but has remained independent since 1865. An occupation by United States Marines, beginning in 1916 and lasting for eight years, was initially instigated to protect United States commercial interests. It was continued, however, because of the threat of European intervention when financial collapse and continuing disorders completely disrupted the economy and government. After the marines withdrew, a freely elected government had a precarious existence for a few years. Rafael Trujillo, in 1930, established his iron dictatorship, which lasted until his assassination thirty-one years later. Vicious political conflict and the threat of a communist takeover followed his death, terminating in complete anarchy in the spring of 1965. To protect American lives, United States troops were landed. These later merged with the military forces of the Organization of American States which had undertaken to protect property and restore order. Free elections were held in 1966 and reasonably orderly government followed, though political unrest continued in the air.

The majority of the population is of mixed Negro and European descent. Most of the sparse Indian population of the island was killed off soon after the Spanish occupation. Whatever remained was assimilated by the new settlers and all traces of the Indians disappeared. There are now an indeterminate number of Haitians in the country, possibly around 25,000, who came over to work in the cane fields and became established as unauthorized residents. The unemployment rate is high, and the population growth rate is said to be 3.5 per cent annually, one of the highest known.

Historically, the economy has been based on sugar. More than 50 per cent of the foreign exchange is now earned by sugar exports. Other agricultural products are cattle, coffee, cacao, bananas, and tobacco. There is some bauxite production, the main nonagricultural export. The United States absorbs 80 per cent of all exports. The basic staples of diet are rice and beans, supplemented with a variety of fruits and vegetables. Very little meat is consumed by the average Dominican family.

Commercial agriculture was developed on the plantation system. There are now many small landowners, but the major production of export items remains in the hands of large landholders. Nearly half of the land is under cultivation or in pasture. Rainfall varies in the farming areas from 20 to 90 inches annually. There is usually a pronounced dry season in the south, starting in November and continuing through April. On the northern coast the dry season is from May through November. Many areas require irrigation for sustained plant growth.

The cattle originally brought to the island by the Spaniards were not an important element in the economy of the early settlers. Frequently

Dominican Republic

they were killed just for their hides. Descendants of the early cattle acquired a high degree of tolerance to the tropical environment through natural selection during the three centuries before the large sugar plantations became established. By the middle of the nineteenth century the cattle population consisted almost entirely of the small, well-acclimated Criollo. The need for a heavier draft animal for plantation work led to the importation of the Zebu from India and probably from other Caribbean islands. Interbreeding, and to some degree planned crossing, of the Zebu with the Criollo cattle produced a superior draft animal. Such crosses were also called Criollo.

As the tractor replaced the draft ox for work in the cane fields, there was a surplus of cattle. The use of oxen is still widespread for cart work, particularly in hauling cane to the plantation rail cars and to the mills. A gradual improvement in the economy since World War II has led to an increased consumption of milk and beef, and the growing of dairy and beef cattle has followed.

CATTLE BREEDS

The commercial cattle population of the Dominican Republic can be placed roughly in three categories—draft, milk, and beef.

Draft oxen and the breeding stock to supply them are estimated at 100,000 head. These are mixed cattle, the result of interbreeding of the original Criollo and the Zebu breeds, principally the American Brahman.

Dairy cattle in commercial operations, that is, those selling milk in an organized pattern, account for around 200,000 head. A few thousand of these are purebred stock of the European breeds, mostly Holstein-Friesian and Brown Swiss of recent importation. Possibly one-fifth of the dairy cattle are of mixed Criollo and European breeding, and the remainder are predominantly Criollo.

The total number of beef cattle is estimated at 250,000 head. These are cattle raised primarily for slaughter as distinguished from discarded draft and dairy stock. A large majority of beef animals carry marked American Brahman influence. Among other breeds represented are some 15,000 head of the Romana Red, either pure or upgraded from the Criollo (the Romana Red is a breed that has been developed on the island. See below). In recent years some European beef breeds have been introduced to be used primarily for crossing. These probably number less than 2,000 head. Most of such cattle are Charolais, though there are some Aberdeen Angus and representatives of the recently

Young Criollo bull, 950 pounds. Owned by a small farmer.

developed breeds of Zebu extraction, as the Brangus, Jamaica Black, Santa Gertrudis, and Charbray.

In addition to the commercial cattle population of 550,000 head, about 450,000 animals are in the hands of the small landowners. Such stock is basically Criollo but often carries Zebu influence. Owned by the essentially subsistence-type farmers in herds of only one to several head, they are used as occasion requires for milk and draft.

Criollo.—These descendants of the original Spanish cattle are found scattered throughout the country. The pure Criollos are the result of natural selection, little attention having been given to their breeding and practically none to selection for a desired characteristic. There is a marked similarity to the pure Criollo as seen in many Central American countries and in the northern countries of South America. This conformity to a common type would logically be expected. Although in different and widely separated areas, the cattle were subjected to similar

Mature Criollo cow, 850 pounds. Owned by a small farmer.

climatic environments and underwent little, if any, artificial selection. It is also known that in the early days of Spanish exploration and settlement cattle were carried from Hispaniola and Cuba to the mainland of both North and South America.

The pure Criollo in the Dominican Republic is a small animal. Mature cows weigh from 500 to 800 pounds, and bulls seldom weigh as much as 1,000 pounds. Color markings are quite distinctive. A nearly solid tan, reminiscent of the Jersey color, predominates. There are, however, wide variations: solid black, mottled black and white, or light tannish grey. All these may have some white undermarkings.

The horns are rather large, upswept, and often with an inward curve at the ends. Conformation varies but is usually quite compact, and the bone structure is fine. The tail is thin and long. A black switch is typical, as are also a black muzzle and black hair around the eyes. The face of the cow has a muscular appearance. Small, tight udders are the rule. The true Criollo has no hump and carries no Zebu influence, but the dewlap

Romana Red bull, ten years old, 1,800 pounds. Central Romana Corporation.

is more prominent than in most nonhumped breeds. Fertility is high; if on a fair level of nutrition, most cows raise a calf every year.

The term Criollo is also used locally in referring to nondescript mixtures of Criollo with the Zebu cattle. Quite typical of these descendants from the Zebu-Criollo cross are the tan-colored oxen with a prominent hump, weighing upward of 1,000 pounds. The mixed Criollo, showing widely varying degrees of Zebu influence, is now much more numerous than the pure type.

American Brahman.—This breed has had the greatest influence on the present-day Dominican cattle. Both purebred American Brahman and strong crosses of these on other cattle are known simply as "Brahman." The general characteristics of the Dominican Brahman are essentially the same as in the American Brahman.

Romana Red.—This recently developed breed is said to have been

Romana Red cow, eight years old, 1,100 pounds.

initiated when one of the large sugar plantations, the Central Romana Corporation, imported 100 head of red, nonhumped cattle from Puerto Rico. Part of the herd was then maintained pure. On the other group, two breeds of Indian Zebu were used, the Mysore and the Nellore. First, a group of the red Puerto Rican cows were bred to bulls of a Mysore-Criollo cross. The resulting progeny were selected for red color. Nellore bulls were later used, followed by periods of back crossing with the nonhumped red bulls of the original line. Selection for a solid red and for good draft characteristic was continued and resulted in large, humped red animals which bred true to type. These were called "Romana Red."

The exact composition of the breed is not known but consists of the red cattle, the Mysore and Nellore Zebu breeds, and the Criollo. As the

An upgraded Holstein-Friesian herd. A nearly pure Criollo cow is in the foreground.

use of draft cattle decreased, the breeding program was modified to selection for a beef type. For a number of years the Central Romana Corporation maintained a closed herd of several thousand head of their Romana Red breed, but recently they have been crossing with Charolais and Charbray. These have been bred artificially from semen brought in from the United States. The objective is to produce a faster-gaining animal for slaughter.

The dictator Trujillo is said to have contributed $1,000,000 to aid in the development of the Romana Red breed.

The color of the Romana Red varies from a bright, medium red to either a much lighter or a much darker shade. Currently the darker shade is preferred. Only minor white markings are permitted on the dewlap and back of the umbilical fold. Horns are of medium size, upturned and quite thick on the bull. The skin is not as loose as on the straight Zebu breeds, and the sheath is quite moderate in size. The dewlap is prominent and loose on both sexes. The tail is very thin with a black switch. A black muzzle and dark hair around the eyes are common. The bone structure tends to be rather fine in the female but is heavy in the

Dominican Republic

male. A beef-type conformation, better than in the straight Zebu breeds, has been developed in the Romana Red breed.

Dairy Breeds.—The two principal dairy breeds which have been introduced are the Holstein-Friesian and the Brown Swiss. Both breeds were imported from the United States and Canada. A few herds are maintained pure, but most are grade animals with either the Holstein or the Brown Swiss influence predominating.

MANAGEMENT PRACTICES

Dairy Cattle.—There are some well-managed dairy farms around Santo Domingo and in other areas of population concentration. These are usually herds of Holstein-Friesian or Brown Swiss cattle. Although a few dairies have milk parlors, milking is generally by hand in open sheds. Cows are on pasture most of the time and are brought in for milking in the early morning and again in the afternoon. The more modern farms put up hay for use during the dry season, employing the method seen in some European countries where the cut grass is spread over slightly elevated wooden slats around a center pole. Such arrangement aids in the curing process during the rainy season, the time when hay must often be cut. A few dairies employ the loose-housing system, keeping the cows in lots around the milking facilities. Their feed is brought to them—green chop in the growing season and hay during those periods when there is little forage growth. A concentrate mixture is frequently fed to milking cows regardless of their level of production.

Male calves, except for the few kept to grow out as bulls, are usually disposed of a few days after birth, often by giving them to small farmers or workmen. Heifer calves for replacements are kept in individual pens, turned to pasture at five or six months of age, and then bred as two-year-olds. The quality of the good dairy herds is maintained by importations of bulls from the United States or Canada.

The large majority of the national dairy herd, however, are Criollo cattle ranging from nearly pure types to those which show a marked influence of one of the European breeds. These cattle typically are in the hands of farmers who rarely own more than twenty head. The cows in milk are maintained on unimproved pastures without supplemental feed. Milk is commonly taken only once daily, and the average production is around 1,500 pounds per lactation. Breeding is usually to any bull that is available in the locality, though a Holstein-Friesian or Brown Swiss sire is used when one is accessible.

In the dairies with herds of 100 head or more, cooling tanks are em-

A cow herd of Central Romana Corporation in process of being upgraded to Romana Red.

ployed for delivery to tank trucks hauling to a pasteurizing plant. In the smaller dairies the milk is handled in cans. Milk produced in the back country is often marketed unpasteurized, being dipped from the milk can in the back of a cart into the housewife's own container.

There is little use of artificial insemination, although an effort was made to introduce this technique several years ago. Lack of an organization to co-ordinate the work and very little interest on the part of the cattle owner appear to be the reasons that its use never became general. A number of individual dairy farms, however, are improving their herds by this method of breeding. The livestock department of the Central Romana Corporation is also using artificial insemination in some of their breeding programs.

Beef Cattle.—Many growers raising cattle for slaughter continue to fol-

A mixed Criollo three-span ox team hauling cane from field to rail cars.

low the same practices that were employed when cattle were used principally for draft. Bulls are run with the cows the year round. There is little grazing control, and mature animals are marketed at the beginning of the dry season. Except for the purchase of Brahman bulls for upgrading when these can be afforded, there is little attention to breeding.

Some of the large sugar plantations, which went into beef cattle production when the need for draft animals decreased, conduct their operations along modern management lines. Upgrading mixed Criollo herds is in progress by the use of purebred American Brahman bulls, or, in the case of the Central Romana Plantation, by their Romana Red and bulls of Charolais breeding.

On these well-managed properties, pastures are maintained in good condition and have been improved by the installation of adequate watering facilities and by fencing for rotational grazing. Brush, weeds, and noxious grasses are grubbed out by hand, often twice a year. There is a limited use of herbicides for this purpose on a few ranches. Pangolagrass and guinea grass have been used in some instances to replace the more prevalent native grasses.

Cattle of North America

On the large sugar plantations the major means of transporting cane to the sugar mills is by railroad. Three spans of oxen however, are used on the carts hauling cane from the field to the points where the rail cars are loaded. Most plantations raise their own draft oxen for this cart work. These are Zebu-Criollo crosses, typical of the first animals which were developed for such purposes. Oxen are first worked as three-year-olds. They are maintained on whatever pastures are available and are given no supplemental feed. As mature five- or six-year-olds, their average weight is 1,100 pounds or more.

Nearly half the cattle in the country are in the hands of the small farmers who own only a few head each. Grazing is on whatever land is available, often in unfenced areas of open country where there is considerable government land. Since practically no attention is given to the welfare of such animals, they depend for survival on their naturally acquired tolerance to the climate and disease. When the owner has a need for cash he sells, usually for slaughter, whatever of his stock he can best do without.

MARKETING

There are no cattle markets as such in the Dominican Republic. Most slaughter cattle are sold by the grower to a trader, a retailer or a wholesaler of meat, or direct to a private slaughterhouse. Price may be negotiated on a per pound basis, but cattle are rarely weighed, so that actually the price paid is simply so much per head. The large grower, one marketing over fifty head, usually sells for export on a per kilogram liveweight basis at the abattoir or, in a few instances, over his own scales. There is no system of grading. The price paid for live cattle depends on a visual estimate of what an animal will dress out.

In early 1970 the top price paid for the best-quality slaughter cattle, two-year-old bulls weighing 800 to 900 pounds, was 19 cents a pound. This was an increase of 3.5 cents a pound over the price paid in 1967 before export of chilled beef to Puerto Rico from USDA-approved plants was started. Prices paid by traders or retailers range downward from the 19-cent level for young bulls to 11 cents a pound for a canner cow in the country. Although the grower has the universal suspicion of the cattleman for the cattle trader, the Dominican trader actually operates on a quite narrow margin. This is probably due to the wide competition in that field.

SLAUGHTERHOUSES

There are four slaughterhouses in the country that have USDA approval

Typical slaughter cattle in holding yard of a private Santo Domingo abattoir.

for export to the United States. Practically the entire output of these plants goes to Puerto Rico. The largest facility, located in Santo Domingo, has a capacity of 120 head per eight-hour day. The captive bolt is used in killing, a mechanical hide puller is employed, and sanitation is excellent. There is no recovery of by-products, however. This plant is municipally owned but leased to a private operator. The other three plants carrying USDA approval are smaller. Two are privately owned and one is a co-operative of cattlemen in the eastern region.

The cities and larger towns all have municipal abattoirs. In addition to these, there are a number of small privately owned plants. There are also many other private operations whose facilities vary from little more than a concrete slab with water for washing down to nothing but a knife and a tree limb. Most of these operators are retailers who purchase an animal for slaughter direct from the grower or trader.

There is a legal requirement that all cattle killed in the municipal plants must be inspected for health, that adequate sanitation provisions are maintained, carcasses weighed, and ownership checked by the police. These same requirements apply to the authorized private plants. The unauthorized slaughter is entirely outside the law. Except in the USDA-approved plants, however, these legal requirements are seldom met. In most other facilities inspection of live animals and vital organs is perfunctory or lacking entirely.

CATTLE DISEASES

The Dominican Republic is free of foot-and-mouth disease, rinderpest, and all of the decimating cattle diseases. Effective prohibition is maintained on the importation of cattle from foot-and-mouth-infected countries. Parasites and common diseases of cattle, however, take a heavy toll of stock because of the lack of protective measures and also because of inadequate nutrition during the dry season. The calving percentages on the exceptionally well-managed sugar plantations are said to approach 90 per cent, but probably averages less than 65 per cent in the beef cattle area in the eastern part of the country. For the small cattle owners with nearly one-half of the national herd, the annual calf crop is in the range of 35 per cent of the breeding cows. There is usually no immunization for the common cattle diseases by the owner of only a few head.

The cattle-tick population varies widely in different areas, depending on the effectiveness of the dipping or spraying programs employed. On ranches where such treatment has been practiced systematically for years, spraying once a month gives satisfactory control. In other areas, dipping or spraying every 10 to 14 days is necessary to avoid serious loss. The major tick-borne diseases are pyroplasmosis and anaplasmosis. Both are major causes of loss in cattle imported from the United States or Canada. Newly introduced animals require unusual care and special treatment until they develop a degree of tolerance.

Brucellosis is widespread except in a few well-managed herds where it has been practically eliminated. The incidence of the disease over the country varies from 5 to 12 per cent. A control program was started in 1967 involving the vaccination of heifer calves at four to eight months of age which called for the treatment of 24,000 head by 1970. The program was unsuccessful. Since there is no control on animals that are found to be reactors and no reimbursement to the owner for their slaughter, effective control is thus impossible.

These same handicaps apply to the control program which was started in 1968 for tuberculosis. The incidence of this disease is estimated to be 1 to 2 per cent of the entire cattle population. There has been some response to the recommendation to sell for slaughter the known brucellosis and tuberculosis reactors as they are detected, but this has not been at a level which would effect control of these diseases in the near future. Voluntary slaughter of infected animals is considerably higher in the beef herds than in the dairy cattle.

The incidence of liver fluke is high among cattle grazed in low-lying areas. Liver condemnations in government-inspected abattoirs run as high as 80 per cent at times. There is some loss in condition of cattle infected by the disease, but other than this and the loss of the livers, it is not considered a major menace to the cattle population.

There have been cases of rabies, but this disease is not considered a serious threat and there is practically no vaccination against it. Gastrointestinal worms are prevalent throughout the country, and the well-managed ranches worm calves at weaning. The screwworm causes no concern.

GOVERNMENT AND CATTLE

There is little evidence of activity on the part of government in the cattle industry other than in the control of the importation of breeding stock. This ineffectual attitude is seen in the lack of enforcement of the laws on cattle slaughter. The large operators, chiefly sugar plantation owners, are well able to improve their programs on their own initiative and are proceeding to do so. The same is true of the large dairies around the principal cities that are owned by business and professional men. The small farmer, with his few head of cattle, has no desire for any government involvement in his operation. From past experience he has come to distrust any official activity involved in his effort to gain a livelihood. It seems that at least for the time being the cattle industry will get along without aid or interference from government.

OUTLOOK FOR CATTLE

The area of land now utilized for pasture is 3,400,000 acres, of which two-thirds is in developed pastures. Such improved land has been fenced for grazing control, protected against weed infestation, and provided with ample stock water. The carrying capacity of the undeveloped pastures could be at least doubled and that of the developed pastures increased by 50 per cent if all the land were brought up to the level of that on the present well-managed properties. Irrigation during the dry season and fertilization would increase even further the carrying capacity of of the land now in pasture.

There are 6,700,000 acres of undeveloped and unutilized lands, a considerable part of which could be developed for grazing if the rank vegetation were first removed.

The current commercial milk supply comes from a national cow herd of 200,000 cows with an average annual production of only 1,500 pounds a cow. With more productive stock, the volume of milk could be doubled with no increase in the total number of cows. This would mean raising the average production to the 3,000-pound level, which is a very realistic goal. The increase would be readily absorbed since milk is now in short supply in the country. The obvious means of obtaining this goal are better nutrition and the use of artificial insemination. The former could be provided by good pasture management, combined with supplemental feeding during the dry season. The latter would, in a relatively short time, improve the genetic potential of the cow herd.

Competent direction and the necessary facilities for such programs do not appear to be in the offing. Such accomplishments would require a cultural change in the attitude of the small farmers who now own most of the dairy cows. They seem content with their lot and are not inclined to go into a major change, especially one that means more work.

The beef industry, as it exists today, is dependent on exports for expansion. Currently, the limiting factor is the quota on the beef which can be sold to Puerto Rico and the United States. (In 1970, this was 11 million pounds). The heavy competition from other Caribbean countries and from most of Middle America appears to preclude much increase in this outlet. If a larger market for beef opened up, the large cattle raisers could soon find ways to increase their production.

All things considered, the untapped resources for both milk and beef production in the Dominican Republic will probably not be utilized for some time to come. Either a major change in the economic level of the republic, which would enable the population to buy more beef and milk, or a pronounced shortage of beef in the world market, will have to precede such a development.

Haiti

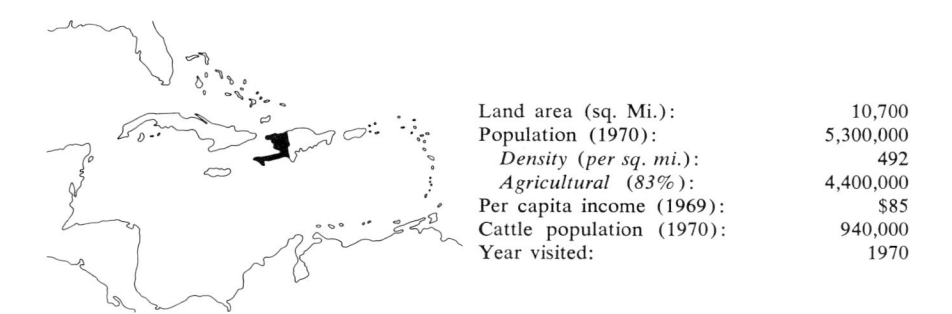

Land area (sq. Mi.):	10,700
Population (1970):	5,300,000
Density (per sq. mi.):	492
Agricultural (83%):	4,400,000
Per capita income (1969):	$85
Cattle population (1970):	940,000
Year visited:	1970

THE REPUBLIC OF HAITI occupies the western side, roughly one-third, of the island of Hispaniola and three adjacent small islands. In area the country is one-fifth the size of Alabama. In the southern part, a narrow peninsula juts out to the west. This area, 150 miles long with an average width of 25 miles, is known as "the South." The northern part of the country is dominated by a central plateau. The major streams rise in the rough country near the border with the Dominican Republic. Much of the land is semiarid with an annual rainfall of around 20 inches. Precipitation, however, increases to 95 inches in the mountainous interior.

Columbus landed on the northern coast on his first voyage to the Western Hemisphere. The entire island of Hispaniola was then controlled by Spain until the end of the seventeenth century. The part that is now Haiti was later settled by the French after the area was ceded to France by Spain in 1697. Less than one hundred years later, seven years of organized revolt among the slave population led to independence from France. Thus, in 1804, Haiti became the second republic in the Western Hemisphere and the oldest in the world with a dominant African population.

In 1822 the Haitians conquered Santo Domingo (now the Dominican Republic) and maintained control for 22 years. Santo Domingo broke away in 1844. There have been frequently recurring intervals of political strife since Haiti became independent. By 1915 the economy had deteriorated to the point where the country was occupied by United States Marines to protect life and property. This occupation lasted until 1934.

Creole cow having the tan color often seen on these cattle.

Although there was then some improvement in the economy and financial responsibility of the government, there was no political continuity until François Duvalier was elected president in 1957. He soon consolidated his position, ruled with an iron hand, and legally became "president for life" in 1964. Duvalier's power was disputed at times, but as dictator he managed to keep law and order in a rigidly controlled police state until his death in early 1971.

Most of the population is of pure African descent. There are small numbers of Caucasians and people of Levantine extraction and a possible 5 per cent of mixed African and European descent.

The economy is almost entirely agricultural. Coffee is the principal export item and the main foreign-exchange earner. Farming is mostly of subsistence crops in the hands of small landowners. The proceeds of any sale of what the land produces go out as soon as received to purchase the bare necessities of life. There are a few large sugar plantations and some sisal production, but these constitute only a small part of the total agricultural output.

The two railroads in the country are used only for hauling cane. One is a narrow-gauge line extending along the coast from 25 miles west of

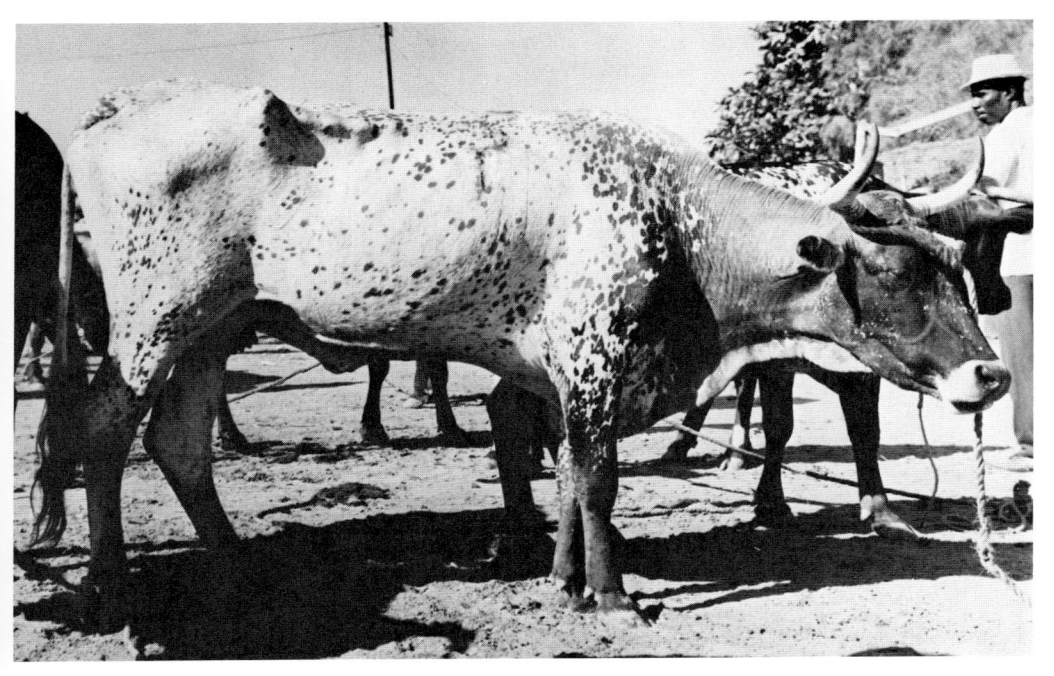

Creole cow with the mottled black and white pattern that is common.

the capital city of Port au Prince to 65 miles north; the other is a short line running along the north coast. The only roads suitable for automobile travel do not extend much farther, and transportation into the interior is largely dependent on trucks or four-wheel-drive vehicles with a high clearance.

Cattle play a small part in the economy. Bulls and oxen are used for cart draft on the few large sugar plantations, and there are a few organized dairy operations. But the cattle population is largely in the hands of the small farmers, who hold the cattle for the most part as the "peasant's purse." These animals are rarely employed for draft, and the hand hoe continues to be the standard method of cultivation. Some milk is taken, but this is not general and the owners rarely slaughter an animal for their own consumption. When a need for cash arises, the animals are driven to the nearest market to be sold for slaughter or to another farmer who can afford them.

CATTLE BREEDS

The Spaniards brought the first cattle to the island in the early days of

Cow carrying headstock used to keep cattle out of fenced plots.

settlement. After Haiti was acquired by France, some cattle may also have been brought from that country. Any such infusions occurred before the days of established breeds in either Spain or France. It appears, however, that the descendants of the Spanish stock had the major influence on the cattle population of Haiti. The pure Creole cattle seen today are very similar to the Criollo cattle of Central America, in which area there was certainly no introduction of French cattle.

Creole.—Zebu types of cattle were not introduced to Haiti until 1949, nearly one hundred years after the Zebu had been carried to most Caribbean shores. The importations of European breeds in more recent years have also been minor in comparison with the numbers which have entered neighboring islands. Both of these factors account for the

One of the few large dairy herds in Haiti, on irrigated Pangola pasture.

Haitian cattle population's being more strongly Creole in type than those in other parts of the Caribbean.

The pure Creole cattle in Haiti are practically identical to the pure Criollo described under "Cattle Breeds" in the chapter on the Dominican Republic.

The mixed Creoles vary widely in size, color, and conformation. The American Brahman–Creole cross is the typical draft type used in the cane fields. This is a moderately humped animal. The mixed Creole, resulting from crossing with the European dairy breeds, is nonhumped. Some of these European crosses have been carried far enough that the characteristics of the dairy breed are sufficient to classify the animals as upgraded.

Dairy Breeds.—Only limited numbers of European dairy cattle have been imported to Haiti, and these have left little impact on the cattle population. At various times in the recent past, Brown Swiss, Jersey, and Holstein-Friesian cattle have been brought in and efforts were made to develop upgraded herds, but very little was accomplished. There is one herd of 25 upgraded Holstein cows of excellent quality. Most of the

Cattle of North America

cows in the few organized dairies are are mixed Creole, with some animals showing the characteristics of one of the European dairy breeds.

MANAGEMENT PRACTICES

The small subsistence farmer usually has from one to three or four head of cattle. Single ownership of more than five head is exceptional. Such cattle are pastured by tethering with a rope over the horns and then secured to a stake in the ground. Less frequently, cattle are permitted to run free with a headstock of forked branches around the neck which prevents the animal from grazing on the enclosed cultivated plots.

During the rainy season there is ample grass for forage. When the rains cease or during one of the frequent drought periods, only dry grass or crop residues are available for feed and cattle go down rapidly in condition. Little attention is given the cow of the small owner other than to see that she is staked out each morning on whatever grassland is available. If he lives near a city or town, the cow owner may sell raw milk to a customer or two, in addition to what he retains for his own household use. Other than this, very little use is made of cattle. When cash is needed, a single animal is led to the local market and sold to a trader. As soon as he can manage to do so, the farmer acquires another animal, possibly a calf to raise to maturity or until there is again need to dispose of it.

The largest dairy operation in Haiti is one with a herd of 200 mixed Criollo cows, some of which could be considered upgraded Holstein-Friesian and Brown Swiss. Around 100 head of these are maintained in milk, run on irrigated pastures of pangolagrass or guinea grass and fed a limited supplemental ration of wheat bran, raw cottonseed, and some cottonseed meal. Milking is by hand in open sheds, and the calves are kept in nearby small pastures. Milk is pasteurized by placing the milk cans in a vat of hot water which is heated by a wood fire. After thirty minutes under heat, the cans are transferred to a cold water tank where the milk is cooled and then put in pint bottles by hand. This method of pasteurization is probably reasonably effective. The milk is sold to consumers at eight cents a pint. Surplus milk is processed on the farm into several varieties of very acceptable cheese. While this is the only dairy operation of this magnitude in the country, there are several dairies milking around 100 head of mixed Creole and upgraded cows.

The major commercial use of cattle, other than sale for slaughter, is in the cane fields of the few plantations that are located in the areas served by the narrow-gauge railroads. Cultivation of cane is now by tractor, but two-span bull teams are used for transportation from the field to

Two-span mixed Creole ox and bull team hauling cane.

the rail cars. Castration of these work animals is haphazard, a poorly applied tourniquet or rubber band being used. Many work bulls are left entire, probably through carelessness. Because they are overworked and on inadequate diet, there is no difficulty in handling the beasts as draft animals. All are of mixed Zebu-Creole breeding. Draft animals are tethered during the periods when they are not worked to graze on such grass or crop residues as are available. No supplemental feed is given them even during the heavy hauling season.

In more remote areas where sugar cane is grown to a limited extent by small farmers, bulls or oxen are used to power the small cane mills. The product of these is either molasses or crude brown sugar that is sold locally.

MARKETING

Probably because of its French colonial heritage, Haiti is one of the few Caribbean countries where cattle are traded in organized markets. There are weekly market days in the larger towns, and in the cattle centers these markets have a sizable livestock section. At the Croix de Bouquets market, as many as 600 head of cattle have been traded in one day. They are trailed for distances of over 100 miles, often requiring three or four days en route. Transactions are usually for one head at a time, between the owner and trader, or occasionally between an owner and another farmer for a cow to go back to the country. Negotiations on a single animal may continue off and on for several hours. The traders buy principally for resale to butchers who do their own slaughtering in municipal abattoirs, or to the one private abattoir in the country located outside Port au Prince.

There are no scales and cattle are sold on a per head basis, priced on an appraisal of what an animal will dress out. In early 1970 a 900-pound bull in fair condition that would yield a 55 per cent carcass was bringing around $100, or 11 cents a pound. A thin canner weighing 500 pounds would sell for $30, or 6 cents a pound. The Haitian American Meat and Provision Company (HAMPCO), a private abattoir slaughtering mostly for export, buys on a liveweight basis from the traders such stock as they were unable to sell locally. The prices involved are 10 to 20 per cent below those paid by the butchers for animals of the same quality. The price paid for bulls is roughly one cent a pound (liveweight basis) higher than that paid for cows. The government tax collectors must be paid a tax of 60 cents a head on all cattle sold before they can be taken from the sales area. All sales by the first owners to the traders are for cash. The traders frequently sell on credit to the town butchers who do not have funds to purchase an animal until the carcass is dressed and sold at retail.

SLAUGHTERHOUSES

HAMPCO, the modern abattoir located at Damien just outside Port au Prince is owned by American interests. The plant has a capacity of 125 head per day, but in 1970 was killing only 9,000 annually because of the scarcity of butcher stock. The facilities have USDA approval for export and, except for a limited luxury trade, most of the output is shipped as chilled or frozen beef to Puerto Rico. A limited number of hogs and goats are also processed by HAMPCO.

The larger towns have municipal abattoirs, the best of which consist of a small, open building, screened on all sides, with a concrete floor and water for washing down. Slaughter and processing are by the individual butchers. There is also much unauthorized slaughter done in the open under a tree. Inspection at the smaller slaughterhouses is limited; the only effective sanitary provisions are at the HAMPCO plant.

CATTLE DISEASES

Haiti has had no history of foot-and-mouth disease, and is recognized by the USDA as a foot-and-mouth-free country. Tuberculosis is not prevalent, and there is very little trouble from screwworms. The tick-borne anaplasmosis and pyroplasmosis are probably the most serious cattle diseases, but these cause no heavy losses because of the acquired tolerance of the native cattle to them. Leptospirosis in recent years has caused considerable calving losses in imported cattle. Many cattle are infected with internal parasites, and 90 per cent of those in marshy areas have liver fluke. There is very little prophylactic vaccination except in a few of the better-managed dairy herds.

The soils are generally lacking in phosphorus, and most cattle suffer from a deficiency of this mineral. Cows often do not conceive until the calf is weaned, and a phosphorus deficiency is considered to be the major cause of this condition. Except in the few well-managed dairy herds, the calving percentage is probably less than 40 per cent annually. Other factors, however, also contribute to this low fertility, particularly a high incidence of brucellosis.

A serious outbreak of anthrax occurred in 1969. A widespread vaccination program was initiated by the veterinary services of the Department of Agriculture, and in early 1970 the disease was considered to be under control. The program contemplated the vaccination of all cattle once a year, but this goal will be impossible to achieve in the large number of widely scattered small herds throughout the country.

GOVERNMENT AND CATTLE

Government participation in the problems of cattle raising is quite nominal. The effectiveness of the anthrax control program is mentioned above. There is a small national agricultural college near Port au Prince. The former plan was to enroll 40 students who would continue as a unit until graduation four years later, after which another class of 40 students would enter. In 1968 this program was changed so that 45 students are now enrolled every year. Most graduates find employment in government service, with such organizations as FAO, or in industry, and thus add little to the development of better management practices in raising cattle in the country.

OUTLOOK FOR CATTLE

Development in all lines of human endeavor in Haiti will have to precede any major improvement in cattle raising. A higher economic level for the mass of the population is necessary to provide a market for the increased production of beef and milk which could be obtained under better management. The limited export market which Haiti enjoys cannot be increased materially. The limitation of the quota restrictions of the USDA which apply to Puerto Rico, Haiti's only export customer, and the strong competition of other Caribbean countries for this market hold Haiti's beef exports within quite narrow limits.

Better management could multiply several fold the productivity of the country's cattle. This would involve development of improved pastures which would require irrigation works for use during the dry season, fencing for grazing control, and provision for stock water. Upgrading the predominantly Creole cattle to more productive beef or dairy animals could also be accomplished under adequate management. Systematic marketing of beef cattle at younger ages, instead of selling for slaughter only the worn-out draft or dairy animal or the odd-lot cow of the small farmer, would add materially to the tonnage of beef now produced.

Such developments as these are for a future day when the mass of the population has climbed to a higher economic plane. For the present, the majority of Haitian cattle will continue much as they have in the past.

Jamaica

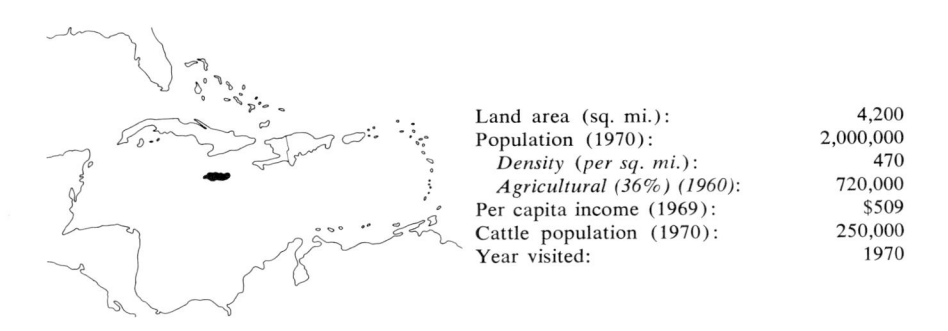

Land area (sq. mi.):	4,200
Population (1970):	2,000,000
Density (per sq. mi.):	470
Agricultural (36%) (1960):	720,000
Per capita income (1969):	$509
Cattle population (1970):	250,000
Year visited:	1970

JAMAICA, A MEMBER OF the British Commonwealth of Nations, lies about 90 miles off the southern coast of Cuba astride the 18th parallel. The land area, 141 miles in length with a maximum width of 51 miles, is one-tenth the size of Louisiana. There are narrow plains on the northern side of the mountainous spine which runs the length of the island, and there are wider plains on the southern side. Short watercourses run to the northern coast and longer rivers to the southern. Some of these are dry during drought periods. The climate, while subtropical with little seasonal temperature variation, is usually quite equable because of the persistent offshore winds. Rainfall varies from a low of 20 inches in some parts of the southern plains to 120 inches and over in the rain forest of the northeastern mountains. Plant growth over much of the island is inhibited during the pronounced dry season from December through May.

Columbus landed on the island in 1494, and colonization was begun by the Spaniards in the first years of the sixteenth century. British forces occupied the island 150 years later, continued the colonization process, and retained direct control until 1958. Jamaica then joined with nine other British island territories in the Caribbean in the formation of the West Indies Federation. This was a step in Britain's program to grant self-government to her West Indies possessions. The effort failed, however, because of the political jealousies among the several national components of the Federation. Jamaica withdrew from it in 1962.

Although over 80 per cent of the population is African or of mixed African descent, the British heritage remains strong throughout the island. Minor elements in the population include European, East Indian, Syrian, and Chinese. Slavery, an essential element in the early agricultural development, was abolished in 1838. Jamaica was among the first European colonies in the Western Hemisphere to take this step.

Government now follows the British parliamentary system with an elected prime minister. The Queen of England is the nominal head of state, represented by a governor-general who is appointed on the recommendation of the prime minister. The upper house of the legislature is the Senate. Members are appointed; the prime minister is responsible for 13 seats and the opposition party for eight of the 21-member body. The House of Representatives is elected by universal suffrage.

Jamaica historically has had an agricultural economy in which sugar has occupied the primary position and bananas second. Other crops of lesser importance were coffee, cocao, and coconuts. Cereals for human consumption have been largely imported. The picture is now different. Bauxite and alumina are the largest foreign exchange earners and tourism is second. Together these accounted for over 60 per cent of the foreign exchange earnings in 1967. Sugar and bananas combined now total only 18 per cent of exports. Jamaica has suffered in recent years both from the effect of droughts on cane production and the low price of sugar in the world market. Industrial production is steadily increasing, though at a rate which is too low to maintain employment at a reasonable level.

CATTLE BREEDS

In the early days of Spanish settlement cattle were brought to Jamaica from Spain and probably from the earlier established Spanish possessions in the Western Hemisphere. Cattle were used initially for milk and draft, but as the cattle population increased, hides became an important product. The descendants of these first bovine inhabitants were called "Creole" by the British plantation owners. With little care or selection, these developed along the same lines as the Criollo cattle throughout Spain's possessions in the Western Hemisphere. They acquired a high degree of tolerance to the tropical climate over a period of 300 years. As the sugar plantations became more productive, a larger draft animal was needed. To this end, Zebu cattle from India were imported in the early 1860's. These, when crossed with the Creole, produced larger and well-adapted draft animals.

These importations of Zebu cattle from India and the resulting admixture with the Creole constituted the major cattle development until the turn of the twentieth century. Then the British colonists, with their

Jamaica

inherent love for good cattle, began to bring in their favored British breeds. Hereford, Shorthorn, Aberdeen-Angus, Devon, Red Poll, Jersey, Ayrshire, and Guernsey were all introduced in varying extent. These breeds were kept in a pure state as well as used for crossing on the mixed Zebu-Creole cattle. Somewhat later there were sizable importations of Holstein-Friesian to supply the increased demand for dairy products.

Because of a foot-and-mouth disease outbreak, the importation of cattle from India was discontinued in 1921. By this time the Creole cow had practically disappeared. The Jersey and Red Poll breeds and their unplanned crosses on the Zebu-Creole cattle seemed to do better in the tropical climate than the other British breeds. Next to the Zebu, these two breeds became the most important factor in the development of Jamaican cattle.

Four breeds new to Jamaica have been developed on the island in recent years. Each of these has its own breed society operating under the wing of the Jamaican Livestock Association. The Jamaica Brahman Cattle Breeders Society was established in 1949; the Jamaica Hope Society in 1952; the Jamaica Red Poll Society in 1952; and the Jamaica Black Society in 1954. The Jamaica Hope is a dairy breed; the other three breeds have been developed for beef.

Jamaica Hope.—In 1910 a small experiment station, the Hope Farm, was established in the plains area in south-central Jamaica. The development of a productive dairy animal was one of the early objectives. The initial breeding work, utilizing some of the more prominent British dairy breeds as well as the Zebu and Creole, led to the conclusion that it was highly desirable to retain some Zebu influence because of the tropical environment.

In the course of the experimental breeding work at the station, Norbrook, a Creole cow possibly carrying some Ayrshire influence, was mated to a Sahiwal bull that had recently been imported from India. The Sahiwal is one of the better milk-producing Zebu breeds. Norbrook was later mated to an imported Jersey bull. The male progeny by these two sires were used extensively on grade Jersey cows in the area. As time went on, there was further use of imported Jersey bulls as well as *inter se* breeding.

A somewhat parallel program, although not as extensive, was followed in the development of a cross of the Sahiwal on Holstein-Friesian cows. In 1950 it was decided to concentrate on the Sahiwal-Jersey breeding. A few high-producing Sahiwal-Holstein-cross cows, however,

Jamaica Hope bull, four years old, 1,500 pounds. Bodles Artificial Insemination Center.

were introduced into the main herd. It was from the mixing of the Jersey, Sahiwal, and, to a minor extent, the Holstein-Friesian, breeds that the Jamaica Hope was formed. The breed is estimated to be 96 to 98 per cent Jersey.

Selection of the Jamaica Hope initially was entirely for type without regard to color. It now embodies milk productivity, fertility, and longevity. Since color was no criterion in the selection, a wide variation exists in any sizable herd. There is a preponderance of the typical, nearly solid tan of the Jersey, but there are also browns, greys, and almost solid blacks. There are individuals in any of these colors with some white spotting. This color range is not too different from that of the Jersey cattle seen today on the Island of Jersey where many individuals do not display the typical tan color. The Jamaica Hope is a somewhat larger animal than the usual Jersey, approaching the size that the Jersey breeders in the United States are now striving for. Cows in a good herd average close to 1,000 pounds, and typical bulls weigh from 1,400 to 1,600 pounds. Conformation is that of a moderate dairy type. The

Jamaica Hope cow, twelve years old, 950 pounds. Bodles Artificial Insemination Center.

loose skin, a characteristic of the Zebu, is noticeable, particularly in the dewlap. There is no evidence of the Zebu hump.

Milk production is exceptional for a tropical breed. The highest-performing cow has produced over 20,000 pounds of 4.9 per cent butterfat milk in a 305-day lactation. Production of the foundation herd at Bodles Animal Production Research Station (successor of the Hope Farm), where the breed is now being improved, has been increased from an average of 7,248 pounds in the 1950–54 period to 9,179 pounds in 1965. Part of this increase, however, is credited to improved nutrition. First-calf heifers now produce slightly more milk in a lactation than mature cows did fifteen years ago. Good commercial herds have average productions of over 5,000 pounds per lactation.

Jamaica Brahman.—The earliest importations of Zebu cattle were from India in 1860 for the development of a better draft animal. These were of the Mysore, Hissar, Guzerat, and Nellore breeds, the last two being the most numerous. Unplanned breeding continued for many

Jamaica Brahman bull, three and one-half years old, 1,400 pounds. Worthy Park Farms.

years. This resulted in a variety of mixtures as the Zebu breeds crossed with the Creole descendants of the cattle brought to the island three centuries earlier.

When the tractor began to supplant the draft ox in the early 1940's, plantation owners began to devote their breeding efforts to the development of an improved beef-type animal. The result was the Jamaica Brahman, an animal quite similar in its general characteristics to the American Brahman. This breed had been developed some years previously in the United States from essentially the same Zebu breeds. Over the past twenty years large numbers of American Brahman bulls have been imported from the United States and introduced into the Jamaica herds. While these have had considerable influence, there is no question that the Jamaica Brahman was developed from Zebu cattle imported from India, independent of the American developed breed.

The Jamaica Brahman has characteristics typical of the Zebu breeds —loose skin, long ears, and pendulous dewlap, sheath, and umbilical

Jamaica Brahman cow with calf, four years old, 1,100 pounds. Worthy Park Farms.

fold. In these features as well as in color, the breed is practically identical to the grey American Brahman. The American breed has somewhat better fleshing qualities generally, but the Jamaica Brahman tends to have a less pronounced slope to the rump. This trait is particularly noticeable in some registered herds. Mature cows weigh from 1,000 to 1,200 pounds, bulls from 1,500 to 1,800 pounds. The breed has gained an enviable reputation in Central America and the northern countries of South America, and sizable numbers of bulls are exported to these areas.

Jamaica Red Poll (Jamaica Red).—The English Red Poll was introduced into Jamaica and other Caribbean islands over one hundred years ago. It was employed for upgrading the native cattle for better milk production and, in some herds, was kept relatively pure. Some crossing with Zebu bulls was practiced which initially contributed to the ability to withstand heat stress. The Red Poll characteristics were generally maintained, however, by the continued use of bulls imported from Britain and their progeny. There is a minor South Devon influence in

Jamaica Red Poll bull, mature, 1,650 pounds. Worthy Park Farms.

Jamaica

the breed as it exists in Jamaica today. This is a result of the use of bulls from a Zebu cross on a South Devon cow in the early Red Poll herds.

Selection was later made for fleshing characteristics as well as for milking quality. When the breed society was organized in 1953, selection requirements discriminated against Zebu characteristics such as the hump and pendulous dewlap. The Zebu influence, which initially was essential to good growth in the tropical climate, was thus largely bred out. The Jamaica Red Poll as seen today is not quite as large as its British counterpart, but it is well adapted to the tropics. Color is a medium red; the body is well muscled and smooth; and, for a beef breed, the cow has a good udder. The average cow weighs 1,200 pounds, bulls up to 2,000 pounds.

Jamaica Black.—Aberdeen-Angus bulls were introduced later than the Zebu and Red Poll breeds for upgrading native cattle. The success of the other island breeds—Jamaica Hope, Jamaica Brahman, and Jamaica Red Poll—led the breeders, who had been using Angus bulls for crossing, to establish in 1954 their own society for the registry of Jamaica Black cattle. These are an indeterminate cross of Zebu and Aberdeen-Angus. The general opinion is that the Zebu influence is of the order of 8 to 10 per cent.

Cows in good herds average 1,100 pounds, bulls 1,400 to 1,600 pounds. The polled characteristic of the Angus has been retained. Color is solid black with minor white markings permitted on the underside, back of the umbilical fold. The Zebu influence is discernible but not prominent since the hump in the male is quite small and seldom noticeable in the female. The dewlap, sheath, and umbilical fold are much less prominent than in a straight Zebu-Angus cross. The Jamaica Black tends more toward the Angus conformation than does the American Brangus, which is also a Zebu-Angus cross. In recent years the Jamaica Black Breed Society has discriminated against animals displaying strong Zebu characteristics.

Holstein-Friesian.—At various times over the past sixty years purebred stock of this breed was imported from both the United States and Canada. Some effort was made to develop heat tolerance by crossing with the Zebu, but the Holstein herds deteriorated and found little place in the dairy industry. In recent years sizable importations again have been made. The incentive for this development was the increase in demand for milk and the prominent world-wide position the Holstein-Friesian holds as a superior dairy animal. For the most part, the Holstein

Jamaica Black bull, mature, 1,500 pounds. St. Jago Properties.

herds that have been established in Jamaica are maintained pure by continuing importations. They are meeting with more success than in the past, but their future in the country as a good dairy animal does not appear to be assured. Generally the herds seen throughout the country, even under conditions of good management, do not show the thrifty condition of the Jamaica Hope. For the same quality of cattle, the milk production of the Holstein in Jamaica is considerably less than is obtained in places with cooler climates to which it is well adapted.

Other Breeds.—Recently some of the larger producers have been crossing Charolais bulls on the recognized beef breeds. This produces a popular slaughter animal. Fed bulls are 100 pounds heavier at 12 to 15 months of age than bulls of the other breeds handled under comparable conditions. Crosses of ½, ¾, or ⅞ Charolais heifers from Jamaica Brahman cows do quite well in the tropical environment when under good management.

Some breeders have used Santa Gertrudis bulls for crossing on Brahman cattle. Reynolds Metals Company has made wide use of Santa Gertrudis in upgrading its herd of Red Poll cattle.

Jamaica Black cow, mature, 1,100 pounds. St. Jago Properties.

Some dairy breeders have been endeavoring to obtain a better young slaughter animal by crossing Brown Swiss bulls on the Jamaica Hope. No definite trend along this line, however, has been established.

MANAGEMENT PRACTICES

In former years the primary use of cattle was for draft. The Zebu-Creole draft animal was well adapted to the environment and could perform its work without receiving much attention. As the draft animals were replaced by the dairy cow and beef animal, it was found profitable to provide better pastures, to control breeding along predetermined lines, and to employ better health-protection measures. Management became the determining factor in the productivity of the national herd, and in 1970 was still improving—from quite elementary practices on some properties to examples of excellent modern methods on others. This is true for both beef and dairy operations.

The practice of leaving bulls with the cows the year round is declining. Many producers now have controlled breeding seasons so that calves will be dropped several weeks before the rains commence. This permits the calf to get a good start before the problems that go with wet weather are encountered. There is also a trend toward breeding cows for the first time at younger ages. Whereas it was formerly common practice to breed cows first at three years of age, progressive producers now breed at 15 to 18 months or when the heifers reach a predetermined weight.

Improved pastures have been established at an increased rate since the end of World War II. Pangolagrass was introduced and has replaced guinea grass as the main pasture plant. Fertilization, principally with nitrogen, became quite general on the better pasture lands along with rotational grazing to ensure maximum production. Since the Jamaican soils are thin and low in plant nutrients, such practices are essential for good production.

The annual rainfall is high enough over most of the cattle-raising area that normally it would be considered ample for good plant growth. It is of such a poor pattern, though, falling mostly from June to December, that a long dry season seriously retards growth. During the latter part of the dry period, only mature dry grass remains for forage. Irrigation is not widespread as yet, but it is increasing. Water for irrigation is obtained from the streams and also from wells. Since many streams are only dry beds after the rains cease and storage reservoirs are meager, wells are the most reliable source of irrigation water.

Marketing of younger animals for slaughter is a growing practice. Two methods of growing out slaughter animals are being followed.

In one, young stock to be grown for beef are kept on improved pastures where good growth is promoted by careful rotation from one pasture to another. With a good quality commercial herd of one of the Jamaican breeds, this feeding method produces a slaughter animal at 30 months that yields a carcass of 500 pounds or more.

Recently a new type of feed-lot operation has been started on some of the larger farms where beef cattle are raised. Animals destined for slaughter, usually bull calves or yearlings, are placed on a full feed for 100 to 200 days. The typical ration consists of 50 per cent dried citrus pulp, 40 per cent brewers' grains, and 10 per cent coconut cake. This is fed along with 4 or 5 pounds a day of chopped green grass. Consumption of the concentrate mixture averages 15 pounds a head daily. Average daily gains of 2½ pounds are obtained. Such an operation produces a carcass of 400 pounds or more at 24 months of age.

Reynolds Jamaica Farms Beef Operation.—The outstanding beef-cattle development in the country is that of Reynolds Metals Company. Baux-

Jamaica

ite mining concessions granted by the government require that, after strip mining, the land must be put back into as productive a state as it was originally. Reynolds has effected a 100 per cent rehabilitation of the land it has mined. This has been accomplished by conserving the topsoil and replacing it after mining has been completed. The large land areas that are held in reserve for future operations are also utilized for stock raising.

Improved pastures of pangolagrass or guinea grass were developed by fertilizing the land. The pastures were cross-fenced for grazing control and provided with stock water where necessary. Such local stock as could be obtained—largely Red Poll, South Devon, and Angus grade stock, along with Zebu crosses—were upgraded by the use of Santa Gertrudis, Brahman, and Charolais bulls. The Santa Gertrudis crosses eventually were determined to be the most productive.

The Reynolds herd now totals 13,000 head and has been selected by weaning weights supplemented by later gains, trueness to type, and fertility. Replacement heifers must weigh a minimum of 425 pounds at 205 days of age, and 800 pounds at two years of age before they can enter the breeding herd. The requirement for herd sires is 450 pounds at 205 days; and on supplemental feed after weaning, 800 pounds at one year and 1,000 pounds at 18 months. Calving is over 80 per cent, and the death loss to weaning is slightly over 2 per cent. Artificial insemination is employed on much of the herd. The cows are inseminated during two heat periods, then turned with pickup bulls.

Slaughter cattle are grown out on grass and finished with a supplemental ration of seven pounds of grain a day. The cattle are slaughtered in a modern company-owned abattoir which also handles a few consignments of nearby producers.

The Reynolds Farms are an outstanding example of what good management can accomplish in the tropics in the way of beef production. It is practically the only beef operation in Jamaica that makes use of artificial insemination.

Dairy Cattle.—Dairy herds in Jamaica are usually pastured the year round and fed a rather sparse ration of prepared supplement. This is about one pound of grain for each four pounds of milk above ten pounds. Practically no hay is put up, and attempts at storing roughage as silage were abandoned a number of years ago because of the problems encountered from excess moisture during the harvesting.

The average yield from the Jamaican dairy cow is 2,500 to 3,000 pounds per lactation. The best commercial herds of Jamaica Hope cattle run as high as 7,000 pounds. While recently introduced Holstein-Friesian cows exceed this production, it has yet to be demonstrated that they can continue to do so over an extended period under practical farming conditions. The room for improvement in milk production per cow, however, is obvious. Better-quality cattle and, probably more important, better nutrition during the dry season will be necessary to increase materially the productivity of the national herd.

The use of artificial insemination was started in 1945 at the government research station at Bodles. Acceptance of the practice was slow although the service was made free by the government in 1951 for the Jamaica Hope breed. By 1968, however, 12,000 cows, possibly one-third of the national dairy herd, were being bred artificially. The practice should be of material benefit in improving the genetic quality of the Jamaica Hope.

MARKETING

There are no organized cattle markets in Jamaica. The sale of slaughter animals is negotiated between the buyer (a butcher or trader) and the farm owner. Delivery to the abattoir is cleared through the Livestock Clearing House, which is under the Ministry of Trade and Industry. Until January, 1970, the government endeavored to hold down the cost of beef by controlling the price of live cattle. The price per pound was regulated but did not apply to fed cattle or to those under 30 months of age. These prices could not be effectively administered owing largely to the subterfuge of selling by the head instead of by the pound.

No system of grading has been established, but there is some recognition of quality in the prices paid for different types of cattle, as determined by age and the feeding procedure. In early 1970 the feed lot cattle, largely bulls under 18 months of age which had been on feed for three months, were selling for 24 cents a pound, liveweight, on the farm. Cattle under 30 months, grown out on improved pastures, were bringing 22 cents a pound. Possibly over 85 per cent of the cattle slaughtered, consisting of older animals which had been on whatever pasture was available and the culls from dairy and beef herds, were selling at 18 cents a pound. This latter group comprised the animals to which price controls had previously applied.

There is considerable demand in Central America, Colombia, and Venezuela for breeding stock grown in Jamaica, particularly the Jamaica Brahman and the Jamaica Red Poll. The quality of the registered stock produced on the island and especially its adaptation to a tropical environment have gained a well-earned reputation. The Jamaica Brahman is considered by some to be better acclimated to these conditions than the American Brahman. It is also preferred because of the less pronounced slope to the rump. The fact that Jamaica is a foot-and-mouth-free area is also an important element in the sale of breeding stock.

Before the Jamaica Hope breed was firmly established, there was some sale of milk cows from Jamaica to other Caribbean islands. These were not the improved type of animal recognized today as the Jamaica Hope, but some of these exportations came to be known by that name. This produced a bad reputation for the Jamaica Hope among some breeders which, although now unwarranted, still persists. The Jamaica Hope is now in good demand on most of the Caribbean Islands.

SLAUGHTERHOUSES

The slaughterhouses of Jamaica, with the exception of the one owned by Reynolds Jamaica Mines, Ltd., are municipally operated facilities. At the Kingston abattoir a charge of $4.50 a head is made by the butcher for processing. The offal and hide go with the carcass. The Kingston plant is the only one in the country producing the by-products, blood meal, meat meal (tankage), and bone meal. There is competent inspection of the live animal before slaughter and of the vital organs and carcass. Capacity of the plant is 120 head of cattle in an eight-hour day. Some 18,000 head are processed annually. Sanitation is fair, although the animals are skinned on the floor without the use of cradles. Stunning is by captive bolt in the stationary killing box, after which the carcass is hoisted for bleeding and eviscerating. It is then dropped to the floor for skinning.

Towns throughout the country have small slaughtering facilities which range from the same type as the Kingston plant down to a mere slaughter slab without a roof.

CATTLE DISEASES

Jamaica has no serious problems with cattle diseases. There was an outbreak of foot-and-mouth disease in 1921. It was stamped out by slaughter, and the island has since been considered free of the disease. None of the more virulent tropical cattle diseases ever gained a foothold. There are strict import regulations which include well-enforced quarantine of imported breeding stock.

Tick-borne anaplasmosis and pyroplasmosis are universal. The carrier tick population is kept in check either by dipping or by the use of spray runs on large properties and by hand spraying on small farms.

The interval between treatments varies from two weeks to two months, depending on the degree of tick infestation in the area. Spray races are generally preferred to dipping vats. It is felt that some Zebu influence is necessary for cattle to tolerate both the tick and heat stress unless acclimatization has occurred over many generations.

Most calves are vaccinated against blackleg at weaning, and sometimes a second innoculation is given to yearlings to be retained in the breeding herd. In some areas vaccination against anthrax is given. It is common practice to worm calves at weaning.

Tuberculosis has been practically eliminated. In the Kingston slaughterhouse only four or five carcasses a year are condemned for this cause. Brucellosis is not considered serious. In some of the better herds it has been eliminated by annual test and the slaughter of reactors. Vaccination of replacement heifers as calves at four to eight months of age is practiced in some herds, but this is not general.

The incidence of liver fluke is quite high in areas with damp soils but not to the extent of seriously damaging infected animals. The principal loss is condemnation of the liver at the abattoir.

GOVERNMENT AND CATTLE

The Animal Husbandry Division of the Ministry of Agriculture and Lands has accomplished marked improvement in the breeding and management of Jamaican cattle. It has been directly responsible for the development of the Jamaica Hope cow and is proceeding with the improvement of the breed along progressive lines. Individuals in unregistered herds meeting recognized standards may be registered. Although the standards employed do not yet include performance requirements, it is anticipated that in the near future milk records will be required for registry. The development of improved pastures through the introduction of pangolagrass, the use of fertilizers, irrigation during the dry season, and proper pasture rotation has progressed to a considerable extent because of Animal Husbandry Division sponsorship. The introduction and growth of artificial insemination and good veterinary service are also largely the result of the leadership of this Division.

There are other fields in which the cattle industry is intimately involved where much could be done by government. There is no grading of slaughter animals—a matter of increasing importance as the practice of concentrate feeding progresses. The lack of a competitive market for cattle places the grower, particularly the small operator, at the mercy of individual buyers. The divided governmental authority under which the farmer operates—the Ministry of Agriculture and Lands for management matters and the Ministry of Trade and Industry in selling—is also a disadvantage.

Actually, the Jamaican cattleman enjoys a wide measure of free enterprise with a minimum of government interference for this day and age. This was evidenced by the termination of price controls on live cattle in 1970.

Sizable incentives are offered to the cattleman in the way of reducing his tax burden. There is no duty on the importation of breeding stock, animal feeds, or agricultural machinery. Considering the high duties on such items in many countries, this is a distinct advantage.

OUTLOOK FOR CATTLE

Even with a very low per capita consumption, Jamaica imports two-thirds of its requirements of dairy products and over one-fourth of its beef. In 1968, per capita consumption of beef and milk was 23 pounds and 100 pounds respectively. Under present conditions of management the lands available for cattle raising are just about completely stocked. The 250,000 cattle population includes something over 100,000 head which can be classified as dairy animals.

It has been adequately demonstrated within recent years that the national herd could be increased by improved management to the extent that the country could become self-supporting in both milk and beef. This would be despite the high rate of population increase, currently almost 3 per cent a year.

The total land available for pasture in Jamaica is estimated as 630,000 acres, of which only a little over one-fourth is now in improved pastures; that is, pastures which have been reseeded to more productive grasses, adequately fertilized, irrigated where necessary, and fenced for proper control of grazing. Generally speaking, one acre of improved pasture will carry a mature animal for one year (not to be confused with the term "animal unit' as applied to a cow and a calf). Unimproved pasture requires two acres or more per animal, depending on the local rainfall.

A modest genetic improvement in the dairy cow that would increase her milk production and an increase in improved pastures to make possible the slaughter of beef cattle at younger ages would go a long way in reducing deficits in milk and beef. Jamaica, with its internally generated income from bauxite mining and the tourist trade, should have the funds available to meet the capital requirements necessary for such expansion in the cattle industry.

Optimistic estimates have been made by some agricultural planners that by the expansion of grain feeding Jamaica could reach a level of beef production sufficient to tap the export market. This goal does not appear realistic at the present level of agricultural technology. The limited number of beef cattle now being fed home-produced concentrates, principally brewers' grains and dried citrus pulp, are consuming the total island production. Thus increasing the feeding of concentrates would involve raising grain on the island or importing feed grain from the United States. Neither of these steps is economically feasible.

The price of beef in Jamaica is already equal to that of beef of a higher quality in the United States. The increased cost of producing additional beef on the island would seem certain to price Jamaica beef out of the export market.

For the immediate future, cattle development in Jamaica will probably involve the improvement of the breeds that have been developed there and better pasture management. This should lead to a more productive national herd and permit a gradual reduction of imports of milk products and beef.

Martinique

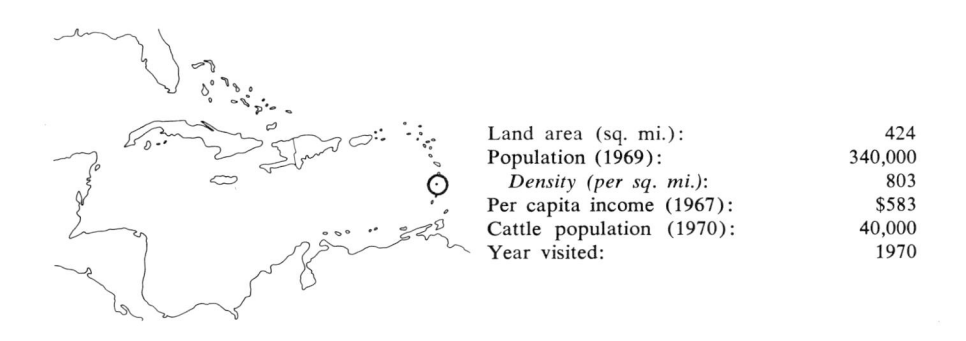

Land area (sq. mi.):	424
Population (1969):	340,000
Density (per sq. mi.):	803
Per capita income (1967):	$583
Cattle population (1970):	40,000
Year visited:	1970

MARTINIQUE, an overseas department of France, is a small, oval-shaped island lying between the 14th and 15th parallels. The Atlantic Ocean is on the east and the Caribbean Sea on the west. The area is three-fourths that of the city of Houston, Texas. From the indented coastline inland the coastal plain becomes hilly and then rises to a mountainous interior with heights up to 4,600 feet. Rainfall varies widely within short distances—averaging 30 inches along the southern coast, 120 inches in the central rain forest, and dropping to 70 to 90 inches on the northern plains. The heavy rains fall from July through November, and there is a pronounced dry season from March until June. The climate is warm, with little seasonal variation, and is moderated by the nearly continuous northeasterly winds.

Columbus is credited with discovery of the island. It was bypassed during the early Spanish explorations and was first settled by the French, whose occupation dates from 1635. The Carib Indians who had migrated to the island some centuries earlier were soon killed off. Except for three brief take-overs by the British during colonial times and a period that approached autonomy in the first years of World War II, the island has been continuously under France. The French influence continues paramount. Although the wave of independence which swept over many of the Caribbean Islands after World War II caused some rumblings in Martinique, the people elected by popular vote to remain a part of France. The economic level is comparatively high for the area—per capita income was $583 in 1967. This is steadily increasing under very sizable French support. The communist element, once a majority, has declined to less than 20 per cent. For the foreseeable future it seems that Martinique will remain as an integral part of France and enjoy stable government.

The large majority of the population is of Negro or mixed Negro and Causasian descent. There are said to be some 2,500 descendants of the early French settlers still on the island. Metropolitan French personnel who have come to the island in recent times as part of the military establishment or to fill government posts constitute the remaining part of the non-Negroid population.

The development of Martinique, like that of the other Caribbean islands, was based on sugar production. Nearly half of the surface area is now cultivated or in pasture; the other half is rain forest or is mountainous with slopes too steep for plowing. In recent years the movement of the people to the city and larger towns has created a labor shortage in the cane fields. This, along with the wide fluctuation in the price of sugar, has caused a rapid change in the agricultural pattern. Cultivation of sugar cane is on a sharp decline, and the land so released has been converted to bananas or vegetable crops for the European market and, on the more hilly lands, to pangolagrass pastures for raising beef cattle.

CATTLE BREEDS

Although Martinique was originally settled by the French, they did not begin to immigrate until nearly 150 years after the Spaniards had established their influence in most of the Caribbean area. The first cattle on the island were probably of Spanish origin. It was the custom of early Spanish explorers to release on the first landfall cattle that were surplus from ships' stores on the voyage across the Atlantic. Such cattle then came to exist in a semiwild state, but were often the nucleus of the cattle holdings of the settlers who eventually followed. It seems that this is what happened on Martinique. In any event, the cattle now seen on the island are similar to the descendants of the cattle on the other Caribbean Islands where Spain was in continuous control for several centuries.

As the need for animal draft power increased on the cane plantations, Zebu cattle from other islands, as well as from India, were imported and crossed with the Creole descendants of the Spanish cattle. This mixture of Creole and Zebu, as well as what remained of the original cattle, were all called "Creole" cattle, a usage which has persisted to the present day. Straight descendants of the Spanish cattle have nearly disappeared. A nearly pure Creole cow is seen occasionally in the more remote parts of the island. These carry the typical characteristics of the Creole: the thin

Nearly pure Creole cow of a farm worker. Mature weight 850 pounds.

tail with a black switch, upstanding horns of fair size, and a dairy-type conformation.

Following World War I the sugar plantations began to mechanize, and the use of cattle for draft declined. This movement continued until the draft ox practically disappeared. The Creole was then grown largely for beef on the larger plantations, usually by continued breeding to American Brahman–type bulls. Within recent years a considerable number of growers have been crossing Charolais bulls on the upgraded Brahman. Red Sindhi bulls have also been used in a few instances in an attempt to increase the milk production of the upgraded Brahman and thus wean a heavier calf.

There was also an upgrading of the Creole to dairy-type cattle in which crossing with the Holstein-Friesian was preferred. Some of the other European dairy breeds were used to a lesser extent, particularly the Brown Swiss which was imported from France and Switzerland.

Mixed Creole cow herd in process of being upgraded to American Brahman.

MANAGEMENT PRACTICES

About half of the cattle that are grown for beef are in the hands of small owners—perhaps a farm worker who has managed to keep a head or two of his own, or a farmer working a few acres on a semisubsistence basis. These cattle are usually tethered by a rather short chain for roadside grazing or whatever grass can be found. They are mostly of mixed Creole breeding, only an occasional animal being seen which could be called an upgraded Brahman or a straight Creole. The owner holds on to his cattle until he meets a need for cash and then sells off what is necessary to satisfy it. Very little attention is given these animals other than to see that they are staked out on the best grazing available. The chains they are tethered with are sold in standard lengths of about 18 feet, and the owners prefer a small animal which can reach sufficient fodder for the day from a circle of this diameter.

The plantation owners who have gone into cattle raising as a means of utilizing their former draft cattle conduct a much more sophisticated operation. Some have developed good pangolagrass pastures with fenced paddocks for grazing control. Under the best management these are rotated every forty-five days. Spraying with 2, 4-d is employed for weed control and to combat brush encroachment. Various exotic breeds have been tried in the past for upgrading, but the only definite trend in such programs has been the use of American Brahman until the recent introduction of the Charolais. Quite a number of growers now import young Charolais bulls direct from France, a procedure readily available

Martinique

to Martinique as a department of France and without foot-and-mouth restrictions on imports. For the most part, however, herds are of upgraded Brahman stock or of mixed Creole cattle.

In the beef herds bulls usually run with the cows the year round. A 70 per cent calf crop is obtained by some of the better growers, but the country-wide average is probably less than 50 per cent. Until recently the general practice was to market two and one-half to three-year-old steers at weights of 800 to 900 pounds. Now, with better pastures and grazing control, there is an increasing tendency to sell young bulls for slaughter when weaned at nine months of age. Brahman calves usually weigh around 440 pounds at this age. Calves that have been sired by Charolais bulls reach this weight at seven months of age.

Commercial dairying is a comparatively recent development. There is one large operation, a 200-cow Holstein-Friesian herd, but most dairies are in the 20-cow bracket. The majority of the herds are either upgraded Holstein-Friesian developed by bulls imported from Canada, or Creole crosses showing a strong Holstein influence. In the larger dairies, cows in milk are fed an imported supplement. Calves not wanted for replacement are sold for veal at two or three days of age.

Very little use is made of artificial insemination. Although a small government bull stud was started in 1960, only 700 cows annually are being bred in this manner. Friesian bulls (the French Pie Noir) have been imported from France and could be used artificially if desired. The progeny of these sires, when bred to the predominantly Holstein-Friesian dairy herd, would produce a calf well worth growing out for beef. Such a practice would fit into the desire to produce more beef rather than to slaughter the dairy calves at a few days of age.

MARKETING

Cattle for marketing are bought directly from the owner by a butcher and slaughtered in a municipal abattoir. The large grower with a sizable offtake to dispose of can obtain the going price, but the worker with only an occasional animal to sell is at the mercy of the buyers. Weaned bulls, just off the cow, weighing 400 pounds were selling for 41 cents a pound in early 1970. A mature bull in good condition was bringing 35 cents a pound, a cow around 32 cents. There is no recognized grading system.

Breeding stock is also sold by the pound. A young grade dairy cow in good condition, weighing 900 to 1,000 pounds, sells for 40 to 50 cents a pound, depending on the buyer's appraisal of her milk-producing ability. The price paid for dairy-type bulls, usually purchased as calves or yearlings, is from 50 cents a pound up.

CATTLE DISEASES

Although there are said to have been no cases of foot-and-mouth disease on Martinique for many years, it is not recognized by the United States as a foot-and-mouth-free area. The island is practically free of tuberculosis; and brucellosis is not considered to be serious.

European breeds are sprayed as often as once a week for the control of ticks. The well-acclimated Creole and Zebu herds are treated much less frequently and in some cases not at all. In some areas, calves are wormed as many as four times before weaning. If inspection indicates the need, cows are also wormed. In extreme cases this may be twice a year.

Martinique has no quarantine regulations and no restrictions on imported breeding stock other than the required health inspection and certification at the country of origin. Importation of breeding stock is actively encouraged by the government, but because there is no quarantine of these animals, the national herd is not adequately protected from infectious diseases.

GOVERNMENT AND CATTLE

Breeding stock and farm machinery may be imported duty free. A subsidy amounting to 40 per cent of the cost of developing improved pastures is given to the farmer. Other than these provisions, the government does not seem to grant much more than lip service to the cattle- or dairyman.

OUTLOOK FOR CATTLE

Martinique currently imports over two-thirds of both its beef and dairy products. The marginal cane lands if converted to grass could readily make the island self-supporting in milk and beef, but it has yet to be demonstrated that such a conversion is economically feasible. The instances where cattle are now raised on land that was until recently in sugar cane have usually resulted from the landowner's desire to get away from the worries of fluctuating sugar prices and labor shortage. It is conceivable, however, that the high carrying capacity of improved pastures in a twelve-months growing season could make cattle raising for either milk or beef competitive with cropping. A carrying capacity of one and one-half to two head an acre, which can be obtained in Martinique on good pasture, is a degree of productivity not often realized.

Trinidad and Tobago

Land Area (sq. mi.):	1,980
Population (1970):	1,100,000
Density (per sq. mi.):	530
Agricultural (21%) (1968):	231,000
Per capita income (1968):	$726
Cattle population (1970):	63,000
Buffalo population (1970):	4,000
Year visited:	1970

THE ISLANDS OF TRINIDAD AND TOBAGO, lying just off the northeastern coast of South America, are a unit of the British Commonwealth of Nations. Trinidad, the larger island, 1,864 square miles in area, was at one time part of the South American continent. It is only seven miles off the Venezuelan coast. The small island of Tobago, 116 square miles in area, is nineteen miles northeast of Trinidad.

The climate of the islands, situated between the 10th and 11th parallels, is warm and humid but alleviated by the steady winds and the surrounding sea. A mountainous ridge with elevations up to 3,000 feet runs across the northern part of Trinidad. This is paralleled by two lower ranges to the south. Tobago also has a mountainous spine which crosses the island from northeast to southwest. The rainy season is normally from June to December with only scattered showers the rest of the year. There is, however, a wide variation in the precipitation pattern from year to year.

These islands were discovered by Columbus in 1498 on his third voyage. Spain did practically nothing in the way of settlement until one hundred years later. Soon after the first Spanish town was established, it was taken by the British under Sir Walter Raleigh. Both the British and Dutch then contested the Spanish possession, and the two islands were more fought over than any others in the Caribbean. They were finally ceded to Great Britain, Trinidad in 1802 and Tobago in 1814. They have been politically allied since 1889 under the British Crown. Both entered the West Indies Federation when it was founded in 1958 but withdrew three years later. Status as a fully independent member of the Commonwealth of Nations was obtained in 1962. Since then the political climate was stable until the mob violence that broke out in 1970.

The majority of the population is of Negro and East Indian descent, 43 per cent and 36 per cent, respectively. The proportion of the East Indian element is gradually increasing. The mixed Negro-Caucasian population is considered to be about 16 per cent; and the remaining 5 per cent is Chinese and European, the latter mostly British and French. Both islands are heavily populated; Trinidad has 300 inhabitants per square mile and Tobago 530. When discovered, Trinidad was inhabited by a few tribes of the peaceful Arawak Indians who had migrated from the mainland of South America. These undoubtedly were soon killed by the Spaniards and have left no trace in the present population.

Agricultural development was slower than on the islands to the north. In the latter part of the eighteenth century, French settlers began the cultivation of sugar cane. Cacao soon became an important crop, and several large coconut plantations were established. African slave labor manned the various plantations until 1834 when slavery was abolished. In anticipation of the abolishment of slavery, indentured East Indians had previously been brought in. Cane production on Tobago never recovered from the labor shortage that followed emancipation of the slaves, and the plantations were gradually converted to coconut groves which required less labor and little animal draft power.

Cattle have played a lesser role in the agricultural development of Trinidad than in the rest of the Caribbean area. The first Spaniards who attempted settlement, around the end of the sixteenth century, were unsuccessful, but they undoubtedly introduced some of the Spanish-type cattle. Commercial agriculture did not really begin until nearly 200 years later. Such cattle as then followed were also of the Spanish type as these were readily available, particularly in Venezuela. The introduction of the Zebu breeds from India to the Western Hemisphere did not reach any great proportions until the middle of the nineteenth century. Around this time they were brought to Trinidad for draft use and were eventually crossed with the Creole.

As the British settlers became established in the agriculture of Trinidad in the 1890's, they introduced some of their breeds. The Red Poll, Hereford, and Ayrshire breeds were principally represented. Of these, the Red Poll was the most successful. None of these breeds made an important contribution to the cattle population of the island although their influence can be seen in scattered individuals of the present nondescript cattle.

An upgraded Holstein-Friesian herd at a Government Experiment Station.

In the 1920's, Zebu cattle were again imported as the demand for draft animals grew. These included the Nellore, Mysore, and Guzerat breeds, and to a lesser extent, the Sahiwal. In this same period, there was also an introduction of Friesian cattle from England.

Following World War II, larger numbers of Holstein-Friesian, mainly from Canada, were imported. More interest in beef cattle also developed, and in recent years the Jamaica Red, American Brahman, Charolais, and Charbray breeds were imported.

Water buffaloes were first imported from India in 1905. The last importation was six bulls of the Murrah breed in 1948.

CATTLE BREEDS

The principal breed which is recognizable in Trinidad today is the Holstein-Friesian. There are a few purebred herds, some of which go back three or four generations. Many of the dairy herds could be classified as upgraded Holstein-Friesian. A greater number of the milk producers, however, have mixed Creole cattle which may show a little Holstein influence.

Creole cow of a small farmer. A nearly pure descendant of the original Spanish cattle.

Except for the cattle of the organized dairies, most of the cattle in the country are nondescript mixed Creole owned by small farmers or village dwellers. Occasionally an individual is seen with characteristics of the original Spanish cattle. There are a few small herds of beef breeds such as American Brahman, Jamaica Red, Charolais, and Charbray.

In the buffalo herds, the river type predominates. The sharply curled horns of the Murrah are much in evidence and, less frequently, the white-spotted forehead of the Nihli.

MANAGEMENT PRACTICES

Dairy Cattle.—The organized dairy farms which are under good management make use of pangolagrass pastures, feed a balanced supplement in proportion to milk production, and often employ milking machines which sometimes are in conjunction with a milk parlor. The common practice is to keep the cows in milk in a shaded loafing yard after the morning milking and return them to pasture after the evening

Buffalo cow herd. Caroni Estates.

Mature buffalo sire showing Murrah breeding. Caroni Estates.

milking. Green chop is fed during the day. A supplemental ration of 18 to 20 per cent protein content consists of corn, wheat bran, brewers' grains, dried citrus pulp, coconut meal, and minerals. All of these ingredients with the exception of the corn are home produced. The use of silage has not been successful since the high moisture content makes it difficult to keep. Practically no hay is put up.

Dairy herds vary in size from 10 to 12 head to some in the 100-cow bracket. Female calves are kept in individual pens until a few months of age, then in group pens until around one year of age when they are turned to pasture. Bull calves are usually disposed of soon after birth. Frequently, they are sold to a small farmer who buys them for $20 to $25 to grow out for slaughter. Milk production from a well-managed herd of good upgraded or purebred Holstein-Friesian cows averages 6,000 to 7,000 pounds per 305-day lactation. The average calving interval is from 14 to 16 months, although in a few herds it has been reduced to 13 months.

Most of the milk on the island is produced in the area around Port of Spain where over one-fourth of the population lives. At the dairies, milk is cooled by a refrigerated coil which is placed in the milk can after it is filled. The cans are trucked to a plant producing sterilized milk. No pasteurized fresh milk is produced in Trinidad except that furnished the employees of the few large concerns which operate their own dairies for this purpose.

The Ministry of Agriculture maintains an artificial insemination center at the experiment farm near Port of Spain. Semen is collected locally or imported frozen. Conception rates are as low as 2.3 inseminations per pregnancy. Such an abnormally low rate is possibly due to extraneous causes such as nutrition rather than to faulty technique.

Texas Star Farm.—An unusual part played by American industry as it enters foreign countries is occasionally seen in the Caribbean area where it has become involved in some form of agricultural development. An example is the Star Farm of the Texas Company, the major oil producer in Trinidad.

Starting in 1959 with a small milking herd to supply its staff with fresh milk at the Pointe a Pierre refinery, the company has developed a highly efficient tropical dairy.

A herd of 100 head of upgraded ¾ to ⅞ Holstein-Friesian cows is currently maintained on pangolagrass pastures. These are on rough, hilly land not readily adaptable to any other use. The stock is in excellent condition, and the average production exceeds 6,000 pounds per 305-

day lactation. This productivity has been obtained by combining three factors: (1) imported semen from proven sires in the United States and Canada; (2) well-kept records on the performance of all cows; and (3) careful selection of the replacement heifers. Milking is done in a modern milk parlor, and the milk is pasteurized and bottled on the farm. The effects of tropical disease and parasites are minimized by good management. Heifer calves are raised to breeding age, and those not required for replacement are sold to nearby farms.

The dairy is only one division of Texas Star Farm. The over-all project has been developed as a demonstration farm for various food crops. A small beef herd is also maintained.

Beef Cattle.—There are a few herds of cattle which are run for beef production. These vary in size from 25 to 150 head. The main herd is usually of a nondescript type where upgrading is being attempted by the use of American Brahman, Jamaica Red, Charbray, or Charolais bulls. The breeding herd and young stock are run on pangolagrass. Steers and, in some instances, bulls are marketed off the grass at from one and one-half to two and one-half years of age, weighing 750 to 900 pounds.

Beef Buffalo.—Trinidad is one place in the world where a systematic effort has been made to develop the water buffalo for beef production. More efficient in the tropics, particularly in their use of coarse roughage as well as being more tolerant of parasites and indigenous cattle diseases, buffaloes have for centuries supplied man with milk and drawn his plow or cart in many countries of Asia. In these countries the flesh is used for food only when the worn-out beast is no longer useful. (The exception is in India and in Egypt, where unwanted male calves are slaughtered for meat at an early age.) The potential as a tropical beef animal, however, has not been developed.

When the need for the draft buffalo began to decline in Trinidad in the late 1940's, Dr. Steve Bennett began to develop a beef type from the buffalo herd on one of the sugar plantations of Caroni, Ltd. The selection criterion was for a well-fleshed beef-type conformation, emphasizing a broad, nonsloping rump, a long, deep body, and a straight back. Fertility and early maturity were also selection factors. In 1970, the herd numbered 300, of which 150 were females of breeding age. Color varies from a dull copper to black, but no attention was given to color in the original selection program. The copper color is now preferred. Horn shape is not uniform but more nearly resembles that of the tightly curled horn of the river buffalo than the long, wideswept

Finished buffalo steers, 20 months old, 1,000 pounds after three months on feed.

horn of the swamp buffalo. Mature cows weigh 1,200 pounds and bulls 1,600 pounds.

In the breeding herd, one bull to 80 cows is used. Bulls are changed every 30 days for a 30-day rest period. Although bulls stay with the cows the year round, most calves are born during the months of August and September. The heat period of the buffalo cow, at least in the Trinidad environment, is seasonal to this extent. Buffalo cows normally calve for the first time at two and one-half to three years of age.

In-calf heifers and yearling bulls sell for $500 for breeding stock. Weaned calves at eight months of age, weighing 500 pounds, sell for $325.

In addition to the foundation herd of Caroni Ltd., there are a number of other beef buffalo breeders on Trinidad and Tobago who have started in recent years by use of Caroni bulls. There still remain several hundred head of draft buffalo in Trinidad working on the sugar estates or doing cart work on small farms. This use of the buffalo is rapidly declining, however, and it is probable that those remaining will be eventually diverted to beef herds.

A noticeable difference between the young buffalo carcass and that of a steer of similar age is the markedly white fat of the buffalo. A number of taste panel teams are said to have compared buffalo and cattle meat with no significant differences being noted. Just what cuts were used in making these tests or the method of preparation of the meat was not mentioned. Generally buffalo meat is tougher and coarser in texture than cattle meat. Such differences could be relatively unimportant in areas where the native methods of preparing meat differ materially from those in the United States.

MARKETING

The small cattle holders, who in total own a majority of the cattle in Trinidad, sell slaughter animals to a speculator on a per head basis. The price is equivalent to something less than 20 cents a pound liveweight. The speculator usually has his animals slaughtered in the municipal abattoir and sells the carcasses to a meat merchant at 45 to 50 cents a pound. The Hindu owner of one or two animals frequently seeks an intermediary to sell to, even though at a lower price than he could obtain from the speculator. This intermediary, after a period, then sells to the speculator at a profit. Thus the Hindu manages to escape the stigma he would suffer if he had sold the animal directly to a man who would kill it. The religious attitude toward bovine life is carried by the Hindu to whatever part of the world he travels. With time, however, it seems that this may relax to a degree in foreign lands.

Large growers, who have systematized their sales to some extent, sell either the live animal or the carcass direct to a supermarket or other large buyer. Young slaughter buffaloes usually bring two cents a pound more, on a liveweight basis, than other cattle.

Except for the few large company-owned dairies supplying their employees, commerical milk producers sell to the one large sterilizing plant. The price paid is based on the nonfat solids content. No. 1 Grade with a minimum solid content of 11.9 per cent was selling for 5.5 cents a pound in 1970, grading down to 4.5 cents a pound for milk with less than 10 per cent solids.

SLAUGHTERHOUSES

The major slaughterhouse is the municipal plant in Port of Spain. Several other facilities are scattered through the smaller towns. There is also one modern USDA-approved packing plant. The Port of Spain slaughterhouse, built in 1910, exhibits a presentable appearance despite its age. The walls are tiled, there are good washing-down facilities and ample holding pens. The captive bolt is used in killing. Carcasses are hung for eviscerating, and the blood is saved for the production of "blood pudding," a sausage-like product made from cooked blood and a bread filler.

The plant was built when live cattle from nearby Venezuela were available as a source of beef. When a foot-and-mouth outbreak cut off these imports, frozen beef from Australia and New Zealand was brought in to supplement the local supply. Currently only 2,400 head of cattle and 18,000 hogs are killed annually. Inspection and sanitation are adequate.

The plant will process the animal of any individual owner at the standard charge of $2.00 a head plus $1.00 for inspection. The small holder of a single animal could realize a considerably higher return if he would avail himself of this privilege, but he seldom does so.

CATTLE DISEASES

Trinidad and Tobago are considered free of foot-and-mouth disease, and the law against importations from a foot-and-mouth country is rigidly enforced. Bat-carried rabies is endemic. It is estimated that 75 per cent of the cattle in the country are vaccinated annually against it. A new vaccine is now being tried which is said to be effective for three years. Periodic spraying for ticks is practiced in well-managed herds. This is done once or twice a month in the wet season and every three months in the dry period. Brucellosis and vibriosis are said to present no serious threat to cattle health. The country is free of blackleg.

Imported northern breeds, particularly the Holstein-Friesian from Canada, are very susceptible to the tick-borne anaplasmosis and piroplasmosis. Serious losses occur unless individual temperatures are taken daily for a period of several months to discover the affected animals. These are subjected to heavy doses of antibiotics. Even where this is done, there is loss in condition and productive ability for several months. Grade herds which go back several generations to a Creole and Zebu base withstand these blood diseases better than British or Continental cattle.

Low fertility exists in many herds, both purebred and upgraded. This is probably the result of seasonal nutritional or mineral deficiency. The pangolagrass pastures require heavy nitrogen fertilization. There is a long dry period when plant growth cannot be sustained. The high nitrogen content of the feed intake at times and again the low nutritional value of the matured grass could be factors contributing to calving intervals of 14 to 16 months or more. The mixed Creole and Zebu nondescript cattle do not seem to be affected and calve regularly.

Twenty years ago tuberculosis was the cause of widespread losses in the draft cattle and buffaloes on the sugar plantations. It was found to be due to the common practice of keeping draft animals under shed roofs to provide for ready collection of manure for fertilization. Separation of feed and manure was inadequate. The pens were cleaned only periodically and the cattle were sometimes in manure up to the belly. They went down rapidly in condition and became readily susceptible to tuberculosis. Incidence as high as 30 per cent was observed, even in buffalo herds. The disease was brought under control by test and slaughter, and the incidence is now thought to be under 2 per cent, based on inspection at the Port of Spain abattoir.

Worming for gastrointestinal parasites is necessary with young calves. It is even desirable to treat some buffalo calves. Other than this, no prophylactic treatment of any kind is employed in the buffalo herds.

GOVERNMENT AND CATTLE

Considerable effort on the part of the government is directed toward increasing milk production. At the Central Experiment Station a program is underway to develop a Zebu-Holstein cross as a future tropical milk type. The goal is roughly a ⅞ Holstein-Friesian and ⅛ Zebu animal to be derived from interbreeding Canadian purebred Holstein-Friesian with Sahiwal cattle. Third-generation cows were in production in 1970, but the program had not progressed to the point where any significant results could be claimed.

The Crown Land Development scheme is a government program to establish small farmers on 25-acre units with 10 to 12 dairy cows. The land being utilized was formerly a United States Air Force base. A unit consists of a house, milking shed, established pangolagrass pastures which are cross-fenced, and the cow herd. The cost, $12,500 a unit, is to be returned to the government on a long term, nominal interest basis by the individual farmer. At the going price of 5 cents a pound for milk, it appears that the unit operator can barely provide for his family, to say nothing of the burden of capital repayments.

OUTLOOK FOR CATTLE

Trinidad currently imports nearly half of the beef it consumes and probably a larger proportion of its milk products. Draft animals seem certain to disappear from the sugar plantations within the next few years in spite of efforts of government-favored labor unions to prevent this change. A logical step would be to divert the draft cattle and the buffaloes to beef production. Although the additional beef so obtained would not fill the present gap between production and consumption, it would be an important step in this direction.

Increasing the number of improved pastures—by the establishment of pangolagrass or other high-producing tropical grasses, fencing for rotational grazing, and irrigation—would add materially to the carrying capacity for both dairy and beef cattle. Considerable areas formerly devoted to intensive cacao culture have been abandoned or are poorly utilized due to the low price of this product. Although not the best land for raising cattle, these cocoa areas could be so utilized to advantage. The future price of sugar also enters the picture, as the favored market that Trinidad now enjoys may be adversely affected when Britain enters the European Common Market.

Factors such as these could make available enough agricultural land to make Trinidad self-sufficient in both milk and beef. A program with such objectives, however, would require a large-scale development with a heavy capital investment. Whether or not this will be undertaken by interests capable of pushing it to a satisfactory conclusion cannot be ascertained.

A unique element in the livestock picture of Trinidad is the water buffalo grown for beef. The possibilities along this line are important to economical beef production in tropical areas throughout the world where beef is in short supply and the economic level of the people is rising. The concept of the water buffalo as a source of good meat is new. Adequately promoted, it could lead to the buffalo's becoming an important element in meeting the future protein demands of the world.

French Guiana

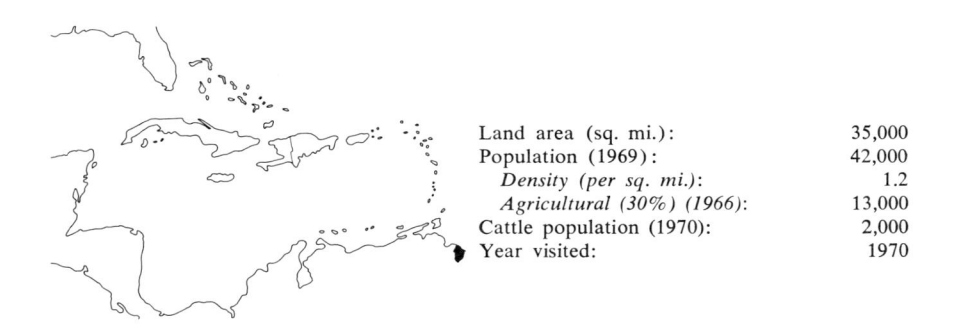

Land area (sq. mi.):	35,000
Population (1969):	42,000
Density (per sq. mi.):	1.2
Agricultural (30%) (1966):	13,000
Cattle population (1970):	2,000
Year visited:	1970

FRENCH GUIANA occupies a wedge-shaped area on the northeast coast of South America. It lies between Surinam on the west and Brazil on the east and south; on the north is a 200-mile coastline on the Atlantic Ocean. The land area is three-fourths that of Mississippi. Located between the 2nd and 6th northern parallels, the climate is tropical with a high humidity although moderated somewhat along the coast by continuous northeasterly winds. The main rivers, rising in the highlands on the Brazil border, run in a generally northerly direction to the Atlantic. They are the only means of transportation into the interior. Except for scattered Indian and Bush Negro villages along the rivers, all the inhabitants live in a few localities on the narrow coastal plain.

The record shows that Columbus traced the coast of the Guianas in 1498 but did not land. Some early Spanish attempts at settlement were made at the beginning of the sixteenth century but were soon abandoned. During the first years of the seventeenth century the French occupied the offshore island of Cayenne and established trading rights along the coast for the area between the Orinoco and Amazon rivers. Over the course of the next century most of this region fell into other hands. France managed to retain possession of what is now French Guiana except for brief occupations by the Dutch (from 1660 to 1664) and the Portuguese (1808 to 1816). After World War II, in the general effort to strengthen the hold on her former colonies, France changed the colonial status of French Guiana to that of a Department of France with an elected representative to the French parliament. This move, however, appears to have accomplished little in advancing the economy of the former colony.

France began the deportation of criminals to French Guiana in 1797, and the infamous penal colonies that were established were used intermittently until finally abolished after World War II. A prosperous sugar plantation economy flourished during the first half of the nineteenth century, but then deteriorated and has now practically vanished. Other agricultural developments, such as rice culture and bananas, which have been successful in Surinam and Guyana, have made no significant progress. When De Gaulle attempted to enter the nuclear race, the French Missile and Space Center was set up in French Guiana, which led to a sizable construction boom and increased employment. For a number of years commercial shrimp fishing has added to the national income. The accomplishments of these various activities, however, do not indicate a particularly bright future for French Guiana economy. Undoubtedly, it will remain dependent on aid from France.

French Guiana is the most sparsely populated country in the Western Hemisphere. Considering the country as a whole, there is only one person to the square mile, and fewer than 20 to the square mile on the coastal plain. Nearly two-thirds of the population are in the environs of the capital city of Cayenne. Most of the inhabitants are known as "Creole," which includes those of African as well as mixed African and European descent. There is also an element of East Indians who have not mixed to any great extent with the other races. In the interior there are thought to be around 1,000 descendants of the original American Indians and a few thousand Bush Negroes. The latter are the pure descendants of slaves who escaped to the jungle and reverted to their original tribal life. There is a small European element, almost entirely of French descent, in government and commerce.

CATTLE BREEDS

Cattle have practically disappeared from French Guiana. When the sugar plantation operations were no longer profitable, a general retrogression in all agriculture set in and there was no longer a need for draft animals. The economic incentive to raise cattle, and people with the ability to care for them, disappeared. Of the 12,000 to 15,000 cattle that were estimated to have been in the country in 1915, there are probably now less than 2,000 head of all ages. Many of the former animals were probably killed off to supply the local demand for beef without any effort made to replace them.

The present cattle are concentrated in the populated areas around

Mature Criollo cow—tan color, 750 pounds. Government Livestock Station.

Cayenne and the other scattered communities along the coast. These are predominantly an entirely nondescript type. Many individuals show the Zebu influence, and others show traces of various European dairy breeds. The noteworthy feature of the cattle of French Guiana, however, is the occasional animal displaying the typical characteristics of the Criollo common to Latin America and the Caribbean islands. This is a small tan or black animal with a barrel-shaped body and dark hair around the eyes. Undoubtedly these animals are descendants of the first cattle brought to the Western Hemisphere by the Spaniards. Where they came from is a matter of conjecture, as Spain established no permanent settlements in any of the Guianas. Yet representatives of the pure or nearly pure Criollo are seen in all three Guianas.

The few cattle in French Guiana are held mostly by small farmers

Cattle of North America

or village dwellers as a form of savings, to be disposed of only when there is a need for money. Milk may be taken incidentally for household use but there is no commercial milk production. One butcher maintains a herd of 180 head down the coast from Cayenne from which he draws animals for slaughter as needed. The government livestock station has 15 head of mixed Creole cattle and an equal number of "pensioned" cattle—those belonging to nearby owners who have sent them to the station for breeding. A Red Sindhi bull, 18 years old and in excellent condition, is used for breeding.

The cattle at the government station are maintained during the day on good pangolagrass pastures and brought in under shed roofs at night where electric lights are kept burning as protection against the rabies-carrying vampire bat. While the station is largely devoted to horticulture, it is also attempting to raise bulls for distribution to the small cattle owners in the vicinity.

French Guiana imports all its milk and milk products as well as 90 per cent of the beef and pork consumed. The total number of cattle killed annually is around 200 head. The price paid by the butcher for the odd animal he purchases is roughly equivalent to 25 cents a pound liveweight.

A private project has been initiated to establish pangolagrass on 2,500 acres located on the coastal plain near the Kouroo. It is planned to run 1,000 head of breeding cows. The program projects the marketing of three-year-old animals off grass. In 1970 this was the only development in cattle raising that was even in the planning stage for French Guiana.

Guyana

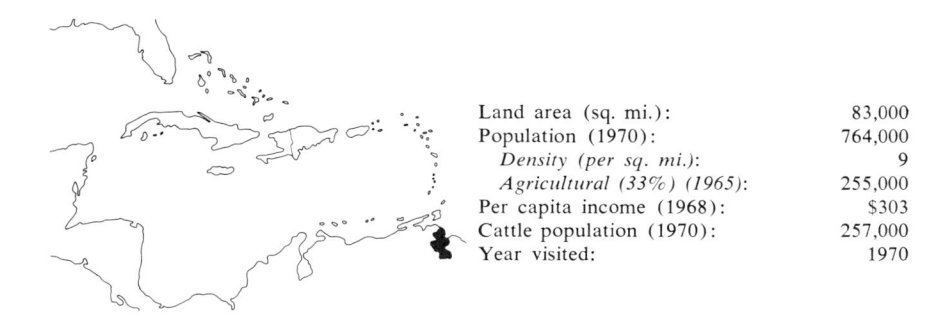

Land area (sq. mi.):	83,000
Population (1970):	764,000
Density (per sq. mi.):	9
Agricultural (33%) (1965):	255,000
Per capita income (1968):	$303
Cattle population (1970):	257,000
Year visited:	1970

THE REPUBLIC OF GUYANA occupies the area in the northeast corner of South America that lies between Surinam on the east and Venezuela and Brazil on the west. The Brazil border also extends along the south. There is a 300-mile coastline on the Atlantic in the northeast. Guyana is nearly one-third the size of Texas. Venezuela lays claim to two-thirds of Guyana's territory, the area between the present border and the Essequibo River, and Surinam claims some 6,000 square miles which lie on the eastern side of the New River in the southeastern part.

A low coastal plain, only the seaward fringe of which is arable, extends to a white sand zone. Beyond this is a mountainous rain forest interior that drops off to a low savannah, 300 feet above sea level, along the Venezuelan and Brazilian borders on the west. There are also other savannah areas, one in the south known as the intermediate savannah, and another one inland from the northern coast. The rainfall of 80 inches along the coast is ample for sugar cane and rice, but it falls in a pattern which requires irrigation at times. Normally there are two rainy seasons, one from mid-May to mid-July and the other from mid-November to mid-January. The periods between are drier in the interior than on the coast. In the western savannah, the Rupununi, the heavy rainy season extends from April to August with 60 to 70 inches of precipitation but less than 10 inches during the remainder of the year.

The climate on the coast is warm and humid, alleviated to a considerable extent by onshore breezes. In the interior it is much more tropical. Practically all cultivated land is on the seaward side of the coastal plains. At high tide this land is below sea level, protected by 140 miles of sea defense. The polder system was started by the Dutch settlers, and earthen dykes were consolidated by planted vegetation and progressively advanced as more land was reclaimed. Large segments of the outermost dyke have deteriorated in recent years and are being replaced with piling and concrete walls.

Spanish sailors touched the coast of what is now Guyana before the end of the sixteenth century, but the first actual settlement was attempted by the Dutch in 1581. The Dutch West India Company then exploited the area and a sizable plantation economy was established by 1620. Dutch control was maintained until 1796 when the British moved in. The country as British Guiana was ceded to Britain in 1814. Status as a British colony was maintained until it became the independent country of Guyana within the Commonwealth of Nations in 1966. Dissatisfied with a freedom which carried even an implied connection with the British Crown, Guyana elected to become a republic. It no longer acknowledged the Queen as head of state but remained within the Commonwealth. This status was assumed in 1970. Since independence the government has evidenced an increasingly socialistic trend.

The two major elements comprising 80 per cent of the population are the East Indian and the African. Nearly 50 per cent are descendants of the East Indians brought in as indentured labor after the abolition of slavery, and 30 per cent originated from the African slaves which preceded them. Some 12 per cent of the total population is of mixed extraction, 5 per cent are American Indians scattered through the interior, and the remainder are either Chinese or Europeans (largely British and Portuguese). The Negro sector has managed in recent years to dominate the political scene despite the numerical superiority of the East Indian group. The basic antipathy of these two races has led to serious social disorders in the past. Although an undercurrent of racial antagonism still persists, the situation has improved in recent years.

Historically, the economy has been the sugar plantation-type prevalent over most of the Caribbean area. About 95 per cent of the sugar is grown on only thirteen large plantations. Paddy rice cultivation followed sugar and is now approaching half the export earnings of sugar. In recent years bauxite and alumina have become the most important elements in the growing economy. In 1969 these products exceeded sugar by 50 per cent in the value of exports.

Cattle were never an important adjunct to the agricultural economy. The unique method employed by the Dutch in developing the cane fields created only a limited need for the draft ox. The fields are generally

Steer showing strong Criollo influence. Rupununi District.

below sea level, and the system of using dykes for protection from the sea provides ready water transportation of cane from field to mill. The method of cultivation adopted, known as the "cambered bed system," supplies both the irrigation essential for a good crop during dry spells and the drainage of low, convex fields that is necessary to hold down the salt content. The cane beds are formed by shaping convex strips 24 to 36 feet wide. These beds are separated by ditches 18 to 20 feet wide, for drainage. Cane is transported from the field by shallow steel barges, called punts, on the major irrigation canals. Formerly cattle or mules were used to move the punts. One ox could draw five punts holding five tons each, thus moving a total of 25 tons of cane. Such a task requires at least five three-span ox teams when sugar cane is moved by ox cart as was the general practice in the Caribbean area.

Cattle may have been brought to the Guyana area by the early Dutch

Cow in a Rupununi herd with black-and-white markings, a typical Criollo pattern.

settlers, but, if so, any trace of such importations of European stock has been lost in the cattle now seen in the country.

CATTLE BREEDS

One major introduction of cattle into Guyana was from Brazil. These cattle were brought across the Takutu River and its tributary, the Ireng, which form the present boundary between the two countries. At first these were Criollo descendants of the early Spanish cattle. They were followed by the Zebu-Criollo cross. (The first Zebu cattle in South America were the ones brought to Brazil from India by the Portuguese.) Representatives of pure or nearly pure Criollo cattle are still seen in the general area that borders on Brazil. This is the harsh savannah known as "the Rupununi." The majority of the cattle, however, show some Zebu influence. The better herds have now been upgraded to American Brahman. Santa Gertrudis bulls have also been used but apparently have not been as popular as the Brahman.

Along the coast in the agriculturally developed belt where swamp areas are utilized for pasture, the cattle are of a completely nondescript type. Zebu influence is seen but is not general. Of the European breeds, the Holstein-Friesian influence is most common, but there are also traces of Guernsey, Jersey, and Hereford. These swamp cattle have resulted from random breeding of the former draft animals with such exotic breeds as were brought in mainly for milk production. Through natural

Upgraded American Brahman herd. St. Ignatius Livestock Station. Rupununi District.

selection they have become remarkably well adapted to their inhospitable environment.

In the vicinity of Georgetown, the capital and major city, the dairies have small herds of upgraded Holstein-Friesians. There are only minor indications of any of the other European dairy breeds.

MANAGEMENT PRACTICES

The nondescript swamp cattle which constitute a majority of the national herd are raised with very little attention from their owners. A village dweller with other employment may keep two or three head. Holdings vary from this small number up to a few hundred head, although there are very few herds with 100 or more. Milk may be taken from some of the cows for household use. Animals are commonly sold for slaughter only when a specific need for cash arises.

The lands utilized for pasture are those ill adapted to growing cane or rice. Forage consists of poor sedges and grasses common to land which is under water much of the time. Cattle are turned out in the morning and are brought in at night, primarily for protection from

Cattle of North America

Slaughter steer on coastal plain, 750 pounds. Kabawer Ranch.

theft. They often graze in water up to their bellies but maintain themselves in fair condition and are unusually hardy and disease free.

A project to improve the native cattle has been underway for several years on the Kabawer Cattle Ranch. This is a holding of 15,000 acres belonging to the major sugar producer and is located on land adjacent to the cane- and rice-growing areas. The herd of 6,500 head consists of 2,700 mother cows and 200 sires. The remainder is young stock. The breeding herd was originally the nondescript swamp cattle, but there is now considerable evidence of the upgrading program that is underway.

The forest, which encroaches on the pasture land, is being cleared and planted to improved grasses. There is good control of grazing as the pastures are rotated seasonally according to whether they are low-lying and under water in the wet season or are higher and drier.

Breeding is selective: American Brahman, Romana Red, and some **Santa Gertrudis** bulls are used. Bulls are run with the cows the year round at the rate of six bulls per hundred cows. Cows do not breed regularly every year, and the average calf crop is around 50 per cent. Steers, normally castrated before one year of age, are slaughtered at three and one-half years at an average weight of 800 pounds. The meat is sold through the company-owned supermarket. Heifer calves are retained in the herd if of reasonable conformation.

With the exception of one operator in the open-range ranching country of the Rupununi, Kabawer is by far the largest commercial cattle enterprise in the country.

The Rupununi District is on the plateau area near the Brazilian border. There cattle are raised in one of the harshest environments that man has ever chosen for them. The elevation is about 300 feet. Part of the area lies on the outer edges of the Amazon basin with the outlet to the sea some 1,500 miles away. The annual rainfall is 80 inches, but 90 per cent comes during the April to August wet season. There are only scattered showers the rest of the year. Poor and shallow soils predominate on the undulating terrain, which has practically no gradient to afford adequate drainage.

The only grasses are of a coarse, fibrous type, low in protein as well as minerals except when in the early growth stage. The intricate drainage system of watercourses is lined with palm trees. Carrying capacity ranges from 40 to 60 acres per head of cattle of all ages. Holdings are measured by the square mile. The largest is a unit of 2,000 square miles with 30,000 cattle, operated by the Rupununi Development Company. From this, holdings range down to 25-square-mile ranches, running a total of 300 to 400 head. These are often held by practically subsistence ranchers whose production averages no more than 40 to 60 head annually.

The total number of cattle in the Rupununi numbered around 60,000 at the time of the "cattlemen's rebellion" in 1969, described below. For the most part these are run under open-range conditions although there is some boundary-line fencing. The region is completely isolated except for air transportation. A four-strand barbed-wire fence costs $1,200 a mile. Fenced paddocks are usually provided around the homestead only to the extent necessary for working cattle.

The area was first settled by unauthorized Portuguese emigrants from Brazil, who brought their Spanish-type Criollo cattle with them. These cattle had undoubtedly come into Brazil from Venezuela, which also corners on the Rupununi just north of the Guyana-Brazilian border. The Portuguese were later followed by British and other European settlers. Exotic bulls were introduced in the early 1900's, mostly of

Zebu types familiar to the Portuguese but also some British breeds, notably Hereford. In recent years upgrading of the predominantly Criollo cattle has been attempted, mainly by the use of American Brahman and a few Santa Gertrudis bulls.

Bulls are run with cows throughout the year, and calf crops vary from 25 to 50 per cent. The low level of nutrition during much of the year is most likely the reason for this poor calving rate as well as for the high calfhood death loss. Bulls not wanted for breeding are castrated at one and one-half to two years of age, the delay allowing the owner to determine the best prospects to keep as herd sires. Most herds are rounded up regularly for branding and vaccination against bat-carried rabies. Both operations are carried out on most ranches by roping and throwing.

Steers are marketed at five to seven years of age, weighing from 800 to as much as 1,000 pounds. The production runs from 6 to 8 per cent annually for the average herd. The small operator of a 50-square mile property and a total herd of 800 head has an average gross income of around $5,000 from the sale of 50 to 60 six-year-old steers. After expenses this hardly leaves enough to live on unless the rancher grows a good part of the family food. Gardens are often established on small irrigated plots rendered fertile by having once been the site of a corral.

Until recently cattle were trailed to Georgetown over a route of 350 miles. Three weeks were required for the drive. As cattle raising in the Rupununi became better organized, an abattoir was built at Lethem in the center of the growing area. Unchilled dressed carcasses can now be flown from there to the Georgetown market in a little over an hour. (An outbreak of foot-and-mouth disease in November, 1969, however, stopped all movement of beef from the area.)

There is only one freehold property in the Rupununi District as large as 50 square miles. Nearly all ranches, from 25-square-mile units up, historically have been operated on "permissions" granted on an annual basis by the government of the British colony. These permissions were automatically renewed for nominal rentals and there was never occasion to question the continuity of an operation. This practice was continued initially by the newly independent government, which came into being in 1966. When questions arose in regard to the loose arrangement, the ranchers were dissatisfied with the answers.

This led to an ill-conceived "rebellion" at the beginning of 1969, in which a majority of the ranchers endeavored to take and hold by force the ranching area of the Rupununi. They were surreptitiously aided in this endeavor by the Venezuelan government, which lays claim to the Rupununi as well as other Guyanese territory. The rebellion was quickly put down by the Guyana police and the military. Many of the ranch headquarters were burned and leveled. The Rupununi ranchmen who had participated in the rebellion escaped by air to Brazil and Venezuela. Their cattle and other possessions left behind were confiscated by the Guyana government. The ultimate effect of this disorder remains to be seen. Operations of the Rupununi Development Company, which runs half the cattle in the region, were not affected.

Dairy Cattle.—There are a few thousand head of dairy cattle around the major city of Georgetown. The Bel Air Dairies, the largest dairy operation in Guyana, is under the same ownership as the Kabawer Cattle Ranch. In 1970 there were 150 head of upgraded Holstein-Friesian cows in the milking herd. The dairy was originally formed by the consolidation of the individual plantation herds of the owner. The cattle are maintained on pangolagrass pastures and fed a supplement consisting largely of home-produced products—rice bran and coconut meal, plus some imported grain, molasses, and urea. Milk production averaging 5,000 pounds per lactation is claimed, although this would not be expected from a casual appraisal of the condition of the herd. A milking machine in an open-type barn delivers milk directly to a trailer-mounted stainless steel tank which is hauled daily to the pasteurizing plant in Georgetown.

Most of the milk produced commercially is from 5- to 20-cow dairies having upgraded, or mixed, Criollo-Holstein herds. These are maintained on native grass pastures without supplement. There is also the incidental sale of milk by the owner of two or three cows to people in the villages.

In 1970 the price paid the farmer for milk was six cents per imperial pint.

There is a government artificial insemination center near Georgetown that supplies Holstein-Friesian semen.

MARKETING

Cattle are generally sold on a dressed-weight basis. In the coastal area the representative of a co-operative society, or a trader, buys at the farm or ranch on the basis of carcass weight. A commission of one cent a pound on the dressed weight is deducted, which roughly is equivalent to 3.3 per cent of the price. In early 1970 the average price of slaughter animals in Georgetown was 30 cents a pound dressed weight, the equivalent of 15 cents a pound liveweight, before deducting the commission. The top price for an animal in better than average condition was a cent or two higher. The butcher selects the animal he wants to

Slaughter cattle at Georgetown abattoir.

buy in the holding yard of the abattoir and deals with the trader for it.

Prior to the ban on shipments out of the Rupununi, a limited number of cattle were exported from the area by river boat to Surinam.

When the government abattoir in Lethem was completed in 1958, all cattle marketed in the Rupununi were sold on a cold dressed-weight basis to the government agency which operated the facility. The price to the producer was equivalent to 10 cents a pound liveweight. Air freight on carcasses from Lethem to Georgetown was four and one-half cents a pound. This partially accounts for the difference between the Georgetown and Lethem prices. Small quantities of warm carcasses have also been exported by air at times to Trinidad, Surinam, and French Guinea.

SLAUGHTERHOUSES

There are two government-owned abattoirs, one in Georgetown and another in Lethem. These handle all the slaughter of cattle under inspection in Guyana. Neither has chilling facilities, and carcasses are delivered warm the day they are killed. Those from Lethem are flown early in the morning to Georgetown so that butcher shops can sell them the same day they are killed.

The Lethem plant was processing 3,500 head annually before the foot-and-mouth outbreak in 1969. The Georgetown facility, an old plant right in the city, without modern equipment, kills from 200 to 300 head per week. Cattle are sometimes held for four or five days in the holding yard, on water and a token feed of cut grass. This delay occurs when the trader or co-operative fails to gauge the market demand accurately.

CATTLE DISEASES

A handicap to Guyana is the lack of quarantine on the Brazilian border, where foot-and-mouth disease has been endemic for years and cattle wander freely back and forth across the river boundary during the dry season. Historically, in this area of the Rupununi there have been sporadic outbreaks of foot-and-mouth disease. The latest, in 1969, was of the O type. Vaccination of cattle in affected areas was attempted by the government veterinary services as a means of control, but considerable difficulty was experienced in getting sufficient supplies of vaccine, even for the cattle in the isolation areas. The coastal area, where over two-thirds of the cattle are grown, was kept free of the disease. Since the outbreak, the only movement of cattle between the two areas has been of dressed carcasses transported by air. Control measures, however, seem to be ineffectively administered and it is doubtful that the Rupununi and the coastal area can be isolated from each other.

The incidence of tuberculosis in the country is thought to be low as very few infected cattle are discovered in either the Lethem or Georgetown slaughterhouse.

Blackleg is encountered in parts of the country and vaccination is then resorted to. Screwworm infestation of young calves is bad, particularly on the navel and gums. The better cattlemen treat more or less systematically for it. Tick infestation varies widely, and spraying or dipping is done as needed. In the coastal area spraying is at 15- to 30-day intervals.

The few large ranches and dairies in the coastal area vaccinate their calves with Strain 19 for brucellosis. Internal parasites are serious in young cattle, and the calves are wormed regularly until they are weaned.

Small operators take practically no precautions against disease or parasites and call for veterinary service only as a last resort.

In the Rupununi all cattle on the organized ranches are vaccinated every year against rabies carried by the vampire bat. The small cattle holdings of the American Indians here, however, go unprotected.

GOVERNMENT AND CATTLE

The Ministry of Agriculture and Natural Resources outwardly displays an atttitude of encouragement toward the cattle industry, but the accomplishments to date have not been impressive. The same bulls have been used at the artificial insemination center so long that inbreeding has resulted in a decline in production of herds using the service. The St. Ignatius Livestock Center at Lethem maintains a herd of upgraded American Brahman and another of Santa Gertrudis and supplies bulls to ranchers on either a loan or a purchase basis.

A USAID project calls for investigational work at the Ebini Livestock Station on beef-cattle breeding and management, along with other agricultural developments. The work, however, was just getting organized in early 1970. Although not as harsh a cattle country as the Rupununi the intermediate savannah where the Ebini Station is situated would seem to be a poor cattle area. The sandy soils are of extremely low fertility so that heavy fertilization is necessary. Although rainfall averages 90 inches annually, there are two pronounced dry seasons and irrigation would be necessary for sustained plant growth. It is questionable whether cattle can be raised profitably here for a market that is currently paying only 15 cents a pound for beef on the hoof.

The policy of the government in granting only annual "permissions" to the cattlemen in the Rupununi has retarded substantial development except on the lands used by the Rupununi Development Company, where tenure seems to have been somewhat more assured. The whole area is stocked at about capacity under present conditions. If the ranchers had reasonable assurance that they could realize a return on a long-term capital investment, they could increase carrying capacity. Wells for stockwater for use in the dry season and low earthen dykes to hold back the flood waters in the wet season, along with some surface irrigation, would allow more cattle to be run and would also increase productivity by providing better forage. Fencing in some areas would also permit better grass utilization.

The socialistic tendency of the newly independent government raised the specter of dispossession of the land the rancher had utilized for years.

Guyana

Development on the Rupununi for cattle raising cannot now be anticipated in the foreseeable future.

In early 1971 the World Bank granted the Guyana Government a $4,400,000 loan for development of the livestock industry in the country. Competent agriculturalists prominent in the sugar industry are interested in the project. It thus appears that as development proceeds it will have the benefit of constructive guidance with the prospect of worthwhile results.

OUTLOOK FOR CATTLE

Such development as the future may hold for either dairy or beef cattle in Guyana will depend on the degree to which private enterprise is allowed to function.

Back of the cultivated coastal areas where the rain forest begins, productive tropical grasses could be established after clearing the jungle. This could provide for both a national dairy and beef herd of a size to make the country self-sufficient in these products. The rising economy of the country could supply the capital expenditure necessary for such development and absorb the higher prices for milk and meat on which expansion would hinge. The political climate is such, however, that such developments are highly problematical. In spite of the optimistic forecasts which are heard, it appears that progress along this line will be slow.

Surinam

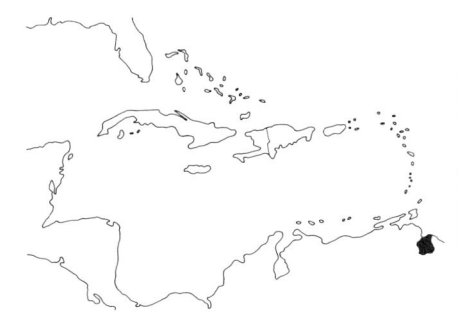

Land area (sq. mi.):	63,000
Population (1970):	403,000
Density (per sq. mi.):	6
Agricultural (25%) (1964):	10,000
Per capita income (1968):	$580
Cattle population (1970):	50,000
Buffalo population (1970):	300
Year visited:	1970

SURINAM, an integral and self-governing part of the kingdom of the Netherlands, lies on the northeastern coast of South America between Guyana on the west and French Guiana on the east. The southern border is with Brazil. A northern coastline on the Atlantic Ocean extends from the Corantijne River (the Guyana boundary) to the Marowijne River (the border with French Guiana). The area of Surinam is not quite as large as that of Louisiana.

All the rivers rise in the mountainous area in the south and flow generally northward to the Atlantic. Most of the cultivated land is on the coastal plain which extends the width of the country. Much of this is below sea level. An intermediate zone of bush savannah and forest, 30 to 50 miles wide, parallels the coastal plain. Farther to the south the terrain becomes hilly and then mountainous with elevations over 4,000 feet on the Brazilian border. Much of this country is in rain forest. Since the time of the earliest settlements, all agricultural development has been on the coastal plain in the vicinity of the mouths of the large rivers. This is where practically the entire population lived until the recent discovery of bauxite. Scattered tribes of American Indians and Bush Negroes are the only inhabitants of the interior.

When the early Spanish explorers were establishing themselves on the continent, they paid little attention to the northeastern coast of South America. Later they took some action in the Guianas, the general term for the coastal area between the Orinoco and Amazon rivers. By the end of the seventeenth century both the British and Dutch had become active in the Caribbean area and the adjacent shores of South

America. An English settlement was founded on the Surinam River in 1630 but was abandoned a number of years later. In 1651 the governor of the British colony of Barbados initiated emigration to the Guianas, and permanent sugar plantations were soon established. When the Dutch moved into the area that is now Surinam, they firmly established their position there by trading New Amsterdam (Manhattan in the United States) for the British rights in Surinam. Subsequently the British retook the area by force during the Napoleonic era, but it was finally returned to the Netherlands in 1816 and has remained in its possession.

The sugar plantation economy originally established by the British in the seventeenth century was based on African slave labor, and this system was continued by the Dutch. During the slavery years a considerable number of slaves escaped to the jungle interior, reverted to their tribal way of life, and their descendants, known as Bush Negroes, are still living in the wilds. When slavery was abolished in 1863 and all plantation Negroes were freed, there was a lack of labor for the cane fields. To fill the labor gap, over the years many different groups of people were brought in.

Chinese coolies, first from the Dutch East Indies and later direct from China, were employed. The Dutch also brought over Malaysian people from the East Indies, and these Indonesians continued to come in as free immigrants in later years. Their last entry was in 1939. Beginning in the early 1870's and continuing until after World War I, East Indians of both the Hindu and Moslem elements entered the country as indentured laborers.

All these different groups from which labor for the plantations was obtained account for the varied yet racially unmixed population of Surinam today. The Negro descendants of the emancipated slaves who live in the populated areas, including those that have intermixed with Caucasians, are all called "Creoles." The East Indians and the Creoles constitute 70 per cent of the population. These two are now about equally divided, though the proportion of the East Indian sector is steadily increasing. Javanese and other Indonesians are maintaining about 15 per cent of the population, and the Bush Negro 9 per cent. The Creoles and the Bush Negroes do not intermingle. The proportion of indigenous Indians, about 2 per cent, is declining. This is also true of the 2 per cent Chinese element. The remaining 2 per cent of the population is mostly Jewish and Lebanese but also includes the numerically small contingent of Dutch civil servants.

The Negro (Creole) element has been in political control since self-government was granted in 1954. Compared with most South American countries, Surinam has had an unusual stability in government. This has been fostered in no small degree by its position as a member of the Netherlands community with the accompanying Dutch influence.

The sugar economy on which Surinam was founded has deteriorated continuously in the past century, and the growing of cane has practically ceased. From 105 plantations in the early 1800's, three remained in 1950. By 1970 only one was left, and it was on the verge of closing down because of labor difficulties.

Rice has supplanted sugar as the major crop. Bauxite and alumina are now the backbone of the economy. The country is rich in mineral resources and has the water-power potential to generate the electricity to exploit them. If the present political stability continues, Surinam's future is industrial rather than agricultural.

Because of the structure of the sugar plantation operation, which depended mostly on water transportation of cane to the mills, the use of cattle as draft animals was never important. Horses appear to have been preferred for hauling instead of oxen. Water buffaloes were brought in at one time for work in the rice paddies, but they were not widely used. Only a few hundred head now remain in the country.

CATTLE BREEDS

The sparse cattle population of Surinam is found on the farms around Paramaribo, the capital city, and the few other agriculturally developed areas. It is impossible to trace the original base from which this population, as seen today, was derived. In any sizable group of cattle, true representatives are seen of the Creole cow that is common to Latin America and the Caribbean area. This indicates that the original Spanish cattle in the New World were the common ancestors of Surinam's cattle. Conceivably the Creole cattle in Surinam could have been remnants of those left by the Spaniards who touched this part of South America at the end of the sixteenth century. Because this early appearance of the Spaniards in the area was never consolidated and has left hardly any traces, a more probable hypothesis is that the early Creole cattle were brought from Barbados by the British when they established their permanent settlements in Surinam at the middle of the seventeenth century.

The Creole cattle of Surinam which show no major influence of other breeds are typical of the type wherever it is found. The light tan shade, nearly solid, is the most common color pattern. Nearly all black and

Cattle of North America

Mature Creole bull of a small farmer (horns have been blunted), 950 pounds.

an occasional mottled black-and-white are also seen. The thin, narrow tail with the black switch, the muscular face with dark hair around the eyes, and the fairly large horns, thin and widespread, are all to be noted in Creole cattle in Surinam. The pure or nearly pure Creole cows vary in weight from 650 to 950 pounds, depending on their level of nutrition. Bulls weigh up to a maximum of 1,100 pounds.

Apparently the Zebu was never brought to the plantations for the breeding of a larger draft animal and was not introduced into Surinam until 1948. In that year there was an importation of American Brahman from the United States with the objective of producing a better beef animal. Practically all of the breeding work aimed at developing more productive beef cattle is done at the government livestock station. American Brahman, Santa Gertrudis, Hereford, and some Charolais bulls have been used, and the influence of all of these can be seen in the station's herd. The American Brahman, however, has been the most widely used breed for crossing.

For some years the Holland Friesian has been employed in upgrading the Creole cow to a more productive milk animal. Recently Holstein-

Mature Creole cow of a small farmer, 750 pounds.

Friesian semen from the United States has also been used at the government station in an effort to increase milk production. A minor influence of other European dairy breeds, particularly the Jersey, is seen.

Most of the cattle in Surinam are still either nondescript Creole or Creole showing a degree of Brahman or Friesian influence. A few dairy herds can be called upgraded Friesian.

The buffaloes in Surinam are of medium size, cows weighing around 1,000 pounds and bulls 1,500 pounds. They are of the swamp type, grey to dark grey in color. Horns are swept back for the most part, although on some individuals there is a pronounced inward sweep toward the extremities. The cows often show a narrow whitish streak of hair across the upper chest just below the neck.

MANAGEMENT PRACTICES

With the exception of a few dairies in the 100- to 200-head bracket and

Buffalo bull. Government Livestock Station.

the herds on the government livestock farms, the cattle of Surinam are in small herds. Around 80 per cent of the cattle are held by owners with three to five head and a limited number having up to twenty head. These owners are small farmers and village-dwelling East Indians. Two-thirds of the latter are Hindus who maintain their animals from religious motives, plus such return as is realized from the milk produced.

The small cattle holder pays little attention to the care of his animals other than to stake them out to as good pasture as can be found. This may be waste patches of a farmer's land, swampy areas, or along the roadside. While quite solicitous of their well-being, he is not oriented to the requirements of good husbandry. The average calf crop is probably under 50 per cent, and the calf loss under one year of age is about 40 per cent. By comparison, an 85 per cent calf crop is obtained on the government livestock farms and the death loss during the first year is 7 per cent.

Many of the cows are milked. The milk is either used in the owner's household or sold in the local village or in Paramaribo. Practically no

Buffalo cow. Note white markings on neck and chest. Government Livestock Station.

supplement is fed. The average production is 2,000 pounds per lactation.

The cattle sold for slaughter are normally unwanted bulls around seven years of age or nonproducing cows. Many are sold only when a need for cash arises.

There is one herd of 50 head of buffalo which is utilized for the production of yearling bulls for slaughter. Buffalo cows are not milked and buffaloes are not used for draft. The few small herds that are seen display the excellent adaptation of this animal to tropical conditions—heat stress, parasites, and disease. They are in better condition than even the best-maintained cattle, evidence of their ability to utilize efficiently coarse, rough forage. The average herd of cattle generally shows lack of adequate nutrition and is usually seen in poor to fair condition.

An artificial insemination center is maintained at the government livestock station. The annual number of services, mostly to cows on small farms, is 4,000. Purebred Friesian bulls, originating in the Netherlands, as well as Holstein-Friesian semen imported from the United States, are used.

MARKETING

The farmer sells either to a trader or a buyer representing a retail butcher. Many of the cattle reaching the abattoir were originally owned by an East Indian Hindu who maintains a cow and calf. In such cases, because of the Hindu's reverence for cattle, a sale for slaughter cannot be considered and the transaction must pass through the hands of an intermediary. He, in turn, then sells the animal to a trader. Such subterfuge is not necessary in the case of the East Indian Moslems who account for one-third of the Indian cattle owners.

In 1970 the price paid the farmer for the odd animal sold ranged from 21 to 25 cents a pound liveweight, depending on condition and estimated yield. The average carcass weight at the Paramaribo abattoir was 290 pounds with a 45 per cent dressing percentage.

In 1970 the farmer was receiving 5 cents a pound for milk at the pasteurizing plant in Paramaribo.

SLAUGHTERHOUSES

The municipal abattoir in Paramaribo handles most of the slaughter in the country. The annual kill averages 8,000 head of which about 3,000 are imported animals. The retail butchers do their own processing in the plant, paying a fee of $2.00 a head for slaughtering and $1.00 for inspection. The killing box and captive bolt are used, and skinning is done on cradles. Carcasses are hung on overhead rails for eviscerating and splitting. Sanitation and inspection are fair. There are refrigerating facilities, and all butcher shops are required to have refrigeration.

CATTLE DISEASES

Surinam claims to be free of foot-and-mouth disease and tuberculosis. Restrictions against the importation of cattle from foot-and-mouth countries are well enforced. Rabies carried by the vampire bat was quite prevalent a number of years ago, but is now controlled by vaccination.

Tick-borne diseases and internal parasites are the most serious plagues of cattle. The government livestock station and the few large farms spray all cattle on an average of once a month. Many small farmers wash their cattle by hand and have been taught to use Bercotox, an insect repellant, in the water. Screwworm is prevalent, and the large farms use Smear 62 systematically on the navels of young calves and on any skin abrasion on older animals.

Calfhood death loss is very high, probably averaging 40 per cent of the calves dropped. This extreme loss is largely due to malnutrition, which lowers resistance to disease and parasites. Except for the small number held on improved pastures, most cattle exist on coarse, native grasses and sedges of the low-lying swampy fields. As a result, a cow does not have adequate milk for her calf and the calf as it grows must get along on the poor forage. Mineral deficiencies are also contributing factors to poor calf health. These nutritional deficiencies certainly contribute to the low calving rate. On the average, cows calve only every two years.

GOVERNMENT AND CATTLE

Surinam imports 20 per cent of the beef, 50 per cent of the butter, and all of the cheese consumed. Although the tropical climate is a major handicap to raising cattle, the country could become self-sufficient in beef and dairy products by the development of improved pastures and the following of reasonable husbandry practices. The work of the government livestock farms, just out of Paramaribo, is demonstrating what can be accomplished along this line.

One of the two government farms husbands a dairy herd; the other is for beef production. On the dairy farm, 125 acres of pangolagrass maintain a herd of 200 upgraded Friesian cattle in which there is considerable Brahman and Creole influence. The 90 cows in milk produce an average of 4,500 pounds per lactation. The annual calf crop is from 85 to 90 per cent of the cows of breeding age and the death loss is 7 per cent of calves up to a year of age. These results are far above the national average and illustrate what can be accomplished at a level of husbandry practical for a commercial dairy. This has further been demonstrated by the establishment of a model farm on five acres, four acres of which have been put into pangola and paragrass pastures. These suffice for the maintenance of six upgraded Friesian cows which have an average annual production of 5,500 pounds.

On the beef cattle unit of the government livestock farm, a herd of 340 head of mixed Creole cattle is maintained. Improved pastures are being established, and it is felt that one acre will carry 1.6 head of cattle of all ages. The breeding program contemplates the use of American Brahman and Hereford bulls to develop an upgraded animal that is three-eighths Brahman and five-eighths Hereford. The cost of establishing improved pastures is said to be $160 an acre, which includes clearing the jungle growth by hand labor, cultivation, fertilization, and seeding.

OUTLOOK FOR CATTLE

Surinam in recent years has made more progress in developing its economic structure than either of the other Guianas, and the same measure of success could also characterize its cattle industry in the future. The grass-roots approach of its experiment station work evidences a practical solution to the problems involved. Businessmen are becoming interested in the possibilities for profit in cattle in a country which now imports a large part of its dairy products and beef. Although climatic conditions are no more favorable to cattle raising than in Guyana and French Guiana, Surinam shows signs of developing a more productive national herd than either of the neighboring countries.

PART TWO: **Middle America**

Middle America

Middle America

Middle America is commonly defined as including Mexico, Central America, Panama, and often the Greater and Lesser Antilles. Both groups of Antilles have been included in the section on the Caribbean Islands. For convenience Panama is here included in Central America, which designation traditionally has comprised only the countries of Costa Rica, El Salvador, Guatemala, Honduras, and Nicaragua, because of the closely related political history of these countries.

The six Central American countries (as here defined) have many similarities in their cattle and the practices in raising them. Because of the small size of the individual countries—Nicaragua, the largest country, is less than one-fifteenth the size of Mexico, and the total area of all Central America is less than one-third that of Mexico—cattle operations within the borders of an individual country tend to be homogeneous. Central America is therefore discussed as a separate section of Middle America, followed by Mexico.

Central America

The total area of Central America is 208,000 square miles, equal in size to the Gulf Coast states of Louisiana, Mississippi, Alabama, and Florida. The terrain is characterized by mountain ranges, generally running from northwest to southeast. Older mountains are dominated by more recent volcanic formations, and much of the area is currently subject to earthquakes and volcanic activity. Coastal plains follow the shores, which are usually quite narrow on the Pacific side and broader and more extensive on the Caribbean. In the lowlands the climate is tropical. The plateau areas enjoy a more equable climate by reason of the higher elevation. The heaviest concentrations of population are usually found in these higher regions. The rainy season, called "winter," lasts for six months, May to October, in the northwest, increasing to seven months in Costa Rica and nine months in Panama. The dry season, from October or November to April or May, is "summer."

Before the arrival of the European in the Western Hemisphere the Mayas had developed in northern Central America what was probably the most advanced Indian civilization on the continent. Some of the other Indian tribes on the south also had an advanced culture, and in the less heavily populated areas there were scattered primitive peoples who had reached a stage comparable to that of the tribal Indians on the north. All were eventually subdued by the Spaniards, whose conquests began in the early 1520's, when they moved in from Mexico on the north and from Panama on the south.

Because the region was lacking in extensive mineral resources, the magnet that most attracted the conquistadors to the New World, colonization and development were not as extensive as in Mexico and South America. Some agricultural settlements, however, were established along with Roman Catholic churches and monasteries.

The rule of Spain in the area north of Panama continued for three centuries. Panama was governed from Colombia until independence was gained in 1903. The other five states came under the captaincy-general of Guatemala until they became independent from Spain in 1821. The fact that they were ruled by a captaincy-general instead of a vice-royalty was indicative of the lesser importance that Spain placed upon these colonies, which yielded little gold or silver. A captaincy-general lacked the prestige of the vice-royalties that Spain established in richer areas of the western Hemisphere.

For a brief period after gaining their independence, the former Central American colonies (Panama not included) were annexed to the newly formed Mexican Empire. When this collapsed in 1823, the Federal Republic of Central America was formed of Costa Rica, El Salvador, Guatemala, Honduras, and Nicaragua. Each of these has remained an independent state since the dissolution of the Federal Republic in 1838. All have been subject to revolution, coups, and armed conflict with their neighbors.

The only European power that disputed the Spanish rule in Central America was Great Britain. Spain in her conquests had bypassed a small, rather inhospitable area of about 9,000 square miles (now British Honduras) that lies northeast of Guatemala along the Caribbean coastal plain. This was a good timber area, and British loggers began moving in about 1638. Eventually some agricultural settlements followed. This kind of activity extended southward along the coast of Honduras to the San Juan River in Nicaragua. In 1862 the Crown Colony of British Honduras was proclaimed, and is today a self-governing British colony. (Guatemala lays claim, with little foundation, to the area.) The British foothold in the isolated eastern part of Nicaragua persisted, and the region was finally proclaimed a protectorate. At United States insistence it was abandoned in 1860.

The present population of Central America, now nearly 15 million, gives a density of 72 per square mile, as compared with 66 for conterminous United States. The people are either pure Indian descendants of the indigenous population or mixtures of these with the Spaniards, varying from nearly pure Indian to nearly pure Spaniard. There is a small minority of Europeans, mostly of Spanish origin. The mixed Indian and Spanish people are sometimes referred to as mestizos. Non-Indian natives who have predominantly European cultural traits may be called

Ladinos, but a pure Indian who assumes the cultural attributes of the European is also a Ladino. Neither term is heard very often in Central America.

In some areas of the lowlands along the Caribbean there are sizable Negro and mixed Negro and Indian elements in the population. They are descendants of slaves the British brought in from Jamaica or Negro laborers introduced more recently by large plantation owners.

From both geographic and ethnic considerations the most logical policy for the five Central American states that were included in the Federal Republic seems to be consolidation into one entity. There have been some forty-five formal attempts to achieve this end since 1846. All failed because of local rivalry and the political jealousies of the individual states. For the last twenty-five years more emphasis has been placed on economic autonomy, and some progress has been made in the establishment of a Central American Common Market. In theory, and to some extent in practice, there is now a common tariff and free circulation of goods among the five states except for such items as food staples and petroleum. Such co-operation could conceivably lead to some form of political union.

The economy of Central America is essentially still agricultural, although over the past few decades continuous increases in industry and commerce have been achieved. Commercial production of crops and livestock is primarily in the hands of large landowners, but the number of small producers is increasing. Most of the people remain occupied in subsistence agriculture. These farmers often meet their immediate money needs by the sale in local markets of small quantities of the products they have raised. Hardly any of these products, however, reach regular channels of commerce. In most states coffee is the major export crop, with bananas and cacao following. Although the production of cotton has increased rapidly in recent years throughout the region, it is a major crop only in Nicaragua.

The basic bovine population of all of Middle America descended from the cattle the Spaniards brought to the region. This movement began in the first third of the sixteenth century and continued for an indefinite period. The cattle reached central America from Mexico and later from the Caribbean Islands where Spanish rule had been established.

Distinct breeds of cattle recognized today had not been developed in Europe at this time. There were undoubtedly local types in Spain which, during the course of many generations, had acquired similar characteristics through natural and, to some extent, artificial selection.

Cattle of North America

These were the source of the original cattle in Middle and South America. Their descendants, either relatively unmixed representatives or interbred with other later introduced types or breeds, are commonly called Criollo cattle throughout Latin America.

The common practice among cattlemen of applying the term Criollo to both the unmixed descendants of the original Spanish cattle and the miscellaneous mixtures of these with either European and Zebu types is confusing. This complication is further confounded by an effort to distinguish between "mixed" Criollo cattle and "upgraded" Criollo cattle. These terms as here used are defined as follows:

Criollo (pure): The nearly pure descendants of Spanish cattle which were brought to the Western Hemisphere before the introduction of other European and Zebu types.

Criollo (mixed): The nondescript progeny which resulted from unplanned crossing of pure Criollo with either northern European breeds or Zebu types of cattle.

Upgraded cattle (of any specific breed): These are the result of the continued use of the bulls of one breed on Criollo cows, and on the female offspring of such breeding, carried to the point where the progeny displays characteristics that are predominantly those of the breed of bulls so used.

The De Lidia, the cattle of the bull ring, are said to have been imported from Spain to Mexico around the middle of the seventeenth century, but it is doubtful that any of them reached Central America. During the latter half of the nineteenth century Zebu cattle from Brazil, and later from the Caribbean Islands, were brought to Mexico, but they found their way to Central America in limited numbers. The same is true of the European breeds, representatives of which followed the Zebu to Mexico.

Later importations of foreign cattle were not begun until well after the turn of the twentieth century. A few Red Poll herds, some representatives of which are still in existence, were established in Honduras and Panama forty years ago. The United Fruit Company introduced the Nellore and Guzerat Zebu breeds to the north coast of Honduras in the late 1930's. About the same time a few small herds of the European dairy breeds, particularly Guernsey and Jersey, were established in Nicaragua and Costa Rica. Except in a few isolated areas, however, none of the introductions had any appreciable effect on the predominantly Criollo cattle population of Central America.

Central America

Since World War II there has been a steadily increasing importation of Zebu-type cattle, which have been used in upgrading the Criollo to a larger and more productive beef animal. As a result the grey color of the American Brahman has been conspicuous in all the Central American countries. The American Brahman, although possibly not as tolerant of such an environment as some Zebu breeds, performs excellently. The attempts of breed enthusiasts to establish such northern European breeds as the Hereford and Aberdeen-Angus have met with less success because of their susceptibility to heat stress and tick-borne diseases.

Along with the gradual but fairly steady improvement in the economy there has been a marked increase in the demand for milk. Although much of the milk production still comes from Criollo cattle, there is a steady increase in the number of herds of predominantly European breeds. The Guernsey and, to a lesser extent, the Jersey were the first to become established. Both are now outnumbered by the Holstein-Friesian, which appears certain to become the milk cow of Central America because of its high milk-producing ability, rather than a greater tolerance to the tropical environment. In most of Central America the concentrations of dairy cattle are in high plateau areas adjacent to population centers where climate is more temperate. The beef herds are usually at lower elevations, where heat stress and tick infestation are a greater handicap to northern breeds of cattle.

The pure Criollo cattle will probably disappear from Central America in the near future. Today Honduras has the largest number. They are found mainly in the mountains and in the more remote coastal areas. The pure Criollo are also seen in Guatemala and El Salvador and occasionally in Nicaragua and Costa Rica, but they have practically disappeared from Panama.

In both Nicaragua and Costa Rica pure Criollo cattle have been selected in a limited way for milk production. Within a period of only thirty years a remarkable strain of climatically adapted, dairy-type cattle was developed. This selection of a milk type from the Criollo illustrates the possibilities of developing more productive cattle from indigenous populations by the utilization of modern procedures in breeding. During a period which extends back over four centuries, the pure Criollo through natural selection has evolved into a type of nonhumped cattle which is even more tolerant of the tropical environment of Central America than are the recently introduced Zebu breeds. It is an excellent draft animal, considering its size. What could be developed from it in the way of a beef type has never been determined. It is lamentable that the pure Criollo is passing out of the picture without its full potential as either a milk or a beef animal having been explored.

The future of beef cattle in Central America hinges on the export market to the United States. In 1968 the United States absorbed nearly one-third of the total slaughter of Central America and is importing more every year. Costa Rica, Guatemala, Honduras, and Nicaragua are the major exporters. Panama produces only a small volume for export and El Salvador exports none directly.

The export industry has been developed to a major extent through United States contributions of capital and know-how by grants, low-interest loans, and technical assistance provided under numerous United States AID programs. Since the major cattle producers are a few large growers in the higher-income brackets, the United States export trade does little to help the poor of these countries other than to stimulate a small increase in employment.

The Central American countries are free of foot-and-mouth disease; yet their proximity to South America leaves them subject to threat. Defenses against the disease are not as thorough as those upon which cattlemen in the United States have come to rely. The Central American Internation Regional Organization of Plant Protection and Animal Health (OIRSA) does what it can along protective lines with such funds as are available to it. But surveillance for the detection and prevention of the disease is not well organized, and in case of an outbreak the boundaries between the countries would provide little impediment to the movement of cattle.

Costa Rica

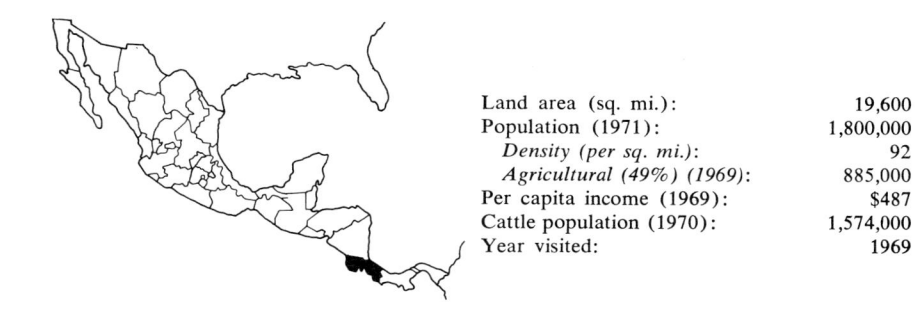

Land area (sq. mi.):	19,600
Population (1971):	1,800,000
Density (per sq. mi.):	92
Agricultural (49%) (1969):	885,000
Per capita income (1969):	$487
Cattle population (1970):	1,574,000
Year visited:	1969

THE REPUBLIC OF COSTA RICA occupies a wedge of land lying between the Caribbean Sea and the Pacific Ocean, bounded by Nicaragua on the north and Panama on the southeast. The land area is somewhat less than one-half that of Louisiana. There are coastal plains along both coasts—narrow on the Pacific but wider along the Caribbean. Most of the country is mountainous with a central plateau area at 3,000 to 6,500 feet in elevation, encircled by mountains on three sides. The lowlands are tropical, with heavy rainfall on the Caribbean side but much less near the Pacific side, which has a pronounced seven-month dry season. The uplands are fairly temperate, and the mountains are cool.

Columbus, on his last expedition to the New World, made a token landing on the Caribbean coast of the area that is now Costa Rica. Settlement by the Spaniards was slow because they found neither the gold nor the silver they sought. Colonization did not begin until the second half of the sixteenth century. What development that did occur was along agricultural lines. The colony was governed by the captaincy-general of Guatemala until 1821. Then, after two years under the Mexican Empire, Costa Rica joined the Federal Republic of Central America. With the dismemberment of the federation in 1838, the country became independent.

When the first Spaniards arrived, they found the area more thinly populated by native Indians than the colonies on the north. Early Spanish estimates placed only 25,000 indigenous inhabitants in the region. The pure-Indian population today is probably less than 2 per cent of the

Criollo (pure) bull, two years old, 1,000 pounds. Reddish tan color (partially dehorned). IICA, Turrialba.

total, and in the remainder there is a much higher proportion of European ancestry than in the other Central American countries. While Spanish influence predominates to the extent of possibly 80 per cent of the people, there is a sizable element of British, German, Austrian, Dutch, and other European descendants in the mixed population.

The economy is primarily agricultural. Nearly all industrial activity is involved in the processing of foodstuffs, other light industry, and commerce. Well over half the people are engaged in farming or stock raising. With the exception of banana production, which is mostly under foreign corporations, the agriculture of the country is in the hands of Costa Ricans. Coffee is still the major crop, bananas are second, and cattle are third. The production of sugar cane and rice has been increasing in recent years. Less than half the arable land is cropped or used for pasture.

Costa Rica is further advanced in cattle raising than any of the other Central American states. Along with Nicaragua, Costa Rica enjoys the highest quotas for beef export to the United States. The cooler climate of the central plateau and a dry season that is relieved by some rain are definite advantages for cattle raising.

Mature Criollo (pure) cow, 950 pounds. Reddish-tan color (dehorned). IICA, Turrialba.

CATTLE BREEDS

Costa Rica, along with Panama, has made notable progress in upgrading the Criollo cattle. European dairy breeds were brought in over fifty years ago. These were followed by importations of Guzerat and Nellore stock from Mexico. The process of upgrading the Criollo to a higher-yielding milk type on the one hand and to a larger and faster maturing beef animal on the other has proceeded continuously and has accelerated rapidly since the late 1950's. Today the American Brahman influence dominates in the beef-cattle herds.

Pure Criollo.—These cattle are rarely seen today in Costa Rica. There is a small herd at the experimental farm of the Instituto Interamericano de Ciencias Agricolas (IICA) at Turrialba, where a consistent effort has been made to develop a strain of milk-type Criollo. The herd, said at one time to have numbered over 300 head, was founded through the efforts of Dr. Jorge de Alba. The base was the best specimen of pure Criollo cattle which could be purchased in the remote areas of the Central American countries. The selection program followed was for milk productivity and trueness to type. This resulted in animals with remarkably uniform characteristics: a light reddish-tan color, fine bone, short hair, a thick hide, a long, very thin tail with a short brush, prominent tail stock, and rather prominent upswept horns. Conformation is distinctly a dairy type. Comment is heard that some Ayrshire or Jersey influence was introduced into the herd, but Dr. de Alba states that this is without foundation.

A small herd of about 50 of these cows still maintained in 1969 at IICA is a noteworthy example of what can be accomplished by systematic selection from indigenous cattle with a high degree of tolerance to the tropical environment. The primary objective for which the herd is now maintained is to furnish pure Criollo bulls and cows for an experimental crossing program with Jersey and Guernsey breeds. While selection of the herd is based on milk production, no effort is made to evaluate the milk producing ability of a bull's progeny.

Under present conditions the average herd production is 4,400 pounds of 4.6 per cent butterfat milk for a 305-day lactation period. The best cows produce 6,600 pounds. Nutrition is ample, and better care is provided against parasites and tropical disease than the naturally tolerant Criollo probably requires. Two-year-old bulls weigh 1,000 pounds, and mature cows in average milking condition weigh 950 pounds. Only bulls from the highest-producing cows are kept, and these are discarded as younger bulls come on. No effort to prove the bulls which are discarded was being made in 1968, and undoubtedly some excellent genetic material was going to the slaughterhouse because of this policy. All the cows are dehorned. In general appearance these selected Criollo cattle are the same as the pure representatives seen in other parts of Central America, although because of a well-balanced ration they are in excellent condition. In the unselected Criollo cattle, however, the coloring is more varied.

A marked similarity is noted in this selected Criollo herd in Costa Rica and the milk-strain Criollo which has been developed in Venezuela. There have been minor exportations of the IICA cattle to Venezuela in past years, but these were not of sufficient volume to have had any marked influence on the type as developed there. It appears that, by selecting for both type and milk production, very similar animals were developed independently in the two countries. The cattle of both Venezuela and Costa Rica definitely trace back to the Spanish cattle brought to the New World over four centuries ago.

Criollo (Mixed).—Even in the more remote parts of Costa Rica the

Purebred Red American Brahman bull, three years old, 1,450 pounds. Home raised on a coastal ranch near Puntarenas.

cattle usually show a pronounced influence of some exotic breed, often American Brahman or Guernsey. Draft oxen frequently may be classed as upgraded American Brahman. The small dairy herds show Guernsey or, less often, Holstein-Friesian influence. Some large ranches where little attention has been given to breeding practices have herds with a good representation of mixed Criollo, particularly in the northwest part of the country near the Nicaraguan border.

Introduced Breeds.—The American Brahman has had more impact on the beef cattle of Costa Rica than any other breed. This is predominantly the grey-colored type, although Red Brahman have increased in popularity recently. Most growers follow a continuous upgrading program using either purebred bulls or highly upgraded bulls in their herds. Many purebred herds are maintained, usually in conjunction with a commercial operation.

A few European beef breeds are maintained as purebred herds to produce bulls for crossing on commercial herds. These include Charolais, Charbray, and a few Aberdeen-Angus and Beefmaster. The influence of these, however, has been negligible on the total cattle population.

Purebred three-year-old Guernsey heifer. Home raised on a modern farm near San José.

In the national dairy herd the Holstein-Friesian probably predominates today, having made rapid progress in the past ten years because of its high milk production. The Guernsey, losing ground to the Holstein, is easily second in numbers, and the Brown Swiss is third. The Jersey, although one of the first dairy breeds brought to Costa Rica, is disappearing. Some excellent purebred herds of all these breeds are maintained, based for the most part on importations of stock from the United States. More recently they have been improved by artificial-insemination programs using semen from the United States. The bulk of the milk production of the country comes from upgraded herds of these breeds.

MANAGEMENT PRACTICES

Beef Cattle.—The major beef-producing area of Costa Rica is the hilly region lying on the Pacific side of the Continental Divide in the northwest part of the country. The annual rainfall there is normally around 80 inches, most of it falling from May through November. This section

Typical upgraded American Brahman–Criollo herd of a large grower. On irrigated pasture in the dry season.

of the country is practically without precipitation during the long dry season. Cattle are generally maintained on the dry mature growth during this period, often without adequate grazing control to ensure sufficient feed. A few of the more progressive ranches are beginning to put up some hay in order to carry stock through the dry season in better condition. This practice should gradually increase.

Beef cattle are also raised on the Pacific coastal plain to a limited extent. There flood irrigation during the dry season can maintain lush green pastures, but the practice is not widespread. Although many of the short, swift streams which flow from the central highlands to the Pacific run nearly dry after the rains cease in November, a few could furnish sufficient water to irrigate sizable areas.

There are examples of well-managed properties in the country with good grazing controls. These utilize permanently fenced pastures, often subdivided with temporary electric fences. There is good provision for stockwater, sometimes with a pipeline system for supplying corner-located water tanks in the pastures. Such practices, however, are not general and most slaughter cattle grow out to market age with little care.

There are many small growers whose annual sale of steers is in the fifty-head range, but most beef production comes from larger ranches which market several hundred head. Both large and small growers follow the practice of taking some milk from a portion of the cow herd to help defray current operating expenses. On a ranch with around 1,000 brood cows, 200 may be milked once a day. The calves are taken off in the late afternoon and returned to the cows as they go to pasture the next morning after milking. The yield runs around eight pounds of milk a head a day. This gives a return of $100 a day for 200 head, a profitable operation in view of the low labor cost involved.

Breeding practices vary widely. Generally bulls run with the cows the whole year, the ratio being three or four bulls to 100 cows. There is some utilization of a nine-month breeding season, spaced so that calving begins at the end of the dry season. Herds under good management obtain a 65 to 70 per cent calf crop, but the country average is said to be around 40 per cent. The most progressive purebred breeders make use of artificial insemination and often have a technician of their own to do the impregnating.

The screwworm takes a heavy toll of Costa Rican cattle. Good management utilizes maternity pastures as the most effective control measure. Pregnant cows are brought into enclosures near the headquarters where Smear 62 can be applied promptly to the navels of newborn calves. They are then watched closely for the first few weeks. Such care materially reduces the death loss.

Slaughter cattle are usually marketed at three to four years of age, when they weigh from 900 to 1,000 pounds. With good management, high upgraded Brahman steers can reach a 900-pound weight at two and one-half years. Some ranchers who are financially able to do so buy young stock, usually weaned calves at nine or ten months of age, and carry them through on grass to three years of age. The profitable United States export market has made this feeding operation quite lucrative to a rancher who has the available pasture.

Dairy Cattle.—Most of the dairies are within a thirty-mile radius of the capital city, San José. There are numerous herds of fifteen to twenty Criollo mixed cows which show a strong influence of Holstein-Friesian, Guernsey, or Brown Swiss. Occasionally a nearly pure Criollo cow is seen in the herds. These small dairy herds are run on pasture.

The main supply of milk in the country comes from larger, better-managed dairies. Some of these have purebred or high-grade Guernsey or Holstein-Friesian herds milking 200 to 300 cows. These are maintained at United States standards except that milking is usually done by hand. Such dairies are located in the highlands at elevations of 4,000 feet or higher, where with good care the European breeds can be maintained in excellent condition. Although comparatively moderate temperatures prevail, the tropical environment still takes its toll of the European breeds both in milk yield and in the size of a mature animal. A Holstein-Friesian cow weighs 100 to 200 pounds less than her genetic equivalent in the United States. Her milk production is only about 75 per cent as great. This is with the very best of care. An average productive age of eleven years is reached in some herds. Production in such herds averages 9,000 pounds of milk in a 305-day lactation.

Most of the large dairies raise breeding stock. They breed only their best cows to dairy bulls; the rest are bred to a beef-type bull to produce feeder cattle. Artificial insemination is employed. In the small dairies calves are killed at one to three days of age because it is more profitable to sell the milk than to raise the calf.

There is a small government bull stud at El Alto near San José in which seven bulls of the dairy breeds are maintained: three Holstein-Friesians, two Guernseys, a Brown Swiss, and a Jersey. Service is furnished gratis, but there is a charge of $3.00 for fresh semen and $4.50 for imported United States frozen semen. Small dairies in the vicinity are the only customers, and the results are not impressive. Around fifty cows are bred daily. The conception rate is 60 per cent with as many as three services.

Feeding and pasture practices vary considerably with the elevation. Often cows are turned to pasture at 4:00 P.M. after milking and brought in again at 4:00 A.M. At 6:30 A.M. they are returned to pasture for about two hours and then brought in again to be kept under shed or in a barn during the heat of the day. Supplemental rations usually contain no grain. Wheat bran, cottonseed meal and hulls, and molasses are a common mixture. Six pounds are fed as a base with a pound additional for each two pounds of milk produced. Green chop or, in some cases, silage is fed during the period that cows are under shelter.

Except in the more remote areas the draft ox has disappeared. The occasional team seen is a cross of American Brahman, Guernsey, or Brown Swiss on the native Criollo. Their use is limited to inaccessible settlements along the coast or to areas in the mountains where the roads are impassable to motor vehicles.

Cattle of North America

MARKETING

One of the few instances where cattle are offered for sale at a common point in Central America is the Montecillos Market in Alajuela, fifteen miles west of San José. The Cattlemen's Association of Costa Rica has recently built a modern sales facility next to a new co-operative slaughterhouse there. It features paved pens with a capacity of 1,000 head of cattle, two sets of scales, an auction ring, and other pens for holding cattle after they have been sold. There are provisions for watering stock but none for feeding.

Auction sales were not being held in 1969, although that is the ultimate objective. The principal buyers are butchers and some traders who buy direct from farmers who have trucked in their cattle. The function of the trader is largely a financial one. He usually pays the seller at once and carries the butcher-buyer for eight days for a commission of 1.4 per cent. This market facility, however, gives the seller the opportunity to deal with more than one buyer and thus offers some element of a competitive market. This is a feature practically unknown elsewhere in Central America.

At the Montecillos Market cattle are brought in on the weekend and sold on Sunday afternoon and Monday morning. All animals are for local consumption. Except for the occasional head bought by a local farmer to take back to the country, they are slaughtered in the adjoining plant. The average run of 1,000 head a week consists of 70 per cent three- or four-year-old upgraded Brahman steers. The rest are discarded cows, bulls, and a few draft oxen. There is very little recognition of quality other than the butcher-buyer's appraisal of the dressing percentage to be obtained from animals he buys. In early 1969 the exceptional 1,000-pound four-year-old steer, grading good by USDA standards, brought 9 cents a pound. A canner cow brought a little under 8 cents. Practically all sales are by liveweight.

Except for the Montecillos Market the cattle trade is oriented largely toward the United States market for frozen beef. Of an estimated 170,000 head slaughtered annually, 70,000 of the heavier and better animals leave the country as boned frozen beef.

The government allots a percentage of the national annual export quota to each of the five plants approved to participate in the trade. The Cattlemen's Association then determines the number of head that can be supplied to each plant from the surrounding area. The government next advises individual growers in the area of the number of head they are to furnish the plant for export, as well as the time for delivery. In

Montecillos Cattle Market.

early 1969 the price of export cattle, regardless of quality, was 15 cents a pound.

The bulk of this trade is three- to four-year-old upgraded Criollo steers at around 1,100 pounds. It is obviously to the grower's advantage to sell his larger and heavier animals through this channel. The remainder of his offtake has to go to the local butchers for 9 cents a pound.

The going price for raw milk in 1969 was 6 cents a pound to the producer. Dos Pinas, a co-operative organization with a modern dairy plant in San José, bottles pasteurized milk and sells a full line of dairy products. This concern handles 30 per cent of the milk produced in the San José milkshed. Sanitation standards are required on the farms of the producers from which it buys, and the whole operation is well above the level found in most of Central America.

SLAUGHTERHOUSES

The co-operative slaughterhouse adjoining the cattle market at Montecillos is the only modern plant in Central America slaughtering exclusively for local consumption. Because of failure to pass USDA inspection for export trade, the plant was relegated to local use. Most of the output will undoubtedly go for export as soon as official USDA approval is obtained and a cattle allotment authorized. Export slaughterhouses sell only a small part of their production locally.

In 1969, Costa Rica's five approved export plants were processing 70,000 head annually. This capacity could easily be doubled if the export quota were increased. The workload is on a seasonal basis because of the scarcity of slaughter animals during the dry season. Sanitation and inspection at these plants are good. Skinning is usually done on cradles, but the carcasses are handled on the rail from there to the cooling room.

As much of the output as possible goes into boned United States–type cuts. The carcass is chilled 24 hours before cutting and is then quick-frozen. These export plants are located in the San José and Puntarenas areas. Since there are no roads connecting these regions with the Caribbean port of Limón, the frozen packaged beef must move either by rail to Limón or by refrigerated truck through Nicaragua and Honduras to a Guatemalan port.

The beef for home consumption is usually processed in the municipal plants located in the larger towns. These are minimal facilities, usually not much more than a small building with a concrete floor, water, and perhaps a few skinning cradles.

CATTLE DISEASES

Constant vigilance against the tick, screwworm, and internal parasites is essential for a profitable cattle operation. In spite of this, a major part of the national herd is run without adequate prevention or veterinarian treatment. The screwworm, as it preys on young calves, is probably the most devastating parasite. Short of eradication, it can be successfully combated only by the use of maternity pastures and daily inspection during calving, practices found on only a few ranches. Calf losses from birth to weaning are approximately 15 per cent for the country as a whole; good practices can hold these losses to under 5 per cent.

Most growers spray by hand for ticks. Depending on the area and season, good control can be obtained by treatment at intervals of two to four weeks. Often, however, spraying is resorted to only when the cattle are observed to be in unseasonably poor condition.

Nearly all commercial growers and all those raising purebred cattle use the three-way vaccine for blackleg, malignant edema, and hemorrhagic septicemia at branding or when calves are three or four months of age. Cattle of all ages are vaccinated against anthrax and rabies in areas where these diseases are prevalent. Neither of these diseases is widespread.

Calves must be drenched for internal parasites at five to six months of age and again as yearlings. Older cattle are not usually seriously affected and are treated only when their condition so demands.

There is a government program for elimination of brucellosis and tuberculosis which calls for testing by government veterinarians and mandatory slaughter of infected animals. The owner is paid one-half of the fair market value of the animals killed. This program has accomplished very little. Small growers are reluctant to accept the loss involved in shipping a reactor and hesitate to become involved with the government. The large ranches and dairies take care of themselves. Well-managed dairy or purebred beef herds are usually vaccinated against brucellosis during calfhood, often at four months of age. Legally this can be done only by a licensed veterinarian.

Satisfactory veterinary service is difficult for the small grower to obtain because there are few competent private veterinarians in the country. Those in government service are allowed to accept fees for private practice, but often they are not available when needed. The larger ranches keep a veterinarian on retainer to make a monthly inspection of the herd, do pregnancy testing, handle brucellosis vaccination, and perform other routine tasks.

GOVERNMENT AND CATTLE

The government takes an ostensibly favorable attitude toward the cattle industry. There are adequate laws covering inspection and sanitation in slaughterhouses approved for export, and the government veterinarians in these plants are competent. Periodic visits of the USDA inspectors ensure that proper standards are maintained. Government is active in its efforts to increase the volume of the "voluntary" quota for export to the United States. Beyond this, however, there is little concrete evidence of government assistance to the cattle industry.

The low calving rate and the lucrative United States market have caused some concern that the cattle population might decrease. Actually it has shown modest increases in recent years. Nevertheless, there is a law prohibiting the slaughter of pregnant cows or those under eight years of age. This law is often ignored in the many small municipal plants.

There are land-reform laws on the books, but practically nothing has been done to implement them. With only 92 inhabitants a square mile, population pressure on the land is not serious. There are practically uninhabited wide jungle areas on the Caribbean coastal plain, most of which are subject to future agricultural development. There are very few extremely large land holdings. These elements temper the threat of agrarian reform for the time being.

OUTLOOK FOR CATTLE

The process of upgrading the Criollo cow to American Brahman in beef herds and to the Holstein-Friesian, Guernsey, or Brown Swiss in dairies appears to be in its final stages. The European and the other Zebu-type beef breeds which have some adherents do not seem likely to become very important.

The local market for beef is easily saturated, and it appears improbable that the quotas for export to the United States will be materially increased in the near future. The world demand for beef, however, is increasing and eventually could result in a considerable expansion in the cattle industry.

Next to endemic diseases and parasites, lack of adequate nutrition during the dry season is the most serious handicap in raising cattle. Cattle deteriorate rapidly in condition as the dry season advances. This could be offset by either irrigating or feeding hay or silage. In many areas water could be made available for irrigation by constructing simple gravity systems. The frequency of rains in the wet season presents problems in haymaking, but a number of progressive ranchers have shown that these difficulties can be overcome. The better nutrition would improve the condition of the cow during breeding season and result in a larger calf crop. Heavier cattle could also be produced at younger ages. These factors, along with better parasite and disease control, could easily double the current production of beef with no increase in pasture lands now used.

The uninhabited jungle areas on the Caribbean side of the Continental Divide are equal in extent to the agriculturally developed land on the Pacific side. Vegetation on the Caribbean side is green practically all year because there is some precipitation during the so-called dry season. The higher land of the region, most of which is now without roads, could provide excellent year-round pasture without irrigation or fodder storage. The possibilities for a further expansion of the cattle industry here far exceeds what could be accomplished by better management methods in the present cattle region on the Pacific side. Participation in the world market for beef, rather than dependence on the limited outlet to the United States, will be necessary for such development.

El Salvador

Land area (sq. mi.):	8,300
Population (1971):	3,500,000
Density (per sq. mi.):	430
Agricultural (60%) (1964):	2,100,000
Per capita income (1969):	$283
Cattle population (1970):	1,350,000
Year visited:	1969

THE REPUBLIC OF EL SALVADOR is the smallest of the Central American countries. The area is one-sixth that of Mississippi. It is enclosed by Guatemala on the west, Honduras on the north and east, and the Pacific Ocean on the south. Although only 135 miles in length, with an average width of 60 miles, the country is divided into three distinct climatic regions because of differences in elevation. Along the Pacific there is a narrow, hot coastal plain; the central plateau and valleys are subtropical, warm rather than hot; and the thinly populated mountainous area in the north is usually cool.

When the Spanish conquerors first entered the region in 1523 from Guatemala, they were strongly resisted by the Indians. These tribes had developed an advanced civilization similar to that of the Aztecs in Mexico. El Salvador was not heavily colonized because it lacked mineral wealth. There were, however, some agricultural settlements by Spaniards who intermarried with the Indians.

Spanish control was established in 1525 and continued for nearly three centuries. Revolt against Spain began in 1811, but it was ten years before Spanish rule ended. In 1838, when the Federal Republic of Central America was dissolved, El Salvador became an independent country.

During the remainder of the nineteenth century the country was in nearly continuous turmoil, the result of conflict between the conservative and liberal elements of the population. After the turn of the century there were nearly thirty years of relative stability in government, followed by a period of revolts and coups, often communist-inspired and led. From 1950 to 1960 strong leaders maintained an orderly government, but it has since been followed by coups and communist-engendered unrest. Nevertheless, the continuity of government has been uninterrupted since the adoption of a new constitution in 1962.

During the periods of political stability El Salvador has made considerable industrial progress. The currency has been stable with the United States dollar since 1934. The country has the only steel plant in Central America, an oil refinery, and a cement plant. The economy is basically agricultural, with coffee the main crop and cotton second. Two-thirds of the population live at a subsistence level. A continuing movement to unionize farm labor was causing serious concern among large landowners in 1969.

The population of El Salvador is homogeneous, 92 per cent being of mixed Indian and Spanish descent. The rest of the people are either pure Indian or pure European. El Salvador is by far the most heavily populated country in Middle America, now more than 400 people a square mile.

The country also has the heaviest cattle density—156 a square mile, or three times that of any other Central American country. Until quite recently the primary use of cattle was for draft; milk was taken regularly but was of secondary importance; and beef was only the salvage product of worn-out work oxen or discarded milk cows. In recent years this situation has begun to change with the importation of Zebu breeds to the higher country between the plains and mountains.

CATTLE BREEDS

There still remains a rather large representation of Criollo cattle in El Salvador. There were no sizable importations of Zebu cattle until after 1960, and the European dairy breeds followed them. Breeding stock of both types, principally bulls for crossing, were brought in first from Mexico and Guatemala and then from the United States. Ships calling at El Salvador ports are not equipped to handle cattle, and most imports from the United States have come by air.

Criollo (Pure).—The Criollo cattle seen in El Salvador show more of the typical characteristics of the descendants of the old Spanish cattle than do those in most of the other Middle American countries.

A mature cow varies in size from 550 to 850 pounds. While there is the usual variation in color—black or nearly all black, some red, and a few black-and-white mottled—the predominant color is tan. This may be a light, almost cream color, sometimes with a rather grey appearance, or a rather dark tan with a decided reddish cast.

Criollo (pure) cow. San Salvador market.

The horns are large, outswept, and upturned, frequently with a lyre-shaped tip. The tail is thin and long, sometimes nearly reaching the ground, and with such short hair that it gives the appearance of having been clipped. The switch is short and thin. Darker-colored hair around the eyes is common. The face is quite narrow and, like the rest of the head, has a lean, muscular look. There is a general tendency for animals to be slightly swaybacked.

The use made of cattle by the small owners for many generations before the recent introduction of exotic breeds may well have led to some elemental selection for a draft and milk type. Many teams of work oxen seen throughout the country are good draft-type animals. They are well boned, with sound hoofs, heavy forequarters, and thin but strongly muscled hindquarters. Some cows are seen with good udders and large teats, though there has been no systematic selection for milking ability. Individual animals are said to give up to 16 pounds of milk a day.

Criollo (Mixed).—This hybrid type of Criollo is rapidly increasing owing to the widespread use of Zebu and European dairy bulls. At the

Criollo (mixed) ox team. Central El Salvador.

present time the mixed Criollo is a nondescript cross of these types of cattle on the Criollo. The American Brahman influence, however, predominates.

Introduced Breeds.—There is little evidence in Salvadorian cattle of any Zebu breeds except the American Brahman. These have been used for crossing to improve fleshing quality, and there are also a few small purebred herds. Some use has been made of Indo-Brazil bulls. Bulls of the Santa Gertrudis, Angus, and Charolais breeds have been introduced, but their impact on the cattle population has been negligible.

Of the European dairy breeds the Holstein-Friesian is by far the most prominent, with the Brown Swiss a poor second. There is one large herd of pure Ayrshires. These pure dairy herds are usually maintained in the higher country between the coastal plain and the mountains. A few Holstein-Friesian herds are the result of several generations of crossing on the Criollo.

MANAGEMENT PRACTICES

Most of the cattle in El Salvador are in the hands of small farmers who have fewer than 20 head of essentially dual-purpose animals. The

An upgraded Holstein-Friesian herd on irrigated pasture.

primary emphasis is on draft, although milk is taken for household use and for sale, often in the form of cheese. Only discarded work oxen or nonproductive cows are sold for slaughter unless there is a pressing need for cash. Unwanted calves are usually killed when a day old. The importance of the work ox in Salvadorian agriculture is indicated by the fact that out of a total cattle population of 1,250,000 head there are fewer than 300,000 cows. Very few young animals are raised for slaughter.

Nearly all cattle are run on such pasture as is available year round. This is either rough or brush-covered land that cannot be cultivated or consists of crop residues on cultivated fields. During the six-month rainy season feed is lush and ample. A month or two after the rains cease, the dry, mature growth supplies barely enough nutrients for body maintenance and practically nothing for the growth of a young animal.

Over the country in general there is little planned breeding except for the limited use of artificial insemination. This service is provided free by the Ministry of Agriculture bull stud in San Salvador. Coconut milk is used as the diluent in the preparation of semen, which is kept at room temperature for use over a four-day storage period. Some 22,000 cows were inseminated in 1969, and the practice increases by 10 per cent annually. The conception rate was 60 per cent on the first service. There are 225 technicians in the country who were trained at the bull-stud laboratory. There is also considerable private use of imported semen from the United States in larger herds. The charge for this is around $2.50 an ampule.

Prophylactic treatment for disease prevention is minimal unless some outbreak gives cause for concern. Fortunately the Criollo not only is a good rustler but also carries a high tolerance to the most prevalent diseases.

Large-scale operations are limited in number. There are only about eight haciendas with 500 or more head of cattle. Some of these are dairy farms in the vicinity of San Salvador which milk 200 to 300 head. There are a few herds of up to 1,000 head of mixed Criollo cows in the process of being upgraded. Both types of cattle operations have frequently resulted from a city-dwelling businessman's desire to diversify his interests by landownership and cattle raising.

A typical large dairy farm has pastures that have been improved by establishing such African grasses as pangolagrass or Africa Star. Hay or silage is put up for use during the dry season, and a supplement is fed to cows in milk. When water is available from wells or springs, pastures may be irrigated for four months of the dry season. This practice will then give a year-round lush grass with a carrying capacity of about one head an acre. There is some use of loose housing; feed is brought to the lots where animals of all ages are kept throughout the year. Both milking machines in milk parlors and hand milking are employed. Bull calves are killed when a day old, but most heifer calves are raised for replacements or for sale as breeding stock.

The larger owner of mixed Criollo cattle is usually upgrading his herd to American Brahman. He often milks a number of the better-producing cows in order to provide a steady income. All cattle are run on pasture, and practically no provision is made for supplemental feed during the dry season. Heifer calves are grown out for replacements, and the poorer cows are sold off as part of the upgrading process. Steer calves are grown out for slaughter at around three years of age at weights of 850 to 950 pounds. Vaccination for blackleg and some spraying for ticks are common practices in this type of operation.

MARKETING

El Salvador has a surfeit of cattle markets, the largest of which is in San Salvador. Most of the commercial cattle of the country eventually find their way to it. There are several other markets in towns east of San Salvador to which owners bring cattle on the regular market days. Cattle frequently move from one market to another and possibly to a third before reaching San Salvador. Transactions are carried out be-

A mixed American Brahman–Criollo herd. Cows are milked once a day.

tween dealer and dealer in such trading. Advantages which should accrue to producers from being in a position to bargain with more than one buyer are lost in the cut taken by traders before the animals are finally sold for slaughter in San Salvador. This system originated in the days when cattle moved on foot from eastern growing areas to the major consumption area around San Salvador. Most cattle are now transported by truck, but the chain of markets still persists.

Slaughter cattle are usually sold by the producer or trader-owner to the store owner, who pays a fee for having the carcasses processed in the municipal slaughterhouse.

El Salvador is the only Central American country which does not export beef to the United States and thus fails to enjoy the price incentive of that outlet. It is estimated that some 25,000 head of live cattle which originate in El Salvador are slaughtered in Guatemala in USDA-approved plants for export to the United States. This outlet may have an indirect effect on the price structure for live animals.

All sales are by the head, and prices increase as the San Salvador

San Salvador cattle market.

market is approached. This market is an open area in the city, fenced off from the street and surrounding property. There is some provision for watering stock but none for feed. All cattle belonging to one owner are held in a group by an attendant, frequently with their heads tied together.

In early 1969 slaughter animals sold in San Salvador at the equivalent of 15 cents a pound. Very little attention is paid to quality. Practically all cattle sold are work oxen and cows, usually marketed because they are no longer useful. There is, however, some sale of young stock, work oxen, and cows in milk to small farmers. Such transactions are often the result of an owner's need for cash.

The sale of milk reveals the hectic state of health regulations and the absence of marketing controls. Price is customarily on a "per bottle" basis, the bottle holding 0.75 liters, or approximately 1.65 pounds. The price normally received by the producer is equivalent to 5.3 cents a pound for genuine whole milk of 3.5 per cent butterfat content or slightly less. Pasteurized and bottled, it retails for 17 to 20 cents a pound. There is a wide sale of raw, diluted milk, from which most of the butter-

El Salvador

fat has been removed, at 8 cents a pound retail. To meet this competition the processing plants sell a pasteurized, diluted product of 1.5 per cent fat content at 9 cents a pound.

SLAUGHTERHOUSES

In comparison with other Central American states, El Salvador suffers in having no USDA approved facilities for the export of frozen beef. With the exception of two small, recently built private plants, all slaughterhouses are municipally owned. The largest of these is the one in San Salvador. Except for overhead rails and a killing box, mechanical equipment is lacking. Killing is done with a long dagger inserted at the base of the brain. Skinning is done on benches. There is no refrigeration, and carcasses are sent warm to the retail outlets. Health inspection is perfunctory, and ample use of water for washing down is the only sanitary provision. There is no by-product recovery.

The rest of the 73 municipal slaughtering plants in smaller towns and cities through the country have even poorer facilities. There is also considerable slaughter of cattle by individual butchers.

CATTLE DISEASES

Tick-borne disease and screwworm are the worst plagues suffered by Salvadorian cattle. There are outbreaks of anthrax, but its incidence is not considered serious, and vaccination is used only when an outbreak occurs. There is also some vaccination for rabies in areas where it is prevalent. Tuberculosis is widespread, as is brucellosis. Some larger owners, through exceptionally good management, have eliminated both of these diseases in their herds. This was accomplished by slaughter of tuberculosis-infected animals and by vaccinating calves for brucellosis.

Vibriosis has not been positively identified, but because the major importations of breeding stock are from the United States and Mexico, it probably exists.

Routine vaccination for blackleg and hemorrhagic septicemia is quite general except in small herds owned by subsistence farmers.

In 1969 there was considerable concern over the wide incidence of a killing disease which was characterized by symptoms similar to rabies. Careful investigation proved it was not that disease. A poisonous weed, escobilla, was the suspected cause.

Pure Criollo cattle have developed a remarkable tolerance to indigenous diseases. Newly imported cattle require exceptional care to survive their first year in the country. This involves frequent spraying for ticks, usually by hand sprayers packed on the back, and prompt detection and treatment of anaplasmosis and pyroplasmosis. After crossing and several generations of upgrading, the European breeds—and to a lesser extent the Zebu—require more care than the straight Criollo.

A serious handicap to animal health can be traced to the theoretically free veterinary services authorized by the Ministry of Agriculture. Under bureaucratic administration the treatment thus provided is haphazard and undependable. It is also a major factor in keeping competent veterinarians from practicing.

Malnutrition resulting from the long dry season is a greater handicap to cattle production than disease. The gains which the cattle put on during the lush pasture period of the rainy season are largely lost during the dry period that follows, unless, as in the exceptional operation, the cattle are on irrigated pastures during this season. The storage of fodder in the form of hay or silage to carry cattle through the dry season is unknown except on a few large farms.

GOVERNMENT AND CATTLE

The government of El Salvador has been little involved with the cattle industry. Land reform at the present time does not seem to be a threat to the cattle raiser. This will probably not change in the near future since so many of the cattle owners are small operators with only a few head. The political uncertainties and the heavy population pressure which continues to increase could conceivably change this situation.

OUTLOOK FOR CATTLE

Although El Salvador already has the densest cattle and human populations of any Central American country, there is still room for a sizable expansion in numbers. Much of the central plateau is occupied by the coffee plantations that are still the mainstay of the country's economy. But large areas in this region are not adapted to coffee culture and could be developed as pasture lands. At higher elevations and with a cooler climate this region is better cattle country than the coastal plain, and the same holds true for the more mountainous, rougher country on the north. As population pressure forces increased production of food crops in the coastal plains, stock raising can be expected to move back into the hills.

As the need for more intensified use of available land increases, production of hay and silage for the dry season and the use of irrigation for pastures will probably increase. The additional beef and milk production which could be obtained by providing adequate nutrition during the six-month dry season could exceed that which has resulted from upgrading the Criollo cow. With an annual rainfall of around 100 inches over much of the country and a twelve-month growing season if water is available, proper management and development of irrigation would enable cattle to make good gains through the dry season.

    The Criollo cow is certain to disappear in the near future. Although the subsistence farmer resists this trend because his Criollo ox team requires no special care and the cow gives more milk than can be expected from her calf by a Zebu sire, he cannot hold on to them much longer. Pure Criollo bulls are seldom seen today, and bulls of the American Brahman, Brown Swiss, or Holstein-Friesian breeds sire most of the calves. The eventual result will be a more productive beef or milk animal when properly cared for, but one that for many generations will be more susceptible to indigenous disease and heat stress.

Guatemala

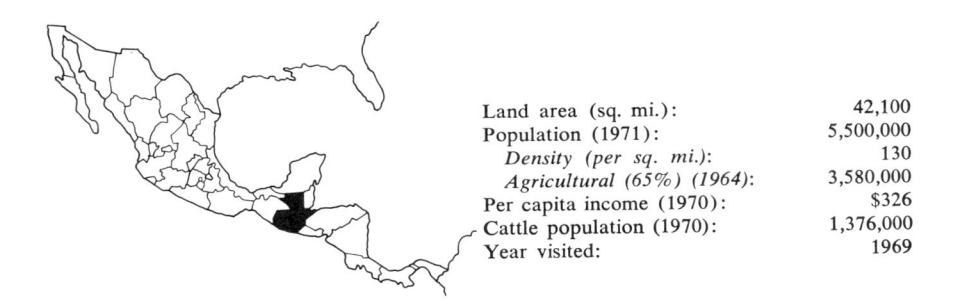

Land area (sq. mi.):	42,100
Population (1971):	5,500,000
Density (per sq. mi.):	130
Agricultural (65%) (1964):	3,580,000
Per capita income (1970):	$326
Cattle population (1970):	1,376,000
Year visited:	1969

THE REPUBLIC OF GUATEMALA is bordered by Mexico on the north and west; British Honduras, a short Caribbean coastline, Honduras and El Salvador on the east; and the Pacific Ocean on the south. Guatemala for years has laid claim to the territory of British Honduras, calling it Belize. This is a rectangular area of 8,000 square miles which lies between the northeastern Guatemalan border and the Caribbean Sea. Guatemala's position apparently has little to support it except the natural desire for more territory and a longer coastline on the Caribbean.

In area Guatemala is seven-eighths the size of Louisiana. There is a narrow coastal plain along the Pacific, and Caribbean lowlands stretch inland for a considerable distance from the east coast. The central highlands—plateau areas cut by deep valleys and bordered by volcanic mountains—cover nearly two-thirds of the surface. Most of the population is concentrated here. The northern one-third of the country is covered by tropical jungle and is thinly populated. This region is available for major development in the future.

The area was invaded by Cortez from Mexico in 1523, and the captaincy-general of Guatemala was established in the following year. This was the seat of Spanish authority for the Yucatán Peninsula, the southern states of Mexico, and all the area southeastward as far as Panama until independence of Central America was proclaimed in 1821. Mexico subsequently annexed Guatemala, along with the rest of Central America, which it lost with the formation of the short-lived Federal Republic of Central America described above. Guatemala

Cattle of North America

became a republic in 1839 and since that date has had a turbulent political history.

Communism had become a feature of Guatemalan political life by 1950, but the communist-dominated government was overthrown in 1954. A military regime assumed power in 1963, and a new constitution went into effect in 1966. The political situation remains far from stable, and control measures approach those of a police state because the insurgent element has moved into the capital, Guatemala City, which harbors nearly one-seventh of the total population.

Most of the people are of mixed Indian and Spanish descent, though 43 per cent are pure Indian. The European element is small. The natives found by the Spaniards were Mayas, and their language is still spoken by the Indians in the more remote regions.

The economy is primarily agricultural, involving 68 per cent of the population. The country is largely self-supporting in the food staples, mainly corn and beans. Coffee has been the largest export crop, with bananas second, but in recent years cotton and sugar have become important sources of foreign exchange. The mineral resources of the country are extensive, but little has been done to exploit them other than a recent development in the production of nickel.

The rainy season is from May to October, and there is some plant growth for another month or two. Most large-scale agricultural development, cattle raising as well as farming, has been along the Pacific coastal plain. Around Guatemala City there are some modern dairies, but the highlands, which in many ways are the best cattle country, have not yet been developed for livestock. The mountainous high-plateau region is inhabited by primitive Indian subsistence farmers, who grow small hillside patches of corn or beans.

The large Guatemalan landowner has not suffered much from land-reform programs. The principal requirement is that the land must be utilized productively, by either cropping or grazing. As long as this condition is met, large holdings remain intact. As a result, commercial agriculture and cattle raising are conducted on a sizable scale by a comparatively few well-established owners.

CATTLE BREEDS

In Guatemala, as throughout Central America generally, the term Criollo is used to designate both the pure descendants of the original cattle of the Spaniards and the unplanned mixtures of these with other more recently introduced breeds. These are mostly Zebu types. When crossing has progressed to the point where the Criollo influence has been

Mature Criollo cow, light-tan color, in better-than-average condition, 850 pounds (horns have been blunted).

practically bred out, such cattle are known by the name of the dominant breed, such as Brahman (for American Brahman). As in other Central American countries, both Criollo (pure) and Criollo (mixed) are considered here.

Criollo (Pure).—There are few cattle left in the country which can be considered as representative descendants of the original Spanish animals brought to Middle America. Such Criollo cattle as exist are found in the southeast near the El Salvador border.

Color varies widely, as in all Criollo cattle. The tan color, ranging from a quite light shade to a decidedly reddish tinge, is probably the most common. This is the same color as that of the Criollo in Costa Rica, which was developed there for milk production. Before the introduction of the Zebu breeds, this same tan-colored Criollo is said to have been quite common throughout Latin America.

Criollo (Mixed).—This type accounts for most commercial cattle, which are varied crosses of American Brahman and Santa Gertrudis bulls on

Mature Criollo cow, black color, in average condition, 750 pounds.

native Criollo cows. The result of this admixture is a larger, heavier-boned animal which reaches market weights at younger ages than the pure Criollo. The Zebu influence is readily discernible in the American Brahman crosses by the long ears, prominent sheath, sloping rump, and grey color. Santa Gertrudis crosses soon attain the general appearance of this breed—a dark-red color, a slight swelling forward of the shoulder, a minor slope to the rump, and a fair beef conformation. Although it would be difficult to make actual weight comparisons, the Santa Gertrudis–Criollo crosses seem to equal or possibly exceed American Brahman crosses on the Criollo.

Introduced Breeds.—The American Brahman and Santa Gertrudis are the most prominent beef breeds which have been brought into Guatemala for upgrading the native Criollo cattle. Both do well on the tropical coastal plains.

A few Aberdeen-Angus have been imported from the United States, and some Charolais have come from Mexico. Neither of these has made much impact on the total cattle population. The first crosses of either breed with American Brahman or Santa Gertrudis can be grown out

Purebred American Brahman bull. Home-raised in southern Guatemala.

satisfactorily in the tropical climate but require more care than the straight Zebu types.

Dairy cattle are raised mostly in the area near Guatemala City at elevations of around 5,000 feet. These are mainly Holstein-Friesian and crosses of this breed on the Criollo. The bulls used were imported from the United States and Mexico.

MANAGEMENT PRACTICES

Beef Cattle.—The main cattle-raising area is along the southern coastal plain bordering the Pacific. Most of this region was originally tropical jungle, but large areas were cleared for the cultivation of bananas, sugar cane, and cotton. More recently cleared land has been utilized for cattle raising by the establishment of improved tropical grasses, such as pangolagrass and Africa Star grass.

Many cattle operations are on a large scale, involving from 1,000 head upward, although there are some smaller growers handling only a few hundred head. Toward the El Salvador border, on the eastern part

Purebred Santa Gertrudis bull. Home-raised in southern Guatemala.

of the Pacific coastal plain, there are farmers bordering on the subsistence type who have small herds of Criollo cattle, sometimes nearly pure but more commonly of mixed type. Practically all commercial cattle, however, come from the larger hacienda owners with holdings of up to 30,000 acres.

Rainfall, which averages 100 inches or more along the coast, is more than ample to maintain pastures in lush condition during the wet season. Plant growth may be sustained for a few weeks after the dry season sets in by the retained soil moisture but then deteriorates rapidly, leaving only the mature dry grass for feed. More progressive ranchers put up some hay to feed during the dry period and may also feed some cottonseed hulls. There is also some irrigation with water diverted from short, fast streams that rise in the coastal mountains. This is usually flood-type irrigation, but some sprinkler systems are used. Where irrigation is employed, carrying capacities of less than one acre an animal unit can be obtained. Without irrigation 1.7 acres is the general average, although good pasture management can reduce this materially.

The common breeding practice is to run bulls with the cows all year.

A Santa Gertrudis herd on irrigated pasture at the end of the dry season. Southern Guatemala.

Cattle are run under fence with pastures cross-fenced for breeding and grazing control. A number of ranchers have started using a five- or six-month breeding season. Some are even breeding for a three-month period, and the cows are pregnancy-tested two months after the bulls are taken out. The cows with calf are then run in a separate pasture, and the open cows are put back with the bulls for another three-month period. After this any open cows are culled. Heifers are usually bred as two-year-olds to calve at three.

The calf crop realized by the large growers averages 60 per cent calves dropped, with the death loss to weaning running 5 to 10 per cent. Exceptionally good management will obtain a calving percentage of 85 to 90 per cent and somewhat lower losses to weaning. Calves are normally weaned at eight to nine months of age, as near as possible to the start of the dry season. Better operators obtain a 400-pound steer at this age.

The steer calves may be either grown out by the first owner or sold as calves to a feeder-type operator. In either case the best slaughter animal reaches a weight of 950 to 1,000 pounds at two-and-a-half years of age. This applies in the case of cattle showing a strong influence of Zebu or Santa Gertrudis breeding and having some supplemental feed, such as hay or cottonseed hulls, during the dry season. The feeder-type operator runs from a few hundred up to several thousand head of such stock annually, all of which he usually purchases as calves.

Recently there has been an increasing interest in feeding out bulls instead of steers. While not extensively practiced in 1969, this practice probably will grow. Most cattle slaughtered are destined for export to the United States as boned, frozen beef. A young bull would have advantages over a steer for such trade, showing faster gains during the growing period and producing a leaner carcass.

On the best ranches modern management methods are employed. Pastures are carefully rotated to obtain maximum usage without overgrazing and are clipped to control weeds. In some purebred herds, and even in a few commercial herds, calves are ear-tagged as dropped and identified with their mothers. Weaning weights are recorded, and selection for breeding stock is based on these records. There is some practice of artificial insemination, with the use of semen from the United States as well as from domestic bulls.

Dairy Cattle.—The principal dairy operations are in the high-plateau area surrounding Guatemala City. Although few in number, these dairies are in the 100-cow range. They usually have Holstein-Friesian herds, the foundation stock of which was imported from the United States. The cow herd is either maintained in yards with all feed brought to it or kept on pasture except when taken in for milking. Sanitation is good.

The European breeds do much better at the 5,000-foot elevation around Guatemala City than on the coastal plains. This is apparent in the condition of most dairy herds. The cattle tick, however, still persists and must be fought constantly. This is much less a problem when the herd is maintained in dry lot continuously. There is considerable use of artificial insemination in the large dairy herds.

Around towns and villages in the more remote areas the local dairymen have small herds which usually show some Holstein-Friesian influence. These herds are on pasture except when brought in for milking. Sanitary conditions are inadequate.

Criollo cows are milked by subsistence farmers both for household consumption and for local sale when there is a surplus.

MARKETING

All cattle sales are individual transactions between owner and buyer.

Guatemala

There are no regular cattle markets. The two major slaughterhouses have buyers who travel through the country to purchase cattle. Sales on the basis of price-per-pound on the slaughterhouse scales are replacing the old practice of selling by the head. Small ranchers often truck a few head into the slaughterhouse, and these are paid for when delivered.

There is no systemized method of grading either live cattle or carcasses, but fleshing quality is recognized to some extent in the prices at which live cattle are moved. This applies mainly to the estimated dressing percentage an animal will yield rather than to the quality of the beef. An old bull usually brings the same price as a young steer if the anticipated dressing percentages are the same.

For pricing purposes the plant ordinarily divides the cattle into three classes as they are weighed. These classes and the prices they were bringing in early 1969 are as follows:

	Liveweight Price per Pound
Distinctly American Brahman or Santa Gertrudis– type cattle of either sex, in fair flesh and weighing from 900 pounds up (in practice these will be mostly 2½- to 3-year-old steers, although some bulls are now sold at this age)	$0.15
Better-fleshed Criollo (mixed) cattle, either sex, weighing 750 to 900 pounds (these are mostly 2½- to 3-year-old steers)	0.14
Thin and old cattle of the USDA canner type, either pure or mixed Criollo, usually weighing under 750 pounds	0.12–0.13

In country transactions between the grower and the feeder-type rancher, a good 420-pound steer or bull calf of upgraded American Brahman or Santa Gertrudis quality was bringing $70 a head in early 1969. Heifer calves of similar quality, bought for breeding stock and weighing around 375 pounds, were selling for $80 a head.

The best purebred breeders average around $450 a head for yearling bulls to be used in the herds of commercial growers.

SLAUGHTERHOUSES

The only modern slaughtering facilities are two privately owned plants, one in Guatemala City and the other 34 miles south in Escuintla. Both of them are oriented toward the export of frozen, boned meat to the United States and are under USDA inspection. This trade has increased rapidly in recent years and in 1967 accounted for over 27 per cent of the total beef production of the country. In 1969 the export trade was estimated at 21 million pounds.

Cattle received at these plants are held in paved yards on water but without feed for 36 to 48 hours. Most are Criollo (mixed) cattle, but the proportion of upgraded American Brahman and Santa Gertrudis is steadily increasing. While the plants lack modern equipment, operations are conducted under satisfactory sanitary conditions. Killing is by captive bolt, and skinning is done on cradles. Boning and packing for freezing are hand operations. Most of the production is transported in refrigerated truck trailers overland to a Caribbean port, where it is placed on ships destined for Florida ports. There it is distributed, and the trailers are returned to Guatemala for reloading.

Both plants also sell meat for consumption in the Guatemala City area. The current price (early 1969) to retailers is 25 cents a pound for dressed carcasses.

Most of the beef for local consumption is processed under unsanitary conditions in small municipal facilities.

CATTLE DISEASES

Guatemala, like all other Central American countries, is free of foot-and-mouth disease. The tick and the tick-borne diseases are the worst plague to the cattle of the country. The pure or nearly pure Criollo has developed a near immunity to these and can withstand heavy infestations and remain in fair condition. The Zebu breeds and admixtures of these with the Criollo require frequent dipping or spraying if they are to do well. The best growers usually treat every 21 days throughout the year, although in some local areas continuous treatment is said to have so reduced the tick population that it is repeated only when close observation shows that the tick is returning. Spray runs are generally preferred to dipping vats.

The screwworm causes heavy losses from navel infection in young calves. Many growers attempt to treat all calves twice but still encounter considerable loss. There is talk of starting a program using the sterile male fly to eliminate the screwworm fly. Since there are no funds to finance such a program however, it is unlikely that it will be attempted in the near future.

Prophylactic treatment is given for blackleg and anthrax in areas where they have occurred. There is also treatment for rabies in areas where there are carrier bats. The large dairies and many of the better-managed beef-cattle ranches vaccinate for brucellosis at weaning, usually at eight months of age.

Tuberculosis is said not to be a serious problem in Guatemalan herds.

Cattle imported from the United States are susceptible to anaplasmosis, and special care is necessary to bring such stock through the first few months in the country. There is some vaccination of entire herds for this disease, but the practice is not general.

GOVERNMENT AND CATTLE

The most notable feature of the Guatemalan government's attitude toward the cattle industry is reflected in the fact that no land reform programs to date have affected the operations of the large cattlemen. The country historically has had to depend on agricultural production to earn foreign exchange. Basic exports have been coffee and bananas, but in recent years cotton and sugar have found overseas markets. This has reduced to some extent the wide fluctuations in income as the price of coffee fell in the world market, but the uncertainties of economic dependence on such crops make further diversification desirable. Such considerations may support the attitude toward land reforms. They probably also account for the interest shown by the government in the cattle industry.

The Ministry of Agriculture has a livestock-development program and has begun importing good breeding stock from the United States to aid in the improvement in the country's cattle.

Export quotas to the United States are fixed by the Ministry of Economy.

Price controls were imposed on beef carcasses in 1968, but prices remained at 25 cents a pound, which, under current conditions, left a fair margin for the grower.

OUTLOOK FOR CATTLE

The future of beef cattle in Guatemala is tied to the upgrading of the Criollo to a Zebu-type animal. Such possibilities as may have existed for development by systematic selection of a more productive type of Criollo have been lost in the heterogeneous mixtures resulting from the unplanned crossing practices of the past. American Brahman and Santa Gertrudis herds appear certain to dominate the scene in the future.

In the sizable dairy herds, the Holstein-Friesian predominates. Such herds are mainly in the higher country where, with good care, this northern breed adapts quite readily.

The small farmer on the coastal plains will probably keep his Criollo animals for some time to come. The Criollo-type cattle are the only ones that can maintain themselves well enough without prophylactic treatment to furnish him with milk and pull his plow and cart.

As the world demand for beef increases, there is room for a major expansion in the cattle industry of Guatemala. In the present cattle country on the coastal plains, land devoted to pastures could be doubled in productivity by the establishment of improved grasses, irrigation during the dry season, and the storage of hay to carry cattle over this period in good condition. Better-quality beef animals obtained by the upgrading process now in progress will further increase production.

Even greater increases in cattle numbers could be attained by utilization of the forest-covered central highlands and the low-lying northern jungles of the state of Petén, which in itself occupies over one-third of the country and is only sparsely populated by primitive Indians. The highlands, despite their rough terrain, constitute what in many ways is the best cattle country in Guatemala. This is particularly true with regard to European breeds. The northern jungles, when cleared, could be made more productive than the coastal plain in the south where most of the cattle are now raised. Both regions have ample rainfall, and, although both have their dry seasons, the one in the north is shorter.

While economic conditions place these developments some time in the future, the cattle population of Guatemala could be doubled or tripled if the political climate continues to be favorable to the large operator.

Honduras

Land area (sq. mi.):	43,400
Population (1970):	2,700,000
Density (per sq. mi.):	63
Agricultural (67%) (1961):	1,650,000
Per capita income (1969):	$246
Cattle population (1970):	1,830,000
Year visited:	1969

THE REPUBLIC OF HONDURAS occupies an area bounded by the Caribbean Sea on the north, a short Pacific coastline on the south, and a longer border with Nicaragua which tapers to a point on the Caribbean in the east. Guatemala and El Salvador form the western boundary of Honduras.

The interior, comprising three-fourths of the land area, is mountainous and cut with deep, winding valleys. The Caribbean coastal plain is wide and contains the best agricultural land; the narrow strip of lowland along the Pacific is drier and not as fertile. The climate of the coastal areas is tropical throughout the year, but inland at higher elevations it is more moderate.

The rainy season generally lasts from April to November, although it changes less abruptly in the north, where there is some precipitation even during the dry season.

Columbus landed in Honduras on his last voyage in 1502. Later the rule of Spain was firmly established and continued under the captaincy-general of Guatemala until 1821. Then in 1838, when the Federal Republic of Central America dissolved, Honduras became independent as a separate country. At that time the British were in control of the Caribbean coastal area and had been since the end of the eighteenth century, when they established their logging operations. Spain had been unable to end this British foothold, but it was gradually relinquished after Honduras became independent.

The people of Honduras are mostly of mixed Indian and Spanish descent. Pure Indians, living in remote, isolated areas, account for less than 6 per cent of the population. There is also a small element of Negroes, possibly 2 per cent, descended from slaves brought to the north coast by the British. The large landowners are of Spanish ancestry and constitute another small minority. They are now engaged in industry and commerce. The working people and small farmers comprise a majority of the population.

Conflict between the wealthy, landed Spanish descendants and the working people smoldered during the three centuries of Spanish rule. The internal political situation became acute following independence. Wars with El Salvador and Nicaragua, violence, and near anarchy followed. Both Mexico and the United States intervened in efforts to restore order. The United States involvement early in the twentieth century was largely to protect its sizable investments in banana plantations. The never-ending conflict between liberals and conservatives has continued down to the present, although since the last military coup in 1963 there has been somewhat more stability in government.

The country is practically self-supporting in food, of which corn is the main staple. Commercial agriculture initially was of the one-crop type, dependent on the export of bananas grown by large United States–owned corporations.

Practically all commercial stock raising and farming are in the hands of large landowners, while most of the agricultural population is engaged in subsistence farming. Small but rather steady gains have been made in recent years in industry and commerce. Transportation is a major problem, and the government is devoting much effort to road building. For years the only railroads were those serving the banana plantations in the north.

More than three-fourths of the farms are under 23 acres in size, and the average is only 8.5 acres. Most of the smaller operations are carried out by farmers only a small proportion of whom derive sufficient cash income from crops or livestock to enable them to aim for a higher economic level. About 20 per cent of the arable land is in what the government classifies as "large farms," those of more than 2,500 acres. The average size of such holdings is 6,600 acres. These large farms are owned by only 0.1 per cent of the farm owners. While the holdings of the large landowners are not as extensive as in some Latin American countries, such operations emphasize the wide gap that still exists between the rich and the poor.

CATTLE BREEDS

Today in Honduras there are many Criollo descendants of the original

Mature Criollo cow, tan color, in good condition, 900 pounds.

Spanish cattle which show little interbreeding with foreign cattle types. The lack of roads and other transportation facilities has kept the interior inaccessible, and that is where most of the marginal farmers with their Criollo cattle live. According to the last available statistics, collected in 1954, the small farmers owned 75 per cent of the total cattle. There probably has been no major change in the ownership pattern since then. Among this cattle population there are still relatively pure strains of the Criollo. It is only since the early 1960's that introduced breeds have gained in sufficient number to make an impact on the cattle population countrywide. Beginning in the 1930's, however, large herds were built up by the United States–owned banana plantations in an upgrading program using Zebu and Red Poll bulls on Criollo cows. Because these herds were isolated along the northern coast, they had no marked influence on the cattle population as a whole.

Criollo (Pure).—There are good examples of pure Criollo in the more remote areas of Honduras. In some parts cows are maintained in fair milking condition without either supplement or prepared forage, even

Criollo cow, black and white, black ears, 750 pounds. In a small mountain dairy herd.

during the dry season. Some cows have very good udders. The better cow of such Criollo herds averages 10 pounds of milk a day. Cows weigh up to 900 pounds. There is frequently a fair-looking Criollo bull with these herds.

In the mountainous areas small herds of ten to twenty Criollo cows are to be seen. Many of these have degenerated over the years from lack of adequate food and from unplanned breeding. Cows in such herds weigh no more than 500 pounds each. Bulls kept by many of these small cattle owners are merely those remaining after the better animals have been castrated for draft oxen.

Two distinctive color markings are seen in the Honduras Criollo. An animal may have black ears and a white body, or a white body with a few small black spots. Although smaller, these cattle have the same

Mature Criollo bull, tan color, 800 pounds. Eastern mountains.

general appearance as the Blanco Orejinegro of Colombia. Or an animal may be black and white, similar in color pattern and in conformation and size to the Chinampo of Baja California. Although these two color markings are seen occasionally throughout Latin America, they appear to be more common in the cattle of Honduras. There are also animals with the tan coloring appearing in Criollo cattle wherever they are found.

Criollo (Mixed).—The hybrid Criollo of Honduras is predominantly the result of unplanned breeding to American Brahman bulls. A few cattle show Brown Swiss influence. The work oxen of the small farmer sometimes show traces of either or both of these breeds.

Introduced Breeds.—Among the initial efforts to introduce exotic types were the importations of the Nellore and Guzerat breeds from Jamaica by the United Fruit Company. The use of these Zebu breeds was followed by the introduction of Red Poll cattle, also from Jamaica. The

Young Criollo bull, black, 700 pounds. Central mountains.

company maintained these herds to supply beef and milk to their employees on the banana plantations along the north coast.

Before the use of the Red Poll on the northern plantations there had been an importation of the breed by individual owners in the Choluteca region on the south coast.

The American Brahman was brought in about 1949, and the Santa Gertrudis came ten years later. In the 1960's upgrading efforts favored the American Brahman on the Criollo to produce a faster-gaining beef animal with a heavier slaughter weight.

Among the dairy breeds the Brown Swiss and the Holstein-Friesian have been used most extensively, although there have also been importations of Guernsey and Jersey stock. The Holstein-Friesian is popular because of its higher milk yields. The Brown Swiss is especially liked by the smaller growers since it appears to be somewhat more tolerant of tropical environment than other dairy breeds. Also, when crossed with the Criollo, a fair triple-purpose type animal results.

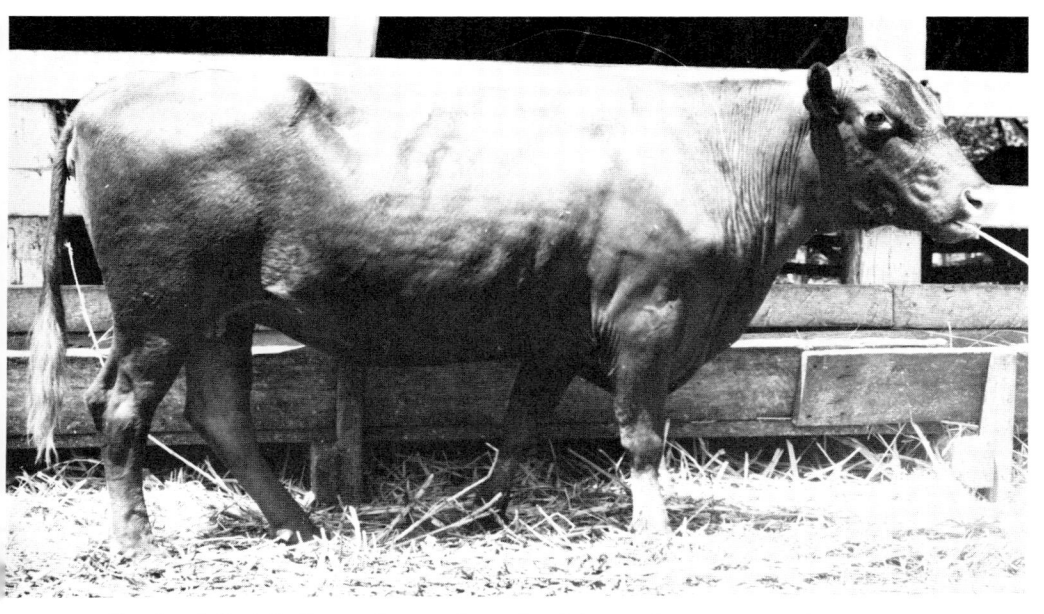

Purebred Red Poll cow from a large herd in southeastern Honduras.

MANAGEMENT PRACTICES

Cattle raising in Honduras often involves dual-purpose use. Medium-sized growers, with herds of 50 to 150 cows, produce both milk and beef. The subsistence-type farmers generally use cattle for draft and the production of milk. Milk in excess of the owner's requirements is often sold either to neighbors or to a small dairy plant. Cattle are sold for slaughter from these small herds only incidentally. Among the large growers, however, there are herds for beef production only. These supply a major part of the slaughter animals for the export market. In the north there are concentrations of strictly dairy herds, some of which are in the 100-cow range.

There are three principal areas for growing cattle in Honduras. On the northern coastal plain in the region around San Pedro Sula are the largest concentrations of both dairy and beef cattle. It is in this area of 80 to 160 inches of rainfall that the banana plantations are located. These plantations run the largest herds in the country. In the central highlands most of the cattle are owned by small farmers who sell some milk locally and an occasional animal for slaughter. This is a region of 40 to 80 inches of rainfall. On the narrow coastal plain along the

Highly upgraded Holstein-Friesian cows of a well-managed dairy in central Honduras.

Pacific, where there is a longer and hotter dry season, cattle are grown for beef and are also used extensively for draft.

Beef Cattle.—Cattle grown for beef are run under fence on lush pastures during the rainy season. During the rest of the year they manage to survive for the most part on nothing but standing mature grasses. There is a small but increasing use of irrigation to maintain growth during the dry period. No stored forage is ordinarily kept.

Mention should be made of the lands that were formerly in sugar cane which have been converted to pasture. Cane diseases have caused large areas to be abandoned that were formerly in this crop. On these farms, either for beef or dairy operations, the irrigation ditches of the former cane fields are now used to irrigate pastures of elephant grass or other green forage crops. This practice effects a marked improvement in the productivity of cattle.

Countrywide, the cattle marketed for beef are predominantly mixed Criollo, although the proportion of American Brahman is steadily increasing. Most heifers are kept for replacements or for sale as breeding stock. The slaughter cattle are largely steers, but there is a growing tendency toward feeding young bulls. The percentage of these in the total kill is still small, however. Marketing age is usually two and one-half to three years. Steers with some Zebu influence reach weights of 800 to 900 pounds at this age. The pure or nearly pure Criollo weighs at least 100 pounds less. Weight is the main objective in feeding for slaughter. Thus there is very little difference in price between that paid for a young, well-finished steer and a poorer-quality older animal as long as the final weight is 800 pounds or more. Because of a definite price break at the 800-pound level (animals over this weight bring one or two cents a pound more than those under it), cattle which, because of bad management, drought, or other reasons, have not gained well are held to an older age.

Artificial insemination has not gained acceptance among the beef-cattle breeders. The usual practice in the beef herds is to run bulls with the cow herd all year. The calving rate on a countrywide basis is less than 50 per cent. The death loss from birth to weaning is around 22 per cent. The best operations, however, claim to get 70 to 80 per cent calves and to hold down the death loss to weaning to 2 or 3 per cent.

Dairy Cattle.—In addition to the many small operations in the dairying region on the Caribbean coastal plain, there are farms with up to 300 head of milk cows. The only pasteurizing plant in the country is located there.

The common practice is to turn cows to pasture in the early morning and bring them into yards under shade for the remainder of the day. Green chop is fed year round, the quantity increasing during the dry season. Pangolagrass and guinea grass are widely used for pasture, and elephant grass is cut for forage. Irrigation is employed during the dry season to keep green feed available. Most dairymen feed concentrate to their milk cows, although some feed only roughage with molasses added.

Milk yield varies widely, depending on the quality of the stock, the feeding practices, and management. A purebred or high-grade herd of Holstein-Friesian or Brown Swiss seldom averages more than 7,000 pounds of milk per lactation.

Several of the larger dairies have some form of milk parlor, but most milking is by hand. Sheds and barns are usually well cleaned, but sanitation generally varies from poor to only fair. Milk is hauled into the pasteurizing plant by the producer.

Many dairies kill their bull calves a few days after birth in order to avoid this drain on the milk production. There is little use of milk substitutes in raising heifer calves.

Holding yard at Cortez Packing Plant in San Pedro Sula. Shows typical slaughter cattle

Artificial insemination is widely used in the dairy herds through importation of frozen semen from the United States.

MARKETING

Cattle are sold direct by owner to buyer for private plants processing beef for export. For local consumption the butcher or dealer buys from owners in the local area and has the animals slaughtered in the municipal facility. In early 1969 cattle over 800 pounds were bringing 13 cents a pound. Cattle under this weight brought 10 to 12 cents. Except for emaciated cattle, practically no attention is paid to the quality of an animal. There are no established cattle markets in Honduras.

In the San Pedro Sula area, the only milkshed in the country, the average price paid by the dairy plant for milk was a little over 4 cents a pound. A few farmers producing milk with up to 4.4 per cent butterfat were getting as much as 5.6 cents a pound.

SLAUGHTERHOUSES

There are only three facilities in Honduras which have reasonably acceptable facilities for slaughtering cattle. Two are privately owned plants in San Pedro Sula. There is another in Choluteca. All three export frozen, boned beef to the United States. Sanitation is good, a primary requisite to pass the periodic inspections by USDA veterinarians. Except for a small volume sold to the premium markets in San Pedro Sula and Tegucigalpa, the output is exported.

Beef for local consumption is killed in municipal facilities in the towns. These are primitive, often consisting of nothing more than a concrete floor and a shed roof.

The better-quality cattle usually go to the export packing plants. These range from mixed Criollo showing Brahman influence to some upgraded Brahman cattle. There is only a sprinkling of cast-off dairy stock and a few nearly pure Criollo. A typical ¼ Criollo–¾ Brahman steer dresses around 57 per cent; the pure Criollo, about 45 per cent.

CATTLE DISEASES

The cattle tick is the worst menace to the cattle of Honduras. In the north, treatment is usually to spray every eight days to two weeks throughout the year. In the south, where systematic treatment over the years has materially reduced the tick population, satisfactory control is maintained by spraying every three or four weeks. Complete elimination of the tick could be obtained by more drastic treatment. This is not, however, considered desirable, for animals raised in such a tick-free area would have no natural tolerance to the tick-borne diseases and could not survive when sold into an infested area unless given special care. Cattle imported from the United States soon succumb to anaplasmosis or piroplasmosis unless given special treatment during their first year.

Internal parasites of both lung and stomach are prevalent. Drenching every three or four months until the age of one year and again at two years is common practice. General practice is to vaccinate calves with the three-way vaccine against blackleg, malignant edema, and hemorrhagic septicemia twice during the first year. Screwworm is prevalent and treatment of the navels of newborn calves and attention to any abrasion of the skin are essential.

Calfhood vaccination against brucellosis is prohibited by law. A few herds under exceptional management test all cows for both brucellosis and tuberculosis every year and eliminate all reacting animals.

The sanitary methods mentioned apply to large and medium-sized beef herds and commercial dairies. Pure or nearly pure Criollo ox teams and the few milk cows of the subsistence farmers receive practically no prophylactic treatment but probably manage to survive more easily in the tropical environment than do the upgraded herds. Calfhood mortality up to one year of age is said to average 22 per cent throughout the country, but it is certainly less than this in the small Criollo herds.

GOVERNMENT AND CATTLE

Government programs for advancement of the cattle industry have not accomplished a great deal other than to secure enactment of laws regulating the export trade to the United States. Veterinary services, which are free, are so inferior that a dairyman in the San Pedro Sula area will pay $10 a service to a private technician for artificial insemination rather than use the government veterinarians.

The prohibition of vaccination against brucellosis may have some long-range desirable features, but, in the absence of any mandatory control for elimination of the disease, it is a severe handicap to the progressive cattleman. There is no effective control program against tuberculosis. As the cattle population continues to increase, losses from both diseases will also inevitably increase until effective protective measures are taken.

As far as land reform is concerned, the government has left the cattleman relatively free. There are agrarian-reform laws on the books, and there has been some redistribution of land purchased from a few large owners. Results have not been satisfactory. The fact that agricultural holdings are not extremely large is one mitigating factor. There are also large areas of entirely undeveloped land available for stock raising and farming. For the immediate future there seems to be no threat to the stockman from land reform.

OUTLOOK FOR CATTLE

As in the other Central American countries, the Criollo will eventually disappear from Honduras. While upgrading to Zebu or to European dairy breeds reached sizable proportions only after 1960, already something over one-fourth of the country's cattle population can now be classified as belonging to one of the exotic breeds introduced.

There are two major incentives for raising cattle in Honduras: the beef-export trade and milk. A Zebu cross on the Criollo and then continued upgrading give a larger, faster-maturing animal for slaughter. For dairy purposes the cross on the Criollo by a Holstein-Friesian or a Brown Swiss bull increases milk production. Either practice results in increased financial returns to the owner despite the additional care required for exotic animals. Probably in the relatively near future the commercial beef herds will be dominated by the American Brahman, and the dairies will have mostly either Holstein-Friesian or Brown Swiss stock.

The large areas in eastern Honduras that are very sparsely populated and almost entirely undeveloped have the capability of supporting large numbers of cattle. As the world demand for beef continues to increase, these jungle areas will probably be cleared for pasture.

The Honduran cattle industry would collapse if the export of beef to the United States were stopped. This would happen immediately in the event of an outbreak of foot-and-mouth disease. It is an ever-present threat as long as a few shiploads of live cattle move each year from Honduras to Chile or Peru and the vessels return to reload. The increasing air traffic between Honduras and the South American countries, in all of which foot-and-mouth disease is endemic, is perhaps an even greater concern. In spite of such hazards, however, Honduras could become an important cattle-producing country.

Nicaragua

Land area (sq. mi.):	50,200
Population (1970):	1,900,000
Density (per sq. mi):	38
Agricultural (60%) (1963):	1,200,000
Per capita income (1969):	$391
Cattle population (1970):	1,700,000
Year visited:	1969

THE REPUBLIC OF NICARAGUA occupies a triangular area where the isthmus which connects North and South America bends southward and begins to narrow. It lies between Honduras on the north and Costa Rica on the south, bounded on the east coast by the Caribbean Sea and on the west coast by the Pacific Ocean.

The dominant features of the terrain are, first, the two large lakes, Nicaragua and Managua, lying a short distance inland from the west coast; and, second, the mountainous interior which lies beyond the lake region and extends to the eastern lowlands. The western coastal plain, as far back as pre-Columbian days, has been the heavily populated part of the country; the eastern plain, much of which is swampland, was originally occupied by scattered Indians and, in later days, by Negro immigrants from Jamaica. The area of Nicaragua is about the same as that of Louisiana.

Columbus, on his last voyage in 1502, landed parties on the eastern coast in order to establish Spain's claim to the area. Exploration began in 1519 when the Spaniards moved in from Panama. Soon afterward settlements, which were primarily agricultural, were established.

Nicaragua became an independent state in 1838. Subsequently the country followed the typical Central American pattern of armed uprisings and turmoil engendered by the antagonism between the landed, conservative element of the population and the poorer liberals. This pattern has continued down to the present day. In addition, however, there was more participation by foreign powers in Nicaraguan affairs than occurred elsewhere in the area. During the eighteenth century the

Criollo (pure) steer, three years old, 825 pounds. Northwestern Nicaragua.

British established themselves on the east coast in what came to be known as the Mosquito Kingdom, an area well isolated from the rest of the country by the mountainous interior, and eventually established a protectorate there. The United States protested and refused to recognize the move, and in 1860, Great Britain abandoned her ambitions in the area.

The pure Indian element of the population is estimated at 4 per cent; 70 per cent are of mixed Spanish and Indian descent. Indian ancestry predominates among the workers, and Spanish ancestry characterizes the upper class. There is a sizable Negro element concentrated on the east coast. These are descendants of the slaves the British brought in from Jamaica. The proportion of Europeans is also high, probably around one-sixth of the total population.

The one-crop agricultural economy that supported the country until recent years was based on coffee. This has now been superseded by

Mature Criollo (mixed) cow, 800 pounds. Northwestern Nicaragua.

cotton, the most important export commodity. There is a small but steady production of gold.

Agricultural development dates back to the large plantations established by Spaniards in the seventeenth century. Commercial agricultural production now flows mainly from the mechanized operations of large landholders. Subsistence farming occupies a large part of the agrarian population, which obtains a meager cash income from the sales of produce in local markets. With government aid the status of the subsistence farmer has shown a small but steady improvement in recent years.

Cattle raising, as well as agriculture generally, is largely confined to the western part of the country. From the narrow coastal plain along the Pacific, this region extends into the rolling and hilly country that lies to the west of the mountainous interior and runs from the northern border with Honduras to the southern border with Costa Rica.

Cattle of North America

Before the export of frozen beef to the United States began in 1959, cattle were raised principally for milk production and for use as draft animals. The continuous improvement in the economy during the intervening years has resulted in an increase in the per capita consumption of beef and an as yet unsatisfied local demand for milk. The small farmer still keeps his ox team for cart and cultivation and maintains a few cows for milk. The larger farmers have materially increased their cattle holdings for both milk and beef production. The bovine population today (disregarding the cattle of the subsistence farmer) could be divided evenly between beef and dairy animals.

CATTLE BREEDS

As in the neighboring countries the native cattle are being bred out rapidly. The commercial beef producer is now building up his herd to a Zebu type. The dairyman is going to one of the European milk breeds. Frequently both will also have a purebred or highly upgraded herd from which they produce breeding stock. Even the marginal farmer prefers to use a bull with some Brown Swiss or American Brahman breeding if he can obtain one.

The large concentration of dairy herds is in the country around Managua. The largest beef-producing area is in the Rivas district, the narrow coastal plain that lies between the Pacific Ocean and Lake Nicaragua.

Criollo (Pure).—Southward from the border with Honduras across the length of the country to Costa Rica, the proportion of Criollo influence in local herds consistently decreases. In the north there are still some pure Criollo ox teams and cows. The influence of Brahman, Brown Swiss, and other exotic breeds is much more noticeable around Managua, and further south in the Rivas area even the draft cattle are not typical Criollos.

The pure Criollo cattle in Nicaragua have the same general characteristics that have been described under El Salvador. The black-and-white mottled and the tan hair colorings are particularly noticeable.

Criollo (Mixed).—These are much the same type of nondescript cattle as the mixed Criollos in other Middle American countries and account for most of the cattle population of the country. The mixed Criollo herds of the larger growers are being rapidly bred up to American Brahman.

Introduced Breeds.—The American Brahman enjoys a strong prefer-

Typical upgraded American Brahman herd with a few mixed Criollo cows. Rivas area.

ence with the Nicaraguan cattlemen and far outnumbers other beef breeds. This is true in both purebred herds and commercial herds being upgraded. There is a small but growing representation of Charolais and Aberdeen-Angus cattle. Neither of these breeds has been used to any extent for upgrading the Criollo, but both produce excellent first-cross feeders.

The Brown Swiss is the predominant breed which has been crossed with the Criollo to increase milk production. This cross produces a better beef carcass than most other dairy-breed crosses when grown out for slaughter.

Guernsey cattle, even though introduced before the Brown Swiss, are not as popular. They are the second dairy breed in number, followed by the Holstein-Friesian and the Jersey. In none of the dairy breeds can size or milk production be maintained comparable to that exhibited in a northern climate.

MANAGEMENT PRACTICES

Beef Cattle.—Most of the herds producing animals for slaughter are

American Brahman-Brown Swiss cross cow, 1,200 pounds.

run under fence with a minimum of attention. There is excellent pasture during the rainy season unless an exceptional drought period occurs. There is practically no forage put up for the six or more months of the dry season. In an average operation carrying capacity varies from 7 to 14 acres an animal unit, depending on soil type, elevation, and pasture management. Bulls are usually run with the cows all year. Calf crops average 45 per cent, with a death loss of between 20 to 30 per cent by weaning time. Such operations vary in size from 50-cow herds to those of a few thousand. Most proprietors live in town, and management is left to the often untrained foreman.

Increasing numbers of cattlemen are following modern management methods. In their operations pastures are rotated to obtain the best use of plant growth and to leave ample grass to carry stock through the dry season. Under this system a two-year-old steer gains nearly a pound a day during the dry period. There is some use of irrigation to maintain pastures, and a few growers put up hay to feed sparingly for two or three months before the rains begin. There is a growing practice of

Nicaragua

employing a five- or six-month breeding season, which begins with the first rains and permits calving to start during the middle of the dry season. The best operations of this kind obtain an 80 per cent calf crop with only a 2 or 3 per cent death loss to weaning.

Some of these well-managed ranches run as many as 1,000 head of brood cows. Pregnancy testing is employed, and the open cows are culled. Blood samples are taken annually from the entire herd for tuberculosis and brucellosis control, and infected animals are slaughtered. Cows are identified by brand numbers, and the weaning weights of their calves are an important factor in culling and selection. Twenty-month-old mixed or upgraded Criollo steers are marketed at 900 pounds. The owners of such progressive ranches often live on the property, an unusual practice in most of Central America.

A few cattlemen, dairymen as well as beef producers, buy steer calves at the beginning of the dry season and carry them through to slaughter weight on pasture. The calves are usually crossbred Criollos or dairy-type animals, anywhere from weaning age to yearlings. When run on irrigated pastures or supplemented with a little hay, this stock can be sold at two or two and one-half years of age at weights of 900 pounds or more. The cost of gains is very reasonable, and the profit potentials are among the best to be found in the cattle business in Nicaragua.

Practically no use is made of artificial insemination in the beef herds.

Dairy Cattle.—Dairy herds vary in size from those of 15 to 20 head of crossbred animals to a few with 100 to 200 head of highly upgraded or purebred cows. The smaller herds are owned by marginal farmers. Their cows are run entirely on pasture and receive only the most meager supplemental ration during the dry season. Milk, other than what is kept for household use, is produced mainly during the rainy season. It is processed for cheese to be sold locally. Some small herd owners manage to produce milk throughout the year, and in these herds production ranges from 1,500 to possibly 3,000 pounds a lactation.

Milking is by hand. There is some use of a milk-parlor arrangement in which the hand milker is on a level below the animals. The cows, held in stanchions, are fed their grain supplement while being milked.

There are some larger and more efficient dairies. These are situated in the area around Managua, the capital and main population center. The producing herd usually consists of Criollo cows bred up to a Brown Swiss type. A grain supplement is fed to producing cows. These generally run on pastures where there are sheds for shade. Hay and in some cases silage are fed when the pastures dry up. Pastures are rotated in both the dry and wet seasons to avoid overgrazing.

Some of the best herds of high-grade Brown Swiss cows or other dairy breeds average 7,500 pounds a lactation, but the usual production ranges from 5,000 to 6,000 pounds. Calves are often weaned at three days, fed milk substitutes to three months of age, and then put on dry feed and grown out for slaughter at two to three years. Most heifer calves are kept for breeding stock, either as replacements in the owner's herd or for sale. Good grade bulls as well as most purebred bulls are sold for breeding.

Sanitation in the larger dairies varies from fair to good. Stainless-steel cooling tanks are utilized in some of them. The three main pasteurizing and dairy plants in the country are in Managua, and milk is transported to them by tank truck. Much of the milk, however, moves in cans to collecting stations and then on to the processing plants.

Artificial insemination is used in all the larger dairy herds. Frozen semen is generally used, and the breeding often is done by a farm workman trained by United States firms distributing semen. A government program to train members of the national guard as technicians was unsuccessful.

MARKETING

Buyers for the export packing plants travel through the country puchasing cattle wherever they can. In the western area, served by the plant at Condega, sales are usually by the head. In early 1969 prices for animals weighing from 500 to 900 pounds ranged from $50 to $70. This was the equivalent of 10 to 13 cents a pound, a little lower than prices in the Managua and Rivas areas because of the greater distance to a port. The two packing plants at Managua, serving both Managua and Rivas, paid 13 to 14 cents a pound liveweight for animals delivered to the plant, or 24 to 29 cents a pound for carcasses. The pricing basis was at the seller's option. There is an increasing tendency among producers to sell on the carcass basis, and in 1969, 60 per cent of the annual kill was sold that way.

Most beef consumed locally is bought on the hoof from the small grower by the retailer, who has the animals processed in the municipal slaughterhouse. Prices involved in these transactions are generally below those paid by the packing plants.

SLAUGHTERHOUSES

There are three packing plants in Nicaragua that kill and freeze beef for export to the United States. Two of them are privately owned. The one

Holding yard at packing plant in Condega.

in the small town of Condega in the northwestern part of the country has a daily kill of around 120 head. The other, in Managua, handles 300 head a day. The third plant, also in Managua, is owned by IFAGAN, a joint operation of the National Cattlemen's Association and the Instituto de Fomento Nacional (INFONAC). It too handles 300 head daily when operating. These export packing plants, however, operate largely on a seasonal basis, since few cattle are offered for sale during the dry season.

Killing is by captive bolt in a killing box. Skinning is done on cradles, after which the carcass is handled on overhead rails. Inspection and sanitation are under the supervision of veterinarians of the Nicaraguan Ministry of Agriculture and are subject to periodic approval by USDA representatives.

The production of the export plants is nearly all boned, frozen beef, much of which is packaged as frozen cuts—tenderloins, roasts, and rib eyes—designed to satisfy the United States trade. Carcasses are chilled for 24 hours before boning. Some of the beef would grade a USDA low-good, although much of it is of lower quality. About fifteen per cent of IFAGAN's production is reserved for the local premium trade in Managua.

Municipal slaughterhouses in the smaller cities and towns throughout the country handle most of the beef for local consumption. These are usually minimal facilities. The retailer delivers the animals he has purchased for slaughter and receives the warm carcasses as soon as they are dressed.

CATTLE DISEASES

The tick and the screwworm are the worst hazards to cattle health in Nicaragua. Herds of mixed Criollo or American Brahman under good management are dipped or sprayed for tick control at least every three weeks during the dry season and every five weeks in the wet season. The screwworm is largely responsible for the country's high death loss in calves up to a year old. The only method of preventing serious loss is careful observation as cows drop their calves and treatment of the navel with Smear 62 or an equivalent within 24 hours after birth. Any skin abrasion in an older animal must be treated promptly. Such treatment is practiced only in the better-managed herds.

Brucellosis is widespread. Vaccination against the disease is prohibited by law. A few of the more progressive growers blood-test their cow herd for both brucellosis and tuberculosis annually and eliminate reactors. Even under these circumstances an incidence of 2 per cent brucellosis reactors in a herd is often encountered.

The incidence of tuberculosis is less. There is no control program. Although testing for brucellosis and tuberculosis is a free service of government veterinarians, it is not widely used. All that is being done along this line is the testing and slaughter procedure adopted by a few individual growers.

Internal parasites are present in young calves, and good control requires drenching twice by the time a calf is weaned.

Vaccination during calfhood against blackleg, malignant edema, and hemorrhagic septicemia is quite general. In areas where anthrax is prevalent, the larger herds are commonly vaccinated twice a year. This practice is followed with cattle of all ages and is continued for life.

Botulism shows up from time to time, but the incidence has been minor, and it has not been considered a serious threat. The unnamed killing disease discussed in the chapter on El Salvador has also appeared in Nicaragua; the poison weed escobilla is thought to be the cause. There are also occasional outbreaks of rabies.

GOVERNMENT AND CATTLE

The major involvement of the government in the cattle industry is reflected by the meat-inspection law and the pressure exerted on the United States to increase the beef-export quota. Without this export market the Nicaraguan cattle industry as it now exists would virtually disappear. Nearly one-half of the beef produced in the country is shipped to the United States.

INFONAC is a quasi-government entity which serves as an advisory and information center for agriculture on a country-wide basis and assists farmers in obtaining financing. Its objective is mainly to help the smaller farmers improve their operations. One of the schemes initiated in 1959 was the importation of Brahman breeding stock to be placed in the hands of farmers on a share basis. One-half of the cattle raised are returned to INFONAC; the other half become the property of the farmer. The females returned are again distributed on the same basis to other farmers, and the steers are sold. The objective of this scheme was to increase the rate at which the Criollo cow could be upgraded to Brahman. While it has had no major effect on the cattle population so far, the program illustrates a practical approach to cattle improvement.

The Institute of Agrarian Reform, another government entity, has organized a number of agricultural colonies. These are co-operative units of at least 10 families. Ten acres are allotted to a family, and some common grazing privileges are provided for livestock. Mechanized farm equipment is furnished on a community basis. An exchange program with Israel provides for kibbutz-oriented Israelis to be domiciled in Nicaragua to advise on the operation of the colonies and for Nicaraguans stationed in Israel to gain firsthand knowledge of how the kibbutz functions there.

The individual farmer is charged with the costs involved in handling his plot, together with the interest on the funds he has had to borrow. When the crops are sold, he is credited with what his land actually produced.

The agricultural colonies have made no significant impact on the agricultural economy so far. If, however, the difficulties inherent in such co-operative ventures can be solved, these colonies could eventually be a major factor in avoiding the deterioration in agricultural production that is the usual aftermath of unplanned redistribution of land. There are large undeveloped agricultural areas in the country, and land reform does not appear to be a serious threat today to the large cattleman of Nicaragua, although the high rate of population increase could eventually put more pressure on him.

The law which now prohibits the vaccination of calves for brucellosis is a major handicap to the health of the national herd. As long as it is in effect, there is no chance of eradicating this disease. The incidence of brucellosis is unknown, but it is certainly high, as are the losses from it.

There are no effective control measures for cattle health other than the prohibition of imports from countries with incidence of foot-and-mouth disease.

There is a legal restriction against the slaughter of cows capable of reproduction. Since this restriction is observed only by the export packing plants, the effect is simply to shunt the cows a farmer wishes to sell to the municipal slaughterhouses.

OUTLOOK FOR CATTLE

For the present the beef-cattle population of Nicaragua appears to be close to a practical ceiling. The country's domestic needs, even with the high rate of population increase, are easily met because of the low per capita consumption of beef. Even this, at 34 pounds per capita, is nearly twice that of any other Central American country.

Only token increases in the "voluntary" export quota of frozen beef can be anticipated for the immediate future. Still, Nicaragua has an excellent potential for raising more cattle. There are large jungle areas in the north and east, now practically unpopulated with either people or cattle. Cleared and developed as pastures, these areas could support several times the present number of cattle in Nicaragua.

The present economy will apparently support a higher production of milk than is currently obtained. This could easily come from improved management, a more productive milk cow, and better nutrition. There will probably be a gradual but continuous increase in the national dairy herd, but this cannot be expected to add materially to the total cattle in the country.

The American Brahman seems certain to become the national beef animal. It is already well established and enjoys a strong preference among the growers, both large and small. The tropical climate is unsuitable to the British breeds. With the higher losses they suffer and the more meticulous care they require, they cannot be competitive with the Zebu, especially when the market recognizes no advantage in carcass quality.

The Brown Swiss is still the preferred dairy breed today. The Guernsey is now in second place, but the number of Holstein-Friesians is rapidly increasing and will probably become the major dairy breed, as it has in many countries.

Panama

Land area (sq. mi.):	29,200
Population (1971):	1,500,000
Density (per sq. mi.):	51
Agricultural (40%) (1969):	1,070,000
Per capita income (1970):	$963
Cattle population (1970):	1,297,000
Year visited:	1969

THE REPUBLIC OF PANAMA occupies the narrowest and most strategic section of the isthmus connecting North and South America. This is an S-shaped area from which the Azuero Peninsula extends into the Pacific. The land extends generally eastward from the border with Costa Rica for nearly 500 miles to the Colombian border. The distance between the shorelines of the Caribbean Sea on the north and the Pacific Ocean on the south varies from 37 to 110 miles. The Canal Zone, United States property by treaty, bisects the country near the center of the S. Panama has served as the overland gateway from the Atlantic to the Pacific from the time of the Spanish conquest of Peru, through the days of the gold rush to California, and until the construction of the Panama Canal.

Spanish explorations of Panama date from 1501. Colonization was successful after initial setbacks caused by disease and hostile Indians. Balboa led an expedition across the isthmus in 1513 and was the first European to see the Pacific Ocean. Panama became the vital link in communication between Spain and her colonies on the west coast of South America, and here the treasures of the Incas passed overland on the route to Spain.

As trade began to flourish, Panama became the seat of activity for pirates, who took a heavy toll both in captured treasure and in ransoms. Sir Francis Drake toward the close of the sixteenth century and Sir Henry Morgan during the last half of the seventeenth century were the most famous of the British buccaneers. They harassed the Spaniards on land and sea, capturing and sacking ports and seriously disrupting commerce. This was a phase of the continuous conflict between Spain and Britain throughout the seventeenth century.

Spanish rule was maintained, however, and Panama eventually came under the Viceroyalty of Colombia. When Simón Bolívar freed Colombia from Spain, Panama also declared its independence. But it soon joined the Colombian Union to become, in effect, a Colombian state.

The decision of the United States at the close of the nineteenth century to build the Panama Canal led to Panamanian independence. French interests had previously negotiated a treaty with Colombia for the construction of a canal across the isthmus. When the attempt failed, the United States purchased from France the uncompleted works. Colombia then refused to conclude a satisfactory treaty with the United States, and there was talk of moving the location of the canal to Nicaragua. This led to revolution in Panama against the Colombian government, and, under the umbrella of United States protection, Panama declared its independence in 1903. A treaty was then concluded with the Panamanian government which gave the United States jurisdiction in perpetuity over the Canal Zone, a strip of land 50 miles long and 10 miles wide extending from the Caribbean Sea to the Pacific.

The construction of the canal, its completion in 1914, and the establishments necessary for its operation and maintenance have provided the basis for a considerably higher economic level in Panama than exists in any of the other Central American states.

One-third of the total population lives in Panama City and Colón, the two cities at the terminals of the canal. The pure Indian element constitutes 10 per cent of the population. Primitive tribes reside in the nearly inpenetrable eastern jungles and the mountainous western regions. There are large representations of Europeans and Negroes, but most of the population is of mixed Spanish and Indian ancestry.

The income derived from the operation of the canal and the commerce engendered by it have provided a livelihood for so many people that agriculture does not play a large part in the national economy. Although half the population is directly dependent on agriculture, it is largely of the subsistence type. The productive potential of the land is still far from being utilized, and the country is not even self-supporting in food production. There are a few large plantations. The principal export crop is bananas.

The major farming area is the coastal plain of the Azuero Peninsula. Cattle raising is more advanced in western Chiriqui Province adjoining Costa Rica. The main concentration of dairy herds is around Panama City and in the hilly country on the west. Throughout these areas the

Criollo (mixed) cow in a Chiriqui herd.

annual rainfall is 70 inches, concentrated in the six months from May to November. North of the Continental Divide, rainfall is much heavier, averaging around 150 inches. This region is mostly jungle, undeveloped, and practically uninhabited.

The people of Chiriqui look on their province as the "Texas of Panama" and, not entirely facetiously, refer to themselves as living in Chiriqui instead of in Panama. Within a distance of 50 miles the land rises from the tropical Pacific coast to elevations of over 4,000 feet, where the climate is quite temperate. The terrain changes from a coastal plain to rolling country and then to hills and valleys which can be used only for pasture. Above 2,000 feet pastures ordinarily remain green for most of the dry season, but there is not sufficient moisture to maintain good plant growth.

CATTLE BREEDS

Cattle in Panama are generally used for both milk and meat production. In some areas they are also used as draft animals. The indigenous Criollo has practically disappeared.

Criollo (Mixed).—Throughout the country many herds of nondescript cattle are seen, indicating that the upgrading process to exotic breeds has

Home-raised three-year-old purebred Brangus bull. Chiriqui Province.

not been completed. These cattle are what have been called mixed Criollo. The draft cattle invariably show mixed breeding.

Introduced Breeds.—In the beef breeds the American Brahman predominates, both in the number of purebred herds and in the mixed Criollo and upgraded cattle of the country.

Santa Gertrudis, Charolais, and Brangus are also represented, but as yet they have had very little impact on the cattle population as a whole. The Brangus breed is growing in popularity, especially in the high parts of Chiriqui. Red Poll cattle were brought in from Trinidad in the late 1920's, and there are still some pure representatives.

Guernsey and Jersey were the first dairy breeds introduced into Panama. Their influence is still seen in the mixed Criollo cattle, but they are largely being replaced by Brown Swiss and Holstein-Friesians. Of these, the Brown Swiss is now the most popular, probably because dairy operators prefer a dual-purpose animal.

MANAGEMENT PRACTICES

Many herds in Chiriqui run up to 1,000 head of mother cows. Most of these are grown principally for beef, although a number of sizable

Holstein-Friesian herd upgraded from Criollo base. Near Panama City.

operations take milk for sale from a limited number of the better milking cows in their herd. When this is the practice, milking is done only in the early morning. The cows, with their calves, are then turned to pasture. They are brought to the barn in the early afternoon and separated from the calves until the following morning's milking. Usually no supplement is fed. When on good pasture, an average milk production of 8 to 10 pounds a cow is obtained.

In the Chiriqui herds bulls are normally left with the cows all year. The usual ratio is 1 bull to 25 cows. In some herds the breeding is cut to ten months in order to avoid calving in the last months of the dry season when feed is short. The better operators employ maternity pastures, primarily to facilitate the treatment of newborn calves for screwworm. Exceptional operations obtain an 85 per cent calf crop, with a 5 per cent death loss to weaning. The average is around 50 per cent, with loss to weaning varying from 10 to 20 per cent.

During the long dry season cattle on most ranches go down badly in condition. Small beginnings have been made at putting up some hay to feed toward the end of the dry period. A few of the best-managed properties also employ flood irrigation. This practice permits two and one-half to three-year-old steers to be marketed at 950 to 1,000 pounds. Animals that are grazed on dry pasture after the rainy season will not reach these weights until a year older. Many of the short streams flowing southward from the Continental Divide to the Pacific carry water throughout the dry season and could make possible a wider use of irrigation.

In the area east of Chiriqui Province cattle operations are smaller, and the management is not as good. The cattle there are often only auxiliary to crop farming. The average weight of the steers and bulls from this area, when slaughtered at the National Abattoir in Panama City, runs from 800 to 850 pounds. Of the kill 40 per cent are bulls.

Scattered through the country are a number of feeders who buy cattle at any age at which they can be advantageously purchased and grow them out on grass to slaughter weights of around 900 pounds. Most of this feeder stock comes from small farmers who sell their young animals when they need cash.

There were only about 15 commercial dairies in Panama in 1969. Most of the milk supply comes from herds grown primarily for beef and from small farmers with 20 or 30 cows. In the dairy herds the bull calves are grown out for slaughter either by the original owner or by a feeder.

There are a few breeders of purebred beef cattle in the country. Some follow good management practices and have outstanding herds. American Brahman herds are the most numerous, but there are a few Brangus breeders. The purebred operation is often carried on in connection with a crossbreeding program.

Artificial insemination is used only to a limited extent and does not appear to be increasing. It is employed in a few of the large dairy herds and in the exceptional beef herd on a special group of cows.

MARKETING

Panama's cattle industry is not nearly as dependent on the United States beef export market as are the other Central American countries. In the early 1960's a government subsidy was necessary to move cattle into export channels because of the high prices paid for cattle for local slaughter. Increased production and higher prices in the United States have since brought local and export prices into line. In 1968, however, frozen beef exports were just under 7 per cent of the total beef production.

There are no established cattle markets. Slaughter cattle for the local trade are purchased in rural areas by cattle dealers, who have them processed on a custom basis and sell the dressed meat to retailers.

Residents of the Canal Zone and the United States military establishment there are the major beef consumers. To supply this trade, the principal slaughterhouse in Panama City buys direct from the producers.

Cattle in holding yard at the Panama City slaughterhouse.

In early 1969 the going price for live cattle for local consumption was 17 cents a pound. All cattle except for those in very poor condition brought the same price. Cattle to be slaughtered for the Canal Zone and for export were usually steers three and one-half to four years old and were moving at a slightly higher price.

The export trade is handled on a bid basis. The buying firms in the United States are advised each year of the date set for the sale of export beef. The quota for the year is then allotted to the highest bidders on a liveweight basis. The bids hold for the next 12 months.

The average milk price to the producer is 5 cents a pound, but the better dairy farms near Panama City get as much as 7 cents. Raw milk, sometimes diluted, retails for 10 cents a pound. Pasteurized milk of a fixed quality is sold to consumers for 14 cents a pound.

SLAUGHTERHOUSES

There are two modern slaughterhouses approved for the export trade.

The one in Panama City kills around 10,000 head annually. The one at David, in Chiriqui, kills 20,000. Both plants also sell to local outlets. These are well equipped and have good sanitation and inspection. The sledgehammer is used for killing.

There are small slaughterhouses in towns throughout the country which together handle about 40 per cent of the cattle slaughtered. These are the usual kind of local facility found in Central America without much provision for good sanitation.

CATTLE DISEASES

It is general practice on the larger ranches in Panama to spray for tick control at intervals varying from three weeks to two months, depending on the season and degree of infestation. Treating the navels of newborn calves for screwworm is also common. If this is not done, losses run high. Calves are drenched for internal parasites at weaning and again when their condition warrants.

Blackleg was prevalent at one time, but now the incidence is low. Some operators vaccinate for it, as well as for hemorrhagic septicemia, but the practice is not general. Outbreaks of rabies occur. At such times teams of three or four farmers, supervised by a veterinarian, vaccinate all the cattle in the area involved.

Vaccination for brucellosis was practiced for 15 years but was made illegal in 1955. Even the importation of cattle which have been so vaccinated is prohibited by law. The ring test is now employed in dairy plants to determine areas where brucellosis exists, and the herds involved are then blood-tested individually. On a national average the incidence of the disease is said to be 0.5 per cent of the cattle actually tested.

It is also illegal to vaccinate cattle for anaplasmosis or to import animals which have been vaccinated against this disease.

Tuberculosis is not considered to be serious, and there is no systematic testing for its control.

So far as is known, Panama has been free of foot-and-mouth disease for many years, and imports of cattle from foot-and-mouth countries are prohibited. Efforts are made to prevent any cattle from Colombia (a foot-and-mouth–infected country) from crossing the border into Panama. The eastern part of Panama along the Colombian border is a wide belt of jungle, only sparsely populated, which separates the cattle-growing areas of the two countries. Thus the danger of introduction of foot-and-mouth disease from this source does not appear to be imminent at present. There is, however, a considerable volume of coastal trade handled by small vessels between Colombia and the northern coast of

Panama. These vessels reach villages as far west as the Canal Zone. This traffic could at any time introduce foot-and-mouth disease to Panama. There is no sanitary inspection of the vessels. This should be of vital concern to all Central American countries. If the disease ever gains a foothold in Panama, it would soon spread throughout Central America.

GOVERNMENT AND CATTLE

The government at times has displayed interest in cattle development which has not always been well advised. The prohibition against vaccination for brucellosis has been mentioned. The unsettled political situation in recent years has probably prevented any long-range planning for the cattle industry.

OUTLOOK FOR CATTLE

The future national beef herd of Panama will almost certainly consist mostly of upgraded American Brahman with some representation of other breeds carrying Zebu influence, such as Santa Gertrudis and Brangus. The practice of taking milk from farmers' beef-type cattle will probably continue, but upgraded Brown Swiss and Holstein-Friesian cows should eventually dominate the national dairy herd.

Development of the jungle area north of the Continental Divide could provide for a sizable number of beef cattle, but, to be successful, such operations would require extremely good management. There is not much room for cattle expansion elsewhere in Panama.

Mexico

Land area (sq. mi.):	761,000
Population (1970):	50,600,000
Density (per sq. mi.):	66
Agricultural (46%) (1969):	23,400,000
Per capita income (1969):	$572
Cattle population (1969):	24,876,000
Year visited:	1969

THE UNITED MEXICAN STATES lies south of Texas, New Mexico, Arizona, and California. It is bounded by the Gulf of Mexico and the Caribbean Sea on the east and the Pacific Ocean on the west. Guatemala and a corner of British Honduras border Mexico where it narrows to the isthmus of Central America in the southeast. Two major peninsulas extend from the mainland: Yucatán in the southeast separates the Gulf of Mexico from the Caribbean; and Baja California in the northwest lies between the Pacific Ocean and the Gulf of California.

The terrain is dominated by high-plateau areas ranging from an elevation of 3,600 feet on the northern border to 9,000 feet in central Mexico. The central plateau region is enclosed by two major mountain ranges, the Sierra Madre Oriental (eastern) and the Sierra Madre Occidental (western). Both ranges fall off to coastal plains which are semi-arid in the north and tropical jungle farther south.

Climate is largely determined by the elevation. It is tropical on the coasts and low inland areas in the south but temperate over the central plateau at elevations above 3,000 feet. Rainfall is heavier on the eastern coast than on the western but gradually increases on both coasts from north to south. Irrigation, however, is necessary in most areas for cultivated crops. In general, the rainy season lasts from May to October.

Although some shipwrecked Spanish sailors reached the Yucatán Peninsula in 1512, the conquest of Mexico did not begin until Cortez led his expedition from Cuba in 1519. The centuries-old Indian civilizations of the Toltecs, Aztecs, and Mayas were at a high level of develop-

Mexico

ment when the Spaniards arrived. The superior weaponry of the conquistadors, however, led to rapid subjugation of the Indian cities, and the Viceroyalty of New Spain was established in 1535. This comprised all of present-day Mexico, as well as the region on the north now occupied by Texas and most of California, Arizona, and New Mexico.

The movement to obtain independence from Spain, which flared spontaneously at the turn of the nineteenth century throughout Spain's possessions in the Western Hemisphere, reached Mexico in 1808. A republic was proclaimed in 1822, after nearly three centuries of Spanish rule. Texas broke away from Mexico in 1836 and was an independent republic until it joined with the United States. The areas of California, Arizona, and New Mexico that were part of Mexico were ceded to the United States following the Mexican War of 1846. Since then the independence of Mexico has been maintained, except for a brief period when Napoleon III of France established Archduke Maximilian of Austria as emperor in 1865. He was overthrown and executed in 1867. A revolution from 1910 to 1917 was successful, and the political party which has since dominated the country was the outgrowth of that revolution. Mexico has enjoyed a reasonable stability in government for more than 50 years and is first among Latin American countries in that respect.

Fewer than one million Mexicans are of pure or nearly pure Spanish descent. Roughly half of the remaining population is mixed Spanish and Indian (mestizo), and the rest are pure descendants of the Indians who originally populated the country. Although in recent years there have been sizable and continuous increases in industrial activities, they have effected only a moderate rise in the economic level of the country owing to a corresponding increase in the population. The annual population increase has averaged slightly over 3 per cent. Mexico is still a land of a few very rich and many very poor.

Mexico is predominantly an agricultural country, with well over half of its people engaged in farming and stock raising. Millions of Indians, living principally in the southern part of the country and on the central plateau, are on little more than a subsistence level. The northern states have irrigated and mechanized farming operations. On the vast grasslands there are large cattle ranches showing excellent management. There are also good cattle ranches on the southern coastal plains, particularly on the Gulf side.

The agricultural pattern of Mexico has been in a transitional stage for the past 50 years. Before the revolution of 1910–17 land was almost entirely in the hands of large owners and was worked by peons. Land-reform programs contained in the 1915 constitution have done little to help the masses and have actually proved to be a handicap to agricultural development. The indefinite provisions of these programs led to subterfuge and political maneuvering on the part of large landowners endeavoring to hold their empires together.

The basic law provided that an owner could keep 500 head of cattle and sufficient land on which to maintain them. The remaining livestock and land were to be distributed among the workers on the land. What constituted 500 head of cattle was not defined. The question, for example, of whether a herd of 500 cows would still be considered within the limit after the calves were dropped was left to the interpretation of local authorities. It was soon apparent that a rigid application of the 500-head-an-owner limitation would have a disastrous effect on livestock production because of the losses which would ensue in taxes and in the foreign exchange generated by cattle exports. In 1934, "Certificates of Ineffectability" were issued to approximately 450 large growers. They permitted the ranchers to maintain the level of their operations on the assurance that their land and livestock would be free from expropriation for 25 years. As these certificates expired, it became general practice for a landowner, as the only recourse, to divide his livestock and acreage among various members of his family, even among other relatives. The future of this procedure is highly questionable.

The *ejido* is the instrument through which expropriated land and livestock have been distributed to the agricultural workers. Although the term "purchase of land" for redistribution is sometimes used, compensation, if any, to the owner is quite nominal. The *ejido* is a village which has perpetual tenure on the acquired land. This land is held collectively for grazing or for distribution in small plots for individual cultivation. The *ejido* farmer has the right of occupancy only as long as he maintains the land in cultivation. These plots vary in size from 13 to 36 acres, depending on the productivity of the area as determined by soil conditions and rainfall.

Since the cattle of the *ejido* farmers are run on common grazing lands, the individual owners have practically no control in their management. The number of cattle sold off the *ejidos* is variously estimated at one-tenth to one-sixth of the total marketed over the country. These are of a quality that is inferior to those sold by surrounding ranches.

Frequently the *ejido* dwellers occupy the former buildings of a large hacienda. The original owner sometimes lives in a house on a separate plot of land, but more often he has moved to town. The *ejido* farmer is still on a subsistence level. The only change in his peon status is that he is now under a political overlord instead of an individual landowner. The average *ejido* is nothing but a government-controlled rural slum.

CATTLE BREEDS

Mexican cattle trace to the Spanish stock introduced by the conquistadors. Typical animals seen today, however, are the result of upgrading or indiscriminate crossing with European and Zebu breeds that have been introduced within the past 70 years. The dairy herds are predominantly upgraded or pure European breeds. Within a 100-mile radius of Mexico City there are dairy herds which were developed from purebred stock imported in recent years from the United States and Canada. There are also purebred beef herds which originated from rather recent importations from the United States and Brazil and, in the case of the Charolais, from France. Annual imports of United States purebred breeding stock currently total some 13,000 head of dairy cattle and between 4,000 and 6,000 head of the beef breeds.

Descendants of the original Spanish cattle, little influenced by modern breeds, are now seen only in the more remote parts of the country. These are generally known as Criollo cattle, although in the state of Sonora the term *corriente* is more common, and in Baja California the word *chinampo* is used. All these terms, meaning "common cattle" or "cattle of the country," are applied to the more or less pure descendants of the Spanish cattle, as well as to the indiscriminate mixtures of these with more recently introduced breeds.

Criollo (Pure).—Probably the purest descendants of the Spanish cattle now to be found in Mexico are the chinampo in Baja California Territory (the southern part of the peninsula of Baja California). They are a degenerate type, the result of lack of breeding control and countless generations of subsistence in an arid land where centuries of overgrazing have reduced the available forage to cactus and other desert plants. These cattle have actually become browsers rather than grazers. They have a high tolerance to desert conditions and can go without water for two or three days. On a few large ranches where forage crops have been introduced through irrigation, the chinampo have been practically bred out by crossing. The few head in the hands of the small subsistence farmers, however, are nearly straight chinampo.

The chinampo has a shallow, barrel-shaped body and long, rather thin legs. Its horns are unusually large and are upswept, often turning outward at the tips. The head is quite narrow. Color is extremely varied —nearly solid black or tan, spotted black and white, and nearly solid

Mature Chinampo cows, about 650 pounds. Baja California.

Criollo (mixed) steer, 1,000 pounds. Western Oaxaca State.

tan and white. There are very few reds. Black-and-white markings, quite similar to those on the Holstein, are seen, although there has been no admixture with this breed in the area. Cows weigh 600 to 800 pounds; and bulls weigh 750 pounds to a maximum of 1,000 pounds for an old animal in fair condition.

Pure Criollo cattle can also be found in other parts of Mexico, more frequently in the mountainous areas of the central plateau. They have much the same nondescript appearance as the chinampo but may be larger because they have not been subjected to such extreme nutritional deficiencies. These Criollo cattle are often used as draft animals and for household milk by subsistence farmers.

Criollo (Mixed).—Most Mexican cattle fall in this category. They are crosses on the pure Criollo where upgrading has not proceeded far enough to fix the characteristics of the new breed.

In the northern states the Hereford influence predominates, though indications of other breeds are seen. These include Aberdeen-Angus, Charolais, and some of the dairy breeds, particularly Holstein-Friesian.

De Lidia bull with cows on pasture. Zacatecas State.

On the low plains along both coasts the mixture has been almost entirely with the Zebu breeds. The Indo-Brazil and American Brahman predominate. Most commercial cattle reaching the slaughterhouses show the influence of one or the other, and possibly both, of these breeds. Little evidence of mixing with dairy breeds is noticed.

In the southern part of the central plateau the European dairy breeds, as well as the various Zebu breeds, enter into the mixtures, which usually defy classification. A greater proportion of the cattle here are owned by small farmers on a near-subsistence level. Oxen are used for cart and field work.

De Lidia.—Following the cattle which the Spaniards brought to Mexico for milk and meat were the "brave cattle" for the bull ring. They are said to have been introduced around the middle of the seventeenth century. Although their numbers are small in relation to the total cattle population, their growers are considered to be the elite among cattlemen.

In appearance the fighting cattle are identical to those in Spain. The color is usually black or dark grey, although occasionally brindle and mixed colors are seen. White markings are not discriminated against. The chest is deep and slopes upward to the flanks. The body is barrel-shaped, and the hindquarters are narrow but strongly muscled. The forehead tends to be quite broad. Horns spread upward either in a single curve or with a slight outward curl at the tips.

The consensus is that the fighting bull of Mexico is not on a par with

De Lidia bulls in holding pens at Mexico City bull ring.

his counterpart in Spain. Whether this is a fact or whether it merely stems from the common assumption that an imported product is better than one produced locally is difficult to determine. Since the ban was placed on imports of cattle from foot-and-mouth countries, no De Lidia have been brought from Spain. The gene pool that exists in a population of over 30,000 head should be sufficient to maintain a reasonably high standard in the Mexican fighting bull.

Recently Introduced Breeds.—In the good cattle areas on the central plateau in the states of Chihuahua, northern Sonora, and down into central Durango, the Hereford predominates. There are a few Aberdeen-Angus herds, but they do not seem to have adapted to the environment as well as the Herefords. Most herds are the result of upgrading Criollo cattle by the continuous use of either Hereford or Angus bulls. Neither breed attains the growth it displays in the United States. This

Purebred Holstein-Friesian cows of a modern dairy. Near Mexico City.

can probably be traced to nutrition deficiency, as well as the lack of genetic improvement. The upgrading process was not begun until after the turn of the century.

Farther south, from southern Zacatecas to Mexico City, excellent dairy herds of Holstein-Friesian, Brown Swiss, and Jersey Cattle are concentrated around the large cities. The Holstein-Friesian accounts for about 90 per cent of the national dairy herd. Large herds of 300 to 400 cows in milk are usually maintained pure. The small herds of 10 to 30 cows are often made up of crosses. Holstein-Friesian bulls are commonly used in these herds, although Brown Swiss are growing in popularity because their calves sell at a better price for slaughter.

Zebu cattle were first brought to Mexico from Brazil in 1884. There were many subsequent importations, several of which were destined for transshipment to the United States. The importation of cattle from Brazil, as well as from other foot-and-mouth infected countries, was prohibited in 1954 following the eradication of the disease in Mexico. The Mexican cattleman retains a strong preference for the Indo-Brazil breed, which was well represented in the different importations of Zebu cattle that reached Mexico. After the movement of cattle from Brazil was cut off, there were sizable importations of American Brahman cattle from the United States. (This breed had been developed in the United States largely from Zebu cattle that entered through Mexico.) In 1969

American Brahman cow herd on irrigated pasture in Veracruz State.

the total number of American Brahman cattle in Mexico was approaching that of the Indo-Brazil. Next to these two breeds in popularity is the Guzerat. There are also a few Nellore and Gyr and one purebred herd of Red Sindhi cattle.

There are representatives of the Santa Gertrudis breed in the tropical areas of Mexico. Although they seem to do well, the breed has not met with wide acceptance by large cattle growers.

Charolais cattle were first brought to Mexico from France during the first decade of the twentieth century. About the same time there was also a shipment from Colombia to Mexico. A man named Silva and the Terrazas Estate are recorded as having imported representatives of the breed before 1910. That year saw the beginning of the revolution, and in the unsettled period which followed, all traces of the first Charolais vanished from Mexico.

Continuity of the breed in Mexico was established by an importation in 1930 by Jean Pugibet, the president of a cigarette-manufacturing company and ranch owner. Pugibet had seen the breed in France while serving in the French army during World War I. There were 2 bulls, 5 bull calves, and 10 heifers in the first shipment from France. Pugibet arranged for two subsequent importations, one in 1931 and another in 1937. In all, he brought to Mexico 8 bulls, 5 bull calves, and 29 cows and heifers. One of the bull calves died shortly after arrival. Seven of the heifers were in calf by French bulls. These were the nucleus of the Charolais breed in Mexico, from which were drawn the foundation animals for Charolais herds in the United States.

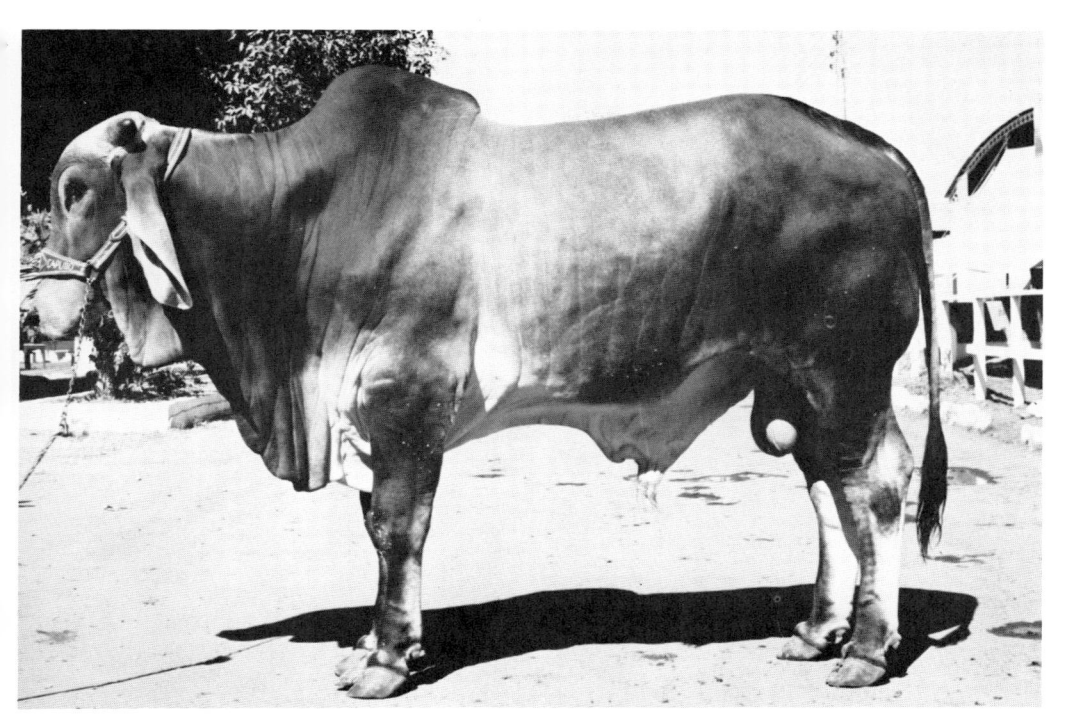

Purebred Indo-Brazil bull at Chiapas State Fair.

The Charolais showed a remarkable ability to adapt to the Mexican environment. This was true not only in the high-plateau area but also, although to a lesser degree, in the coastal regions. The ravage of the cattle tick is felt along the full length of the coastal plains on both the Pacific Ocean and the Gulf of Mexico. Yet Charolais herds, with good care, can be maintained there in thrifty condition and show the desirable attributes of the breed in beef conformation. They do not, however, attain the large size that the breed does in cooler climates.

The so-called purebred Charolais herds generally consist of individuals upgraded to the point where the Criollo influence is not discernible. There are also herds of Charolais crosses on both Criollo and Zebu cattle.

MANAGEMENT PRACTICES

There are two different types of commercial beef-cattle operations in Mexico: those on the high northern plateau and plains, where the up-

Guzerat cow, 1,100 pounds, on a ranch in Veracruz State.

grading of the Criollo to European beef breeds has been accomplished in many herds; and those on the coastal plains, where the upgrading process is principally to the Zebu breeds. The major dairy operations are around the large cities on the central plateau.

Beef Cattle—Northern Plateau and Plains Region.—In the states of Sonora, Chihuahua, most of Durango, and northern Zacatecas, ranch

Young Charolais bull at a fair in Chiapas State.

operations are carried on by the large growers much as they are in the southwestern United States. Many large holdings have been maintained through family division of land tenure and by the Certificate of Ineffectability.

Cattle raising in this region is oriented toward the United States export trade for young feeder steers and frozen, boned meat. A typical ranch operation involves a unit of 25,000 to 40,000 acres in a 10- to 15-inch rainfall area and carries a herd of 1,000 to 1,500 mother cows. Many ranches are even larger. Most cattle are now run under fence.

Stocking rates vary widely depending on the productivity of the land, which is affected by past overgrazing, soil types, local variations in rainfall (often pronounced), and management practices of the individual owner. In good areas 18 to 25 acres an animal unit are sufficient, while 40 to 80 acres are required where there is serious encroachment of semiarid vegetation and where little attention is given to grazing control.

In the southern part of the plains area the average rainfall is around 16 inches, and good pasture management gives a carrying capacity of 8 to 12 acres an animal unit. This requires grubbing out the mesquite and acacia, rotating pastures to avoid overgrazing, and providing adequate stockwater. Small areas of cactus are kept as a form of drought protection. The spines of the plant can be burned, leaving the rest of it edible. This provides sufficient nutrients to maintain cattle over periods of several months when severe droughts are encountered.

In the past the universal practice was to run bulls with cows on a year-round basis. This is still followed in many old, established herds. The more progressive ranches, however, have a five- to six-month breeding season, usually from March to August, which gives a calving period from January through June. In such a program, where one bull to 15 or 20 cows is used, 75 to 82 per cent calf crops are obtained. The average for the state of Chihuahua, however, is 65 per cent. Heifers are normally bred at two years of age to calve as three-year-olds.

Planned breeding programs permit the best utilization of the new plant growth, which lasts from June to the end of August, the period in which 85 per cent of the precipitation occurs. Over the rest of the year cattle must be maintained on the standing mature grasses. Some growers, however, feed cottonseed cake for two to four months of the dry (winter) season, at the rate of 1 to 1.5 pounds a day for a productive cow. Such feeding gives an increased conception rate in the breeding herd and results in stronger calves.

Calves are commonly weaned in December at eight to ten months of age at weights of 300 to 320 pounds. Mexican law requires that a minimum of 20 per cent of a rancher's offtake must be sold for domestic consumption. Also a group of young females must be transferred gratis to *ejido* dwellers, this number currently being 2 per cent of the cattle exported. Most of the steer calves, however, are exported at weaning to points along the United States border. Most shipments include a few short yearlings that were too light to ship in their first year. Good-type calves sometimes bring slightly higher prices than United States–grown calves because of the rapid compensatory gains which can be obtained on good feed. Most grade good to low-good by USDA standards; a few make a choice grade.

The export of live female cattle to the United States is prohibited by Mexican law. Heifer calves not wanted for replacements as breeding stock are normally grown out on pasture for a year. On grass they reach weights of 475 to 500 pounds as yearlings, when they go into the feed lots. The average feed-lot period is 75 days, and gains of 2 pounds a day

A Chihuahua feed lot, northern Mexico.

are common. Slaughter weights are 625 to 650 pounds at a grade equivalent to USDA good. The principal outlet for this class of cattle is Mexico City.

Culled cows and discarded bulls, along with some Criollo cattle which find their way from the mountain areas to the feed lots on the plains, are usually put on a short feed of 30 to 45 days before slaughter. Most of this production goes into 60-pound packages of frozen, boned meat and is exported in refrigerated trucks to the United States.

Feed lots are owned for the most part by the slaughterhouses or by large ranchers. They are frequently well-run operations, similar in construction to those in the United States. A few lots handle 5,000 to 12,000 head annually, but units of this size will not increase rapidly because of the favorable United States market for feeder cattle. The feed-lot operations are run largely by hand labor because of the easy availability of workers and the relatively low wages. Balanced rations are fed with sorghum as the base concentrate and cottonseed meal for protein. In cotton-growing areas, cottonseed hulls are used as roughage. In the southern part of the area there is some sorghum silage. Most of the corn grown in Mexico goes for human consumption and is not generally used for cattle feed.

Beef Cattle—The Coastal Plains.—On the plains along the Gulf of Mexico, large numbers of Zebu-type cattle are raised. The Zebu influence varies from representatives of pure breeds, such as American Brahman or Indo-Brazil, to obvious Zebu-Criollo crosses. In the state of Veracruz alone, most of which has a tropical climate, the estimated cattle population is 4.5 million head. This is one-sixth of the national herd. Most of these cattle, particularly in the Huasteca area from Tampico to Veracruz, are produced for the major market in the Federal District.

Steers for the Mexico City market are grown out on grass to two and one-half or three years of age. The straight Zebu breeds weigh up to 1,300 pounds at market age, and Criollo crosses weigh less. There is also some transshipment of yearling cattle from the extreme southwestern coast in the states of Oaxaca and Chiapas to the Veracruz area for growing out before final shipment to Mexico City. Some use of feed lots is made in these movements to improve condition by a short warm-up feed before shipment.

In the coastal areas cattle are on pasture on a year-round basis, although only mature standing grasses are available for forage during the dry season from December into May. The growing season could be 12 months long if adequate moisture were available. There is some use of irrigation for cropping, but such practice has only a minor application in stock raising. There is very little feeding of grain to beef cattle.

Pangolagrass is one of the best forage plants. Under good management, even without irrigation, it has a carrying capacity of slightly over one animal unit an acre. Inedible tropical growth, however, is persistent, and careful control and maintenance are necessary to obtain such rates. The average stocking rate is 1.7 acres an animal unit in the Veracruz-Huasteca area.

Dipping for tick control is necessary in all the coastal areas. This is done usually at intervals of 30 days in the dry season and 15 days in the rainy period. Even small farms running 150 head have their own dipping tank. The old practice of year-round breeding is gradually giving way to a five- or six-month breeding period. This program is limited to the larger, better-managed ranches which use one bull to about 35 cows.

Although some holdings on the coastal plains have been reduced in size in the land-reform program, there are still many large operations. In the state of Veracruz, an estimated 70 per cent of the cattle are on units running 500 head or more, 20 per cent on ranches in the 200-head range, and 10 per cent on the *ejidos*.

Dairy Cattle.—South of the northern high plains dairy cattle become more numerous. The highest concentrations are in the area from southern Durango to Mexico City. All the dairy breeds do well at the higher elevations in the region, although much of it lies within the limits of tick infestation.

Dairies maintaining 300 to 400 cows in milk are quite common. These large herds are of excellent Holstein-Friesian, Brown Swiss, and Jersey cattle. The production of these units is destined principally for Mexico City.

The entire herd is often maintained in exercise yards with shade provided under thatched roofs. Milking is done in modern milk parlors. Silage, haylage (silage of lower moisture content), and green-chop sorghum are fed along with dry roughage composed of cottonseed hulls, pea hulls, or hay. The concentrate is usually based on sorghum grains and cottonseed meal. Some large dairies turn their cows to pasture during the day, but more generally, green chop is brought to the lot.

Considerable use is made of artificial insemination.

Among these larger dairies it is a common practice to kill the bull calves as soon as the colostrum has been consumed. Heifer calves are usually grown out as replacements.

Near the towns and villages in the central plateau there are many small dairy herds of 10 to 30 cows. These are predominantly Holstein-Friesian grades or crosses of this breed with the Criollo. Brown Swiss crosses are also common, and there are some Jerseys. Cows are usually turned to pasture after milking, and grain supplement is fed sparingly, if at all. The milk is not pasteurized.

In the tropical areas of the coastal plains dairy herds with some Brown Swiss influence are preferred. Brown Swiss crosses on Criollo cattle or Zebu crosses on Brown Swiss cows are quite common. There are also herds of straight Holstein-Friesian, Jersey, and Brown Swiss, but they must be given special care to withstand tick-borne diseases.

In the mountainous areas of central Mexico and in the southern part of the central plains, cattle are frequently used for draft in addition to

American Brahman–Brown Swiss dairy cow near Veracruz.

producing milk for the small farmers. The animals are of the mixed Criollo type. These are usually the result of crossing with the European breeds in the high country and with Zebu breeds in the lower plains.

The Art of Breeding the De Lidia.—The raising of fighting bulls, or "brave bulls," is a breeding art practiced in Mexico along much the same lines as in Spain. Breeding, for the most part, is confined to the high-plateau area some distance north of Mexico City in the 14-inch rainfall belt. Some De Lidia growers also raise beef cattle as well, but the two operations are distinct. Selection for aggressiveness in the breeding stock is the basic concept in producing the bull for the ring. A minimum of handling is considered essential, although ear tagging at

Criollo–Zebu ox team. Chiapas State.

birth for identification, prophylactic innoculation for blackleg, branding, and the testing procedure for yearlings are necessary.

Both bulls and heifers are tested for aggressiveness when a year old. Two recognized methods are employed: *campo abierto* ("open field") and *en plaza* ("in bull ring"). When tested in the open, a rider on horseback upsets each animal, and its spirit and aggressiveness are noted when it regains its feet. *En plaza*, as the name implies, takes place in a small bull ring which is maintained on his property by the De Lidia grower. Each animal tested is turned into the ring individually and confronted by a man on horseback. He is equipped like the picador at a regular bullfight, but his lance has only a small steel point, about one-half inch in effective length (a six-inch pointed spear of larger diameter is used by the bull-ring picador). The rider's feet and legs are protected by metal stirrups and shields, and the horse is covered with a well-padded cloth. The animal under test is lanced in the shoulder. To qualify for acceptance, he must charge the horse. The number of such charges is considered a measure of the degree of aggressiveness of the individual. Four or five distinct charges are considered excellent.

The manner in which this test is met determines the classification into which the owner places each individual. There are three such classifications for the bulls.

Matador: To be grown out to 4 years of age to go into the ring at 1,000 to 1,200 pounds against experienced matadors.

Novillero: To be used by novice bullfighters, usually at 3 years of age weighing 770 to 880 pounds.

Ametures: The discards, usually sold to small clubs; a few are exported to enthusiasts in the United States.

In a good herd 50 per cent of the annual bull crop places in the *Matador* class; 30 per cent rank as *Novilleros*, showing less fight; and 20 per cent are classed as *Ametures*.

After the trial with the picador, the yearling heifers which have shown proper aggressiveness are tested with the *muleto*, the cape. This is done in the same manner as bulls are fought in the bull ring, a man holding the cape at arm's length and advancing to encourage the charge. Improper reactions, such as raising the head or not charging the cape, are points against the heifer, and she is a candidate for discard. Heifers from which selection is to be made for breeding stock are tested with the cape a second time before they are finally chosen to enter the herd. A minimum of 20 per cent in a year's crop of females are discarded for not showing proper aggressiveness.

Bulls destined for the ring are never tested with the cape, because this maneuver, if repeated, might lead the bull to attack the man rather than the cape.

Bulls for breeding are selected both for their pedigree and for their performance in the aggressive test. In addition to the testing procedure, the owner keeps detailed records on the brood cows and on the progeny of the sires. This includes the performances of all offspring in the tests and all bulls that reach the public bull ring.

Following their test as yearlings, bulls and heifers are grown out separately in fenced pastures, always stocked with ample feed. For about five months before shipment to the bull ring, *Matador* bulls are fed a supplemental grain ration of about 13 pounds a day. The *Novilleros* are so fed for about three months.

Heifers are bred at two years of age. An effort is made to breed the bulls with the best record to the best cows. Common practice is an eight-month breeding season using 1 bull to 50 cows. Under these conditions 65 per cent is considered a good calf crop.

Fighting bulls are sold by the grower direct to the bull-ring operator in groups of six for the customary six events of an afternoon, or at times in groups of seven in order to provide one spare in the event that an entry lacks spirit or is faulted after entering the ring. The price a head for bulls grown by the officially recognized top breeders was $800 in 1968; for the next classification of growers, $540 a head. Such prices as these, even $800 for a four-year-old animal, can hardly compensate the breeder for the cost and care he has given in preparing a bull for the ring. His reward is the acclaim he receives when his bulls perform well and the satisfaction of succeeding in an old and respected line of cattle breeding.

MARKETING

There are no cattle markets or auctions at which cattle are sold, and there is no recognized system of grading. Sales are between owner and buyer or their representatives.

In the northern cattle country the large grower usually sells on his ranch to the feed-lot operator or to a slaughterhouse which operates a feed lot. Owners of small herds either sell to a nearby rancher who grows out the animals on better grass or truck them to a feed lot. Sales are by weight except in those instances where an owner may sell one or two animals by the head to another rancher. Most of the marketing in this northern area, however, is to the United States. In 1968 over 700,000 head of Mexican steer calves and short yearlings and 63 million pounds of boned, frozen beef were exported to the United States. The frozen beef represented possibly another 190,000 head, largely of Criollo and culled cattle.

Hereford or Angus steer calves in early 1969 were bringing 30 to 33 cents a pound delivered across the border to United States buyers. This, after taxes, duty, and transportation charges, netted the Mexican grower 22 to 25 cents a pound on his ranch. These figures are for thin ten-month-old calves weighing 320 pounds, which will make rapid gain on good feed.

Cattle from the coastal areas in the south find their largest outlet in the Mexico City area. The owner, who may be the rancher or a trader, retains possession through the municipal slaughterhouse and sells the carcass direct to a retailer. In early 1969 carcass prices were running from 25 cents a pound for a canner-type animal to 36 cents a pound for the best quality, which goes to the supermarkets. This latter price applies

Holding yards, Mexico City slaughterhouse.

to a steer carcass, probably nearly straight Zebu, with a fair fat cover. There is no grading system, but in the trade an animal producing this quality carcass is called "supreme," the equivalent of a USDA low-good to high-good. Such animals bring around 20 cents a pound liveweight and weigh 1,000 to 1,200 pounds. Lighter animals of poorer quality but with considerable Zebu influence sell for 18 cents a pound liveweight; and those of canner quality sell for 14 to 16 cents a pound. All these lower quality animals are termed "Criollo" by the trade.

SLAUGHTERHOUSES

The most modern abattoirs in Mexico are in the northern states of Hermosillo and Chihuahua. These serve the area approved by the Mexican government to export beef to the United States. USDA inspection of the facilities results in sanitary practices on a par with United States standards. Such plants have holding yards and usually operate adjacent feed lots in order to provide a reasonably constant work load. Although mechanized to some extent, they rely on hand labor. Killing is usually done by the captive-bolt method. Meat for local sale is chilled one day before distribution. Most of the plants slaughter hogs as well as cattle.

The Federal District, which contains Mexico City, is the largest beef-consuming area in the country. The principal slaughterhouse is a municipal plant operated by the military. Inspection is by the Secretariat of Health and Public Welfare, and representatives of the Secretariat of Industry and Commerce stamp each carcass with the legal price at which it can be sold. Processing is on a custom basis for the account of the owner of the cattle. The kill is made in a steel box. The *puntilla*, a short dagger with a slightly curved six-inch blade, is expertly thrust in at the base of the brain. Carcasses are processed on an overhead rail, but, in general, operations are not conducted very efficiently. The dressed carcasses are displayed on the rail under roof and are usually sold warm as soon as processing is completed. For the supermarket trade, however, carcasses are chilled for four or five days. Many retail shops now have refrigeration.

Annual slaughter at the municipal plant runs around 300,000 head of cattle, 200,000 hogs, and 350,000 sheep. There are holding yards under roof with space for over 1,000 head of cattle.

In addition to this large municipal plant there are several small operations with slaughtering facilities scattered throughout Mexico City. These are quite primitive. Their daily kill is 20 to 40 head of cattle.

Throughout the country generally, cattle are slaughtered in municipal facilities. These are mostly hand operations, under roof and with concrete floors. Ample use is made of water for washing down, but otherwise there are few provisions for sanitation. Meat, a side or a quarter, is sold as soon as dressed. It goes to the meat stalls in the markets, where it is purchased by the housewives before noon.

CATTLE DISEASES

In 1947 an outbreak of foot-and-mouth disease spread rapidly over the country. Through the joint efforts of the Mexican government and the United States Department of Agriculture, it was finally stamped out in 1954 after all infected animals were slaughtered. There has been no known recurrence. A joint commission of the two countries continues a rigid surveillance. Every report of a suspected case is investigated on the spot. Importations of cattle from foot-and-mouth countries are prohibited by Mexican law.

For years a joint effort has been made by the USDA and the Mexican government to eliminate the screwworm by means of the sterile-male technique. Sterile male flies are produced by the millions in the United States and released in infected areas. The end result is the elimination of the screwworm population, as has been accomplished in the entire southern United States (see the section "Cattle Diseases" in Part IV, "The United States."). While some progress has been made in reducing infestation in certain areas of Mexico, the desired goal of complete elimination is probably years away. If the screwworm could be eliminated in Mexico, the present patrol of the 2,000-mile border with the United States that is maintained by the USDA could be relaxed. In 1972 the Mexican and United States governments agreed upon a joint program to eliminate the screwworm from Mexico and to maintain a sterile-fly barrier at the Isthmus of Tehuantepec in southern Mexico.

Northern Mexico, with the exception of the low-lying coastal plains, is tick-free as far south as the states of Durango and Zacatecas. This region has no serious disease problems, and European breeds do as well as the limitations imposed by the long dry season permit. It is usual practice to vaccinate during calfhood for blackleg. In the southern part of this plains area there is vaccination for brucellosis in the larger dairy herds. Screwworm is still prevalent in much of the region, but serious losses can be prevented by prompt treatment of any skin breaks and the navels of newborn calves.

In southern Mexico, except in the mountainous regions where the subsistence farmers have a few head of draft animals, the cattle tick is prevalent. Even with the Zebu types, either upgraded Criollo or pure Zebu, periodic dipping or spraying results in a marked improvement in condition. Dipping at intervals of 15 to 21 days is considered desirable. Cattle from the tick-infested lowlands cannot be exported to the United States. Calves in this area are commonly vaccinated for blackleg and anthrax at around three months of age. In some parts, rabies, carried by bats, is a menace, and cattle are vaccinated annually against the disease.

GOVERNMENT AND CATTLE

The largest influence of the government on the cattle industry has been the effect of land-reform programs and the establishment of the *ejidos*. The indefiniteness of the law and the inconsistencies in its administration have cast a cloud over the entire industry.

The secretary of agriculture of the federal government allots the United States export quotas for live cattle and frozen beef to the various Mexican states that have been approved for this trade. The Ministry of Agriculture co-operates with the USDA on joint commissions established for control of cattle diseases.

The National Confederation of Cattlemen (Confederación Nacional Ganaderos) is composed of the regional cattlemen's unions of each state. This is a powerful organization which works in close co-operation with both federal and state governments. The state unions combine the functions of United States cattlemen's associations and county agents and, in addition, furnish supplies and veterinary services. These state unions allocate the state's export quotas of live animals and frozen beef to the individual ranchers and packers. Practically all owners of cattle, except for the small farmers with their span of draft oxen or incidental stock, are members of their state union.

OUTLOOK FOR CATTLE

The potentials for increased cattle production in Mexico are sizable in both genetic improvement and better management. The plains and valleys of the northern states produce Mexico's best beef cattle and in considerable number. Most of the herds are of medium-good Hereford type, upgraded from the native Criollo over the past 30 to 40 years. While they reflect a marked improvement over the production that can be obtained from the Criollo, there is still room for betterment. The 320-pound calf at nine or ten months of age now being produced in the good cattle country of Chihuahua is only a fair start.

There is also opportunity for increased production from better range and pasture management. Most cattle are now run under fence, and an increasing number of ranchers make good use of their land by rotation of pastures. As the result of overgrazing in the past, however, there are large areas so overgrown with semiarid, inedible plant growth that there is little, if any, grass.

On the coastal plains in the south the crossbred and upgraded Zebu types are much more productive than the Criollo stock from which they were derived. The Zebu, however, despite its adaptation to heat and its tolerance to tick-borne diseases, is not one of the most productive types of beef animals. Some breeds which have developed from combining European beef breeds with the Zebu are more productive in hot countries than the pure Zebu.

In addition to genetic improvement, there are other channels for increased beef production in the coastal areas. Improved pastures, better control of grazing, and the use of irrigation could increase the carrying capacity of these grasslands. Although there is ample rainfall on the southern coastal plains, it generally occurs in the period from May to October or November. Pastures hold fairly well through the first two months of the dry season but then become dormant, with only the mature growth left for feed. It has been demonstrated that irrigation not only increases the quantity of available forage but also improves the quality. Even without irrigation the present average stocking rate of 1.7 acres an animal unit could be increased by the introduction of improved grasses and better pasture rotation.

For the immediate future development of the cattle industry along such lines as have been mentioned will be confined to experiment-station work and the individual efforts of a few progressive cattlemen.

The small cattleman has neither the means nor the education to avail himself of modern techniques that could lead to increased production. The *ejido* dweller lacks interest and incentive for improvement in his cattle. He has been given access to a plot of land, a few head of cattle, and the right to graze them, all without effort on his part. Although he works the land and milks the cows as a means of subsistence for his family, he makes no plans for the future. While he owns what he produces, the property he utilizes is held only by community tenure.

The cattle industry of Mexico will remain at about the present level, to the extent that the larger cattlemen are able to continue to stave off the objectives of the land-reform programs. Despite a growing demand for beef, large-scale development of the existing potentials probably will not be soon undertaken.

PART THREE: Canada

Canada

Land area (sq. mi.):	3,851,800
Population (1970):	21,681,000
Density (per sq. mi.):	6
Agricultural (28%) (1970):	6,070,000
Per capita income (1969):	$3,670
Cattle population (1971):	13,660,000
Beef	10,547,000
Dairy	3,113,000
Year visited:	1971

CANADA occupies the northern half of the North American continent with the exception of the Alaskan peninsula in the Northwest. All the adjacent islands along both the Atlantic and Pacific coasts are Canadian territory except two tiny islands, St. Pierre and Miquelon, which lie off the southern coast of Newfoundland. These two islands are the only remnants of New France which have remained French territory.

The native beauty of Canada varies from the rolling hills of the Maritime Provinces, which have the charm of the English countryside, to the grandeur of the Rocky Mountains in Alberta and British Columbia. Heavily timbered areas along the Atlantic Coast stretch inland to prairie and parkland before the mountains are reached. Here highplateau areas cut by deep valleys lie between four parallel mountain ranges, the last of which gives way to a narrow coastal plain. Except for the semiarid prairies east of the Rocky Mountains, the country is well watered with clear lakes and fast streams in the mountains, large rivers on the plains. Uncounted thousands of lakes throughout the central lowlands made travel by canoe an easy means of transportation in the days of the fur trade.

Efforts by European nations to obtain a foothold in North America were made earlier in Canada than in the region that became the United States. The first Europeans reached what are now Canadian shores five centuries before the time of Columbus. Recently uncovered artifacts have disclosed that the Vikings were on the northern tip of Newfoundland about A.D. 1000.

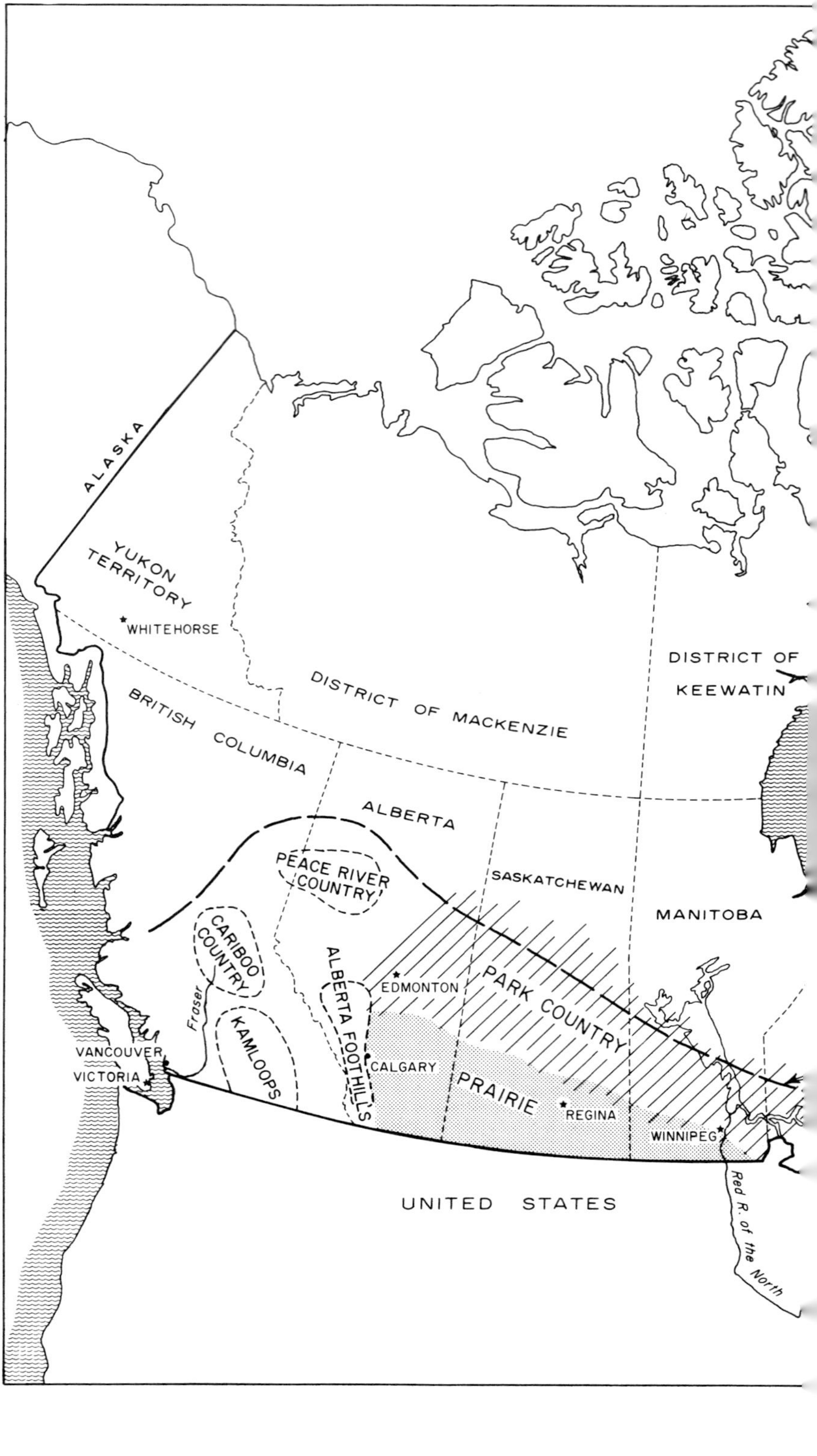
ALASKA
YUKON TERRITORY
WHITEHORSE
DISTRICT OF MACKENZIE
DISTRICT OF KEEWATIN
BRITISH COLUMBIA
ALBERTA
SASKATCHEWAN
MANITOBA
PEACE RIVER COUNTRY
CARIBOO COUNTRY
KAMLOOPS
Fraser
ALBERTA FOOTHILLS
EDMONTON
PARK COUNTRY
CALGARY
PRAIRIE
REGINA
WINNIPEG
VANCOUVER
VICTORIA
UNITED STATES
Red R. of the North

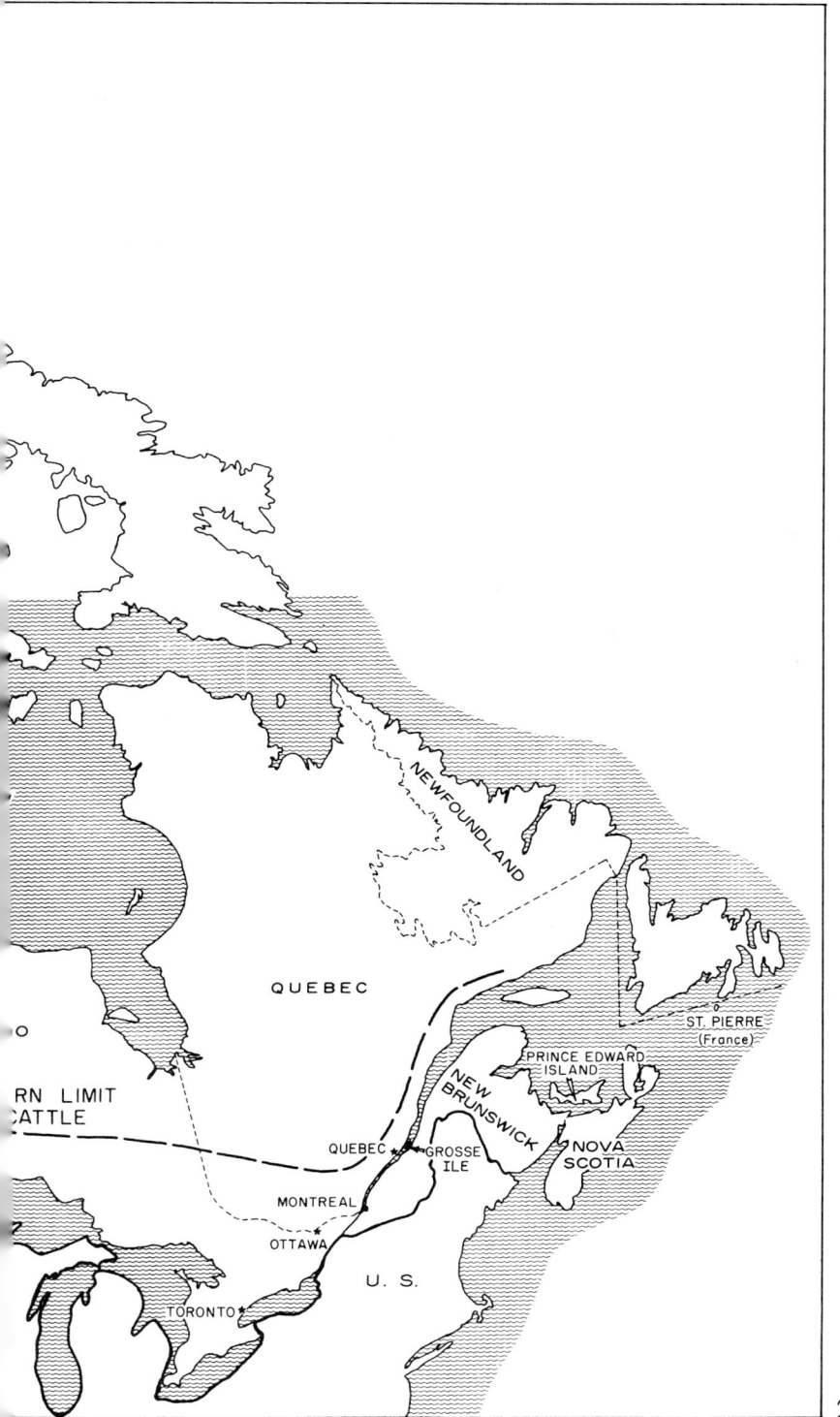

Canada.

Cattle of North America

John Cabot landed a crew from England on Newfoundland in 1497 and claimed that island for Henry VII. Cabot, on the same voyage, touched Cape Breton, which was also claimed for Britain. In 1534, Jacques Cartier sailed up the Gulf of St. Lawrence, landed, and claimed for King Francis I the region that became Quebec. All these early explorations or attempts at settlement had only a short continuity.

In 1600, French fur-trading posts were established along the St. Lawrence River. Samuel de Champlain is credited with founding in 1603 the first French settlement which was more than an outpost for bartering fur from the Indians. It was established at Port Royal (now Annapolis Royal) in Acadia, the name given by the French of that day to the land that is now Nova Scotia. The first buildings on the site of Quebec went up in 1608. Catholic missionaries went to Port Royal in 1611 and to Quebec in 1615. From that time on, Frenchmen were continuously on the soil of New France.

Development of the new land was slow. In the first of many similar interludes the British took the Quebec settlement in 1629 but returned it to France in 1632. Fur trade with the Indians was the principal means of livelihood. It was the pursuit of free individuals over whom there was no control. Traders taking advantage of the situation used brandy in bargaining and offered worthless goods in trade. Antagonisms thus engendered with the Indians severely handicapped efforts at colonization, and, as late as 1660, when New France was made a royal province, there were only 2,300 European inhabitants in the territory.

During the remainder of the seventeenth century New France became stabilized, and by the opening of the eighteenth century warfare with the Indians had subsided. In 1710, however, Britain retook Nova Scotia, and somewhat later war broke out in Europe between Britain and France. British interest in Canada continued to grow, fostered to a great extent by the success of the Hudson's Bay Company in the fur trade. The population of New France had reached 55,000 by 1754. Four years later Quebec fell to the British, followed by Montreal in 1760. Thereafter Canada remained a part of the British Empire until attaining dominion status in 1867.

Throughout the century and a half that France had been laboring at the development of the St. Lawrence Valley, the Hudson's Bay Company, entering through the back door of Hudson Bay, had built a vast backwoods empire. Two pioneering Frenchmen, having failed to interest their own government, after several years of promotional effort sold Charles II of England on the plan of financing a major fur-trading enterprise. The operation proposed was to exploit inland Canada by way of Hudson Bay and thus gain access to the vast virgin fur-trading area on the south and west. In 1670, Charles II granted a royal charter "to the Governors and Company of Adventurers of England Trading into Hudson's Bay," which made them "the true and absolute lords and proprietors of the plains, lakes, forests, and mountains" of western Canada. The "Company of Adventurers" was made up of eighteen stolid English merchants, squires, and army officers, who contributed their share of the finances necessary to get the project under way but never saw a trap or skinned a beaver.

Initially the Hudson's Bay Company was the two Frenchmen, Médart Chouart and Pierre Espirit Radisson, who had interested the king of England in the project. Strategically located trading posts were established from Hudson Bay south to the Great Lakes and west to the Pacific Coast. The early governors of the company saw the necessity for the Indian population to continue in its normal way of life in order that the supply of furs would not diminish. Trading relations were kept on a friendly basis, and the Indians were protected from the white man's aggression to an extent unequaled elsewhere on the continent.

The success of the Hudson's Bay Company eventually led to the establishment of its major rival, the North West Company. The inspiration for this organization was the individual trader working out of New France who was granted a "permission" by the local authorities to carry on his operation in a specified area. This led to rivalries, shameless exploitation of the Indians, and finally hostility on their part. These freebooting traders and trappers were grubstaked by the merchants who bought their pelts. When they left for the woods, they were furnished supplies to be paid for when they returned with the season's take of furs.

The outfitters and merchants in Montreal saw their fur trade diminishing because of the rivalries among the lone trappers and also the success of the organized trading operations of the Hudson's Bay Company. In 1783 a group of Montreal merchants banded together and formed what they called "The North West Company." This company was not a corporation and was never chartered; it is best described as a loose, informal partnership. "Permissions" were obtained for the traders whom they financed, and a degree of order was brought to bartering procedures with the Indians. A unified control over the operations was effected as they expanded westward. Soon after its organization, the explorer Alexander McKenzie became the guiding light of the partnership.

Inevitably a confrontation of the two fur-trading giants occurred. It began in the Red River of the North Basin, which now forms the border of North Dakota and Minnesota before entering Canada. The region was a long-established stronghold of the Hudson's Bay Company. The conflict spread from there through the Northwest Territory. Although the Hudson's Bay Company by its charter had exclusive rights throughout the region, there was no government police force to protect it. Armed conflict between men of the North West Company and the Hudson's Bay Company frequently occurred with the North West men often gaining the advantage.

Eventually both companies realized the burden of continuing their operations under such conditions. Legal authority began to carry more weight after the opening of the nineteenth century, and there the Hudson's Bay Company had a distinct advantage. In 1821 the North West Company merged with the Hudson's Bay Company, which then came into complete control of all of Canada west of Ontario.

In 1791, Canada was granted representative government under a governor general and divided into two provinces: Upper Canada, which was essentially Ontario and British, and Lower Canada, which was Quebec and French. Nova Scotia, which at that time included both New Brunswick and Prince Edward Island, had become British territory in 1713. The Province of Prince Edward Island was separated from Nova Scotia in 1769, and New Brunswick obtained separate provincial status in 1784. Cape Breton Island, off the northeast coast of Nova Scotia, was ceded by the French to Britain in 1758 and became a part of Nova Scotia. British Columbia had been British territory since 1791. Vancouver Island became a crown colony in 1849 and British Columbia in 1858; they were subsequently united, and in 1871 became the Province of British Columbia.

The separation of Canada from Britain dates from the Act of Confederation of July 1, 1867. The provinces of New Brunswick, Nova Scotia, Ontario, and Quebec united in the federation at that time. British Columbia, Prince Edward Island, and Newfoundland and Labrador did not enter the union until later. All the rest of Canada remained as the Northwest Territory with the Hudson's Bay Company in control. After the company surrendered its territory, Manitoba was separated as a province in 1870; Saskatchewan and Alberta in 1905.

After confederation the Hudson's Bay Company came to occupy the position of an empire within an empire. The company's authority was unquestioned even by government. The extent to which its power was recognized was illustrated when the Brtish government was arranging to aid the construction of the railroad across Canada. The Crown requested the Hudson's Bay Company in 1862 to grant a right-of-way

Canada

for the project, which the company reluctantly did. However, the days of the all-powerful giant, which had contributed so much to the westward movement of civilization, were numbered. In 1869 the Hudson's Bay Company agreed to "surrender" its vast territory to Canada. The company, now called "the Bay," continued on and became the mercantile giant of Canada as well as something of an early conglomerate. Because of the land acquisitions the company fell heir to in the surrender negotiations, its financial position was well secured.

During the years of its domination the Hudson's Bay Company had effectively maintained peaceful conditions in its territory through its own efforts, for it exercised the only police power. The vacuum in authority which resulted when the company surrendered its domain produced some incidents of lawlessness. They were soon eliminated by the establishment in 1873 of the Northwest Mounted Police. Except for that brief interval, there was a continuity of law and order throughout the period of development, when western Canada was expanding from a wilderness to its present stage in agriculture and industry.

There is a deep undercurrent in the relationship between Canada and the United States which is much stronger than temporal political activities sometimes indicate. Canada is the one country where the United States citizen can cross an international border hardly realizing that he has done so. In many walks of life the parallelism between the Canadian and American way of doing things appears so natural that it goes unnoticed. This is particularly true of agricultural pursuits, including many aspects of the cattle and dairying industries. Variations in the cattle populations of the two countries follow the same pattern.

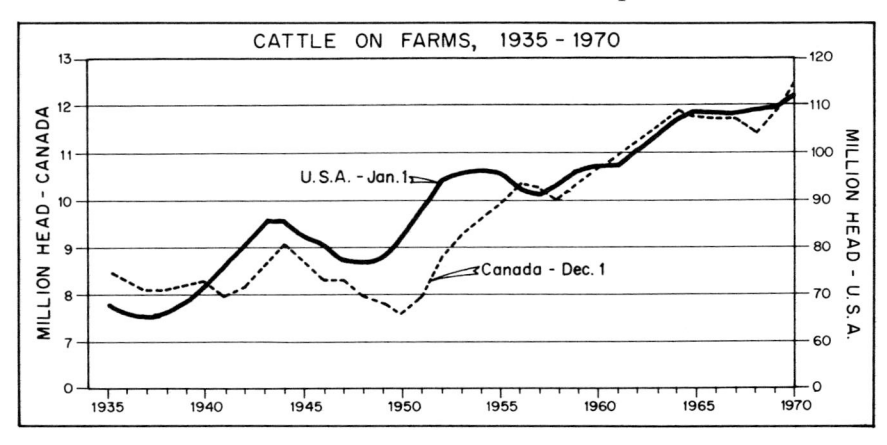

Comparison of cattle populations of Canada and the United States. Courtesy of Canada Department of Agriculture.

Cattle of North America

Canada's vast mineral wealth has been a major factor in its rapid advancement from an agricultural to an industrial economy. The development of oil production that accelerated rapidly after World War II advanced Canada to one of the major oil countries of the world.

CATTLE BREEDS

Canada is the home of the only breed of cattle that was brought to the Western Hemisphere as an old landrace and then husbanded as a domestic animal and developed into a breed. Canadian Cattle are the direct descendants of the prebreed cattle that were brought from Brittany and Normandy in France to the St. Lawrence area in the seventeenth century. Formerly called French Canadian, the name was officially changed to Canadian Cattle in 1930. As here used, Canadian Cattle refers to this breed only. Bovines entering generally into the Canadian picture are designated as "cattle of Canada" or by some similar term. For three and a half centuries the hardy Canadian Cattle have withstood the hazards of upgrading and breeding out by more fashionable breeds as these have reached Canadian shores.

The early cattle which the Spaniards introduced to the West Indies, and which soon reached both North and South America, have been bred out except for a few pockets and some isolated selections for type. The predecessors of the Devon cattle of England and the other early prebreed types that were brought from Europe to the British colonies eventually disappeared in the mainstream of United States Native cattle as these emerged during the first half of the nineteenth century. The Canadian Cattle, however, are direct descendants of the early cattle of Brittany and Normandy and as pure a strain as the Jersey and Guernsey on the Channel Islands.

During the exodus of Loyalists from the British colonies at the time of the American Revolution, cattle of mixed antecedents were taken to Nova Scotia and New Brunswick as these emigrants sought to establish themselves in their new homeland. There was certainly interbreeding of these cattle with the Canadian Cattle that were already in the area. This introduction of cattle from the United States continued into the early years of the nineteenth century with the importations consisting of Native cattle. The principal concentration of Canadian Cattle in Quebec, however, continued to be husbanded in a pure state by the French farmers.

The Hudson's Bay Company made various attempts to introduce cattle to its outposts in Canada. They were initially used for provisioning purposes for its stations, but eventually the company's herds helped supply settlers with a nucleus for their cattle holdings. There is record of some cattle being taken to York Factory on Hudson Bay in 1693, but they probably did not survive. In 1823 the company sponsored a drive of 300 head of Native cattle from the United States to the Red River of the North country. Many of them found their way into the hands of settlers in that area. At about the same time two cows and a bull were taken to Fort Cumberland in Saskatchewan. The company also took cattle from the United States to its fort on Vancouver Island in the 1840's. Some dairy herds were eventually established from this nucleus, which came from California and descended from Spanish cattle. By the middle of the nineteenth century all the forts of the Hudson's Bay Company were stocked with cattle.

The earliest recorded drive of cattle from the United States to Canada took place in 1822, when a French Canadian took a herd from Mississippi to the Red River country. This accounted for a cattle population of 93 head reported to be in the Winnipeg and Portage la Prairie settlements at the end of that year. This movement of cattle to the area appears to have been independent of that of the Hudson's Bay Company, which arrived a year later.

The gold rush to the Cariboo and Fraser River country in British Columbia was the occasion for cattle drives to the region to supply meat to the mushrooming population. There is record of 500 head of cattle being driven up from Oregon in 1863 and another movement of 300 head in 1865. There undoubtedly were other drives of smaller scale during the peak period of prospecting.

When the gold excitement faded, most of the population disappeared, and the market for cattle was lost. The stock, left in the hands of the pioneers who remained, thrived, and by 1876 there was a sizable increase in cattle numbers.

Thaddeus Harper, one of the more enterprising early stockmen, had heard of the fabulous cattle market in Chicago and organized the first cattle drive out of Canada with that market as the ultimate destination. Starting from the Fraser River country with 800 head, he picked up another 400 on the way. They were to be driven to Salt Lake City, Utah, and then taken by rail to Chicago. En route Harper learned there was a good market for cattle in San Francisco, and delivery was finally made to that point. Subsequently there were a few other movements from this pocket of cattle along the Fraser River to other United States markets.

This early backflow of cattle from the wilds of western Canada to the United States was far from representative of the trends of that day. Cattle moving south could almost have met United States herds moving north to stock the plains of the Prairie Provinces. The end of the trail of the Texas Longhorns was the Bow River country on the prairies of Alberta. In 1879 about 1,000 head of range breeding stock were driven from Miles City, Montana, to Alberta. In the years that followed, there were other substantial drives of cows to Canada. One deal called for 3,000 head to be delivered at the United States–Canadian border for $16 a head.

The earliest of these cattle movements contained representatives of straight Longhorns, but the later drives were made up mostly of crosses of Shorthorn and Hereford bulls on Longhorn cows. Thus a strong representation of the Longhorn found its way into the early herds of cattle in the Prairie Provinces.

The development and expansion of cattle breeds in Canada in many respects paralleled what occurred in the United States. Because of the unique similarity of the two countries in environment, history, and human activities, the story of the cattle industry in Canada automatically invites a comparison with that in the United States.

The major breeds which have contributed to the cattle population of the two countries have been the same, and selection for desired characteristics have generally proceeded along similar lines. The "Dutch" Black and White cattle were introduced into the United States in 1852 and into Canada in 1881. In both countries they were selected for milk production and developed into specialized dairy-type animals instead of the dual-purpose, milk-beef type that is still the goal of European breeders. Thus the Holstein-Friesian is now the major dairy breed in both Canada and the United States.

Among the beef breeds of Canada, as of the United States, the Hereford is the most popular, with the Angus next. The Shorthorn is now fourth, having been displaced from third place in recent years by the Charolais.

In the introduction of the new breeds, Canada has been able to move faster than the United States because of its ability to import live animals. The maximum-security quarantine station on Grosse Île in the St. Lawrence River was supplemented in 1969 by a second unit on the French island of St. Pierre off the coast of Newfoundland. Both facilities now enable Canada to import cattle from such European countries as are approved by the Canada Department of Agriculture.

USDA health regulations prohibit the direct importation of cattle to the United States from any country which has not been declared free of foot-and-mouth disease by the Animal and Plant Health Service. Live animals which have passed the Canadian quarantine stations for free movement in Canada are, however, eligible to enter the United States. This same rule applied to semen from imported bulls. In the late 1960's this situation led to a very remunerative export trade in semen of European beef and dual-purpose breeds to the United States from Canada.

When the Grosse Île quarantine station was commissioned in 1965, the demand for Charolais cattle was at its peak in the United States. Efforts of buyers to pass cattle through this quarantine station and then reship them to the United States were soon countered by an embargo by the Canada Department of Industry, Trade and Commerce. There has been no restriction, however, on the exportation of semen.

The tremendous demand for Charolais semen in the United States soon led to the introduction of other European breeds. Pie Rouge de l'Est (Simmental) cattle from France and Simmental from Switzerland were brought to Canada in 1966; Limousin, from France in 1968; and Maine-Anjou, also from France, in 1970. In 1971 the Fleckvieh (German Simmental), the Italian Chianina, the Gelbvieh (German Yellow), and others were awaiting admission to the mainland from the quarantine stations.

Various terms have been employed to designate these recently imported breeds, such as "exotics," "continentals," and "new breeds." All these terms are open to valid objections, but "new breeds" will be used here and should be considered as a contraction of "breeds new to the North American continent."

The Canada Department of Agriculture has maintained a strong interest in the new breeds and has purchased many representatives for work at government stations. The major incentive, however, for the rapid expansion in imports has been the marked propensity of cattlemen in the United States to breed to any cattle for which the promoters hold out a promise of increased growth rate and efficiency. The major introductions of new breeds to the United States through 1971 were the results of importation of semen from Canada.

Presumably the Canadian government will eventually issue export permits for live animals which have come through the Grosse Ile and St. Pierre quarantine stations. When this is done, it should widen the potential of the new breeds. Presently progeny of the new breeds in Canada and the United States have a somewhat narrow genetic background. Most of these animals have been sired by a relatively limited number of imported bulls.

The now well established British and European breeds were introduced to Canada during the nineteenth century, in most cases a number of years after the breeds had become established in the United

States. Importations to Canada, however, were usually direct from the homeland.

There were two basic reasons why the initial importations of cattle came from across the Atlantic rather than from the United States: transportation was more easily arranged from England, and the early Canadian cattlemen wanted their stock from the fountainhead instead of from what might be an inferior and possibly diluted source in the United States. Transportation in both countries was developed from east to west, and it would have been difficult, and in some cases impossible, to have arranged for a shipment of cattle from an inland point in the United States to an inland destination in Canada. Although the distances were greater, it was much easier to bring cattle from England by ship down the St. Lawrence River and then overland to a Canadian destination than to arrange a rail shipment from the United States. The Holstein-Friesian, however, was an exception in the general pattern of the introduction of purebreds. The foundation herds of this breed were based mainly on importations from the United States.

During the last quarter of the nineteenth century and well into the twentieth, representatives of the British breeds were taken to Canada by men of means who had developed an interest in the livestock sector of agriculture. An import from the mother country carried more prestige than an animal from the States. This attitude has accounted for the meager movement of purebred breeding stock from the United States until recently. There is now considerable movement of registered cattle in both directions across the border, as breeders see particular characters they desire in animals on the other side.

The individual breeds of cattle in Canada are discussed under the following headings: "Dairy Breeds," "Beef Breeds," "Canadian Developed Breeds," "New Breeds," and "Other Breeds."

Dairy Breeds

Canadian (French Canadian).—Canadian Cattle, formerly known as French Canadian, are the oldest breed of cattle in the Western Hemisphere. They are the descendants of the old landrace that was indigenous to Brittany and Normandy in France. Cattle from this part of France were brought to Sable Island off the eastern coast of Nova Scotia in 1518 by Baron de Lery in the first attempt by France to establish a foothold in North America. The effort failed, but the cattle are reported to have become feral and multiplied. In the early seventeenth century an unsuccessful attempt was made by the British colonists on the

Atlantic Coast to capture the wild cattle on Sable Island. Eventually all trace of them was lost.

In 1541, Jacques Cartier, the first explorer and settler in the St. Lawrence area, made his third voyage to the settlement which he had founded in what is now the Province of Quebec. On this passage he brought a number of cattle that had been loaded at St.-Malo in Brittany. There is also a record of a cargo from Brittany being brought to the St. Lawrence area in 1601. It is not known whether any progeny of these earliest introductions survived. Regular importations of cattle from Normandy were made by Samuel de Champlain from 1608 to 1610. The French settlements on the St. Lawrence then had a slow but substantial development, and it is certain that the descendants of these cattle furnished the pioneers with draft power, milk, and meat and spread throughout the region as farming expanded. In 1660, King Louis XIV directed his minister Jean Colbert to send to New France some of the best cows from both Normandy and Brittany, but the numbers were small. There is record of a total cattle population of 3,107 head in Canada in 1667, all of which appear to have originated from the animals brought to the settlements from northwestern France.

These indigenous cattle of Brittany and Normandy also furnished the base stock which founded the Jersey and Guernsey breeds on the Channel Islands. Natural and artificial selection has resulted in some differentiation of characters between the Canadian Cattle and the Jersey and Guernsey breeds, but there are marked similarities. This is particularly noticeable when the Canadian and the Jersey are compared. While the hardiness of the Canadian and its ability to utilize coarse forage exceeds that of the Jersey, the two breeds are much alike in most physical characteristics.

Down to the early 1800's the Canadian Cattle were the major bovine representatives in Canada other than the native bison of the western plains. For more than three and one-half centuries in what is now Quebec Province these cattle had been subjected to the harsh environment and bitter winters, to say nothing of the poor forage during much of the year. Survival of the fittest was supplemented by some elementary selection for milk production as farmers retained their better-yielding cows for breeding. There also must have been a decided preference for the black color which has predominated in Canadian Cattle for many generations.

Most Canadian Cattle today are black but shading to brown on some individuals. There are also solid-brown cows, although a brown-colored bull is rare. Black bulls are preferred. The occasional brown

Canadian bull, four years old, 1,700 pounds. Experimental Farm, Deschambault, Quebec.

cow approaches the fawn color of a dark Jersey. Nearly all Canadian Cattle are now dehorned, but before dehorning became the common practice, the horns were upspread and curled forward or inward at the ends, similar to those of the Guernsey. The average cow in a good herd weighs around 1,100 pounds. Bulls weigh 1,500 to 1,900 pounds. These are about the same liveweights as were recorded in 1934 and appear to be the result of selection for a larger animal as well as improved nutrition owing to better feeding practices. The first herdbook of the Canadian breed, published in 1909, gives the weight of cows as "from 700 to 900 pounds" and that of bulls as "about 1,400 pounds." "Brown or faun" is specified as the preferred color, but it also states that "black is not to be discriminated against." The preference for the brown color at that time seems to have been a temporary fashion, for most earlier references to color mention black as predominating, usually a solid black, with some browns appearing. Canadian Cattle were often

Mature Canadian cow, 1,050 pounds. St. Marc, Quebec.

called "Black Jerseys" and also "Black Canadians." Calves at birth are brown but usually turn black or a darker brown after a few months.

The conformation of the Canadian is the dairy type, intermediate between that of the typical Jersey and Guernsey. There is a marked similarity in the conformation of the large-type Jersey now being developed in the United States and the Canadian cow in the better herds. Milk production of the Canadian is comparable to that of the Jersey in Canada. The better herds average 11,000 pounds of milk for 305 days. The butterfat content, however, is somewhat lower than that of the Jersey, averaging around 4.5 per cent. The average milk production for the breed is given as 7,800 pounds. High-producing individuals yield over 17,000 pounds in a lactation.

The Acadians took their Canadian Cattle with them when they emigrated to Louisiana. They had also been trailed into the Mississippi Valley during the early French explorations in that region. These early introductions eventually disappeared, but the breed reappeared in the United States at the beginning of the twentieth century. At this time some of the breed (then still known as French Canadian) were brought to the northeastern states. Herds were established in New York, Pennsylvania, Massachusetts, and Missouri, and an association was formed, but eventually the breed faded from the American scene.

In the middle of the nineteenth century the Quebec Provincial Board of Agriculture made a concentrated effort to breed out the Canadian strain. Some Ayrshire cattle had been imported in 1821 and became popular with breeders who had diversionary interest in cattle. Considerable progress was made by these owners in upgrading their own herds to Ayrshire, and in 1883 the Council of Agriculture declared there was "no longer a Canadian breed." This appears to have been essentially a political attitude dictated by the influential Ayrshire breeders. The small farmers, who accounted for most of the agricultural production in Quebec, had kept their Canadian Cattle pure not only because of loyalty but because the Canadian cow outproduced the unacclimated Ayrshire on the limited feed available. At the same time as the decree of the Canadian Council of Agriculture, other sources reported that 85 per cent of the cattle in Quebec were of the Canadian breed.

An organized effort to keep the Canadian pure was initiated in 1880, and in 1886 the legislature set up rules for the registration of the pure breed. The foundation herdbook was opened the same year. The Canadian Cattle Breeders Association, whose name was subsequently changed to Canadian Cattle Breeders Society (Granby, Quebec), was founded in 1895, and the perpetuation of the breed was assured. In retrospect it seems that little if any Ayrshire influence ever entered the Canadian breed.

Selection of the Canadian cow for milk production paralleled that of the other dairy breeds in Canada. For nearly three centuries after they were first brought to the St. Lawrence area, these cattle were of prime importance to their owners as draft animals. However, the milk produced was a welcome addition to the diet; and after its days of usefulness were over, an animal was butchered for home consumption. The farms of the Quebec settlers were small; and the Canadian oxen, of moderate size but hardy, were ideal draft animals for the small fields.

During the latter part of the nineteenth century milk production became of commercial importance in Canada, initially for butter and cheese and later for fluid milk. Comparison with the Ayrshire increased interest in both the quality and the quantity of milk which the Canadian cow produced. In 1886 competitions were organized for the production of butter and milk from individual cows, and in 1925 the Canadian Breed Association established a record of performance. Herd production averages, based on official tests, were recorded, just as was done for the other dairy breeds.

At the Pan-American Exhibition in Buffalo, New York, in 1901, a six-month test of five of the best cows of nine dairy breeds placed the Canadian first in profit derived from the amount of butter produced per dollar of feed consumed. The supporting data for this award were translated into the following interesting comparison, as written up in the No. 1 Herd Book of the Canadian Cattle Breed:

Breed	Value of Feed Eaten	Number of Cows Which Could Be Kept on Feed Required by 100 Holsteins
Canadian	$113.10	146
Dutch Belted	132.32	124
Guernsey	136.99	120
Jersey	137.78	120
Red Poll	138.03	119
Ayrshire	140.98	117
Brown Swiss	147.26	112
Shorthorn	162.12	102
Holstein	164.69	100

It is of more than passing interest that here, 70 years ago, was a realistic endeavor to evaluate the productive ability of various breeds, and the Canadian cattle placed first.

The Provincial Government of Quebec, which might well be expected to be the chief guardian of Canadian Cattle, has at its experiment farm at Deschambault what is reportedly the oldest and best herd of the breed in existence. In 1971, however, plans were underway to cross Brown Swiss bulls on the farm's cow herd. The objective is to obtain a larger animal and to increase milk production. While some F_1 calves were on the ground, the future breeding program had not been defined. It would be lamentable if this small, irreplaceable gene pool, derived directly from one of the old European landraces, should be bred out of existence.

Ayrshire.—Early Scottish settlers of 1625 on the lower St. Lawrence River brought their cattle with them from southwestern Scotland. These were probably the prebreed stock from which the Ayrshire was subsequently developed in the county of Ayr. These early Scottish imports were soon replaced by the more prepotent Canadian Cattle which had previously arrived in the same area.

Some 200 years later, in 1821, the governor general, Lord Dalhoosie, imported a few head of the early Ayrshires from Scotland. There followed scattered importations over the next 30 years, but the heavy infusion of the breed came in the 1850's. In the succeeding years Ayrshire herds in Quebec gained popularity with prosperous businessmen who had a leaning toward agricultural pursuits. A number of Ayrshire herds were also established in Ontario.

During the years following the establishment of the breed in Quebec, a concerted effort was made to breed out the less impressive Canadian Cattle of the small farmer. Although this move failed, the Ayrshire eventually outnumbered the Canadian Cattle ten to one.

In recent years the Ayrshire has been second to the Holstein in numbers, except for a period in the 1950's when it was equaled by the Jersey. While in 1970 the Ayrshire was again safely in second place, the number of its registrations was only one-tenth that of the Holstein-Friesians registered. The Ayrshires are located in every province, though 85 per cent of them are in Quebec and Ontario.

The Record of Performance initiated by the Department of Agriculture was adopted by the Canadian Ayrshire breeders in 1905. In addition to meeting the 365-day standard for milk and butter production, a cow was required to calve within fifteen months after the beginning of her test. Bulls, to be recognized, were required to have sired four Record of Performance daughters. The selection exercised by the many small herd owners who began testing their cattle in order to qualify Record of Performance daughters did much to improve the Ayrshire breed. Many Ayrshire owners of that day were industrial and professional men, who were quicker to adopt production testing than the average farmer.

The Ayrshire Importers' and Breeders' Association of Canada was formed in 1870 in Quebec. Two years later a rival organization, the Dominion Ayrshire Association, was started in Ontario. Both organizations established herdbooks, and there was considerable antagonism between them. A particular bone of contention was the Dominion Association's policy of registering $15/16$ upgraded females and $31/32$ bulls. Several efforts at the consolidation of the two organizations were attempted, but none were successful until 1898, at which time they were

Ayrshire cow on a southern Ontario dairy farm.

combined in the Ayrshire Breeders Association of Canada (Ottawa, Ontario).

The weight of mature Ayrshire bulls in Canada is given as 1,400 to 2,000 pounds, and cows from 1,000 to 1,250 pounds. Average production per lactation is 10,148 pounds of 4 per cent butterfat milk. There are no noteworthy differences between the Canadian and United States Ayrshires other than perhaps the excellent udders seen in many Canadian herds.

Brown Swiss.—The first representatives of the Brown Swiss breed were taken to the eastern townships of Quebec Province from the United States around 1888. There were subsequently other small importations from time to time, and the Brown Swiss made some progress until 1945. Interest in the breed then seems to have disappeared but was renewed with the reorganization of the Canadian Brown Swiss Association in the early 1960's. Registrations of purebred cattle increased from 580 in 1966 to 741 in 1970. While these numbers are small, it may be significant that while the other dairy breeds, except the Holstein-Friesian, were declining, the Brown Swiss was increasing. Since 1968 three Brown Swiss bulls have been imported from Switzerland. Most of the Brown Swiss cattle are in the provinces of Ontario, Alberta, and Manitoba.

The Canadian Brown Swiss Association (Brighton, Ontario) was organized in 1914 and incorporated in that year under the Live Stock Pedigree Act. The organization was inactive from 1945 until 1963, when it was revived. In 1971 there were 118 members.

Guernsey.—The Guernsey was first brought to Canada in 1878 when Sir John Abbott, of Montreal, imported a number of head direct from the Guernsey. Interest in the breed developed in the Maritime Provinces, and the next importations moved to Nova Scotia from the United States. Additional shipments were made from both Guernsey and the United States. The Guernsey did not become as popular in Canada as the Jersey, and in recent years, Guernsey registrations of purebred cattle have run at roughly half of those of the Jersey breed.

For many years the Maritimes were considered the homeland of the Guernsey in Canada. Now the largest representation of the breed is in Ontario, where 70 per cent of the purebred registrations originated in 1970. Guernsey herds were established in British Columbia in the early years of the twentieth century and are still maintained there in considerable numbers.

The weight of mature Guernsey bulls is 1,400 to 1,800 pounds, and mature cows range from 1,000 to 1,250 pounds. The average milk production of mature cows is given as 9,195 pounds with 4.8 per cent butterfat content.

The Canadian Guernsey Breeders Association (Guelph, Ontario) was founded in 1905. The association adopted production testing in 1909 and has kept pace with the other purebred breed organizations in its progressive attitude toward artificial insemination and modern breed-improvement practices.

Holstein-Friesian.—The Holstein-Friesian was first brought to Canada in 1881 from the United States. In that year a purebred cow and a bull were purchased by Archibald Wright and taken to Fort Garry (now Winnipeg), Manitoba. Two years later a hundred head of Friesian cattle were imported directly from the Netherlands and distributed in the Province of Ontario. There were subsequent shipments from the Netherlands, but from the records it appears that the major source of the Holstein-Friesian national herd was the United States. The influx took place during the latter part of the nineteenth century.

The early development of the Holstein-Friesian in Canada was by the white-collar farmers, who were intrigued more by the novelty of this new breed than by its productive ability. At the time the Holstein was introduced, draft power was still a major function for cattle with the dirt farmer. Such a farmer was not interested in an animal without knowing what its capability would be under the yoke. Also, the consumer was oriented to the rich milk of the Ayrshire and Jersey, and the low fat content of Holstein milk was the subject of considerable ridicule. A common comment of the time was that it was so thin it would rust the bottom of the milk pail.

It was only the perseverance of the Holstein breeders and their influence in agricultural circles that eventually won wide recognition for the breed. The cattle gained a prominent place in the shows of the day, and the large milk yields drew increased attention as the demand for fluid milk grew in the towns and cities.

The Holstein-Friesian breeders were among the first of the Canadian dairymen to employ production records for the improvement of their breed. There was widespread use of production testing for both milk and butter yields in Ontario herds during the last years of the nineteenth century. Then, in 1901, the Canadian Holstein-Friesian Association adopted a Record of Merit system. This was a noteworthy step in the use of production records for the selection of breeding stock. A two-year-old heifer was required to produce 8 pounds of butter in one week, a mature cow 14 pounds, in order to qualify for a Record of Merit. A bull qualified when he had four Record of Merit daughters. These standards were eventually changed to those adopted by the Canada Department of Agriculture for 305- or 365-day production of milk and butterfat. The Holstein breeders themselves, however, had taken the initial step in selection based on production records.

Throughout the period when production testing was gaining acceptance, the breed association continued to register any animal offered solely on the basis of pedigree. There was criticism of production testing by many purebred breeders, who claimed that selection based solely on these criteria was destroying the true breed character. To advance both type and performance, the Association adopted an Advanced Registry for bulls in 1925 and for cows in 1927. This registry required an animal to meet a minimum conformation standard on a scale of points, as well as a minimum production standard. Bulls had to have progeny which would also meet both standards.

The Holstein in Canada has developed independently but along similar lines to her sister in the United States. Both have arrived at comparable high levels of production. Variables in environment and nutrition make it difficult to place the representatives of the breed in one country ahead of those in the other. This statement, though, would be strongly rebutted by Canadian breeders, who feel that they have developed a better "milking machine." It should be noted here that for several decades there has been interchange of high-producing breeding stock between the two countries.

The similarity of the Canadian and United States Holstein is obvious, although the Canadian animal tends to have a more rounded and less angular rump.

In 1971 there were more Holstein-Friesian cattle in Canada than those of any other breed, beef or dairy. The national dairy herd was 80 per cent Holstein-Friesian. The breed is well represented throughout the country, though its greatest stronghold is in Ontario, where the herds are concentrated in the main agricultural areas of the southern part of the province. The Canadian Holstein-Friesian has gained an enviable reputation in many countries of the world, and live cattle as well as semen are widely exported. Although the United States continues to receive more than one-half of the purebred exports, Mexico, Central America, and the Caribbean area all take sizable numbers.

The Holstein-Friesian Association of Canada (Brantford, Ontario) was formed in 1884, but a Canadian herdbook was not established until 1891. Before that year Canadian breeders registered their animals in the United States herdbook. The association has always taken a progressive attitude toward breed-improvement programs. As mentioned, it was the first in Canada to adopt production testing. It was also a leader in the acceptance of artificial insemination.

Red and White Holstein-Friesian cattle were allowed registry in the Alternate Herd Book in 1969. Starting in 1971 black-and-white offspring with a red parent could be registered in the regular herdbook.

The Holstein-Friesian Association of Canada elected to perform the registration function for their breed themselves instead of having this work done by the National Live Stock Records (see the section "Government and Cattle" below).

Jersey.—The first recorded Canadian importation of purebred Jersey cattle was in 1868 by R. H. Stephens to St. Lambert, Quebec. The shipment consisted of 2 bulls and 15 cows which originated in the royal herd at Windsor in England. Three years later another bull and 2 cows were added to the herd. This constituted the founding of the St. Lambert herd, which was for many years the most prominent nucleus of Jersey cattle in Canada. In 1871 a few Jersey cows and bulls were taken from the United States to New Brunswick. There were subsequent importations from both Jersey and the United States over the next three decades.

A private herdbook for the registration of purebred Jersey cattle was established in New Brunswick in the 1870's. During the early years of the breed, however, most purebred breeders recorded their animals in the herdbook of the American Jersey Cattle Club in the United States. The Canadian Jersey Cattle Club (Toronto, Ontario) was organized in 1901. There were 500 members by 1920 and over 1,600 in 1940.

Quebec and New Brunswick were for many years the homeland of the Jerseys in Canada. They gradually spread to the other Maritime Provinces and in later years to Ontario and, in a limited way, to the western provinces. Ontario now has the largest concentration of Jersey cattle, over 60 per cent of the registrations being in that province.

The Jersey population is declining somewhat more rapidly than the decrease in total dairy-cow numbers. The peak of Jersey registrations was in 1950, when nearly 15,000 were recorded. In 1970 only 7,383 animals were recorded.

Apparently no effort has been made to increase the size of the Canadian Jersey cows in recent years as has been done in the United States. Bulls in the artificial-breeding centers usually range in weight from 1,250 to 1,500 pounds. Cows in representative herds weigh 800 to 1,000 pounds. Other than this difference in size, the Canadian Jersey has the same general characters as the breed in the United States. The average milk production of purebred herds is given as 8,500 pounds of 5.2 per cent butterfat content.

Red Poll.—Representatives of the Red Poll breed were imported from England in the early 1880's. There is record of a small herd in the possession of the government of New Brunswick in 1883. Some of these cattle eventually entered private herds. During the 1890's the Ontario Agricultural College brought over a few Red Poll cattle from England, and there were a few importations from the United States. After the turn of the century purebred herds were established in the Prairie Provinces from stock purchased from the United States. Some years later the breed reached British Columbia.

The Red Poll breed has been maintained pure in Canada continuously since its introduction. Although widely scattered over the country, it never became very popular. In the hands of small farmers herds usually numbering only a few head were held for milk production, and surplus stock was sold for slaughter. In the years following World War II there was new interest in the breed, and registrations doubled over what they had previously been. Now this trend has been reversed. In 1970 there were only 206 registrations of purebred Red Polls as compared with over 800 annually in the 1950's.

A 14,000-pound Holstein-Friesian herd in southern Ontario.

Red Poll breeders registered their cattle in the United States herdbook until the enactment of the Live Stock Pedigree Act in 1905. In that year the Canadian Red Poll Association (Francis, Saskatchewan) was organized, and registrations have since been handled by the Canadian National Live Stock Records.

Beef Breeds

Aberdeen-Angus.—Some of the prebreed cattle from which the Aberdeen-Angus was developed may have been taken to Nova Scotia when Scottish emigrants settled the province. There is a record of a purebred bull and heifer being shipped to Quebec in 1860 after the breed had been well established in Scotland. Neither of these introductions, however, left any permanent trace. Continuity of the Angus breed in Canada started in 1876 with the importation of a bull and two cows from Scotland by the Ontario Agricultural college at Guelph. The shipment was arranged by one Professor Brown, a Scotsman who was familiar with the breed in its homeland and had selected the animals from the best stock available. Two more bulls and two cows were selected by Brown during the next four years and added to the Guelph herd.

This nucleus of Aberdeen-Angus attracted wide attention among cattlemen in the United States, as well as in Canada. Record prices were

Angus cow and calf in a Peace River herd of cows averaging 1,200 pounds.

paid for some of the progeny of this herd that were exported to the United States. There were many other importations from Scotland between 1879 and 1886, and from the Guelph herd and these other importations the breed became well established throughout Ontario, the rest of eastern Canada, and, later, in the western Province of Alberta. Importations of purebred stock from Scotland continued well into the twentieth century. In the early 1900's the registrations of Aberdeen Angus exceeded those of Herefords. As the whiteface cattle became popular on the western range, however, they soon outnumbered the Angus. The current ratio is around three Herefords to one Angus on a country-wide basis.

By the late 1880's, Angus cattle were being introduced into the western provinces. Purebred herds were established first in Alberta, and from there the breed was carried to Manitoba. Interest was confined mostly to registered cattle, and it was well after the turn of the century before Aberdeen-Angus bulls were used to any great extent in upgrading the range herds of the Prairie Provinces.

The Angus in Canada conform to the typical character of the breed in other countries. In many of the purebred herds they suffered loss in size in the years after World War II as selection for the compact type followed the trend prevalent in both Scotland and the United States. The commercial herds also eventually suffered in the same manner. In recent years there has been a definite reversal of this trend, and good herds of large purebred Angus cattle now exist. Some of these herds have cows averaging 1,200 pounds. Commercial herds have also improved in scale, although in many herds, both purebred and commercial, 900- to 1,000-pound cows are still representative.

The Canadian Aberdeen-Angus Association (Guelph, Ontario) was organized in 1906. Before that, starting in 1885, purebred cattle were recorded in what was known as the Dominion Polled Angus Herd Register, but these records were later destroyed by fire. At the time of the organization of the Aberdeen-Angus Association only those animals passing an inspection for set standards of quality were accepted for registration. Registration procedures are now handled by the Canadian National Live Stock Records.

Charolais.—The first cattle of Charolais breeding in Canada were a ⅞ bull and a ¾ cow taken from the United States to Spring Coulee, Alberta, in the spring of 1953 by Wayne and Max Malmberg. There were other subsequent importations of percentage Charolais from the States over the next 12 years, but the real establishment of the breed in Canada began in 1966. In May of that year 30 bull calves and 79 heifer calves, which had been shipped from France, were released from the maximum-security quarantine station established the previous year on Grosse Ile in the St. Lawrence River. All the 1966 importations were calves under eight months of age at the time of selection in their homeland, and, as required by Canadian law, they had not been vaccinated against foot-and-mouth disease.

By the end of 1967 a total of 107 bull calves and 434 heifer calves of the Charolais breed had arrived. Some of the animals reached the United States before the Canada Department of Industry, Trade and Commerce prohibited their exportation. The others remained at Government experiment stations or in the hands of Canadian breeders.

There have been numerous subsequent importations of Charolais cattle from France, but because they were also calves, the 1966 and 1967 introductions comprised the principal foundation of the national Charolais herd in Canada as it existed in 1971. The number of full French Charolais cattle in the country was nearly 2,000 head by that time.

The Charolais is not considered here as one of the new breeds. The breed was introduced to the United States in the late 1930's, and it was the success story of the Charolais which led to the subsequent importation of Simmental, Limousin, Maine-Anjou, and various other breeds.

The Canadian Charolais are usually dehorned; otherwise they are representative of the large, white, well-fleshed cattle of France. On the average, however, Charolais in their homeland are more growthy and have a better beef conformation. Some of this difference may be attributed to the fact that the Canadian imports were selected as calves. The Conception to Consumer program, which was recently established by the Charolais Association in Canada (discussed below) could well effect a reversal of this comparison in the years ahead.

Charolais bulls have been widely employed in upgrading the British breeds to the point where the progeny will be officially recognized as "pure" Charolais (an animal of $^{31}/_{32}$ registered ancestry is a pure Charolais).

Limited numbers and strong demand have maintained the prices for Charolais breeding stock at fantastic levels. A market has existed for just about all but the most obviously unacceptable individuals, and quality has suffered. About the only Charolais that have found their way to the packing plant have been a few steers from experiment-station herds slaughtered to obtain carcass records.

In 1970 there were nearly 1,500 members of the Canadian Charolais Association, most of them purebred breeders. The number of herds that included cattle of Charolais breeding was many times that number.

The very profitable business of exporting Charolais semen to the United States has been a major factor in the expansion of the breed in Canada. The owner of a Charolais bull places his animal in one of the artificial-insemination centers and either leases him to the center or receives a fixed price per ampule for the semen.

The limited quarantine facilities have restricted the importation of full French Charolais and generally relegated them to exploitation by a few individuals. At the same time, this has cast an aura of glamour about the breed, stimulated the demand for the cattle, and led to the introduction of the new breeds.

Regardless of the highly profitable promotional efforts, though, the ultimate success of the Canadian Charolais will depend upon their merits as sound beef producers.

The Canadian Charolais Association (Calgary, Alberta), formed in 1960, has been a moving force in the growth of the breed in Canada. The efforts of the association followed for a time the customary lines of broad claims made by all purebred societies. Then, in February, 1968, an outstanding bull-evaluation program, labeled Conception to Consumer, was adopted.

Cattle of North America

SIRE	REG. NO.		NATURAL BIRTH %		% CALVES DIED AT BIRTH		GESTATION PERIOD		BIRTH WEIGHT		INDEX (Adj. 205) WEANING WEIGHT		INDEX A.D.G. ON FEED		INDEX (Adj. 365) YEARLING WEIGHT		INDEX LEAN GROWTH (Steers)
			M & F		M&F		M & F		M & F		M & F		M & F		M & F	
AIGLON	FMC 14	M	85.7	92.9	.0	.0	285.6	285.2	92.1	87.8	98.1	101.4	99.2	100.6	98.8	101.3	100.7
		F	100.0		.0		284.8		83.5		104.1		102.0		103.7		
AMIRAL	FMC 2	M	81.8	90.9	.0	.0	284.6	283.4	96.2	93.4	97.9	99.4	99.0	101.9	98.4	100.7	99.2
		F	100.0		.0		282.1		90.5		100.2		103.5		101.9		
AMOUR	FMC 17	M	66.7	78.8	6.7	7.9	286.7	285.6	100.9	95.9	103.7	101.6	95.7	94.4	100.4	98.2	100.4
		F	90.9		9.1		284.5		90.9		98.0		92.2		94.5		
ARTISTE	FMC 5	M	70.6	81.4	5.9	2.9	287.8	288.0	95.0	90.5	100.4	101.9	100.8	101.1	100.7	101.9	101.8
		F	92.3		.0		288.2		86.1		103.6		101.5		103.4		
BACCHUS	FMC 65	M	100.0	96.2	.0	.0	288.7	287.9	93.9	90.5	100.5	101.2	104.8	106.2	102.1	103.0	98.9
		F	92.3		.0		287.1		87.1		102.0		107.9		104.1		
BALEZE	FMC 59	M	86.4	93.2	4.5	2.3	288.5	287.7	95.8	92.2	99.7	99.5	103.4	102.2	101.1	100.4	96.5
		F	100.0		.0		286.9		88.7		99.3		100.6		99.4		
BASILE	FMC 38	M	78.9	85.3	5.3	2.6	289.8	287.8	98.0	93.4	97.1	95.3	102.1	98.8	99.2	96.5	106.2
		F	91.7		.0		285.7		88.7		92.4		94.9		93.2		
BELPHEGOR	FMC 52	M	95.2	91.0	.0	.0	286.6	284.3	92.1	90.9	102.0	103.1	103.9	104.1	103.1	103.5	104.6
		F	86.7		.0		282.0		89.7		104.6		104.4		104.1		
BERGER	FMC 61	M	93.8	96.9	6.3	3.1	284.9	284.8	92.8	88.8	100.7	100.1	95.9	97.2	98.9	99.1	100.2
		F	100.0		.0		284.6		84.9		99.2		99.1		99.3		
BEVON	MC 22	M	100.0	97.8	.0	4.3	287.5	287.0	86.9	88.2	96.9	96.5	97.9	94.7	97.3	96.4	97.2
		F	95.7		8.7		286.6		89.5		96.1		92.2		95.6		
BIEN AIME	FMC 62	M	68.8	79.1	12.5	8.9	287.7	286.7	93.8	91.8	99.0	101.4	104.3	101.3	101.2	101.3	99.1
		F	89.5		5.3		285.8		89.9		103.4		98.8		101.4		
BIGARREAU	FMC 60	M	100.0	100.0	6.3	3.1	283.7	282.9	89.4	85.4	103.7	100.2	97.8	95.1	101.0	98.3	99.0
		F	100.0		.0		282.0		81.3		96.9		92.6		95.7		
BIJOU	FMC 49	M	92.9	93.8	7.1	8.8	285.4	285.7	98.1	96.7	105.6	103.8	101.9	103.7	103.8	103.8	104.6
		F	94.7		10.5		285.9		95.2		102.4		105.1		103.8		
BOBINO	FMC 42	M	69.2	76.9	7.7	7.7	286.8	285.7	95.9	92.2	103.1	101.8	103.8	104.2	102.8	102.6	102.9
		F	84.6		7.7		284.5		88.5		100.1		104.6		102.3		
BOURGEOIS	FMC 51	M	92.3	96.2	.0	.0	285.5	285.5	92.7	93.2	99.0	100.6	100.2	106.2	100.8	102.9	93.9
		F	100.0		.0		285.4		93.6		101.7		110.1		104.2		
BOUVREUIL	FMC 66	M	88.2	94.1	5.9	2.9	286.6	287.5	93.0	90.7	98.7	99.0	96.5	96.5	98.3	98.2	100.1
		F	100.0		.0		288.4		88.3		99.3		96.5		98.1		
EL FORTIN	M 8289	M	83.3	85.0	5.6	2.8	285.2	284.2	87.3	84.0	97.6	96.0	95.5	94.6	96.5	95.3	97.4
		F	86.7		.0		283.3		80.8		94.2		93.6		93.9		
LAZY JR ARDEN 22Y	MC 891	M	100.0	100.0	.0	.0	287.5	286.4	85.0	84.4	TOO FEW PROGENY TO INDEX						
		F	100.0		.0		285.3		83.7								
LAZY JR 12Y	MC 887	M	100.0	100.0	.0	.0	288.2	286.8	87.7	85.1	96.7	97.9	98.6	100.6	96.7	97.7	93.4
		F	100.0		.0		285.4		82.5		100.5		105.8		100.3		
McGINNESS	MC 293	M	94.4	94.9	.0	2.3	284.1	282.3	93.8	86.4	102.0	101.1	100.6	98.6	101.9	100.6	105.4
		F	95.5		4.5		280.5		79.1		100.5		96.8		99.4		
OVERALL SUMMARY		M	87.3	91.1	3.8	3.3	286.7	285.7	93.3	90.3	100.0	100.0	100.0	100.0	100.0	100.0	100.0
		F	94.9		2.7		284.8		87.3		100.0		100.0		100.0		

1st 25% 2nd 25% 3rd 25% 4th 25%

Conception to Consumer program. Courtesy of the Canadian Charolais Association, Calgary, Alberta.

The program provides an evaluation based on a contemporary comparison of the bull's progeny. The Agricultural and Vocational College at Vermilion, Alberta, conducted a Pilot Test for beef sires from 1965 to 1968. This test placed the major emphasis on the records of the progeny of the sire. The current program follows lines similar to the Vermilion test but has been expanded to include parameters for calving difficulty and carcass quality on the progeny of the bulls.

To begin with, the bull must be accepted for the test by the Charolais Breed Improvement Committee. By means of artificial insemination he is then bred at random to approximately fifty cows in several participating commercial Hereford and Angus herds. Information relating to ease or difficulty of birth, gestation period, birth weight, weaning and yearling weights, average daily gain on feed, and complete carcass data on steers slaughtered at one year of age are the basic data obtained. This information is processed by computer and adjusted for breed of dam and herd differences to give an index on all the bulls in a given year's program.

The first test, 1968–70, evaluated 20 bulls; the 1969–71 test, 24 bulls; the 1970–72 test, 21 bulls. In the 1971–73 test 39 bulls are entered. A summary of the 1968–70 test is shown in the chart on the opposite page.

An outstanding feature of the Conception to Consumer program is the publication of the results on all bulls tested in the final report on the year group to which they belong. The published reports reveal the bull's genetic worth. Several high-priced, widely advertised bulls have left artificial-insemination centers as a result of the test.

With some minor exceptions, since July, 1971, no calf produced by artificial breeding has been eligible for registration unless sired by a bull which has either completed or is entered in the Conception to Consumer Program. Calves resulting from natural matings can be registered without the sire's having been tested.

The Canadian Charolais Association did not elect to come under the Canadian Live Stock Records and handles all registrations itself.

Galloway.—One of the first British beef breeds brought to Canada was the Galloway. The first importation appears to have been made in 1861 by Thomas McCrae, a Scotsman who settled in the Guelph area of Ontario. Purebred herds were then established in that province and spread to Alberta in the late 1880's. For a time considerable use was made of Galloway bulls for crossing on the western range cattle. The initial popularity of the breed, however, soon faded as the other British beef breeds came into general use for upgrading in the Prairie Provinces.

The Galloway influence in western herds was soon lost, although the breed has persisted, mainly in purebred herds. During the past fifteen years it has had a steady but small growth. The strongholds of the breed are Saskatchewan and Alberta, which accounted for 70 per cent of the 599 registrations of purebred Galloways in 1970.

The Ontario Galloway Herd Book was started in 1874 for the registration of animals bred in both Canada and the United States. In 1882 this function was taken over by the North American Galloway Breeders Association, which continued to register cattle for both countries. The United States breeders later withdrew from this organization, but it continued under the same name, functioning entirely as a Canadian entity. It was incorporated in 1905 under the Live Stock Pedigree Act. The name of the Association was eventually changed to the Canadian Galloway Association (Gull Lake, Saskatchewan).

The Canadian Live Stock Pedigree Act authorizes the recognition of only one breed when there are various types within a breed, and only one breed organization can be incorporated. For these reasons all Galloway cattle—black, dun, and even the belted type—are recorded in the one herdbook.

A few Belted Galloways were introduced to Canada about 1955 from the United States. Later there were a few shipments from Scotland. In 1971 there were said to be about 400 head of Belted Galloways in the country.

Hereford.—F. W. Stone, a Shorthorn breeder of Guelph, Ontario, introduced Hereford cattle into Canada. On a trip to England several years after he had established a purebred Shorthorn herd, Stone saw Hereford cattle in their homeland. He was so favorably impressed that he proceeded to purchase a few head to stock another farm south of Guelph. His first importation, arriving in Ontario in 1860, consisted of one bull and eight heifers. These were carefully selected stock from prominent English breeders. Subsequent importations and careful breeding resulted in the Stone herd's becoming one of the most noted sources of Herefords in North America.

A number of other purebred herds were established in eastern Canada in the following years, but the movement of the Hereford to the western provinces did not come until 20 years later, when purebred herds were established in Manitoba, Saskatchewan, Alberta, and British Columbia.

The Hereford influence spread gradually throughout the range country of the western provinces, and the takeover from the Shorthorn began in the opening years of the twentieth century. In 1906 the Shorthorns outnumbered the Herefords 9 to 1. By the end of World War II, numbers of the two breeds were about equal, but by 1970 the Herefords outnumbered the Shorthorns 7 to 1. The Hereford is the most numerous beef breed in Canada and will maintain this position for some time to come regardless of such inroads as may be made by the new breeds and their crosses.

The Polled Hereford was founded through the efforts of Mossom Boyd, of Bobcaygeon, Ontario. Polled Hereford bulls, imported from the United States in 1902, were used on horned purebred cows. These bulls were of the Single Standard, that is, the polled character had been induced by crossing with polled bulls of another breed. Boyd, however, soon went to the purebred strain of Polled Hereford, the Double Standard, obtained from polled mutation Herefords. The polled animals which Boyd used in his herd also came from the United States. The Single Standard Polled Hereford never gained a foothold in Canada, but the purebred Double Standard strain was accepted by a number of breeders and expanded rapidly during the first quarter of the twentieth century.

The Canadian Hereford has the same general characters as its counterpart in the United States. Selection was made for the compact type during the years that this fad was popular in the United States and some herds today still show the results of this negative type of selection. The emphasis of most purebred breeders for a number of years, though, has been on a larger, growthier animal. Excellent purebred herds are now seen, and the artificial-insemination centers have outstanding bulls.

The Canadian Hereford Association (Calgary, Alberta) was founded in 1890, and the first herdbook was published in 1899. Before that time purebred animals were registered in the United States herdbook.

Highland (West Highland).—Highland cattle were brought to western Canada in the early days of the cattle industry on the assumption that their hardiness could be utilized in the development of a type ideally suited to the Canadian winters. The reputation of the Highland breed for hardiness, undoubtedly emphasized by its rugged appearance, attracted the attention of a number of pioneer ranchers of the western prairies.

The first Highland cattle were imported around 1885 by Glen Lyon-Campbell to the Riding Mountain country, north of Strathclair, Manitoba. There is record of an importation to eastern Alberta in 1898 and a shipment of bulls to Lloydminster in southern Alberta in 1907. During

Highland bull, mature, 1,150 pounds. Photographed near Truro, Nova Scotia.

the early years of the breed in Canada, Highland bulls were widely used by the ranchers on their local cattle, and a considerable demand developed throughout the Prairie Provinces for the resulting crosses.

Highlands did not reach the United States until 1922, when the Walter Hill herd was founded in Montana. Representatives of this herd were subsequently sold to Canadian stockmen.

The Highland cattle of Canada have the same characteristics as those in their native Scotland. The color is brown to a light dun with an occasional brindle. The hair is long and shaggy, often nearly hiding the eyes. The horns are large, widespread, and curve upward on the cow and to a lesser degree on the bull. Highlands are slow to mature and are smaller than most British breeds. Mature cows weigh 900 to 1,050 pounds, and bulls 1,100 to 1,500 pounds. The legs are noticeably short, even allowing for the small size of the breed. The Canadian animals, however, appear to be somewhat longer in the leg than their counterparts in Scotland.

In spite of the attention the Highland attracted in the early days of the cattle industry in western Canada, the breed never became popular and is now of little economic importance. Purebred herds have been maintained, however, by their loyal supporters. The Canadian Highland Cattle Society (Duncan, British Columbia) was founded in 1964, and registrations are handled by the National Live Stock Records. There were 55 members of the breed society in 1970 and 124 registrations of purebred animals.

In 1971 the Canadian Highland Cattle Society recognized F_1 Highland cross females as eligible for registration. Such females must be the progeny of a purebred Highland sire and a registered dam of a recognized beef breed. Performance data are required on the individuals submitted for registration. A second-cross female, the progeny of any purebred beef-breed sire and an F_1 registered Highland cross, is also eligible for registration. These registrations are recorded in a separate section of the Highland Herd Book.

Santa Gertrudis.—Santa Gertrudis were introduced into Canada by J. Grant Glassco in 1952, when a few head were taken to Cold Creek Farms at Woodbridge in southern Ontario. Subsequent small importations have come from Texas and Colorado.

The Santa Gertrudis as seen in Canada have the same characteristics as the breed in the United States (see the section "Cattle Breeds" in Part IV, "The United States"). The breed organization is incorporated under the Livestock Pedigree Act but has not affiliated with the Canadian Live Stock Records. Classification of animals to be registered has been handled by Santa Gertrudis Breeders International (Kingsville, Texas). The Canadian Santa Gertrudis Association (Toronto, Ontario) was formed in 1967. Starting in 1972 classification will be carried out jointly by the Santa Gertrudis Breeders International and the Canadian organization, with the latter organization taking on classification on its own in 1973.

The breed has attracted very little attention in Canada.

Shorthorn.—Cattle carrying considerable Shorthorn breeding probably reached Nova Scotia and New Brunswick from the United States during the early years of the nineteenth century. The limited number involved, however, left no notable impression on the Canadian cattle population.

The first authenticated importation of purebred Shorthorns numbered four bulls purchased by the New Brunswick Board of Agriculture in 1825. In 1832 a Shorthorn cow and a bull calf from the United States were bought by Judge Robert Arnold and shipped to St. Catharines, Ontario. Over the next 20 years there were frequent importations of purebred Shorthorns from the British Isles and a few from the United States. The British imports, however, were more numerous and can be considered to have been the base stock of the Canadian Shorthorn.

The early Shorthorn in Canada was considered too slow and sluggish for a good draft animal. While raised principally to produce bulls for the range herds, the cows proved themselves good milkers when used in the dairy herd, and some selection of Shorthorn cows for milk production was probably made in localities where dairy products were at a premium during the early days of the breed.

Eventually both a beef type and a milk type were developed, though the line between the beef and milking Shorthorn was not rigidly drawn by the organization of separate breed societies as was the case in the United States. All Shorthorn cattle, including the polled as well as the horned beef and dairy types, come under the same breed society in Canada and are registered in the same herdbook. The major development of the breed, however, has been for beef.

During the years when the Single Standard Polled Shorthorn (see "Polled Shorthorn" in the section "Cattle Breeds" in Part IV, "The United States") was being developed in the United States, this new type failed to make much of an impression on Canadian breeders. The lack of interest was due largely to the refusal of the Canadian breed society to register a Single Standard Shorthorn since it was not 100 per cent Shorthorn. Later, when the Double Standard Shorthorn was developed in the United States from purebred parents, the polled strain gained a foothold in the Dominion. Some excellent Polled Shorthorn herds were subsequently established, particularly in Manitoba and Saskatchewan.

When cattle were introduced to the Prairie Provinces during the last quarter of the nineteenth century, the Shorthorn was the first breed used extensively for upgrading. A Shorthorn bull calf was brought to Manitoba in 1868, and five years later a sizable herd was driven from St. Paul, Minnesota, to the McKenzie Ranch, near Portage la Prairie, Manitoba. Purebred importations increased rapidly during the following years, and wide use was made of the Shorthorn bulls in upgrading the Longhorn and mixed cattle that were moving into the western provinces. Many purebred herds were also established in the eastern provinces to supply bulls to the range country. The demand for Shorthorn bulls seemed endless, and the use of home-raised grade bulls was widespread because of the scarcity of purebreds.

In the early years of the twentieth century Shorthorn registrations exceeded the Herefords 9 to 1. After World War II the Herefords overtook the Shorthorns and now greatly outnumber them. Islands of both horned and Polled Shorthorns remain in the cattle country, and there are many excellent herds of the breed. Loyal Shorthorn breeders are continuing to maintain their herds in spite of such inroads as may be made by the new breeds and the trend to crossbreeding.

The Canadian Shorthorn Association (Guelph, Ontario) was organized in 1886.

Canadian Developed Breeds

Hays Converter.—The Hays Converter is the only breed of cattle which has been developed on Canadian soil in modern times. In 1952, Senator Harry Hays, of Calgary, Alberta, a former mayor of Calgary, minister of agriculture, and a senator in the dominion government, began selecting the individuals which were utilized in developing a highly productive beef animal. The objectives were a 2,000-pound cow, a 3,000-pound bull, and a finished 12-month-old steer weighing 1,100 pounds.

A group of Hereford heifers was selected from the reputation herd of a neighboring rancher in the Turner Valley of southern Alberta. As a former dairyman, Senator Hays had been impressed with the growth potential of the Holstein-Friesian. Thus inspired, he selected 8 sons of Fond Hope (a 3,000-pound Holstein-Friesian champion) as his foundation bulls. These were from dams with exceptional udders and whose milk had a minimum butterfat content of 4 per cent. These cows were also better fleshed than the typical Holstein-Friesian. Fond Hope himself, with well-fleshed rear quarters, also tended more to the European Friesian than to the Canadian Holstein-Friesian type. The select herd of Hereford females was pasture-bred for two years to these 8 Holstein bulls. From the resulting progeny 159 Holstein-Friesian–Hereford heifers were selected as the foundation cow herd of the new breed.

The second step was a backcross to the Hereford. The 159 heifers were bred artificially to the Hereford bull Silver Prince 7P, a certified meat sire whose mature weight was 2,440 pounds. From the resulting progeny 5 of the fastest-gaining bulls were selected to go back into the herd of 159 Holstein-Hereford females.

The third step was the introduction of Brown Swiss influence. Four bulls, all from dams weighing 1,800 pounds each and which traced to the famous Brown Swiss cow Jane of Vernon, were bred to 100 selected Hereford cows. For several years the best of the resulting female progeny, ½ Brown Swiss and ½ Hereford, were introduced into the original foundation cow herd, which at the time was estimated to approximate ⅔ Hereford, ⅓ Holstein-Friesian. The highest-gaining bulls from this combined group of Hereford-Holstein and Hereford–Brown Swiss crosses were then selected and bred to the females with the best udders. Breeding continued from then on in the closed herd that was so formed,

Hays Converter bull, eight years old, 2,780 pounds. Hays Farm, Turner Valley, Alberta.

mating "the best to the best." The selection procedure followed emphasized weaning weights, yearling weights, and, especially, the udders on the replacement heifers. No attention was paid to color.

Today there is only one herd of pure Hays Converters, that of 600 brood cows on the Hays Farm in Turner Valley. In 1971 these ranged from the third to sixth generation of the Hereford-Holstein-Brown Swiss breeding. They are a large, rugged, well-fleshed type of cattle, long-bodied and with good rumps. Strong legs and good hoofs are characteristic. Particularly noticeable are the excellent udders of the cows. The body is usually black with some white markings or, less often, red and white. Mature cows in the breeding herd weigh around 1,300 pounds.

All cows have calved at two years of age and have continued to drop a calf every year. All open cows are culled on pregnancy test every fall. A mature dry cow which has lost her calf and is kept on pasture through the summer weighs 1,800 pounds by fall. The lighter weight of the brood cow reflects her calving as a two-year-old heifer, and regularly thereafter, on a rather low nutrition level. Bulls weigh up to 2,800 pounds in breeding condition. Steers put on feed after weaning weigh around 1,100 pounds at 12 to 15 months of age.

Hays Converter bulls are leased to ranchers for upgrading their herds.

Canada

Some bulls have also been sold for export. Semen is available for artificial breeding both in Canada and in the United States and is being used to produce crossbred calves from other beef breeds. There is no breed association as yet.

The foundation herd remains in the hands of Senator Hays.

New Breeds

Nearly all the new breeds have followed a pattern similar to that of the first Simmental imported to Grosse Island in October, 1966. (The Canadian quarantine procedure required for an animal originating in a country that has not been declared free of foot-and-mouth disease is described in the section "Cattle Diseases" below.) Upon release from farm quarantine for free movement throughout Canada, the new-breed bulls are placed in the artificial-insemination centers, at the federal and provincial experiment stations, and on private farms. Upwards of three-fourths of the total semen produced in the artificial-insemination centers from new-breed bulls is exported to the United States; the remainder is used almost entirely in Canada. The control over export of the European breeds to the United States is explained at the close of this section.

Several federal and provincial experiment stations are conducting extensive investigations on crossing the new breeds on the long established British beef breeds and are raising purebred stock from imported parents. Little if any of this work has been in progress long enough to afford meaningful comparisons among the new breeds or their crosses on the various British breeds.

Both purebred breeders and commercial cattlemen have made considerable use of the new breeds through artificial insemination. Upgrading to what is called purebred status is recognized by most of the new-breed associations for either a ¾ or ⅞ female and a ⅞ or $^{15}/_{16}$ male. Some commercial producers see in this program an easy and comparatively inexpensive way of getting into the purebred business, if the time element involved is disregarded. Though many of them will abandon the program before the purebred goal is achieved, they make a sizable addition to the ranks of the new-breed enthusiasts for the time being.

An increasing number of crossbred steer calves of the new breeds are reaching the feed lots, although many crossbred male calves are retained for breeding stock even though their progeny are ineligible for registration. Almost any F_1 female was commanding a premium as breeding stock, and trailer loads were being exported to the United States at the close of 1971.

France and Switzerland were the first countries of continental Europe

Cattle of North America

approved by the Canada Department of Agriculture for export through the maximum-security quarantine stations. In 1970, Germany, Austria, and Italy were approved, and cattle from these countries were in the Canadian quarantine stations in 1971.

Both Britain and Australia, officially recognized as free of foot-and-mouth and other serious cattle diseases, have also made their contribution to the new-breed gene pool. Britain exports live cattle to both the United States and Canada through the regular quarantine stations. Because of the incidence of pleuropneumonia in Australia, however, no live animals were admitted from that country until 1972.

The numbers of new-breed cattle from continental Europe that have entered Canada by way of the maximum-security quarantine stations through 1971 are given below.

Breed	Country of Origin	Released from Quarantine	Male	Female	Total
Simmental (Pie Rouge)	France	1967	30	58	88
Limousin	France	1968	55	72	127
Maine-Anjou	France	1969	24	19	43
Simmental	Switzerland	1969	75	207	282
Parthenay	France	1970	1	...	1
Blonde d'Aquitaine	France	1971	2	...	2
Chianina	Italy	1971	23	6	29
Simmental (Fleckvieh)	Germany	1971	2	...	2

In 1972 one bull calf of the Pinzgauer breed from Austria, one bull calf of the Tarantaise breed from France, and 14 head of Gelbvieh calves from Germany reached Canadian quarantine stations.

The following new breeds have been imported from England, Australia, and the United States in recent years. Originating in countries recognized as free of foot-and-mouth disease, these importations did not have to pass through the maximum-security quarantine stations.

Breed	Country of Origin	Year of First Entry	Male	Female	Total*
Lincoln Red	England	1966	9	60	69
Devon	England	1968	2	10	12
Welsh Black	England and U.S.	1968	..	..	100
South Devon	England	1969	60	30	90
Murray Grey	Australia	1969	Frozen semen only		

*Totals include some animals born in Canada from imported parents.

Export Control of European Cattle to the United States.—The Department of Industry, Trade and Commerce administers the export regulations applicable to cattle of European origin leaving Canada. The control measures involved in the export of live animals to the United States are as follows (up to the end of 1971):

September 27, 1967: The export of all purebred cattle of European origin was prohibited. At the time this rule applied only to Charolais cattle.

June 20, 1970: Charolais animals which had remained in Canada for 36 months were permitted to be exported, but such exports could not exceed 50 per cent of the total number of such cattle from any single herd. Also permitted was the export of male animals that were the progeny of two imported animals of European origin or the progeny of purebred parents, one or both of which had been born in Canada.

April, 1971: The export of male progeny of purebred Simmental (including Pie Rouge) was permitted.

September, 1971: The April, 1971, regulations for the Simmental breed were made applicable to the Limousin.

Chianina.—The first importation of Chianina cattle arrived from Italy in the fall of 1971. The shipment included 22 bull and 7 heifer calves. Semen was available early in 1972.

Description of the Chianina as seen in Italy is given in Volume I, page 191.

The Italian White Cattle Association is situated in Edmonton, Alberta.

Devon.—Between 1855 and 1865 a few herds of Devon cattle were established in New Brunswick and Ontario and are known to have flourished for over 40 years. The Provincial Herd Book of New Brunswick, published in 1893, carried 145 entries of Devon cattle. Entries from Ontario were exhibited at the World's Fair in Chicago in that year. There is a record of Devon bulls being taken to the cattle country in southern Alberta back in the 1880's.

Eventually the breed disappeared from the annals of Canadian cattle. With the advent of the new breeds the Devon breed re-entered the Canadian scene and has therefore been placed in this section.

The Shaver Beef Breeding Farms, which sponsored the recent introduction, currently has a herd of 5 male and 17 female Devon cattle at Galt, Ontario.

The foundation stock was two males and eight females imported from

Mature Devon cow, 1,200 pounds. Shaver Beef Breeding Farms, Galt, Ontario.

England in 1968. An earlier importation of two females in 1964 does not appear to have contributed to the re-establishment of the breed in Canada. There is no Canadian breed association.

Limousin.—The first Limousin to reach North America was the bull calf Castor (later renamed Prince Pompadour), which arrived at the Grosse Île quarantine station in 1967. The following year 11 bull calves and 5 heifers reached Canada. By the end of 1971 a total of 60 males and 93 females had been brought to Canada from France.

The Limousin is one of the few European breeds of cattle which was developed strictly as a beef-type animal. The color is a light red, shading to fawn at the extremities. Also characteristic are a long body well off the ground and an excellent beef conformation. Mature cows in a good herd weigh 1,350 pounds and bulls 2,400 pounds. In France, Limousin cattle are raised under more rugged conditions than most European breeds; they are left in the open the year round, even during the calving season.

Some publicity has been given to calving difficulty encountered by females of the British beef breeds with Limousin-sired calves. From

Limousin bull, Western Breeders Service, Regina, Saskatchewan.

the information available it is hard to evaluate the magnitude of this trouble. In their native land Limousin heifers are bred to calve when three years old. Possibly it is asking too much of a two-year-old heifer of one of the British breeds to deliver such a large calf.

The Limousin breed association is presently giving consideration to a sire-evaluation program which will emphasize carcass quality.

The Canadian Limousin Association (Midnapore, Alberta), organized in 1969, was an offshoot of the North American Limousin Foundation in the United States. The latter was the first organization to import Limousin cattle from France to Canada.

Lincoln Red.—Lincoln Red cattle were imported to Canada direct from England in 1966, with the Shaver Beef Breeding Farms, the principal sponsors of the breed. This company is active in the introduction of Simmental, Maine-Anjou, and Devon, as well as the Lincoln Red.

The Lincoln Red is described in Volume I, page 306. It is essentially a large Shorthorn selected for a deep-red color. A polled strain was

Lincoln Red cow and calf. Shaver Beef Breeding Farms.

developed by the introduction of both Red and Black Angus influence in a few English herds. Some of the Lincoln Reds in Canada are naturally polled.

The Canadian Lincoln Red Association (Rocky Mountain House, Alberta) was organized in 1969 and incorporated under the Live Stock Pedigree Act. It is affiliated with the National Live Stock Records.

Maine-Anjou.—The first representatives of the Maine-Anjou breed in Canada, 3 bull calves, were released from quarantine in April, 1969. These were imported by the Shaver Beef Breeding Farms. By the end of 1971, 19 male and 16 female calves had been received from France.

The Maine-Anjou, reputed to be the largest of the French breeds, was the result of crossing Shorthorn bulls on one of the early cattle types of Brittany in France. Originally the Shorthorn cross was employed to obtain a heavier draft animal, but later selection was for a dual-purpose type. Bulls weigh up to 3,000 pounds and cows 2,000 pounds; average weights are 200 to 300 pounds lighter. Color is predominantly dark red with some white markings. Horns in both sexes are rather small and extend outward.

In 1962 the Maine-Anjou and the Armoricaine breed societies in France formed a confederation. The owners of both breeds were dissatisfied with the milk production they were getting and were also worried

Maine-Anjou cow. Shaver Beef Breeding Farms, Galt, Ontario.

about the effects of inbreeding in the relatively small populations. A program was adopted of crossing Meuse-Rhine-Yssel (MRY) bulls of the Netherlands and Rotbunte (German Red and White) bulls on Maine-Anjou and Armoricaine registered cows. (The MRY and German Red and White are the same breed, one domiciled in the Netherlands, the other in Germany, both often referred to as Red and White or Red Pied Lowland cattle.) The two crosses, Red and White Lowland–Maine-Anjou and Red and White Lowland—Armoricaine, were then to be combined to form a new breed to be known as Rouge de l'Ouest (Red of the West).

When the search of Canadian importers for more and newer new breeds led to the Maine-Anjou and the first importation of calves of the breed in 1969, Maine-Anjou breeders in France realized the possibilities in developing this foreign market and changed their goal from a dual-purpose breed, slanted toward milk-producing ability, to a beef breed. The Maine-Anjou–Armoricaine union was dissolved in 1970, and each breed again has its own society and travels its own road. For the Maine-Anjou this has been a lucrative route, for the breed has placed high in exportations to Canada.

The extent to which the influence of the Red and White Lowland cattle affected the Maine-Anjou is not definable, but in some herds it had a marked effect. The Red and White Lowland itself is a large breed and a considerably better milk producer than the Maine-Anjou. Some eight years of crossing has led to a wide variation in the Maine-Anjou breed. Within the range of Maine-Anjou–Red and White crossbreds, there was reported a definite double-muscling tendency (such animals are not eligible for registration).

Well-organized promotional effort has advanced the Maine-Anjou breed to a position of considerable popularity in both Canada and the United States. Size has been an important element in this regard. Reports on calving difficulty show the Maine-Anjou high in comparison to most of the new breeds.

The Canadian Maine-Anjou Association (Winterburn, Alberta) has affiliated with the Canadian National Live Stock Records.

Murray Grey.—The Murray Grey breed was introduced to Canada in 1969 when Lord Roderic Gordon and his wife, of Bentley, Alberta, imported the first semen. The first F_1 crosses were dropped in April, 1970. Until 1972 the incidence of pleuropneumonia in Australia prevented the importation of live cattle to Canada from Australia and limited the introduction of the breed to the use of frozen semen.

Toward the end of 1971 the Health of Animals Branch of the Canada Department of Agriculture approved arrangements for the direct importation of bull and heifer calves by air from Australia. A 30-day quarantine is required in Australia and another 90 days in Canada before animals can be released for free travel in Canada.

The Murray Grey cattle had their origin in the Upper Murray region of Victoria, Australia. In 1905 the chance mating of a light-roan, nearly white Shorthorn cow to a purebred Angus bull resulted in a calf that was silver-grey in color. The same cow produced in all 12 calves, all grey, by various Angus bulls. Eight of these calves were heifers, and over the next 40 years the color remained dominant in their progeny although Angus bulls were used exclusively.

Starting in 1940 an effort was made to maintain the silver-grey cattle by *inter se* breeding. A number of breeders in New South Wales, as well as in Victoria, became interested in the "new breed," and it gained considerable recognition on the show circuit.

Tasmania also has its own line of silver-grey cattle. Although of the same genetic origin as the Australian Murray Greys, the Tasmania Greys, as they are called, are not recognized by the Australian breed society. A silver-colored calf, resulting from the mating of an Angus bull and a white Shorthorn cow, is said to be the foundation of the Tasmania Greys, although there has been much use of Murray Grey bulls in the Tasmania herds. Canadians do not as yet recognize the Tasmania Greys, although they appear to be identical to the Murray Greys.

Currently the objective of Canadian Murray Grey breeders is the upgrading of herds to purebred status, together with the production of crossbred feeder cattle. Females of ⅞ and bulls of $^{15}/_{16}$ Murray Grey breeding may be registered as purebred. The initial females employed in upgrading may be of any breed, either grade or purebred, except the dairy breeds. All cows in an upgrading program must display a good beef conformation.

The Canadian Murray Grey Association (Bentley, Alberta) was founded in 1970 and has affiliated with the Canadian National Livestock Records.

Simmental.—The first bull of the new breeds brought to Canada was a Simmental (Pie Rouge de l'Est) which arrived at the Grosse Île station from France in 1966 and was released the following spring. In 1968, 20 head of Simmentals from both France and Switzerland were released from quarantine. By mid-1971 over 300 head of Simmental cattle had been admitted to Canada.

The Simmental breed was first introduced to North America in 1886 but disappeared after a number of years (see the section "Cattle Breeds" in Part IV, "United States"). The recent importations to Canada have generated a wide interest in the breed, both for upgrading other breeds to Simmental and for the production of F_1 females and feeder steers. Extensive crossing investigations at Canadian experiment stations have been underway since the first Simmental bull became available. The extensive demand for Simmental semen in the United States has been a major factor in the promotion of the breed in Canada.

The popularity of the Simmental in North America is due in large measure to the color pattern of red or tan on white and the white face. When crossed on the Hereford, the Simmental causes no material change in the typical Hereford appearance, a trait highly acceptable to the longtime Hereford breeder.

The Canadian Simmental Association is situated in Calgary, Alberta.

South Devon.—South Devon cattle were introduced to Canada in 1969 direct from England by Big Beef Hybrids, Inc. They were taken to the

Lacombe Adonis, the first Simmental born in Canada, 2,130 pounds at two years of age.

company's ranch in Manitoba. A total of 60 bulls and 30 cows were imported, some of which have been distributed to breeders in British Columbia and Saskatchewan.

The South Devon is described in Volume I, page 316.

There is no breed association for the South Devon in Canada.

Welsh Black.—The first Welsh Black cattle in Canada were 4 head purchased in the United States in 1968. Later there were several importations direct from Wales until now there are around 100 head of purebred Welsh Blacks throughout the country.

The Canadian Welsh Black Cattle Society (Taber, Alberta) is following the standard line of promotion of the other new-breed organizations. Both upgrading and crossbreeding are advocated, and there are a few bulls in the artificial-insemination centers. Where the Welsh Blacks will fit into the Canadian cattle picture is not yet clear.

Other Breeds

American Brahman.—Two American Brahman bulls and two heifers

South Devon bull, two and one-half years old, 1,600 pounds. British Columbia Artificial Insemination Center, Milner, British Columbia.

were taken to Rorketon, Manitoba, in 1951 by Walter Tuer. They came through the Canadian winters in good condition and reproduced satisfactorily. They were reported to have grown a much heavier hair coat than normal as the cold weather came on. The breed has never attracted commercial attention in Canada.

Cattalo.—Canadian experiment stations did much work in bison-cattle crosses during the first half of the present century in an effort to develop an animal which would be more cold resistant than cattle. Some bison-cattle crosses were purchased from Mossom Boyd for experimental breeding by the experiment station at Scott, Saskatchewan, in 1915. Boyd was the enterprising breeder who had developed a line of Polled Herefords by breeding to Angus bulls and had experimented with buffalo-cattle hybrids in the early years of the twentieth century.

The crossbreeding project was transferred to Buffalo Park at Wainwright, Alberta, a year later. The bison-cattle crosses failed to reproduce even to the limited extent which had been expected. Bison calves were then raised on Hereford cows in order to establish a new base for the breeding experiments, and cattle-yak, cattle-bison, bison-yak, and ¼ bison, ¼ yak, and ½ cattle crosses were produced.

Extensive crossbreeding of bison and three breeds of cattle, Hereford, Shorthorn, and Holstein-Friesian, was continued from 1937 to 1950. The first crosses of bison and cattle were called "hybrids," and any progeny with less than ½ bison breeding was termed "cattalo," signifying a mixture of cattle and buffalo, the generally used designation for bison.

In 1950 the crossbreeding herd was transferred from Wainwright to the Research Substation at Manyberries, Alberta. The yak was dropped from the program, and the work continued with cattle and buffalo. Both the hybrid and cattalo males were subject to sterility or a very low fertility; the females usually bred at a low level of fertility. Between 1951 and 1957, 1,115 matings produced 762 cattalo calves. The pounds of cattalo calf produced per cow bred was equal to the pounds of calf produced from straight Hereford breeding. It was concluded that the cattalo were more resistant to cold. The project was finally abandoned in 1964 primarily because of the infertility of the crossbred progeny.

One difficulty was found to be that excessive amounts of fluid in the fetal membranes were responsible for the loss of many calves and some cows in the crossing of bison bulls with domestic cows. This did not occur in the reverse cross, and so domestic bulls were mated with bison cows in the later stages of hybridization at Wainwright. The yak–domestic-cattle crosses were not accompanied by the high mortality found in the bison-cattle cross, though the crossing of domestic bulls on yak and yak hybrids gave a greater percentage of normal calvings than the reverse cross.

The major cattalo breeding now in progress in North America is that of Jim Burnett at Luther, Montana. See the section "Cattle Breeds" in Part IV, "United States."

Yak.—A male and a female yak were presented to the Zoological Gardens in Ottawa in 1861 by C. M. Robinson. Another pair were purchased by the gardens in 1885. The source of both pairs of these yaks is not known. All the yaks are reported to have been productive, although the number of offspring was limited.

The yaks at the gardens were subsequently sent to the Dominion Experiment Farm at Brandon, Manitoba, where experimental breeding

Canada

was started about 1909. The animals did not reproduce satisfactorily there, and were moved to Rocky Mountain Park at Banff, Alberta, and later to Buffalo Park at Wainwright. Reproduction was satisfactory at the two latter locations. Between 1919 and 1930 the Alaska Experiment Station of the USDA obtained from the Canadian breed three bulls and six females in several different shipments.

A few yak-cattle and yak-bison crosses were produced at Buffalo Park under the supervision of the Canada Department of Agriculture in the 1920's. The experimental breeding was abandoned before any conclusive results were obtained owing to reproductive difficulties in both yak hybrids. The yaks have now disappeared from the Canadian scene.

MANAGEMENT PRACTICES

Eastern Canada includes the Atlantic provinces of Newfoundland, Prince Edward Island, Nova Scotia, and New Brunswick. Central Canada includes the provinces of Quebec and Ontario. Western Canada begins at the boundary of Ontario and Manitoba and includes the three Prairie Provinces, Manitoba, Saskatchewan, and Alberta, all of which lie east of the Continental Divide, and British Columbia on the western side. The agriculturally developed area in Canada extends in a narrow belt above the United States border from the Atlantic to the Pacific. The northern limit of cattle raising in eastern Canada follows an irregular line which does not extend above the 51st parallel. In western Canada the line where cattle raising stops runs northwest from the 51st parallel in Manitoba to the Peace River country at the boundary of Alberta and British Columbia and then across the Rocky Mountains in a southwesterly direction to the Pacific Coast. Newfoundland and Labrador, the Northwest Territories, and the Yukon Territory lie north of this arbitrary demarcation of the cattle population and are unimportant in cattle raising. There are also large areas below the cattle line that are heavily timbered, mountainous, and covered with lakes and swamps, lands which are uninhabited by man or cattle. Only 174 million acres (6.5 per cent) of Canada's total land area are in farms and ranches.

In eastern Canada the cattle brought in with the first settlers increased as needed to supply draft power, milk, and meat for the growing human population. The western provinces were the territory of the Hudson's Bay Company, an organization primarily interested in fur trading. The trading posts of the company were often stocked with cattle to augment the food supply, but little effort was made to foster a cattle industry.

The transfer of the territory of the Hudson's Bay Company to Canada

Cattle of North America

in 1870 paved the way for the agricultural development of western Canada. This was the initial step which made possible a range-type cattle industry similar to that which had begun in the western United States a decade earlier. Large numbers of cattle were soon grazing on the western prairies, but the only market for the beef was in the east. This new source of cattle naturally had its effect on stock raising in the eastern provinces. Wide variations in climate, cultural differences among the various nationalities scattered throughout the provinces, the growth of large concentrations of population, and the part which government came to play in agriculture were factors which had their effect on the expansion of cattle husbandry across the country.

Maritime Provinces.—Lying northeast of the state of Maine, the provinces of New Brunswick, Nova Scotia, and Prince Edward Island are largely surrounded by waters of the Atlantic and enjoy an equable climate. Grass is the natural crop on much of the rolling and often heavily wooded countryside and provided the base for heavy shipments of grass-fat cattle from the marshlands of New Brunswick to England during the nineteenth century.

As the population increased, dairying became a major farm enterprise, although today much of the best land susceptible to cultivation is devoted to row crops, especially potatoes. Cattle of the beef breeds make up less than one-third of the total bovine population, but in recent years there has been a small, steady increase in beef cattle and a decrease in the dairy herds.

Dairy farms are almost entirely family operations with herds of 20 to 40 cows and usually include some row-crop production as well. There are only a few farms in the 100-cow range and only one farm presently milking 200 head. The price squeeze has affected the small dairy farmer, however, and is forcing him to choose one of two courses: to expand his operation or convert to raising cattle for slaughter as a side line to some other operation.

Thus the 100-cow herd has become the aim of those farmers who want to continue with their milk cows. This size herd provides a reasonable standard of living for two families and enables a son to join his father in the enterprise. It demands increased investment, however, in forage-harvesting equipment and in milking and housing facilities. It also involves either switching land from cash crops to forage production or buying out a neighbor. While only a few dairy farmers have so far reached the 100-cow level, the size of the dairy herds is increasing every year.

The number of dairy farmers taking the other alternative, converting their small herds to beef production, is also increasing. When the small farmer obtains outside work to supplement his income, he lacks time to handle the dairy chores and breeds his cows to a beef-type bull. There is usually a ready market for his crossbred calves. The farmers taking this route continue to live on the land, fulfilling a desire still strong with many descendants of original settlers in the Maritime Provinces.

The Holstein-Friesian breed accounts for 75 per cent of the dairy cattle; the remainder are mostly Ayrshire and Guernsey, with a few Jersey representatives. Good Holstein-Friesian herds average 11,000 to 13,000 pounds of milk per lactation.

The practice of raising bull calves to sell as yearling feeders is growing. There is also some feeding out of steers on the farm to 900- or 1,000-pound slaughter weights. Only a small proportion of the bull calves, however, are handled in this manner, since most unwanted calves are sold when a few days old. Heifer calves are normally raised for replacements either for the home herd or for sale.

There are a number of purebred herds of the dairy breeds in the Maritimes. As the trend to large farms continues, the demand for high-producing cows and heifers increases. The widespread use of artificial insemination however, has curtailed the demand for all but the best bulls.

The most common practice among dairymen is to keep the cows in the barn or under a loose housing arrangement for five months during the winter. Increasing numbers are making use of free stalls. The herd is then on pasture from May 15 to November 15 with no hay or supplement fed during June and July when pastures are at their best, and a grain supplement is fed from July on. In the case of some high-producing dairies a grain supplement is fed throughout the year. Some dairies also keep their herds confined throughout the year and feed green chop for roughage during the growing season.

Grass, alfalfa, and corn are the principal forage crops. All these crops are put up as silage, although currently corn is preferred for this purpose. Making hay is a major problem. Good drying weather is practically unknown in the Maritimes because of the heavy rains during July and August and the prevailing cool temperatures. Most hay is baled while high in moisture and is of poor quality. While the use of barn-type hay driers has been advocated by the experiment stations, the high cost has delayed their acceptance. The type recommended is a slatted tunnel on the barn floor over which loose hay is piled. Air is forced through the tunnel by a large-volume blower powered by a gasoline or diesel motor.

The established beef herds in the Maritimes are predominantly cow-calf operations. The calves are fed in small local feed lots or are sold through the local auctions to Ontario feeders. There are, however, some farmer-feeders who finish their stock to slaughter weights. Some also purchase stock from their neighbors to feed. One such enterprise in New Brunswick, which has been finishing 1,200 head annually, recently increased its capacity to 2,000 head. This farm of 3,000 acres is highly exceptional in the Maritimes and feeds out ten times as many cattle as the next largest unit.

The beef-cattle herds range from 20 to 80 head, only a handful having more than this number. The only large beef-cattle operation is the recently established Dundas Farms on Prince Edward Island. This complex, embodying some 6,700 acres, was put together by the consolidation of 35 farms in a good cropping area. Three thousand brood cows will eventually be used here in a crossbreeding evaluation program if the objective of the owner is realized. Several of the new breeds are being used for crossing on Hereford, Angus, and Shorthorn cows. The ultimate goal includes the production of purebred breeding stock of the Charolais, Chianina, Limousin, Maine-Anjou, and Simmental breeds. F_1 females will also be produced. As in the United States, these are in demand by commercial cattlemen, who breed them to a bull of a third breed for the production of feeder cattle.

Traditionally beef cattle in the Maritimes have been kept inside during the winter, but this practice has been changing in recent years. The cow herd maintained in the open where there is ready access to timbered areas does as well as one provided with winter shelter and requires less labor. This method of wintering, however, must be arranged so that feeding is convenient. Winter feeding normally lasts from 180 to 200 days. Recently there has been some feeding of silage to beef cattle. Usually two pounds of grain per day are fed as the calving period approaches. The better operations breed so that calving starts the first of March and most of the calves are dropped during the month. Although colder, March has less damp, rainy weather than April, and calving losses are lower.

The Hereford predominates, although there is a sizable representation of Angus, and there are a few Shorthorn herds. Artificial insemination is little used except by farmers who have recently converted from dairying to beef-cattle operations. Many of these farmers are utilizing the bulls of the new breeds because of the high value of the resulting calves.

A number of small purebred herds supply most of the local demand for herd bulls used in natural service.

Quebec.—There has always been a cultural distinction between Quebec and the rest of Canada. The province has a harsher winter climate than Ontario or the adjacent Maritime Provinces. Major differences in cattle raising also exist between Quebec and the rest of Canada. Many of the 40-hectare (99-acre) farms of the early French settlers are still intact and harbor small herds of cattle. Dairy cattle predominate, outnumbering the beef breeds 10 to 1 in the purebred herds. The Canadian Cattle, which served the sturdy French immigrants admirably and with a minimum of care for two and one-half centuries, began to give ground to the Ayrshires toward the middle of the last century. During the opening years of the twentieth century the Holstein crowded out the Ayrshire and now outnumbers 3 to 1 all the other dairy breeds combined. The Ayrshire, however, reached a high level of development in Quebec, and today excellent herds are seen made up of cows with exceptionally good udders.

Most of the people in the province live in the area between the United States and a line 100 miles north of the St. Lawrence River. Both the dairy and the beef cattle are found in this belt.

The usual dairy herds have 20 to 50 cows in milk. The trend to larger farms is apparent but is developing much slower than in the Maritimes. Very few dairies have as many as 100 cows.

Most of the purebred herds are bred by artificial insemination, as are also many of the better-managed grade herds. Old-style barns house many herds, although some have been converted to loose-housing arrangements. Cows are usually kept inside for seven months of the year. A few dairymen are beginning to confine their cows the year round and feed green chop during the summer. This practice is most common among the farmers who have recently increased the size of their operations.

High-producing herds are fed a purchased grain ration throughout the year except for the first 30 days on pasture in the spring. The grazing season starts around the middle of May and lasts until the first part of October.

Corn, grass, and alfalfa silage are fed extensively, usually along with some hay. Haymaking is often a problem because of poor drying weather, though the situation is not as serious in Quebec as it is in the Maritimes.

Bulk tanks for the delivery of milk are in general use throughout the province, but most milking is by machine into the pail.

Slaughter cattle in Quebec are mostly culled dairy stock and calves. Some dairymen sell their bull calves for export to the United States at three days of age. Others raise them for veal.

The beef herds are of little economic importance. There are a few purebred herds of the Hereford, Angus, and Shorthorn breeds, some of which are of very good stock. These enterprises, however, are mostly the avocation of the city farmer.

Ontario.—In many ways Ontario has been the focal point of the expansion of the cattle industry through that part of Canada lying west of the Maritimes. Ontario has one-third of Canada's human population, one-third of the national dairy herd, and one-sixth of the total beef cattle. When the British beef breeds were introduced in the latter part of the nineteenth century, they usually gained their first foothold in Ontario. The development of the dairy breeds, with the exception of the Canadian and Ayrshire, also centered here.

The seat of this growth of the livestock industry has been concentrated in an area the size of the state of Ohio lying between Lake Ontario and Lake Erie on the east and south and Lake Huron and Georgian Bay on the west and north. Toronto grew to be the "Chicago of Canada"—the major cattle market and the center of the packing industry. The rich farmlands of southwestern Ontario supplied the fat cattle for slaughter, much as the corn-belt states in the United States were the source from which the Chicago packers obtained their cattle. At the same time the dairy industry was expanding, and southern Ontario came to be the "Wisconsin of Canada."

In the years following World War II the expansion of the packing industry was westward. This was followed by new and larger feed lots, where the animals were fed for slaughter. The movement of the packing plants and feed lots to the areas where cattle were produced and the feed was grown paralleled a similar trend in the western and southwestern regions of the United States.

The dairy farmer in Ontario is holding his own, however, and from all indications will continue to do so. Dairy cows outnumber beef cows more than two to one. Ontario is recognized as the center of the development of the Canadian Holstein-Friesian, which has gained a world-wide reputation as a superior dairy animal. Over 85 per cent of the dairy cows in the province are of this breed.

The dairy farms fall into two classes: those which produce milk mainly for fluid sales (about a third of the production), and those which produce manufacturing milk that goes into butter, cheese, and other dairy products. In the Toronto milkshed there are a few industrial-type dairies in the 1,000-cow range and one which milks 2,000 cows. But the 40-to 60-cow farm is still the source of most of the fluid milk. The farm producing manufacturing milk usually is located farther away from large population centers and has a herd of about 30 head. Both types of farms are following the trend to a larger size dictated by the economics of the business. The unit with 100 cows or more is still exceptional, but dairies in the 60- to 80-cow bracket are increasing as the small units go out of business or convert to beef herds.

The dairy herd is commonly held on pasture from late April until sometime in October. A growing number of well-managed units keep their cows in confinement throughout the year and feed silage or green chop during the summer months. Some farmers also feed hay to cows on pasture, starting in July when the nutritive value of the grass begins to decrease. Corn silage is the principal roughage fed in the winter, usually along with a small quantity of hay, and there is also considerable use of home-raised grains in the supplemental ration.

The Ontario dairy farmer makes wider use of modern equipment than do his counterparts in the other provinces. There is more pipeline milking, loose-housing and free-stall arrangements are more common, and the newer types of forage-harvesting equipment are used on practically all the larger farms. The larger size of the Ontario dairy unit makes the increased investment possible. The quality of the cattle is also generally higher than in the provinces on the east. Holstein-Friesian herds with a 16,000-pound production average are frequently encountered.

A start has been made in feeding Holstein steers for beef. While the practice is limited, it could develop into a sizable outlet for dairy calves. The strong demand for veal in the Toronto area, however, diverts many a Holstein calf from the feed lot.

The calf-nursery method of raising veal calves has made considerable progress in the concentrated dairy area southwest of Toronto. In 1971 about 50,000 head were raised to heavy veal weights (approximately 275 pounds and up) in these enterprises. Some of the farm-type nurseries finish 100 or more veal calves annually. The industrial-type nurseries, not farm-connected, have capacities ranging from 1,000 to 4,000 calves annually.

In the typical nursery calves are purchased from a trader who specializes in picking up newly born calves, both at local auction sales and at the Toronto stockyards. He has perfected his arrangements so as to give quick delivery to his customers. Calves reach the nursery at four to

Robot calf feeder in a southern Ontario calf nursery.

five days of age, are immediately given their antibiotic shots, and are started on their nipple-pail routine.

The pens holding individual calves are situated on each side of an alley, which runs the length of the nursery building. Provision for good forced-air circulation and maintenance of a fairly constant temperature of around 70 degrees summer and winter are essential. A milk replacer, manufactured from a powdered milk base, is mixed with water and fed each calf from a nipple pail every few hours.

The plant of the Wellandport Feed Mill, Ltd., at Wellandport, Ontario, is one of the larger calf nurseries. The Robot calf feeder, manufactured by Domaf of Holland, has been in use there for two years. It consists of a small, electric-powered cart which moves automatically along a track down the alley between the rows of calf pens. Two nipple buckets are so positioned on each side that two calves on either side can reach them. After a stop long enough to permit the calves to obtain a measured quantity of fluid, the cart is automatically propelled to the next four calves in the alley. A mechanism is provided for automatic mixing of the milk replacer with water and refilling the nipple buckets from the master tank before they become empty.

Holstein calves, which are practically the only breed fed for veal, weigh an average of 100 pounds when they are received and are fed to 300 to 350 pounds for the Toronto market and 200 to 250 pounds for Quebec. Daily gains of 2½ pounds are obtained with a conversion of 1 pound gain for 1½ pounds of feed. The prepared feeds used, basically powdered milk fortified with minerals and antibiotics, cost around 23 cents a pound. The five-day-old calves cost $40 delivered to the nursery in 1971—a substantial advance from the $28 they cost a few years earlier when off-the-farm calf feeding began.

The "white veal" which the nurseries produce brings a premium of around 5 cents a pound over ordinary barnyard veal. Finished calves sell for 50 to 55 cents a pound liveweight. This works out to be a fairly profitable enterprise, provided the highly unpredictable death loss can be controlled. Death loss varies from 4 to 12 per cent, but with good management can be held to the lower figure.

The white veal has become known to the trade as "Provimi veal," named for the first manufacturer to make the milk-replacer feed.

Most of the commercial beef cattle raised in Ontario are grown on family-type farms of 160 to 240 acres with 15 to 30 brood cows. The owner usually has a mixed farm operation, raising corn, small grains, or tobacco. Cattle are carried to utilize the plant growth on wasteland which is not easily cultivated and such crop residues as are available after harvest. The Hereford is the most numerous breed, and the Angus is next. Herds other than of these two breeds are mostly mixed—a few dairy-type animals, possibly a few Charolais crosses, or various crosses of the beef breeds.

Calves may be either sold in the fall or held over until the following year and marketed as yearlings. The time of disposal often depends on the availability of winter feed, as well as on the farmer's cash needs.

The cattle are on pasture and crop residues for six months. During the rest of the year they are fed hay and sometimes home-raised grain. Most hay is native or improved grasses, although some alfalfa is grown, and its use is increasing. Very little silage is fed to the small beef herd, unless the farmer is conducting a regular feeding operation. All stock is kept in the open but with access to a shed roof for protection in bad weather.

There are still many farmer-feeders, the precursors of owners of commercial feed lots. The cattle and grain entering this kind of operation may be home-grown, but both cattle and grain are often purchased. The amount of home-grown roughage determines the size of the operation.

Feeder cattle, either calves or yearlings, are purchased in the fall, carried through the winter on a growing ration, and finished off the following summer or fall after a nearly full grain feed for 90 to 120 days. The number of cattle handled in this manner by a farmer-feeder is 50 to 200 head, probably averaging under 100.

Ontario has always been a stronghold of the purebred breeder. It was the area from which the purebred business grew in Canada, and it still holds the dominant position in purebred dairy cattle. The development of the new breeds, by contrast, has centered in Alberta, where many of the imported bulls are found.

Breeders of purebred cattle are feeling the effect of artificial insemination on their bull sales. This breeding practice has spurted with the introduction of the new breeds. It is not as yet widespread among the established beef breeds but is increasing every year.

The Prairie Provinces.—The vast region which was to become the Prairie Provinces was open country under control of the Hudson's Bay Company until 1870. In that year the territory was released to the Canadian government. While Manitoba had some earlier settlements in the south along the valley of the Red River of the North, there were only widely scattered outposts throughout Saskatchewan and Alberta. In the years after 1870 development proceeded rapidly: the Northwest Mounted Police were organized in 1873 and brought law and order to the region; the Homestead Act of 1875 enticed farmers from the east to the virgin land; and the Canadian Pacific Railway, completed in 1885, provided an outlet for grain and cattle, which were being produced in larger quantities than could be disposed of locally. All these factors contributed to the pattern of cattle raising in the Prairie Provinces during the last quarter of the nineteenth century.

The open-range cattle operations began to spread from Montana across the border into southern Alberta in the 1870's. Both men and cattle moved up from the United States, and large ranches were established, utilizing both government and railway lands for grazing. The early cattle were the Longhorns, mixed in varying degree with Shorthorns and Herefords, which had already been driven from Texas as far north as Montana.

A smaller addition to the cattle population of the Prairie Provinces came to Manitoba and Saskatchewan from the eastern provinces, principally Ontario. Straight Shorthorn and Hereford herds were established and supplied the stock for upgrading the large holdings of mixed Longhorn cattle that had increased rapidly in Alberta. Wheat farming followed soon after the cattle ranches became established, the better-watered and more level land going to wheat and the drier and rougher areas used for grazing.

For a number of years during this period of crop development there was a great demand for draft horses from the railroad-construction contractors, as well as from the wheat farmers. Some ranches converted entirely from cattle to horses. Then, as transportation and marketing facilities for cattle became better organized and the market for horses declined, grain farmers began to raise a few cattle. Rough areas that were difficult to cultivate could be pastured, and stubble fields were grazed. Eventually the cattle herd was increased by using some of the cropland for growing hay for winter feed.

After World War II the large ranches in southern Alberta which grazed herds of up to 20,000 head on holdings of 100,000 acres or more began to be broken up. Only a few remnants of these old ranches remain, now materially reduced in size.

The agricultural pattern which has evolved in the Prairie Provinces now encompasses grain farms, farms raising both grain and cattle, the ranch-type operation devoted entirely to cattle, the farmer-feeder, and the feed lot.

What is known as the "prairie region" begins in southwestern Manitoba and extends across southern Saskatchewan and Alberta to the foothills of the Rocky Mountains. This area of lower rainfall, particularly the shortgrass country of southwestern Saskatchewan and southern Alberta, where the rainfall is less than 12 inches, was the historic rangeland of Canada. It is still mainly used for cattle raising, although in smaller units than formerly. The grain farms are generally north of the rangelands. Many farm units are grain-cattle-type operations.

Where the prairie lands approach the mountains, the "foothills region" begins. Here the rainfall increases to 15 or 20 inches, and in many ways this is one of the best cattle areas in Canada. Here grain production is subordinate to cattle raising.

The "parkland" is a belt lying north of the prairie that extends from southwestern Manitoba through Saskatchewan and Alberta into the Peace River country. In the parkland the soils are poorer and precipitation is heavier. The region is covered with forests interspersed with open grasslands. Stock raising and grain farming are generally combined throughout the parkland area.

Cattle of North America

In the early days in Alberta all stock was wintered on whatever dry growth remained on the range in the fall. In a hard winter severe losses were accepted as an unavoidable hazard. Even today some stockmen in more favored areas of the prairie region carry their cattle through the winter without hay, feeding only about two pounds of grain a day during periods of bad weather. The more general practice throughout the prairie and parkland regions is to feed hay regularly for three to five months during late winter and early spring. Hay is usually baled and piled in locations handy for winter feeding.

Most of the cattle and the cattle-grain farms in the Prairie Provinces are now family operations, outside help being employed only for short times in critical periods. In the parkland the average farm ranges from one to two sections and has 40 to 75 head of cattle. In the prairie region the farms are somewhat larger, since more land must lie fallow in the summer in order to raise grain. The cattle-grain farmer usually sells his calf crop in the fall, keeping only the heifers he wants for replacements.

While a number of large ranches running several thousand head of cattle still remain in the shortgrass country, the typical operation has from 100 to 200 brood cows on 8 to 15 sections of land. Steers are usually marketed as yearlings, although the number marketed as calves is increasing every year. The size and type of operation in the foothills is similar to that in the shortgrass region, except that the cow-calf operations outnumber the yearling spreads.

The last agricultural region to be developed in the Prairie Provinces was the Peace River Valley. This valley, mostly in Alberta but extending into British Columbia, is predominantly brush and timberland, though there are areas of green parkland and some of muskeg. The initial development was small-grain farming in the open grasslands. After this land was utilized, additional areas were cleared for cultivation. Raising alfalfa and fescue for seed was found to be more profitable than growing wheat. The next step was the introduction of cattle to utilize the plant residues.

The Peace River Valley probably is subject to more severe winters than any other region where cattle are raised. Some herds are maintained as far north as Fort Vermilion at 58 degrees latitude. Here temperatures of minus 40 to minus 50 degrees for two weeks at a time have to be anticipated, although such extreme weather is not encountered every year. (In Alaska and the Scandinavian countries cattle are maintained at more northerly latitudes but under less severe climatic conditions.)

Of the Alberta beef-cow herd 5 per cent, some 85,000 head, are now held in the Peace River country, and excellent herds of both Hereford and Angus have been established there. The Angus cow herd at the Fairview Agriculture College averages 1,300 pounds. The valley cattle are normally fed for 200 to 210 days during the year and are turned out around the first of June. Good brush cover or open sheds situated close to the hay storage suffice for winter protection. Silage, made from an oat-pea mixture, and hay are utilized for winter feed. Two and one-half tons of hay, or the equivalent in silage, is said to be ample to carry a cow through the winter. Because of the long days pasture growth is lush and rapid by the end of spring and normally lasts until October.

British Columbia.—The Hudson's Bay Company established several herds of cattle at their outposts in British Columbia during the 1830's. By 1836 there was reference to the company's owning 1,000 head of cattle west of the Rocky Mountains. This was the beginning of the cattle industry in the area. As the settlements grew, so did the cattle population.

The discovery of gold on the Fraser River in 1856 brought more cattle into the country from the Oregon Territory. As timber operations developed, there was movement of population north from the Vancouver area, and the agricultural development that followed included the husbandry of both horses and cattle. The Shorthorns (then known as Durhams) dominated the cow herds throughout the nineteenth century. The Herefords came later.

The best-known cattle-ranching areas of British Columbia are the Cariboo country and the Kamloops area. Small-scale cattle operations also extend from Alberta across the line into British Columbia in the Peace River Valley. The Cariboo country is the high-plateau area lying between the Cariboo Mountains on the east and the Coast Ranges on the west and is drained by the upper waters of the Fraser River. High-mountain ranching effectively utilizes the excellent summer grazing. Cattle are wintered in the valleys where hay can be put up.

The major cattle country of the province, however, centers around Kamloops and the area extending from there south to the United States border. It is a dry land with 10 inches of rainfall and elevations down to 1,100 feet. Back from the valleys the intervening mountain ranges provide summer range from June to October. Hay—formerly of native grasses but now with a high percentage of alfalfa—is fed for three to four months during the winter. Much of this hay land is irrigated, either by sprinkler or by flooding. Practically all of the hay is baled, and most ranches of any size have modern hay-harvesting equipment. Spring and fall pasture is provided by the benches and hilly land surrounding the valley meadows where the cattle are wintered.

Profitable operations depend upon leased government land for summer pasture. Many ranches run 250 to 300 brood cows and market either calves or yearlings. Larger spreads range up to several thousand head. Typical of the large ranch is an 800-cow outfit with 5,000 acres of deeded land and 15,000 acres of leased land.

Years ago only three-year-old steers were marketed. Now most ranchers ship yearlings, although a few sell two-year-old steers, and some have gone to calves. Nearly all cattle leave the ranches as feeders, supplying primarily the Vancouver feed lots with the surplus going to Alberta and even to Ontario.

Good-quality Herefords predominate in British Columbia, but the number of Angus herds is steadily increasing. Very few Shorthorns are left. Some crossing is seen, principally Angus bulls on Hereford cows to get a white-faced black feeder steer, which is very popular with the feeders. This type of crossing, practiced for years in a limited way, is now increasing. There are also the occasional Charolais bull and some of his progeny. As yet there has been only a limited interest in the new breeds. Beef cattle outnumber dairy cattle two to one.

The dairy herds of British Columbia were gradually upgraded to Jersey, Guernsey, and Ayrshire after the opening of the twentieth century. The Holstein-Friesian movement began after World War II. Preference for the black-and-white cattle grew so rapidly that they now account for 70 per cent of all dairy cattle in the province. Jersey and Ayrshire herds, however, generally continue to hold their own. There are quite a number of purebred dairy herds maintained more as a matter of pride than for the sale of breeding stock. Because of the wide use of artificial insemination in the dairy herds there is now only a limited market for purebred bulls.

Most of the dairy cattle in British Columbia are found in the lower Fraser River Valley and on the southern half of Vancouver Island. There are small herds near the smaller centers of population, but the main concentrations are around Vancouver and Victoria, where more than one-half of the population of the province lives.

Values of the land occupied by dairy farms, because of its proximity to the large population centers, have increased so rapidly that milk production has become a highly specialized activity. The farm is usually just large enough to meet the roughage requirements—mainly alfalfa, hay, and corn silage—of the milk herd. Supplemental feed is purchased in the form of prepared products.

Many dairies are one-family operations, milking 40 to 60 cows each, but there are increasing numbers with as many as 100 cows each. The larger dairies require a full time hired man, or they may be two-family arrangements, a grown son having gone into business with his father. All dairies selling fluid milk have bulk tanks, and most of them have pipeline milking machines.

There are a number of local calf nurseries of the farm type which furnish a good market for young calves. A five-day-old Holstein bull calf sells for $50 and a Guernsey calf for $30. They are raised to veal weights on milk replacer in nurseries that handle from 12 head to 100 or more.

Artificial Insemination

The use of artificial insemination began in the dairy herds in southern Ontario in the mid-1930's and soon spread to the other provinces. By 1971 this method of breeding the milk cow appeared to be leveling off except in Quebec. The preponderance of small herds in Quebec caused a slower acceptance of artificial insemination there. Today, with the size of the dairy herds increasing, management is improving, and greater use is being made of artificial insemination. In 1971, 39.9 per cent of the national dairy herd was bred artificially—31.6 per cent to sires of the dairy breeds and 8.3 per cent to beef sires. In Quebec only 24 per cent of the dairy cows were bred artificially, and very few of these were by beef bulls.

Before the introduction of the new breeds and the increased interest in crossbreeding, the use of artificial insemination in the beef herds was only 0.7 per cent of the total number bred annually. In 1971 the per cent bred by artificial insemination was 4.1, or 160,000 beef cows out of a total of 3,337,700. The new breeds accounted for one-fifth of this total. Artificial insemination will undoubtedly increase in the beef herds in the future—how extensively will depend on the outcome of the varied cross-breeding and upgrading programs now underway.

There are now 19 commercial artificial-insemination centers throughout Canada. There had been a steady decline in the number of centers as smaller units combined with centrally located and larger, more efficient facilities. In 1969 there were only 14 operating centers. The additional 5 units operating in 1971 were established primarily to cater to the heavy United States demand for semen. All the insemination centers are modern facilities meeting the standards established by the Canada Department of Agriculture. With the exception of the one in Quebec, the units are operated by co-operatives of farmers or by individuals. The Quebec center is under the provincial Department of Agriculture. The bulls of the dairy breeds in active use in the various units have good

Feed lot, 75-head capacity. Charlottetown, Prince Edward Island.

progeny records behind them and include a number of exceptional individuals. Excluding the new breeds, the beef sires generally have performance records, and some have progeny data as well.

To gain their place in the sun, all the stations have managed to obtain a few bulls of the new breeds and some Charolais representatives as well. Most of the bulls, however, are dairy animals of the breeds most widely used in the area where the center is located. There is free interchange of semen between stations to meet any special demand.

Export permits, allowing animals of the new breeds to enter the United States, were not granted until 1971 and then only on a very restricted basis. This situation led to an extremely profitable semen-collecting and merchandising business, almost reminiscent of the gold-rush days of the past century. The promotional efforts of the growing number of new breed sponsors have built up an apparently insatiable demand for semen both in the United States and in Canada.

Calgary, Alberta, is the center of the new-breed semen trade. The five new bull studs devoted primarily to this business are housed there. Many of the bulls in these stations are owned by an individual or a syndicate for whom semen is collected on a custom basis. Tales are told of bulls that have produced $1 million worth of semen and of individual animals that have sold for as much as $175,000. All these glamorous animals were selected in Europe when they were less than eight months old with few, if any, performance records behind them.

Feed Lots

The feeding of cattle for slaughter has followed much the same pattern in Canada as in the United States. Over the 20-year period from 1951 to 1971 the number of cattle going to Canadian packing plants doubled, but the number of grain-fed cattle more than quadrupled. The percentage of fed cattle in the total kill increased from 29 to 65. The small-herd owner in the eastern provinces is still finishing steers on the farm, but the large increase comes from the new commercial feed lots in the west.

The first feed lots were established in the corn-producing area of southern Ontario which was close to the stockyard and packing-plant concentrations. They were small operations compared with those of today, the largest handling only a few thousand head annually. As the packing plants and public markets moved westward to the areas where cattle were raised and most of the feed grains produced, the modern mechanized feeding facility was a logical development. Feed lots sprang up across the Prairie Provinces from Manitoba to Alberta. British Columbia, lying west of the Rocky Mountains and isolated to a degree from the rest of Canada, had always enjoyed a more self-contained economy, including the cattle industry. Larger feed lots also came into existence here.

Another element which contributed to the rapid growth of feed lots in Manitoba and Saskatchewan was the large surpluses of wheat which accumulated during the 1960's. Cattle feeding became recognized as an alternative method of marketing grain in which considerable control could be retained in the producers' hands. Methods of including wheat in the grain ration were developed, and barley, a more satisfactory cattle feed, replaced wheat in many farm operations. Grain farmers with no previous interest in cattle pooled their assets and built feed lots. The facilities for feeding cattle in the Prairie Provinces are now probably overbuilt to some extent and often operate at less than capacity because of lack of feeder cattle.

Alberta has now overtaken Ontario as the major cattle-feeding and slaughtering center of Canada. Nearly 40 per cent of all fed cattle in the country are marketed from there. One feed lot, that of Lakeside Feeders, Ltd., at Brooks, Alberta, has a pen capacity of 25,000 head and has put through 50,000 head in a year. It is the largest lot in Canada. There are a number of operations handling 12,000 to 15,000 head

Representative pen of yearling steers. Slatted fence for wind protection in background. Western Feedlots, Inc.

annually, but most units handle 5,000 to 10,000 head. The larger facilities have automated feed-mixing equipment and bulk storage for the principal components. Smaller units, those under 10,000-head capacity, usually buy prepared feeds from a nearby mill. Stilbestrol was universally used by the commercial feed lots until incorporating this hormone in cattle feed was made illegal in 1972. This followed closely on similar action in the United States.

The typical arrangement for all but the small farm-type lot is two rows of pens on either side of a wide roadway which serves both for hauling in feed and for hauling out manure. Feed bunks, running the full length of the pens, are filled from a truck as it travels down the roadway. The quantity of feed to each individual pen is weighed and recorded, and the total cost of feed consumed by each lot of cattle is thus obtained. Lots in the Prairie Provinces have a windbreak of high slatted boards along the windward side but no other protection from the weather. In the Vancouver area of British Columbia some operations have shed roofs as well as windbreaks. They give protection from the heavy rainfall and also from periods of extremely hot weather.

The practice of custom feeding is increasing, and many of the larger lots operate almost entirely on this basis. Cattle are fed for growers who wish to market their own cattle as finished animals, as well as for other interests who seek the tax shelter offered by cattle feeding. In the latter case the feed-lot operator buys the feeders and sells the finished cattle for the account of the owner. The usual arrangement is for the investor in a lot of cattle to pay for the feed consumed plus a fixed charge a head a day for the facilities provided and the cost of whatever veterinary expense is incurred. The practice of feeding cattle on a fixed charge per pound of gain is not widespread in Canada.

The cattle fed are today predominantly yearling steers, of which 70 per cent grade choice. The number of calves entering the feed lots is small but is steadily increasing as producers change from yearling to a cow-calf operation. Yearling heifers are also fed when they are available, and there are isolated instances of lots of bulls, calves, or short yearlings being fed to reach good and choice grades.

In 1970 the government established quality grades for young bull carcasses with the proper conformation and finish. This was a drastic change in policy from the previous practice of disqualifying even superior young bulls and has encouraged the feeding of young bulls.

Fed cattle are now sold directly to the packers from the large feed lots. This method of marketing is also increasing in the smaller units. The central markets and auction sales are largely patronized by the smaller farmer-feeder.

MARKETING

The practices in marketing cattle in Canada closely parallel those in the United States. Prices of equal-quality slaughter animals in the two countries are closely tied together after adjustment for a tariff differential of 1.5 cents a pound and the prevailing foreign exchange difference, which at the close of 1971 was close to par. This permits the ready movement of cattle in either direction when price levels between the two countries vary. There is also some movement of dressed beef, although this is confined mainly to manufacturing beef. The close price relationship between finished cattle on the Canadian and United States markets is shown in Figure 8.

The Public Stockyard.—There are nine public stockyards scattered through the provinces. They perform a function similar to the stockyards or terminal markets in the United States. Historically the large packing plants were located near the population and transportation centers in the eastern provinces. The public stockyards grew up on their doorsteps. The Canadian Pacific Railroad was the major factor in

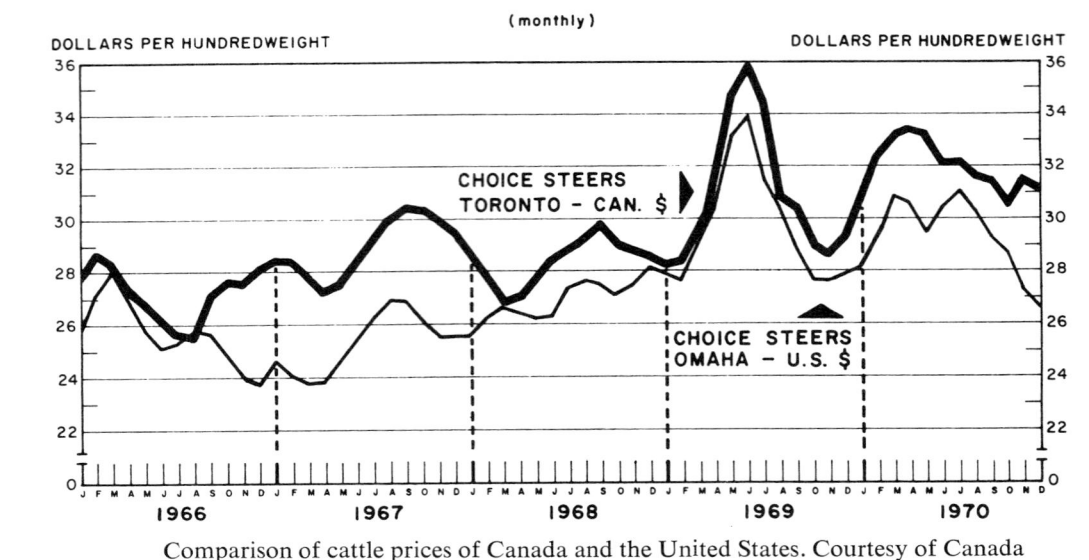

Comparison of cattle prices of Canada and the United States. Courtesy of Canada Department of Agriculture.

the establishment of stockyards, and, through its affiliates, it still owns some of the larger units.

The current trend is for the packing plants to move to the cattle- and grain-growing regions. The volume of trade in the stockyards has naturally followed this westward movement. Three of the largest public stockyards—those at Calgary, Lethbridge, and Edmonton—are in Alberta and handle one-third of the total cattle sold in Canada's public markets. Toronto, however, still handles the largest volume of any single unit. On a large day 5,500 head of cattle and upward of 500 calves will be sold.

The Canadian yard is much better oriented to the producer-seller of cattle than is the United States stockyard. Sales of cattle consigned to the Canadian commission firms, which in a few instances are co-operative producer-owned organizations, are made through an auction ring. The salesman passes out for future disposal any lot on which he decides that the highest bid is not in keeping with his appraisal of its value. The private-treaty method of selling by commission firms was replaced by auctions in 1950, and the change has definitely prolonged the life of the commission firm and the public market. This method of selling permits the rapid handling of small numbers of cattle and also keeps the farmer with only one or a few animals to sell coming to the stockyards, because he has more confidence in the auction method. In Edmonton, where annual sales are 350,000 head, the average draft is one and one-half animals. The major yards are at Calgary, Edmonton, Winnipeg, and Toronto, which together handle more than 200,000 cattle a year. Smaller public yards are located at Lethbridge, Regina, Saskatoon, Prince Albert, and Montreal. The Canadian Public Stockyards sold 40 per cent of all slaughter cattle in 1970 and 30 per cent of the replacement and feeder animals.

Direct sales from the feed lots and producers to packing plants is steadily increasing as is also the volume of cattle handled through the Canadian auction markets. The number of cattle sold through the Canadian Public Stockyards, though steadily declining in recent years, is now showing a tendency to stabilize, and it appears that this time-honored landmark of the cattle industry will continue to function in Canada for some time to come.

The Auction Market.—The auction markets are scattered throughout Canada wherever cattle are raised, in dairy areas as well as where beef cattle are grown. They provide a ready and convenient sales outlet for the small cattle owner. He can load a few calves or a culled cow in his pickup, take them to town, and return with his money the same afternoon. He also sees his offering sold—a privilege every owner enjoys. Sales in the auction markets are held one or sometimes two days a week, depending on the volume offered. Buyers include packing-house representatives, traders, and farmer-buyers seeking replacement stock. The auction markets handle 50 per cent of replacement and feeder cattle and 10 per cent of the total slaughter cattle, the latter being mostly culled cows and bulls.

Direct Sales to Packers.—Direct sales of fed cattle to packing plants increased rapidly as the feed-lot development moved westward. Such transactions are handled by two methods: private treaty between the feed-lot manager or the owner and the buyer of the packing company, or by a number of packer-buyers who visit the feed lot and make offers for a lot of cattle which they have been notified will be offered for sale at a specified time. Over 55 per cent of fed cattle are now sold direct to the packers. The pricing may be on either a liveweight or a carcass-weight basis. Selling on the carcass-weight basis has been increasing in recent years.

Ranch Auctions.—In the Kamloops area of British Columbia the practice is growing of selling feeder cattle at ranch auctions. The sale is advertised, and the offerings, usually a few hundred head divided into

Yearling steers moving into corral at ranch auction in Kamloops area.

lots by estimated weight and grade, are sold in the home corral. The method offers the advantage of direct delivery to the buyer while maintaining competitive prices.

Calf Clock Auction.—At the Toronto Stockyards the Dutch clock method has been adapted to the veal calf market.

The Dutch method of selling is on a descending-bid basis, the auctioneer asking for what he estimates to be the highest possible bid to start and descending from that point. Thus the first bidder becomes the buyer. For example, if calves are selling in the $50 to $53 range, the auctioneer may start the bidding at $53. If there are no takers at that price, he asks for $52.95, then $52.90, and so on, until a bid is made and the calves are sold.

The prime argument in favor of the Dutch method is that it speeds up the bidding. To facilitate matters further the clock is used in most markets where this method is employed. The clock is actually a large dial fitted with numbers descending from 100 to 05 in increments of

Calf auction clock, Toronto Stockyards.

five. A sweep hand moves at a steady pace around the dial once the mechanism is set in motion, going from 100 to 95, to 90, and so on, or, in the case of actual bidding from, say, $10.00 to $9.95, to $9.90, and down.

A desk equipped with a key by which the clock may be stopped is provided for each buyer. When the sweep hand reaches the point where the buyer decides to bid, he presses the button, the clock stops, and an electronic device identifies him. At the Toronto yards the commission man starts the clock and controls the sales. He has the option of refusing the price at which the clock is stopped if he considers it too low. Many

329

Cattle of North America

single units can be moved in a remarkably short time with this method.

Beef Grading.—The Canadian grading system for beef, first inaugurated in 1929, historically has recognized only the quality of a carcass as the determining factor in arriving at the grade to be assigned to it. On September 5, 1972, new regulations respecting the grading of beef carcasses as approved by Parliament became effective. These regulations superimposed yield grade standards on the two highest of the old quality grades, choice and good, now A and B under the new system. Yield grading is now required on all A and B carcasses.

The application of the new grades is handled the same way as in the past—by a federal grader in the packing plants who is an employee of the Livestock Division of the Production and Marketing Branch of the Canada Department of Agriculture. Each carcass is stamped by the grader with the grade number as he determines it, and it must be ribbon-branded ("rolled" in United States trade terminology) under the supervision of the grader to show the applicable grade.

Quality grade was formerly determined mainly by the maturity of an animal as evidenced by the degree of ossification of the cartilage, the conformation, and the general appearance of the external fat cover. Color and texture of the muscle tissue and the degree of marbling in the rib eye, observations that could not be made previously because the carcasses were not "ribbed" (cut between the eleventh and twelfth ribs to expose the rib eye), now enter into determination of quality grade.

The sole factor in the Canadian yield grading system is the fat thickness over the rib eye as related to carcass weight. All carcasses of A and B grades must be ribbed before grading. Carcasses falling under C grade may require ribbing to determine quality.

The table on the opposite page sets forth the general requirements to meet the standards for quality and, where applicable, the standards for yield of the newly established Canada grades. The chart is self-explanatory except for Grade D, Class 4.

Grade D, Class 4, is a catchall for inferior carcasses that are considered to be of manufacturing quality—that is, destined for sausage or canned-beef products. The official description is that this grade includes "poorly muscled females and steers and may also include excessively fat mature carcasses and young carcasses failing to meet Canada C Class 2 grade."

The Canadian quality grades do not correlate closely to the USDA quality grades. The USDA quality grade prime had no equivalent in the old Canadian grading system. The high degree of marbling in the

CANADA CARCASS GRADES

Quality Grade Requirement	Grade Designation	Yield Grade Requirement Carcass Weight, Lb. 300–499 / 500–699 / 700+
Grade A applies mainly to young steers and heifers but may include young bulls. A slight degree of marbling required in rib-eye, which must be bright red, lean, firm, and fine-grained. Fat cover is white and uniform over body.	Red stamp and brand Canada A 1 Canada A 2 Canada A 3 Canada A 4	Fat thickness in inches at twelfth rib .20–.30 .20–.40 .30–.50 .31–.50 .41–.60 .51–.70 .51–.70 .61–.80 .71–.90 over .70 over .80 over .90
Grade B applies to young steers and heifers and may also include young bulls. There is no minimum marbling requirement. Fat cover may be pale yellow and somewhat lacking on hips and chucks.	Blue stamp and brand Canada B 1 Canada B 2 Canada B 3 Canada B 4	.10–.30 .10–.40 .20–.50 .31–.50 .41–.60 .51–.70 .51–.70 .61–.80 .71–.90 over .70 over .80 over .90
		Class-determining Factors
Grade C in general applies to carcasses of young animals that fail to meet standard in some requirement for grades A and B. No yield grade is applied. Ribbing may be required to determine quality.	Brown stamp and brand Canada C Class 1 Canada C Class 2	Carcasses which fall just under Grades A or B in some requirement. Applies to carcasses from young animals that have a marked deficiency in muscling.
Grade D applies to mature females and old steers and to young animals falling below Canada C Class 2 in some requirement.	Black stamp and brand Canada D Class 1 Canada D Class 2 Canada D Class 3 Canada D Class 4	Good and excellent muscled mature females and steers. Medium muscled mature females and steers. Fair muscled mature females and steers. Manufacturing quality.
Grade E applies to mature bulls and stags.	Black stamp and brand Canada E	

rib eye and the related thick fat cover of USDA prime were never acceptable to the Canadian trade. The new grading system likewise made no provision for such excessively fat carcasses except to relegate them to the inferior grade, Canada D, Class 4. Neither does the new grading system have a close approximation to the USDA choice grade, although Canada A grades would encompass some carcasses that would fall in the lower range of USDA choice (referred to in the United States trade as low choice).

Canada A and B grades would fall mostly in the range of USDA good. A few Canada C, Class 1, carcasses might fall in USDA low good grade. The bulk of Canada C, and the Canada D and E grades, have little correlation to the USDA lower grades.

The yield grades are designated by the numerals 1, 2, 3, and 4 following the quality grade designations and are determined by the relationship between the warm carcass weight and the fat thickness at the twelfth rib of the carcass as indicated. The yield grade 1 designates the highest proportion of salable meat to warm carcass weight, and 4 the lowest. In general the designation 1 in the Canadian yield grading system compares roughly to the yield grade 1 in the USDA system; Canada 2, to USDA 2; Canada 3, to USDA 3 and the upper part of USDA 4; and Canada 4, to the lower part of USDA 4 and USDA 5. The principal differences between the USDA and the Canadian yield grading systems are that the area of the rib eye in relation to carcass weights and the percentage of internal fat enter into the USDA standards, and neither of these factors is involved in the Canadian system.

In Canada all carcasses coming under federal inspection must be ribbon-branded with the quality grade. The yield grade is also indicated for carcasses grading A and B. In the United States rolling (or ribbon branding) the quality grade on a carcass is optional with the individual packer and is applied only to prime and choice carcasses. Yield grading is also optional and when used is indicated only by the grader's stamp.

PACKING PLANTS

Packing plants meet exceptionally high standards of sanitation and animal health inspection. All plants processing meat which may enter interprovincial commerce are under inspection by the Health of Animals Branch of the Canada Department of Agriculture. This covers 90 per cent of the beef produced in the country. The remaining 10 per cent is processed in small local facilities that come under either provincial or, in a few cases, municipal inspection.

Canada

Although only a limited number of Canadian plants export beef to the United States, the Canada Department of Agriculture requires all plants under its jurisdiction to be able to qualify for export to the United States. This requires an annual inspection of 104 Canadian plants by USDA officials.

The decentralization of the packing industry which was concentrated around Toronto began in the years after World War II. This has resulted in the installation of modern equipment in all the new and larger plants, together with wide-scale modernization in the older plants.

The plant of Canada Packers in Toronto, the largest Canadian meat processor, has a capacity of 129 head of cattle an hour and averages 1,000 head a day, plus about 2,000 hogs. Some plants break carcasses into institutional cuts, but most beef goes to the retailer as whole sides. A full line of by-products is manufactured in all the larger plants.

There are many small slaughtering facilities throughout the country serving local communities where home-grown beef can be processed at a cost which competes with that of products shipped in from the large centrally located units. Many of these local processors have a capacity of 50 to 200 head of cattle a week and in addition often handle hogs. Their facilities consist of a small holding yard; a small, substantial building; an overhead rail for dressing; and usually a cooler. Their output goes to local retailers or into the frozen-food lockers of their patrons. Cattle are either purchased and slaughtered by the plant operator for the wholesale or retail trade or are processed on a custom basis. The provincial and municipal inspection ensures sanitary operations.

The major supply of fed cattle to the packers now comes direct from the commercial feed lots. Culled cows and Bologna bulls for manufacturing beef are bought in the public stockyards and auction markets.

There is little or no packer feeding of slaughter cattle. In recent years the practice of buying on a yield basis has increased. Of the total fed cattle killed, over half are now sold on a basis that involves carcass yield to some extent in the formula that has been arrived at between buyer and seller. As an example, the buyer may agree to pay an extra $1 per hundred pounds of liveweight if the average dressing per cent of a lot of cattle is 62 or higher.

CATTLE DISEASES

Canada's cool climate provides a healthy environment for cattle. Government efforts to control, eradicate, and prevent diseases of livestock have been highly successful, and this is especially true for cattle. The Health of Animals Branch of the Canada Department of Agriculture supervises all foreign or interprovincial movements of cattle. The provincial departments of agriculture direct intraprovincial animal health activities.

The only foot-and-mouth disease outbreak in Canada in the twentieth century was detected in February, 1952. It was quickly eradicated, and the country was again declared free of the disease in August of that year. An immigrant farm worker discarding pork brought from Europe was thought to have caused the outbreak.

Canada has been considered virtually free of bovine tuberculosis since 1961. This happy circumstance can be attributed to a rigid test and slaughter program.

The incidence of brucellosis is now so low that calfhood vaccination is no longer recommended. A testing program was started in 1957, and the first nationwide test was completed in 1966. The ring test on milk deliveries to processing plants and the back tagging of culled dairy cows is now proving an effective means of further reducing the incidence of brucellosis. Currently, 75 per cent of the counties in Canada are considered brucellosis-free. The remaining 25 per cent are in the "certified" category, that is, not over 5 per cent of the herds in a county harbor reactors, and less than 1 per cent of the total cattle are considered to be infected.

The incidence of the venereal diseases of cattle is not considered serious. Vaccination for vibriosis has been approved by the veterinary services. The extent to which this disease affects the national herd is not known.

It is customary throughout Canada to vaccinate for blackleg, just as it is in the United States. Often the three-way vaccination against blackleg, malignant edema, and hemorrhagic septicemia is used.

Quarantine Stations.—The Canada Department of Agriculture provided the first maximum-security station in North America for livestock importations. This permits the movement of cattle into the country from any area reasonably free from foot-and-mouth, rinderpest, and other serious cattle diseases without subjecting the national herd to undue risk.

The first of these stations was completed in 1965 on Grosse Île, in the St. Lawrence River 40 miles east of Quebec. A second station was built in 1968 on the French Island of St. Pierre off the southern coast of Newfoundland. Operation of the St. Pierre station is directly under the Health of Animals Branch of the Canada Department of Agriculture, although the ownership is French. Grosse Île has facilities for 250 cattle, and St. Pierre can handle 200. Both stations will probably accommodate 900 head annually from Europe as long as the demand for the new breeds persists.

The quarantine and control provisions, as defined by law for the admittance of cattle to Canada from countries that have not been declared free of foot-and-mouth disease, are a prime example of precautionary measures for protection of animal health. They are as follows:

1. The country of origin must be approved by the veterinary director-general of Canada (VDG). An import permit for the specific animals to be shipped must be obtained from the Canada Department of Agriculture.

2. Import cattle must be held 30 days in an approved quarantine station in the country of origin.

3. Import cattle must not be over nine months of age on arrival at the Canadian quarantine station. Actually this means that an animal is less than eight months old at the time of selection in order to allow a sufficient interval for the 30-day quarantine in the European station and the voyage to Canada.

4. All food and litter used by import cattle in the European station must have been shipped direct from Canada (or other approved source).

5. Movement, maintenance, and inspection of import animals are at all times supervised and approved by a veterinary officer of the Canada Department of Agriculture.

6. Shipment to Canada must be by surface vessel, and all food and litter must have come direct from Canada.

7. After movement to the Canadian quarantine station, import cattle must remain under inspection 90 days.

8. After final inspection and release, import animals must go direct from the quarantine station to a farm in Canada having a susceptible herd of cattle and approved by the VDG.

9. After 90 days of departmental quarantine on the farm, import animals are given a final inspection and released for free movement.

10. The importer pays all costs for feed and care at the quarantine station, transportation costs, and $10 a day quarantine fee. Total import costs run to over $2,000 a head.

The Canadian quarantine procedures covering the importation of cattle from such countries that have not been declared free of foot-and-mouth disease are the most thorough of any country in the world.

GOVERNMENT AND CATTLE

Before confederation in 1867 government involvement in the cattle and dairy industries was limited to action by the agricultural boards of the provinces. Among the first of these organizations was the Ontario Board of Agriculture, established in 1843. In the eastern provinces the boards established registers for purebred stock well in advance of the herdbooks of the breed associations. The provincial boards were instrumental in the introduction of the "new breeds" of that day—the British breeds which had recently been developed in England and Scotland. With the establishment of the Canada Department of Agriculture in 1867 the activities of the provincial organizations ceased to expand, and in some areas were absorbed by the federal agency.

The major livestock function of the Canada Department of Agriculture (CDA) in the years immediately following its organization was the control of disease and the providing of quarantine facilities for imported animals. The Federal Experimental Farm system was not authorized by act of Parliament until 1888. From that time on the activities of the department rapidly increased.

The executive head of the CDA is the minister of agriculture, an elected member of Parliament. He is the only official in the department who changes with the political winds. All the other employees, including the deputy minister, who is the acting head of the department, are civil servants.

Federal government control and participation in matters involving cattle come under two branches of the CDA, the Production and Marketing Branch and the Research Branch. The functions of these organizations are summarized below.

Production and Marketing Branch.—

LIVESTOCK DIVISION. The functions of this division are the establishment and supervision of grade standards for meat products; the administration of stockyards; the production testing of dairy and beef cattle, which service is performed in co-operation with the provincial departments of agriculture; and supervision of the registration of purebred livestock.

HEALTH OF ANIMALS DIVISION. This division is charged with inspection of all meat entering interprovincial or export trade; the prevention and control of animal diseases, including quarantine procedures; and research on animal diseases.

Research Branch.—

ANIMAL RESEARCH INSTITUTE. The research stations and experimental farms of the federal government are operated by this division.

Canada

Cattle-breeding, productive-ability, and nutritional investigations are conducted at 12 establishments at various locations throughout the country.

Each provincial government has an agriculture department whose livestock divisions are active in all phases of the cattle industry as adapted to local circumstances. Dairy cattle receive major attention in the Maritime Provinces, where milk production is more important than beef production. In the Prairie Provinces the reverse is true, and the principal work of the experiment stations is with beef cattle.

Historically the provincial departments of agriculture, which replaced the boards of agriculture, exercised control over the registration of all purebred livestock. The regulations applying to registration procedures and issuance of certificates of registration varied widely from province to province. This led to confusion as purebred stock was moved from one province to another and even greater difficulty in the export of breeding stock to the United States.

After several years of negotiation with the provincial departments of agriculture and the breed associations, the CDA obtained legislation to correct this situation in the form of the Live Stock Pedigree Act of 1905. The act is administered by the CDA and "provides the authority for incorporation of Breed Associations for purebred livestock to register and transfer animals, affiliate with the Canadian National Live Stock Records, and to conduct the affairs of the Associations." Other livestock —horses, sheep, poultry, and dogs—are also covered by the act.

The Canadian National Live Stock Records (CNLSR) is the operating organization that was given authority under the Live Stock Pedigree Act to perform the functions of registration, transfer, and issuance of certificates for animals of any breed. The CNLSR is in effect a nonprofit co-operative of the participating breed associations. Participation is voluntary, and a breed association may elect to do its own recording. Only two of the major breed organizations, the Holstein-Friesian and the Charolais, have elected to remain outside the CNLSR and do their own recording.

Registration and transfer certificates issued by the CNLSR carry the seal of the CDA. This has sometimes led to the assumption that the CNLSR is an arm of the federal government. At its inception the organization was subsidized to some extent by the government, but such aid is now limited to the furnishing of office space.

No breed is eligible for registration by the CNLSR unless its articles of incorporation are approved by the Canada Department of Agricul-

Cattle of North America

ture, and different types of one breed (polled and horned, for example) cannot be registered as separate breeds under the Canadian system.

Meat inspection in small slaughterhouses which are involved only in local killing and whose production does not enter interprovincial commerce is usually handled by provincial inspectors. Extension work directly with farmers comes under the provincial department of agriculture. There are both federal and provincial experiment farms, and they often engage in parallel enterprises. The Federal Livestock Division maintains the Record of Performance on purebred dairy herds; the provincial departments of agriculture perform the same function for commercial dairy herds.

OUTLOOK FOR CATTLE

Canadian economists estimate that the consumption of beef during the decade of the seventies will increase 40 to 50 per cent from the 1969 level. If the Canadian and United States supplies are consolidated, however, the increase for Canada in 1980 over 1969 is put at 36 per cent, assuming that the increase in the United States production after supplying the United States demand will be available to Canada. These forecasts are based on expected population increases and higher per capita consumption of beef. Increased beef production in Canada will surely mean a substantial increase in the national beef herd. While the dairy industry will probably supply more steer calves for the feed lots, most of the increased beef production will have to come from the cow-calf man.

Some increase in the productivity of the beef cow can be expected from breed improvement. Heavier weaning weights and faster feed-lot gains are certain to come. The new breeds and crossbreeding will also play a role; how significant remains to be seen. With all this, though, the national beef herd will have to be increased to meet the anticipated demand.

With a breeding herd at the 1972 level of close to 3⅓ million beef cows and 2⅓ million dairy cows, there is ample room for expansion. The dairy herd has been steadily declining during the past decade as the quantity of milk produced per cow has increased. The rate of increase in the annual production of the average dairy cow is now declining and further decreases in dairy-cow numbers will be slow. On the average, beef cattle numbers in recent years have increased at a rate which just about compensates for the decrease in dairy cattle, holding the total cattle population fairly constant. A larger national herd, however, now appears a necessity. The increase will be in beef cattle.

Canada's potential for increasing cattle numbers and production is

Canada

ample to meet the demands of the foreseeable future. The northern limit at which cattle can be profitably raised can be extended beyond the present line as shown on the map at the beginning of Part IV. Experience in the Peace River Valley has shown that cattle can be raised farther north than was thought possible a few years ago. Innumerable small dairy herds in eastern Canada, from the Maritimes to Ontario, will move into the production of feeders. As has been indicated, this trend is already well under way.

Throughout the Prairie Provinces grain farmers will increasingly graze herds of cattle on wasteland and plant residues. Cattle husbandry will probably edge farther north in the parkland, while the ranchers of Alberta and British Columbia increase production as they move from yearling to cow-calf operations. The larger ranchers in the Alberta foothills and in the high plateau and the mountain valleys of British Columbia will probably achieve no large increase in cattle numbers in the years ahead. As a matter of fact, the competition between recreation and grazing rights on the public domain may force a decrease in size of herds to an extent impossible to overcome by better management. All things considered, however, Canada should have no trouble increasing her beef herd to meet whatever demands are put on the country in the years ahead.

The dairy breeds are now well stabilized, with the Holstein-Friesian constituting 80 per cent of the national herd, followed by the Ayrshire, Guernsey, Jersey, and Canadian, in that order. The annual increase in the Holstein population is now so small that it seems certain that the other four dairy breeds will continue at their present levels.

The Hereford accounts for over 55 per cent of all beef cattle. The Angus is next, followed closely by the Charolais if its various crosses are included. The Shorthorn, which includes a small number of milking Shorthorns, is in fourth place among the beef breeds. The major question in the future of the individual breeds is the effect that crossbreeding and the new breeds will have on the cattle population. These are currently highly unpredictable elements in the cattle picture.

For many years the western cattlemen have practiced crossbreeding to a limited extent. This has consisted mostly of using Angus bulls on Hereford first-calf heifers to achieve easier calving of two-year-olds. This practice has increased with the recent emphasis on crossbreeding.

The new breeds have not been available long enough to establish a trend for their future place in crossbreeding. Practically the only systematic crossbreeding utilizing the new breeds up to 1972 was the work done in the experiment-station herds. The ubiquitous Charolais or new breed crossbred seen among herds across the country is usually the result of the owner's curiosity about what the crossbreeding talk is all about.

For the immediate future considerable effort will be devoted to upgrading cattle to one of the new breeds. The creation of his own purebreds by top crossing to the fourth or fifth generation is a temptation the small cattleman cannot resist. While some breeders will develop purebred herds in this manner, most will probably give up along the way.

PART FOUR: United States

United States

Land area (50 states) (sq. mi.):	3,615,210
Population (1970):	203,185,000
Density (per sq. mi.):	56
Agricultural (4.6%) (1970):	9,400,000
Per capita income (1970):	$3,910
Total cattle population (1972):	118,287,000
Beef cows:	38,789,000
Dairy cows:	12,392,000
Total cows:	51,181,000

THE UNITED STATES of America occupies the central part of the North American continent from the Atlantic Ocean to the Pacific Ocean. The southern boundary is Mexico and the Gulf of Mexico, and on the north are Canada and the Great Lakes. The United States lies for the most part between the 30th and 49th parallels, and the climate is generally temperate, although the northern tier of states and extensive areas of high elevations in the Rocky Mountains are subject to severe winters. The southern states bordering Mexico and the Gulf of Mexico approach subtropical conditions during the summer months. Rainfall generally is adequate for seasonal plant growth east of the Rocky Mountains. In the West precipitation varies from 10 to 20 inches in the mountain areas to arid or desert conditions in the extreme Southwest. Along the northern Pacific Coast rainfall is ample.

The first permanent English settlement on the continent was established at Jamestown, in what is now Virginia, in 1607. The Plymouth Colony was founded in Massachusetts in 1620; in the following year the Dutch settled on Manhattan. Settlements along the Atlantic Coast then followed rapidly, and by 1700 all the original thirteen British colonies except Georgia had been organized. The colonies were outgrowths of land grants by the British Crown to private interests—companies, groups of farmers, and influential individuals. Eventually each colony came under the control of a governor appointed by the Crown, although the inhabitants of a colony had a voice in selecting a representative body which was advisory in nature.

In 1776 by a unilateral declaration, the thirteen colonies proclaimed their independence and after eight years of war with their homeland succeeded in establishing their sovereignty by treaty in 1782 and 1783. The United States at that time consisted of the lands lying along the Atlantic Coast and extending to an undefined western boundary, which eventually was recognized as the Mississippi River. The Great Lakes and a disputed region along the St. Lawrence Valley lay to the north, and Spanish-occupied Florida was the southern boundary. From this nucleus the country expanded over the next century by purchase, cession, and voluntary amalgamation to its present territory.

New York, New Hampshire, and Massachusetts claimed regions of the present state of Vermont. These claims were eventually withdrawn, and Vermont entered the Union as the fourteenth state in 1791. The thirteen original colonies claimed lands extending westward to the Mississippi River. In time these lands were ceded to the federal government or to other states.

Following are the acquisitions which, together with the area of the original colonies, make up the United States today.

(In the section below the number preceding each paragraph refers to and explains the numbered area marked on the accompanying map.)

1. Louisiana Purchase.—At the close of the eighteenth century Spanish claims in North America equaled in area those held by Spain in South America. All of the present area of the conterminous United States lying west of the Mississippi River, as well as Florida, was claimed by Spain. In 1800, Napoleon gained from Spain the area between the Mississippi River and the Rocky Mountains. A few years later, because of his mounting troubles in Europe, he sold the region to the United States for $15 million.

Robert R. Livingston and James Monroe were authorized to purchase New Orleans from France for $2 million. In the course of the negotiations in Paris, all of the territory of Louisiana was unexpectedly offered to them for $15 million. Although not authorized to do so, Livingston and Monroe accepted the offer, managing to arrange for $3 million of the purchase price to be allowed as indemnity for damages by France to United States ships. The net terms submitted to Congress were $12 million for the vast area. There were strong objections to paying "such a large price for a worthless tract of land." Congress, however, after much heated debate, approved the treaty for the Louisiana Purchase.

2. Red River of the North Basin.—Possession of the fur-trading area in the fertile Red River of the North Basin was in open dispute between the United States and Great Britain until the end of the War of 1812.

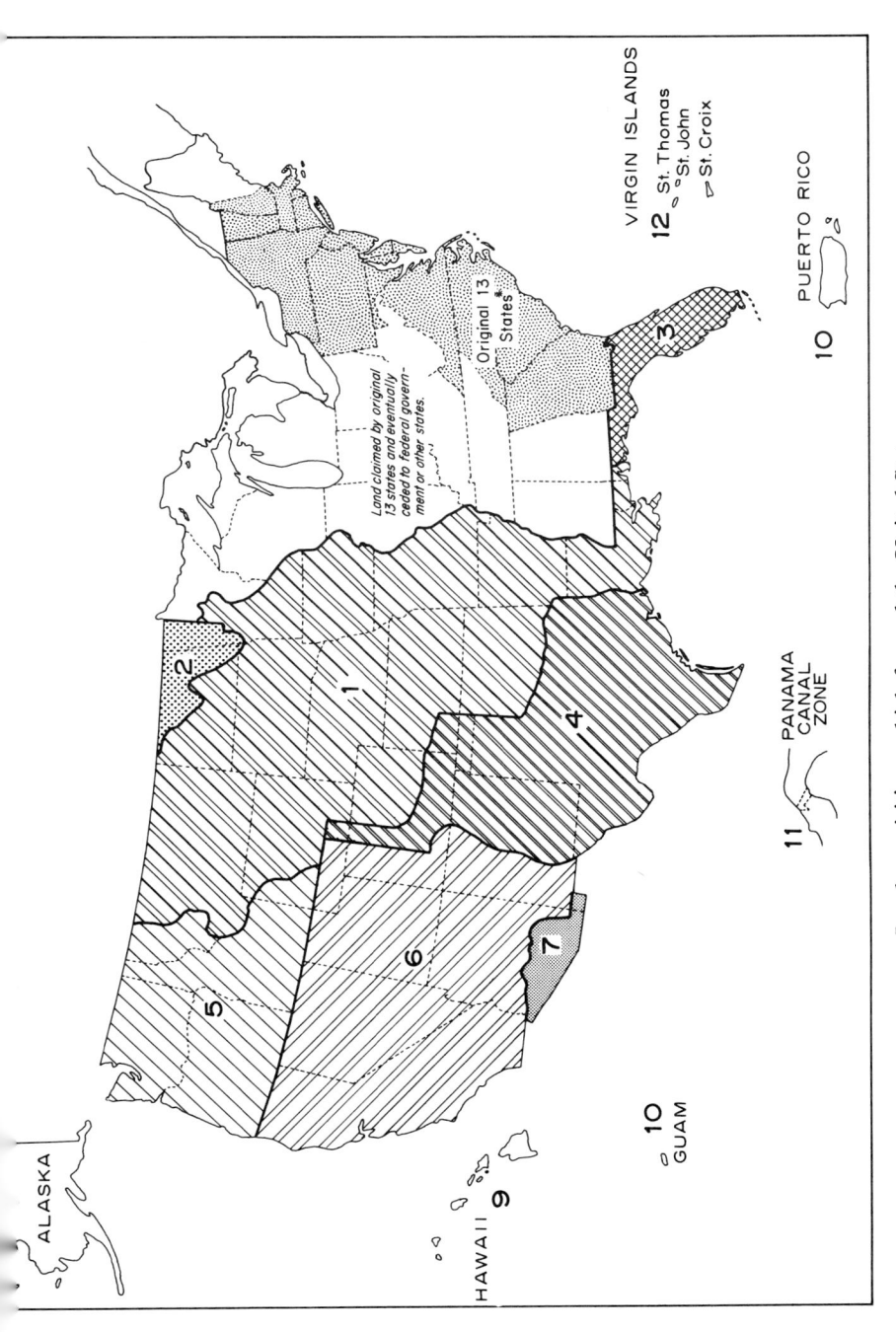

Land acquisitions which formed the United States

Britain then gave up claim to the area, and in 1818 the boundary between the United States and Canada in that region was fixed at the 49th parallel.

3. Florida Cession.—In 1819, Spain ceded Florida to the United States. In return, the United States assumed $5 million in claims of American citizens against Spain.

4. Texas Annexation.—Texas, which gained independence from Mexico in 1836 and was a sovereign state until the end of 1845, was annexed to the United States in the following year at its own request, to become the twenty-eighth state. The original territory of Texas as annexed included the eastern half of New Mexico, the Oklahoma Panhandle, the southwestern corner of Kansas, and a strip north through Colorado and into Wyoming ending at the 42d parallel. All of the area now excluded from the present boundaries of the state of Texas was purchased from that state by the federal government in 1850.

5. Oregon Region Treaty.—The vast area north and west of the Louisiana Purchase that was not held by Mexico was a wilderness traveled only by fur traders in the opening years of the nineteenth century. The boundary between the United States and Canada was in continual dispute. The "Oregon Question" became a major political issue in the United States. In 1846 a settlement by treaty between Great Britain and the United States placed the boundary at the 49th parallel.

6. Mexican Cession.—The lands west of the original boundaries of Texas to the Pacific Ocean remained under the control of Mexico until they were ceded to the United States in 1848 at a cost of $15 million. This agreement was an aftermath of the Mexican War. The purchase included what are now the states of Arizona, California, Nevada, Utah, the western parts of New Mexico and Colorado, and the southwest corner of Wyoming.

7. Gadsden Purchase.—In 1853, 30,000 square miles south of the Gila River in Arizona were purchased from Mexico for $10 million. The negotiations concluded Mexico's claims for depredations by the Apache Indians living on United States soil and the United States' desire for a favorable transcontinental railroad route.

8. Alaska Purchase.—Alaska, over which Russia had claimed dominion since 1841, was sold to the United States in 1867 for $7.2 million.

9. Hawaiian Annexation.—After years of seeking annexation by the United States in order to avoid domination by one of the European powers, the Hawaiian Islands succeeded in being accepted as United States territory in 1898.

10. Spanish Possessions Treaty.—Following the Spanish-American War in 1898, Spain relinquished the last remnants of her once vast domain in the Western Hemisphere, ceding Puerto Rico, the Philippine Islands, and Guam to the United States. Cuba attained independent status under United States protection. The total consideration for all the areas was $20 million. The Philippine Islands were subsequently granted independence; Puerto Rico was granted commonwealth status; and Guam remains an unincorporated territory of the United States.

11. Panama Canal Zone.—The Panama Canal Zone was acquired from the Republic of Panama in 1903. The United States by treaty acquired full sovereignty in perpetuity for a strip of land 10 miles wide and 50 miles long across the Isthmus of Panama between the Atlantic and Pacific oceans.

The international situation involved in the acquisition of a canal site became critical as negotiations with Colombia were nearing a close. The area that is now Panama was at the time a part of Colombia. The Colombian senate unexpectedly refused to ratify the treaty. The Panamanian politicians, fearful of losing the canal, sought United States support in seceding from Colombia. This support was given unofficially but effectively in the form of naval presence. On November 3, 1903, Panama proclaimed its independence, and on November 18 the treaty ceding the Canal Zone to the United States was concluded.

12. Virgin Islands Purchase.—In 1917, during World War I, the islands of St. Thomas, St. John, and St. Croix in the Virgin Islands group in the Caribbean were purchased from Denmark for $25 million. These islands were important as a security link in the outer United States line of defense. They now comprise an unincorporated territory of the United States.

The entry of the white man into what is now United States territory was made when Coronado crossed the Río Grande in 1540. His explorations reached as far north as present-day Colorado and extended from western Kansas to Arizona. During the seventeenth century missions were established in this area. The Spaniards also entered Florida nearly a century before Jamestown was established, although no permanent settlements were founded until years later. Then in the eighteenth century, as expeditions crossed the Río Grande from Mexico, numerous missions were founded by Spanish religious orders from Texas to California. Others were established in Florida. However, these outposts

Cattle of North America

contributed little to the economic growth of the rapidly expanding country except for the part played by Texas in establishing a base for the cattle industry.

During the course of the nineteenth century the United States developed from an agriculturally oriented economy into a major industrial complex. The trend continued after the turn of the century. During colonial times there was hardly a household which did not contribute in some measure to its own sustenance: a milk cow pastured on the village common or an extensive green garden. In 1865, 85 per cent of the population lived in the country. By 1910 the farm population had decreased to 33 per cent of the total and continued at about the same rate until the beginning of World War II. The farm population then dropped rapidly. Today it makes up less than 5 per cent of the total population. Coincident with the decrease in farm labor, there has been an increase in farm size and a corresponding decrease in the number of farms. In 1935 the number of farms in the United States was the highest on record, 6,812,000. By 1970 this number had decreased to 2,895,000 farms. Smaller and noneconomic units were combined with larger farms, or farm operations were abandoned.

The trend toward larger and more highly mechanized farms sharply accelerated around 1950. Larger fields, specialized fertilizers and herbicides, and more powerful and more highly automoted machinery are factors which continue to dictate an increase in the size of the farm unit. The raising of crops and the production and feeding of livestock are taking on more and more of the aspects of the factory and production line. The utopian image and the romance of the typical American farmsteads of past generations are rapidly passing from the agricultural scene.

The three areas beyond the limits of the conterminous states—Alaska, Hawaii, and Puerto Rico—have individual historical and husbandry aspects in the handling of cattle which differ materially from those of the rest of the United States. These three areas are covered at the end of the section "Management Practices" below.

BACKGROUND OF THE CATTLE INDUSTRY IN THE UNITED STATES

There were no cattle (*Bos*) in the Western Hemisphere until introduced by man. The only member of the bovine family indigenous to this part of the world is the American bison—the buffalo—which is a separate branch of the bovine family (see the chart "Origin of Today's Cattle," *World Cattle*, II, 1026–27).

Buffalo cows and calves. Wichita Mountains Wild Life Refuge, Cache, Oklahoma.

The buffaloes, among the most spectacular populations of wildlife that the world has known, were barely saved from extinction by conservation forces in the United States at the close of the nineteenth century. Millions of these animals ranged the vast grasslands between the Missouri River and the Rocky Mountains until the hunters who preceded the settlers began to exterminate them. Many Indian tribes had long derived most of their sustenance from the buffaloes. These they systematically hunted but without noticeably decreasing the population.

Government reserves now harbor sizable herds, and there are many private herds which are held in what approaches a domesticated state. Various attempts have been made to develop a cross of the bison with *Bos Taurus* (humpless cattle) without much success (see "Cattalo" in the section "Cattle Breeds").

SOURCES OF CATTLE

Many kinds of cattle have been brought to the territory that is now the United States at various times over the past 360 years and have intermingled in many different ways. The cattle of the United States as seen today generally trace to three major sources, which over the years have contributed in varied ways to the melting pot. These sources were

Cattle of North America

the Spanish cattle that were carried to the West Indies; northern European cattle, including those from both the continent and the British Isles; and the Zebu cattle of India, which reached the United States direct from India or by way of Brazil and Mexico.

Beginning in 1966 a wave of introductions of breeds new in the Western Hemisphere began and in 1972 was continuing at an accelerated rate. Various terms have been used in referring to these breeds as a class, such as "exotics," "continentals," and "new breeds." None of these are aptly descriptive. Because they are new to North America, "new breeds" is used here in referring to them.

Two breeds of African cattle have entered into the composition of two minor United States breeds. These are the N'Dama, which contributed to the Senepol breed, and the Africander, which contributed to the Barzona breed. There have been other minor introductions of foreign cattle, but they have left little, if any, impression on the cattle population of today. The Canadian (French Canadian) Cattle, the Normandy of France, the Red Sindhi, and the Red Danish were brought to the United States and have now disappeared except for minor remnants of the last two breeds. The Himalayan Yak and Siberian cattle reached Alaska but are now gone. A few descendants of the British White Cattle, brought in during World War II, remain in Texas. To complete the record, all these are mentioned briefly under the heading "Other Breeds," in the section "Cattle Breeds."

The chronology of cattle movements involved and the parts which the different types played are discussed below.

Northern European Cattle

Cattle imported from northern Europe, including the British Isles, have made the largest contribution to the United States cattle population. There have been four major waves of these importations. The first began at the time of the founding of the British colonies; the last is still in progress and probably has not yet reached its peak.

The prebreed types arrived during the first half of the seventeenth century and were practically the sole basis of the cattle population for the next 200 years. Introduction of the newly developed British breeds began early in the nineteenth century. Purebred cattle from continental Europe were also imported. This influx of purebred cattle reached a peak toward the close of the century and then declined as registered cattle in the United States were able to meet the demand for seed stock. The Charolais movement started in the late 1930's when representatives of the breed were brought from Mexico to the United States. Charolais importations then continued as rapidly as quarantine regulations permitted. Finally, in 1967, the influx of the "new breeds" began through use of imported semen.

The Prebreed Types.—The cattle supplied to the settlers in the British colonies in the early years of the seventeenth century were obtained from whatever stocks were available near the ports of embarkation. Often they were among the possessions of the emigrants. They were needed to serve as draft animals and to furnish household milk. Meat was more economically obtained from wild game, sheep, and hogs. Little is known of the physical characteristics of the cattle as the only mention usually made of them is in regard to their number and sometimes their color. To the historians of the time, cattle were mostly just "cattle." There were no breeds of cattle as they are known today, even in England.

There were, however, both on the British Isles and on the Continent, local concentrations of cattle of a particular type or color. They were the result of some measure of natural selection, which possibly had also been influenced by a local farmer's preference for a color or a shape of horn. While these local groupings were not sufficiently differentiated in type to warrant calling them "breeds" in the modern sense, they were definitely the forerunners of the breeds that are known today.

The first cattle of which there is continuity of record to become established in the British colonies were shipped from England to Jamestown in Virginia in 1609. An earlier attempt to establish a settlement had brought cattle to the Atlantic Coast in 1585, for the settlement attempt organized by Sir Walter Raleigh on Roanoke Island. When a supporting ship returned the next year, the Roanoke settlers had disappeared, and no trace of the cattle was found. In 1610 a few head of cattle were imported to Jamestown from the West Indies. There were also some early shipments to the colonies from continental Europe. Most of the importations, however, both before and after the American Revolution, originated in the British Isles.

Within a few decades after their introduction the varied types of cattle that were husbanded by the colonists were able not only to maintain their numbers but also to afford a modest increase for export. Importations from the Old World had practically ceased by 1650. The cattle population of that day became the melting pot from which emerged the "native" cattle of America, as they were known two centuries later. They are referred to in the 1866 Report of the Commissioner of Agriculture as follows:

Cattle of North America

It thus appears that the original stocks of cattle brought into the colonies were mostly English, with the few exceptions made by the Dutch, Danish, and Swedish importations, and perhaps a few French cattle introduced by the Huguenots to South Carolina, and possibly by some of the Huguenots who early came to New York. These cattle spread along the coast and into the interior, with the colonists, and as the latter intermixed their settlements, their herds became intermixed also, and in time made up that *conglomerate* race which has since spread throughout the United States, and is now known as "native cattle."

Although the term Native as applied to United States cattle has now disappeared and, in fact, was always a misnomer, Native will be used here to designate the mixed cattle population of the United States down to the days following the Civil War when the purebred influence became a dominant factor. Neither a breed nor even a type, these Native cattle were the foundation on which a major part of the United States cattle population was built.

As the population of Native cattle expanded, it did not become homogeneous throughout the country. In New England there were often large red animals showing the characteristics of the Devonshire cattle. The southern seaboard states had smaller and more nondescript cattle. There were, at the beginning of the nineteenth century, a few islands of cattle showing a strong Shorthorn influence in Massachusetts and Kentucky. But there were no herds that could be said to be of one breed. All were called Native cattle until the individual breeds became established later.

Mention should be made of the French Canadian cattle that were taken from Quebec around the great Lakes and down the Mississippi River to Fort St. Louis on the Illinois River (see "French Canadian Cattle," in the section "Cattle Breeds," in Part III, "Canada"). In 1765 French Canadian cattle were also taken by the Acadians to Louisiana, where they were maintained for some time as dairy animals. Both these small islands of a distinct cattle type eventually disappeared but are worthy of note as having been an element of the United States cattle population at one time.

The Purebreds.—During the closing years of the eighteenth century the efforts of the first purebred breeders in Britain began to attract attention. Over the next 75 years the landed gentry and prosperous farmers of both England and Scotland developed their now well-known breeds and established herdbooks. Interest in the improved types of cattle soon spread across the Atlantic. The man of industry who had reached a level of prosperity at which he could indulge in the countryside and livestock; the more successful farmers who had been raising cattle; the trade-minded entrepreneur ever-present when there is movement of cattle; and, somewhat later, the western stockman wanting to upgrade his Longhorn herd—all developed an absorbing interest in the British purebred stock. This interest initiated the second wave of cattle importations from Britain and northern Europe to the United States.

The most favored beef breeds were, first, the Shorthorn, followed by the Hereford and, later, the Angus. The five dairy breeds that became popular were the Ayrshire, Brown-Swiss, Guernsey, Dutch Black and White (later called Holstein-Friesian), and Jersey.

The first purebred stock brought to the United States were a few head of the new English Shorthorn breed in 1793. From then through the first half of the nineteenth century there were frequent though small importations of the British breeds as these gained prominence in Great Britain. Except in the case of the Shorthorn, the numbers involved were too small to register much impact on the United States cattle population until after the Civil War. From 1865 to 1895 the importations of purebred stock increased rapidly. It was during this period that the Dutch Black and White and the Brown Swiss were introduced to the United States.

As they became established in their new home, all the foreign breeds were maintained in a pure state, as well as being used widely for upgrading. Most of the cattle which had been in the country before these purebred importations, principally the Native cattle in the areas east of the Mississippi River and the Longhorn in the Southwest, were eventually upgraded to one of the British breeds. Sometimes the breed characteristics were significantly modified by selection, as in the conversion of the Friesian to the Holstein-Friesian and the segregation of the polled Shorthorn and Hereford strains of these two breeds. In the main, however, the British breeds raised in the United States were maintained close to the standards of the time in their homeland.

Several of the many breeds that developed in continental Europe were imported to the United States, but continuity of only two breeds of commercial importance has been maintained up to the present day. These were the Dutch Friesian, known at the time as Dutch Black and White, from Holland, and the Brown Swiss, from Switzerland. At various times small herds of Simmental, Normandy, Red Danish, and French Canadian cattle were established, some of which had their own breed society.

Cattle of North America

The Charolais.—This was the last of the imported European breeds which has had an opportunity to exercise a major influence on the United States cattle population. The Charolais is considered separately because of its more recent introduction, as well as the nature of its impact on the United States cattle population. The purebred cattle from Britain and the Continent which entered the country in large numbers during the last third of the nineteenth century, in addition to propagating their kind, were utilized largely in upgrading the Native and, later, the Longhorn cattle. The Charolais, coming some 50 years later, were used primarily on the already improved strains and even on the purebreds to produce stock of high-percentage Charolais influence.

Although Charolais cattle had been introduced to Cuba in the early 1900's, the breed did not reach the United States until the late 1930's. They were in a way the forerunner of the new breeds which began to come into the United States in the late 1960's. The Charolais is now third in numbers of all the beef breeds in the United States.

The New Breeds.—The importation of frozen semen on a mass scale for artificial insemination was an innovation in the introduction of foreign cattle breeds to the United States which began in 1967. Encouraged by the wide acceptance of the Charolais breed, as well as a newly awakened interest in crossbreeding, promoters began introducing breeds that until recently were practically unknown to United States cattlemen.

Quarantine and import requirements have forced the use of artificial insemination for the introduction of most of the new breeds. With minor exceptions this breeding has been accomplished by the use of semen imported from Canada. Most of the new breeds originate in European countries affected by pockets of foot-and-mouth disease. With an exception to be noted later, USDA animal health regulations for years have prohibited the importation of both live cattle and "live tissue" from such countries. "Live tissue" by definition includes semen. Canada, however, has elaborate quarantine and health inspection facilities which are recognized as satisfactory by the United States. Cattle which have passed the Canadian quarantine and inspection requirements are eligible for entry in the United States, but such movement was prohibited by Canadian law in 1967. Semen, however, could be exported freely. Thus the situation was created for the lucrative trade in new breed semen from Canada.

In 1971 a significant new route was opened which made possible the direct importation of semen from Europe to the United States.

United States

An exception to the USDA health regulations, which permitted the importation of frozen semen from approved countries not free of foot-and-mouth disease, had been in effect since 1965. Surprisingly, importers did not avail themselves of the opportunity thus offered to import semen until six years later. The requirements are that bulls be placed in strict quarantine under supervision of USDA veterinarians in their homeland. Semen can then be collected and shipped to the Clifton, New Jersey, quarantine station, where it is held until examination at the Plum Island Laboratory of the Animal Health Division of the USDA proves the shipment to be free of foot-and-mouth and other diseases and to be otherwise acceptable. The semen is then released for free distribution in the United States.

In 1971 several commercial artificial-insemination centers which did not manage to obtain a favorable position in the Canadian import program were receiving frozen semen via Plum Island direct from France, Switzerland, Germany, and Italy. This source of new breeds to the United States seems certain to grow and can be expected to make heavy inroads on the importations of semen from Canada.

At the close of 1971 it appeared that the United States would get its own maximum-security-guarantee station that would provide for the importation of live animals from countries not declared free of foot-and-mouth and other contagious diseases. The construction of such a facility had been authorized by Congress in 1970, but no site had been selected for actual construction. Toward the end of 1971 the USDA obtained from the Navy Department an island, Fleming Key, off Key West, Florida, with the object of using it for the authorized quarantine station. It was considered possible that the station might be available for use in 1974. The station was designed for the importation of 800 head of cattle annually, to be received in two lots of 400 head each.

Spanish Cattle

There is no authentic record of the kind of cattle which were brought from Spain to the Caribbean Islands and whose progeny soon reached the continental shores. During the period of this movement there were no cattle breeds as such in Spain (nor, as has been noted, anywhere else in Europe).

Columbus' second and third voyages left from the western coast of Andalusia, Spain. The cattle that were initially taken to the Indies, as the newly discovered lands were then known, undoubtedly came from this part of the Iberian Peninsula. Nothing is known of their physical characteristics except that they were adapted to a warm, dry climate. The frequent references to black cattle in the literature probably resulted from a generic use of the word "black," much as cattle in old writings are often called "neat" cattle. Coronado, in his 1541 expedition as far north as Colorado, described buffalo calves as "red in color, like our own cattle." In *The History of Cuba*, W. F. Johnson says of the Spanish cattle that were taken to Cuba:

The original cattle were of a type peculiar to Spain in the sixteenth century—rather small, well-shaped and handsome animals of a light-brown or dark Jersey color, similar to the wild deer in shade, and usually carrying a dark streak along the spine, with a rather heavy cross of black at the shoulders.

There is no basis for the common assumption that the original Spanish cattle were black; probably they were of varied color. The reference, however, to a "light brown or dark Jersey color" is noteworthy. Throughout Latin America, from Central America and Mexico to Barbados and Trinidad, wherever relatively pure Criollo (Creole) cattle can be found the small Jersey-colored cow will be seen.

These early Spanish cattle entered the North American continent at two main landfalls, the Mexican coast in the vicinity of Veracruz and the Florida Peninsula. During the seventeenth century there was some movement of Spanish cattle from the West Indies to the southern colonies, but they were absorbed in the local cattle population. In the early years of the eighteenth century a few cattle reached New Orleans from Cuba and were the foundation of the local herds in Louisiana. Toward the end of the century descendants of the Spanish cattle which had reached California were taken to Hawaii.

To the Southwestern United States by Way of Mexico.—There were probably no cattle included in the expedition which Cortez landed on the Mexican coast in 1519 to initiate the conquest of Mexico. There is definite reference to cattle in a supporting movement two years later. Cortez departed from Cuba, where some cattle had already arrived from Hispaniola (the island of present-day Haiti and the Dominican Republic). It was on Hispaniola that Columbus had first introduced cattle to the Western Hemisphere on his second voyage to the New World in 1493.

As the conquistadors and the Catholic Church became established after the Aztecs were conquered, cattle became an important element in the economy of Mexico. They followed the continuously advancing frontier as the population moved northward with the establishment of military forts and missions. Records show that Coronado's expedition

United States

of 1540 brought 500 cows across the Río Grande. They were probably included as food supply. There were other, smaller movements of cattle to the same area before 1539, but it is unlikely that any of these cattle survived in what is present-day United States territory.

The increase in cattle numbers throughout northern Mexico was phenomenal. In 1576 one cattleman of that area was mentioned as having a herd of 20,000. While the accuracy of this count might be questioned, the fact is well established that large numbers of Spanish cattle roamed the high-plateau lands of Mexico and eventually spread into what is now Texas and New Mexico.

Little use was made of the beasts, and they multiplied rapidly in the favorable climate. Those that were slaughtered for their hides more than met the requirements for beef, since beef tongues often were the only parts of the carcasses kept for food. Only a small number were required for draft, and very few were milked. Except for those animals held for such day-to-day needs, cattle ran in a semiwild state, and breeding was uncontrolled. Eventually completely feral herds broke off from those of recognized owners and became as "wary as the deer."

During the eighteenth century Spanish military expeditions, and the religious element accompanying them, took cattle into present-day Texas, New Mexico, Arizona, and California in sufficient numbers for permanent herds to be established. There is record of a permanent settlement in southern Arizona with cattle that had been brought in by Jesuit missionaries in 1698. In 1750 at San José, Texas, south of San Antonio, there was a herd of 2,000 cattle cared for by 200 Indian converts. Later in the century the original cattle holdings of some missions grew to larger herds, one mission reportedly having 40,000 head. A Spanish exploratory expedition starting from Arizona in 1775 carried 355 head to a mission in what is now San Diego County, California. Most of such cattle eventually came to exist in a semiwild state and then to join the increasing herds of entirely feral cattle which roamed the plains of the Southwest at the close of the eighteenth century.

These Spanish cattle were the progenitors of the Longhorn, which has now disappeared except for a number of park and private herds maintained in an endeavor to perpetuate the type.

To Florida from the West Indies.—The second significant introduction of Spanish cattle to United States territory took place in Florida. Spanish efforts to establish settlements in Florida began in 1513. The Indians, however, were unusually hostile, and later there was conflict with the French, who were also seeking to control the peninsula. It was not until the latter decades of the sixteenth century that permanent settlements were made and the first sizable importations of cattle arrived. At that time, cattle raising had been going on for nearly a century on some of the islands of the Caribbean. These herds were the source of the Spanish stock brought to Florida. The Florida Scrub cattle which descended from these Spanish cattle had nearly disappeared by 1971.

To Louisiana from Cuba.—In the opening years of the eighteenth century, as the French were establishing themselves at the mouth of the Mississippi River, a few small shipments of Spanish cattle reached New Orleans from Cuba. These furnished the foundation for a range-cattle operation in western Louisiana which persisted until the end of the century. The Spanish cattle in Louisiana remained isolated from other cattle until they eventually were absorbed in the westward movement of the Native cattle.

To Hawaii from California.—Mexican cattle that were the pure descendants of Spanish cattle reached California during the eighteenth century. They were taken to the island of Hawaii in 1793 and were the progenitors of the Hawaiian wild cattle which existed unmixed with any other breed for half a century.

Zebu Cattle

There is a record of a few Zebu cattle from India arriving in the United States before the days of the Civil War. All trace of these was lost. A few subsequent importations aroused considerable interest because their progeny were observed to withstand the ravages of the heavy insect infestation and the hot, humid climate of the Gulf Coast. This generated a growing demand for Zebu cattle which were then imported direct from India or from Brazil by way of Mexico. These were the two sources of the breeding stock employed in establishing the Zebu and the Zebu crossbreeds in the United States.

EARLY CATTLE POPULATIONS

Long before the days of the herdbooks, three basic cattle types had emerged in the United States. These were the Native cattle of the eastern United States, the Longhorn of Texas, and the Florida Scrub cattle. On these three types the present-day cattle populations have to a large extent been built.

All the glamour that time and nostalgic literature have bestowed on early American cattle has fallen on the Longhorn. Two earlier types were never accorded the spot in bovine history that was their due. One was the Native American cattle, mentioned above and described in more detail below; the other was the Florida Scrub, which adapted itself over three centuries to the heat, parasites, insect scourges, and coarse feed of the region. The Native cow is gone, and only a few remnants of the Florida Scrub are left, but both are due recognition as basic American types that made their contribution to the present cattle population.

A parallel to the Longhorn and Florida Scrub, having the same antecedents, was the Hawaiian wild cattle. Remnants of these in nearly a pure state are found on the island of Hawaii today.

Native Cattle.—The now-forgotten Native types comprised nearly the entire cattle population east of the Mississippi River before the Civil War. They were of no distinct breed but represented a varied composition of several European bovine types.

Although the original Spanish cattle, the ancestors of the Longhorns, were the first *Bos* on the North American continent, the progenitors of the native cattle in the eastern half of the country were the first cattle to become established in what is now United States territory. After the end of the Civil War these cattle served as the base stock for the upgrading to the various European breeds of beef and dairy cattle. Similarly, the Longhorns furnished the base on which the Shorthorn and Hereford herds were built in the West. The Longhorns, though showing a wide variation in such characteristics as color and size of horn, were a distinct type of cattle. The same cannot be said of the Native cattle. Derived from several European types, they varied widely in size, conformation, and other characteristics.

The seventeenth-century colonists brought with them the cattle that were in the areas from which they embarked. These were types which had not been influenced to any great extent by artificial selection. They came from the many local cattle populations which were to serve as the foundations from which the present-day breeds were to be selected a century and a half later.

The red cattle that the Pilgrims brought to the Massachusetts colony from Devonshire were the source from which the Devon and South Devon breeds were eventually developed. The Dutch, in settling Manhattan in the 1620's, introduced the Black and White lowland cattle from Holland, the genesis of the Friesian. They also brought some of their Red and White cattle. In 1633 the Danes brought to New Hampshire what are frequently referred to in agricultural writings of the time as their "large Yellow cattle." They were a type of cattle native to lower Slesvig. Cattle from that area in Denmark have been described

Cattle of North America

as being of a solid "red" or "yellow-red" color. Combined with the Angel cattle, the Slesvig cattle formed the Red Danish breed during the nineteenth century. Cattle from Sweden came to the Swedish settlement in Delaware in 1638. There is even mention of Hungarian cattle arriving in the colonies in the seventeenth century. These and subsequent importations brought to the colonies the basic types of British and continental cattle which were to combine into the Native stock of North America.

As "cowpen" people (discussed in the section "Management Practices" below) moved into the Piedmont area of the Carolinas, they found wild cattle which had become established there. These may have been descendants of Spanish cattle which had been brought from the West Indies in limited numbers in the earlier days of the southern colonies. Another theory advanced is that the first Spanish cattle in Florida had become feral and eventually found their way to the Piedmont region. If the feral cattle of the Piedmonts were of Spanish descent, this influence was soon bred out by the Native cattle which the cowpen people brought with them.

The major importations of European cattle to the colonies ceased well before the end of the seventeenth century. One historian records that the "mass movement of neat cattle ended in 1640." The New England settlements even had a surplus of beef for export at times after 1650. Certainly the cattle population of the colonies as a whole could have been self-sustaining from the middle of the seventeenth century on. The base stock for the Native cattle was thus on hand well before the opening of the eighteenth century. In the *History of Agriculture in the Northern United States 1620–1860*, Bidwell and Falconer state: "It appears, therefore, that the colonial cattle were derived from four main stocks—English, Danish, Dutch, and Swedish—which by process of intercolonial trade soon became indistinguishably blended."

Cattle continued to be brought over from the Old World to satisfy the whims of a few individuals or perhaps for a specific breeding purpose. Importations of Spanish cattle from the West Indies also continued in a small way. The numbers involved in these movements were minor during the last half of the seventeenth century and throughout the eighteenth century. Further introductions of cattle from Europe began with the importations of purebred stock in the first half of the nineteenth century. There was thus a period of nearly two centuries during which the four European types intermingled.

All these types, interbreeding to the extent permitted by the limitations the terrain imposed on movement, entered into the composition of the Native American cattle. These were described by one writer as "in color extremely various." They were a very useful although often nondescript type. In colonial days, however, and even into the nineteenth century, a negative type of selection frequently occurred. If the farmer needed cash, the better heifers went to market. When an ox was needed, the largest and most vigorous animal was chosen and castrated, and a poorer one left intact to do the breeding.

During the two centuries that elapsed between the time that European cattle were introduced to the Atlantic Coast and the major importations of the British breeds which started after the Civil War, cattle were just cattle over most of the country with only an occasional island of a recognizable type. The Devonshire cattle of the Pilgrims were raised separately for many years in areas of New England. The "Yellow cattle" of the Danes were concentrated in New Hampshire. Later, a cross of the Devonshire and the Yellow cattle was called the Old Rock Stock and was highly favored in New Hampshire and Maine as a draft animal.

Cream Pots was a name used at times for various strains of cattle which had undergone some selection for milk and butter production. Starting in 1820, a line of high-producing milk cows was developed on the farm of William H. Slingerland in New York State from Shorthorn importations. Genetically they were probably as pure Shorthorns as existed at the time. Though called Cream Pots, this line was unrelated to, and should not be confused with, the Cream Pots developed by Samuel Jaques of Charleston, Massachusetts.

Development of the Jaques Cream Pots started in the mid 1830's with the purchase of two cows of Danish origin, one chosen because she was "a large milker" and the other as "an extremely fine butter cow." These two cows were bred to a Shorthorn bull, and through several generations the progeny were selected for milk and butter production. Production of 75 pounds of milk a day and 18 pounds of butter a week was recorded. These were extremely high records for the time. Jaques referred to his cattle as a breed and, although they were eventually lost to posterity, he could well have had the nucleus of a new milk breed if his line had continued. The Jaques Cream Pots were probably the development nearest to a cattle breed founded on the Native cow.

Lamenting on the fact that no American breed of cattle had emerged, the *Report of the Commissioner of Agriculture for 1866* stated:

The establishment of an American breed, commencing by judicious selection of the heaviest native milkers, and careful breeding with short-horn bulls of families of established reputation as dairy stock, should suffice to satisfy the

Cattle of North America

ambition of any enlightened breeder in the country. It would be the work of a lifetime, however, and few have either the skill or the patience for such a work.

In post-Revolutionary days American breeders developed the Morgan horse in Vermont and, later, such swine breeds as the Chester White, Poland China, and Duroc. No American breed of cattle, however, emerged from the broad genetic band of the Native cattle. Distinctive cattle breeds of a later day that were developed in the United States, such as the American Brahman, Brangus, and Santa Gertrudis, were twentieth-century additions to the bovine population and entered the scene long after the Native cattle had been bred out of existence.

During the eighteenth century relatively small local concentrations of cattle having similar characteristics coalesced and might well have led to the development of actual breeds. This was the same kind of situation which led to the segregation of distinct breeds in England and on the Continent during the latter part of the eighteenth century and the opening decades of the nineteenth.

In Britain the landed gentry had the time for this accomplishment. In the United States in those days the farmer with livestock had little time to devote to them after performing the daily chores involved in gleaning a living in the new land. This was before the day of the "city farmer" or professional man with an interest in livestock. When the farmer eventually felt the need for more productive cattle, he found the British breeds readily available.

Successful businessmen began to take an interest in agriculture, particularly in cattle, as the industrial age got underway at the opening of the nineteenth century. The early Shorthorns were imported from England by businessmen of Baltimore and Philadelphia. Other breeds soon followed—first the Hereford, then the Jersey and the other dairy breeds, and finally the Angus.

The dirt farmer was slow in seeing the advantages of better stock and also lacked the means to obtain such animals. As more capital flowed into agriculture, the man with cattle began improving his herd by using purebred bulls. After a few generations of top crossing on his Native cows, he had the working equivalent of whichever breed suited his fancy. Some successful farmers then started in the purebred business.

Cattle continued to be an element in the farmer's power requirement until well after the turn of the twentieth century. From New England across the country to California draft oxen were in use until the close of World War I. Although for some time horses had been replacing the

Three-year-old steers on a Virginia farm in 1904. Native cattle in eastern United States. National Archives Photograph.

last of the draft oxen, these faithful animals did not completely disappear from the American scene until the 1920's.

The Native cow was the foundation on which the grade-cattle population of the eastern United States was developed, and it was well into the twentieth century before the type became unknown.

The Longhorn.—The Mexican cattle run by the early cattlemen of Texas came to be known as Longhorns when they were trailed to the northern markets. Previously they had been just "Mexican" cattle. These and the Florida Scrub cattle were the nearest approach to indigenous cattle that can be claimed for the North American continent. Descended from the stock which the Spanish military, missionaries, and colonists had brought to the New World, they had been subjected to more than three centuries of nearly natural selection before entering the trade channels of the United States.

Whether the Longhorns can properly be called a breed may be open to question. Perhaps they are best referred to as, in the words of a writer of an earlier day, "a kind of cattle." They stemmed from the same root as the present day Criollo cattle of Mexico and of all Latin America.

Their progenitors reached the plateau region of northern Mexico

Draft oxen on a New York farm in 1909. Native cattle in eastern United States. National Archives Photograph.

during the first half of the sixteenth century and were running feral there in the 1550's. Even those retained in a domesticated state by the missions and later by the hacienda owners were not subjected to any particular breeding control. Natural selection in the environment of northern Mexico was therefore the major factor influencing the character of both the domesticated and the wild cattle of the area.

During the seventeenth and eighteenth centuries the Spanish cattle became established in the area that is now Texas and to a lesser extent in New Mexico.

When Mexico gained her independence from Spain in 1821, Texas, then a part of Mexico, became an attractive region for many inhabitants of the southern states who were seeking greener pastures. Florida Scrub cattle, also descendants of the Spanish cattle, were trailed to East Texas by such settlers. The number of Spanish cattle reaching Texas in this manner was small, however, compared with the uncounted herds of wild and semiwild cattle which had descended from those that had

Longhorn steer, nine years old, 1,200 pounds. Wichita Mountains Wild Life Refuge, Cache, Oklahoma.

begun moving up from Mexico a century earlier. The Florida cattle could have had little, if any, influence on the Mexican cattle that had spread widely over Texas.

Thus the Longhorn evolved. As first known to the Texas cattlemen, the strain was in the initial stage of development as a landrace acclimated to both the semiarid plateau regions of northern Mexico and the hot coastal plains of South Texas. Natural selection had not proceeded far enough to establish a fixed type when man abruptly stopped the process to begin the upgrading practice, breeding the Longhorn cow to Shorthorn and Hereford bulls during the last quarter of the nineteenth century.

The Mexican cattle, as first known to the early United States stockman, were rangy animals with long faces, narrow heads, and widespread

Longhorn cow. Wichita Mountains Wild Life Refuge, Cache, Oklahoma.

horns usually with some upturn. The horns were exceptionally long, but in retrospect the length has been exaggerated as the Longhorn has faded into the past. On some nine- or ten-year-old steers there were horns with up to nine-foot spreads, but four to five feet would be more typical of the animals that were trailed north. The horns on the bulls and cows were not as long as those seen on the old steers.

Color varied widely: patched, speckled black on white or red on white, brindle, and occasionally a nearly solid red, tan, white, or black. Conformation was slab-sided and raw-boned, with light hindquarters and heavy front quarters. Weight varied widely depending on the available feed. Mature cows in average condition weighed 650 to 750 pounds; bulls up to 1,000 pounds. Six- or seven-year-old steers would weigh up to 1,200 pounds if on good grass. The Longhorn was an inferior beef-type animal and yet found a ready market in the North in the years following the Civil War because of the shortage of beef.

Longhorn cattle were exceptionally hardy, withstanding equally well

Longhorn bull. Jack Phillips ranch, West Columbia, Texas.

the hot, arid climate of the plains and the tick-infested humid coastal areas of South Texas. The cows were unusually fertile, considering the low levels of nutrition on which they existed at times. Although the often-repeated comment that the Longhorn cow calved regularly every year is probably exaggerated, she certainly calved with sufficient regularity to populate rapidly the South Texas plains. The herd now maintained by the United States Department of the Interior in Oklahoma averages a 90 per cent calf crop, although admittedly it is held on better pastures than were the Longhorns on their native range in the nineteenth century.

In 1927, with a $3,000 Congressional appropriation, 1 bull, 19 cows, and 1 bull calf, all showing typical Longhorn characteristics, were selected for propagation by representatives of the Wichita Mountains Wild Life Refuge at Cache, Oklahoma. They were chosen after an inspection of some 30,000 head of Longhorn-type cattle that were scattered throughout the hinterlands of South Texas and northern

Longhorn steer. From photograph taken on a ranch in South Texas about 1900. National Archives Photograph.

Mexico. The foundation herd of 21 head has been expanded to 900 head, which are now maintained at the Cache Reserve and at the Fort Niobrara National Wild Life Refuge in Nebraska. There are several privately owned Longhorn herds in South Texas, as well as a few herds in state parks. They were gathered from such remnants of the species as could be found in the Southwest 40 years ago. One notable source of original Longhorns from which several of the present herds were founded was the herd of Graves Peeler, an old-time Texas rancher whose Longhorns had been maintained pure from the days of Mexican cattle.

The service which the Longhorn gave to the cattle industry of the United States in its formative years was invaluable. The Longhorn cows made possible the rapid expansion of the cattle herds of the western-

Longhorn-Hereford cross steer. From same photograph as the one on the opposite page. National Archives Photograph.

range country, and their hardiness and fertility under adverse conditions contributed immeasurably to the cattle industry of the western United States.

A breed society, the Texas Longhorn Breeders' Association (West Columbia, Texas), was founded in 1964 to save the Longhorn from extinction.

Florida Scrub Cattle.—The cattle which populated Florida in 1819, when the area was purchased by the United States, came to be commonly known as "scrub" cattle. Like the Longhorns of Texas, they could perhaps be best described as "a kind of cattle." The word "scrub" is often used to denote inferior cattle of mixed breeding. It was applied to the Spanish cattle because they were small and nondescript in appearance. These survivors of the early Spanish cattle became accli-

Florida Scrub cow, 450 pounds, and two-year-old calf, 250 pounds. 1904. National Archives Photograph.

mated to the harsh environment through natural selection. While their hardiness alone entitled these beasts to more consideration, they never acquired a more distinguished name and, therefore, are here simply called Florida Scrub cattle.

Numerically, the Florida Scrubs did not equal either the Native or Longhorn populations. They never achieved the fame of the Longhorn and lacked both the size and the romantic legend of those animals. They were, however, just as distinctive a derivation of the Spanish cattle as the Longhorns. The Florida Scrubs served as the foundation stock on which the bovine population of Florida was founded, although they have never been accorded recognition for the part they played in the adaptation of cattle to the inhospitable range.

During the first half of the sixteenth century cattle were taken to Florida by early Spanish explorers attempting to colonize the area. They came from the herds which had been established on the islands of the West Indies. The early colonizers had great difficulty gaining a foothold because of bitter Indian opposition. Farm enterprises were frequently destroyed by the Seminoles, and the cattle were either slaughtered or driven to other areas. The precise date when cattle became firmly established in Florida is therefore not known.

In 1565, however, there were substantial numbers of Spanish cattle in the herds that the Seminoles had gathered, as well as on the farms of surviving Spanish settlers. Under the sparsely populated conditions it is safe to assume that there were wild or semiwild cattle in the timbered areas in addition to the cattle the Indians maintained in domesticated herds.

When conditions in the area stabilized toward the end of the sixteenth century, the descendants of the early Spanish cattle became the native stock. They underwent nearly three centuries of what amounted to natural selection before the influx of settlers from the neighboring states began. The Florida climate may have its salubrious attraction for the northern vacationer, but before the fever tick and screwworm were eliminated and some provision made for winter feed, it was a rough country for any kind of stock.

Not for some years after the United States purchase of Florida were outside cattle taken to the state in significant numbers. When farmers from the states to the north began moving into Florida, the population of small Spanish cattle met their immediate needs. Later, as they became well established on their Florida homesteads, they brought in stock from the adjoining states. There are records of many cattle being trailed to Florida from Georgia, Alabama, and the Carolinas between 1830 and 1850. This seems to have been the first effort to upgrade the Florida Scrub cow.

As the purebred British breeds increased in the northern states following the Civil War, efforts were made to introduce these highly publicized

Scrub cow ill with salt sick. Central Florida 1932. Photograph courtesy R. B. Becker.

animals to Florida. There is a record of a Devon bull being brought to the Alachua area in 1861. A few Ayrshire bulls were imported in 1870 and some Jerseys in 1873. The Red Poll breed was represented by a number of cows reaching the Alachua area in the late 1890's.

There are frequent references to the heavy losses encountered with these newly introduced cattle. Even today without special care the northern breeds deteriorate rapidly when they first enter southern Florida. These first importations, moreover, were made at a time when the infestation of the fever tick and screwworm was universal over the state. Northern cattle could hardly have survived under the prevailing conditions in sufficient number to exert any appreciable influence on the population of hardy Florida Scrubs. The herds of Florida at the beginning of the twentieth century were thus composed of essentially pure descendants of the stock originally brought to the area by the Spaniards.

Same cow shown on the facing page one year later after receiving iron supplement. Photograph courtesy R. B. Becker.

The cattle introduced to Mexico were from more westerly islands of the West Indies than those that had entered Florida and arrived on the continent about 50 years earlier. This factor alone could not have caused any noteworthy differences between the cattle of the two areas. Yet after three centuries of natural selection the Florida Scrub and the Longhorn emerged as two distinct types, though retaining some marked similarities.

The Florida Scrub was a smaller animal than the Longhorn. Its horns were unusually large but not nearly the equal of those on the Longhorn. The Florida Scrub carried a heavy middle, undoubtedly acquired through the necessity of consuming large volumes of forage with a high moisture content. The much trimmer Longhorn lived on drier and more nutritious grasses. On the other hand, both types had narrow faces, and their color patterns were similar.

Florida Scrubs varied widely in size among themselves, depending on the quality and amount of forage available. Mature cows on rough palmetto land would weigh 450 pounds; but on prairie ranges when not overstocked, they would reach 650 pounds. As in all the Spanish-type cattle, there was a wide range of colors: blacks, reds, some fawn or Jersey, mottled black on white, an occasional brindle, and sometimes a white top line. The Florida Scrubs were remarkably long-lived and hardy. The cow was not especially fertile, usually calving every other year, probably because of the exceptionally low nutritional level at which she existed.

Mineral deficiencies over wide areas of Florida probably contributed to the loss in size of the Scrub cattle. Florida cattle owners have feared "salt sick" in their cattle from the days of the early settlers. Loss of condition owing to acute anemia often occurred after the wet season, and many cattle died in certain areas. The only remedy known was to move the sick animals to a different part of the country which experience had shown was free from salt sick. Work at Florida State University in 1932 showed that a deficiency of iron was the principal cause of salt sick, although lack of copper and some trace minerals was also involved. The photographs on the opposite page show the effect of giving iron to a Scrub cow with salt sick. Many generations of suffering from insufficient iron could account for the small size and inferior conformation of the Florida Scrubs.

The Scrubs dominated the Florida scene until well into the twentieth century, but continual crossing with the British breeds or Zebu-type cattle has now practically eliminated them. In 1971 a few cow herds of pure or nearly pure Scrub cattle still remained in southern Florida, but they were being bred to British and Zebu bulls.

Some consideration has been given to establishing an exhibition herd of Scrub cattle in one of the Florida state parks along the lines of what was done with the Longhorn in Oklahoma. In 1971 a bull and several cows were presented to the Florida Department of Agriculture by J. C. Bass, who had a herd of Florida Scrubs in the southern part of the state. These are being maintained by the state near Tallahassee but there are no concrete plans for perpetuation of the type.

Hawaiian Wild Cattle.—In 1793 or the following year Captain George Vancouver is reported to have carried five cows and one bull of Mexican cattle from California to the island of Hawaii. They were the descendants of the cattle which the Spaniards had introduced to the New World, for at the time no other members of the bovine family had reached the

Florida Scrub cows. James Durrance Estate 1971, Basinger, Florida.

Wild cattle bull, 1,000 pounds. The bushy tail is seen on some individuals but is not representative. McCandless Ranch, Hawaii.

West Coast. In the years immediately following there were other movements of cattle from California to Hawaii. All the early cattle were placed under a taboo by the king of Hawaii and therefore multiplied rapidly, migrated to the sparsely inhabited parts of the island, and became completely feral. Later the taboos were relaxed, and beginning in 1825 cattle were captured and killed for their hides and tallow. The wild-cattle population continued to increase, however, those killed being mainly near cultivated areas. For over 50 years they existed as a pure strain of Mexican cattle, descending from the same source as the Texas Longhorn and the Florida Scrub cattle.

Around the middle of the nineteenth century a few representatives

Wild cattle cow, 700 pounds. McCandless Ranch, Hawaii.

of the newly developed British breeds were introduced to the islands. The Durham was brought in first, followed by some of the dairy breeds and then the Angus and Hereford. Every effort was made to keep the domesticated cattle separated from the wild Mexican animals both to keep the better grass for the tame stock and also to avoid undesirable interbreeding. Stone walls were built to enclose the one type of cattle and keep out the other.

As time went on, mixing of the Mexican and British cattle became inevitable. In many operations interbreeding was practiced intentionally by taking off the Mexican bulls and using those of the British breeds in a systematic upgrading program. There was also unavoidable intermixing of the British breeds with the still feral herds. In Hawaii "wild cattle" had always been the designation of the Mexican cattle, and, as interbreeding with the British breeds became more common, "wild cattle" encompassed any animal that was running uncontrolled, whether it was a pure Mexican descendant, or half Shorthorn or Hereford cross.

As cattle raising increased in importance on the larger ranches, elimination of the wild cattle was looked on as a necessity for a profitable operation. Boundary fences were built, and the wild cattle were killed or collected and sent to market. The year 1925 marked the time when systematic elimination of the wild cattle began. Today there is only one ranch on which any large numbers are found. The McCandless Ranch on the southwestern part of the island of Hawaii has an estimated 2,000 head of wild cattle scattered over nearly 40,000 acres of virgin forest and brush land. Only 22,000 acres of the ranch lands, which run from the coast to an elevation of 8,000 feet on Mauna Loa, have been put into developed pastures. The remainder of the 60,000-acre property is utilized solely by the wild cattle and wild sheep and goats. Many of the cattle carry traces of the British breeds, but pure or nearly pure descendants of Mexican cattle can still be seen.

These wild cattle are of varied colors. Black with some white markings is common, and a nearly solid reddish-tan color is also seen. The white face of the Hereford predominates on individuals that carry evidence of the British breeds. Cows weigh around 700 pounds and bulls up to 1,200 pounds. All are of rangy conformation, slabsided, and with thin hindquarters. The horns are long and thin, widespread and upturned. The wild cattle are hardy and have developed a high tolerance to prevalent internal parasites. Young animals are caught by roping and used for rodeo work.

CATTLE BREEDS

The many breeds which enter into the composition of the present United States cattle population are described in the following pages under the headings, "Introduced Breeds—Beef Type," "Introduced Breeds—Dairy Type," "American Developed Breeds," "New Breeds," and "Other Breeds."

No classification of dual-purpose breeds is given because the tendency in the United States has been to select from the European dual-purpose breeds for either a beef type or a dairy type. The Red Poll, sometimes called a dual-purpose breed, is now being selected for beef characteristics. The Milking Shorthorn, actually only a specialized type of Shorthorn, has been developed largely for milk production and is therefore classified as a dairy type.

Introduced Breeds—Beef Type

Angus.—It is probable that representatives of the Scottish cattle that were the forerunners of the Aberdeen-Angus breed were taken to New England in colonial days. References are occasionally found in writings of the time to "black" cattle in New England long before the breed was established and named in Scotland. There is a record that a young bull and a heifer from the Earl of Southesk's herd of noted Angus were shipped to Montreal, Canada, in 1859, but they left no trace.

The first noteworthy introduction of the breed to the Western Hemisphere took place in 1873. In that year George Grant, a Scottish silk merchant who had retired and established residence in Victoria, Kansas, imported four Angus bulls. (This enterprising Scotsman is said to have cornered the crepe supply in London when he heard of the critical illness of Prince Albert and on his death made a fortune from the mourning Londoners.) The four Angus bulls in Kansas were used on Longhorn cows retained from herds being driven from Texas to the Kansas railheads. When grown out to slaughter weights, the Angus crosses were naturally heavier animals and dressed out more meat. Within a few years purebred herds of Angus cattle had been established, principally in Illinois and New York.

The first breeding herd in the United States was established by two Scottish businessmen of Chicago, James Anderson and George Findlay. They imported one bull and five heifers to Lake Forest, Illinois, in 1878.

There were continuing importations of Angus to the United States for many years, and the breed finally became as well established here as in its native Scotland. Importations declined in later years as the United States herd grew to proportions that far outnumbered its representatives in the British Isles and finally represented only a few bulls, which served mainly as promotion leaders for purebred breeders.

As a beef type the Angus was improved during the early years by such selection techniques as were known at the time. These consisted of visual appraisal of conformation and mating "the best to the best." Later, pedigrees and show-ring performance became the major guidelines in the breeding programs of the purebred owners.

The Angus is a dominantly polled breed. It is either solid black or solid red with only minor white markings permitted behind the navel. It has an excellent beef-type conformation characterized by smoothness. Size varies over a considerable range, depending on whether an animal is of the large or small type.

Angus bull, four years old, 2,205 pounds.

In some herds mature cows weigh around 900 pounds. Such herds usually trace back 30 or 40 years, when the compact conformation was preferred. The Angus is somewhat smaller than the Shorthorn or Hereford, and breeding for the small type perhaps had a more noticeable effect on the blacks than on the other breeds. When the industry began to look with favor on larger animals, commercial Angus herds seemed to be slower recovering their size. Most of the purebred breeders went back to breeding larger cattle some time ago. Where the commercial cattleman has taken advantage of the larger breeding stock now available, he has been able to bring his cows up in weight to the range of 1,000 to 1,100 pounds and in instances to 1,300 pounds. Bulls of the large type weigh 2,000 pounds or more.

The occurrence of dwarfism in some lines of breeding was serious at one time, but this problem has now been overcome.

A small proportion of the Angus breed carries a recessive red gene. The mating of two such individuals can produce a red calf. Red offspring of purebred parents have always been accepted in the Scottish Herd Book, but in 1917 the American association closed its herdbook to cattle of red color. The Red Angus Association, established in 1954, has since built up a highly productive strain of these Reds which genetically are just as pure Angus as the black cattle (see "Red Angus" below).

The Angus was something of a latecomer to the American cattle scene, having been preceded by both the Shorthorn and the Hereford. These breeds were widely and successfully employed in upgrading the Longhorn before the Angus became well known. The Hereford established such a reputation for hardiness and fertility that by the time Angus bulls became available grade Hereford herds strongly predominated in the range country.

The reputation which Angus cattle acquired over the years for excellent carcass and for bringing a somewhat better price for either feeder or slaughter animals was built in the barnyards of the Middle West. This recognition was not, however, sufficient to offset the preference of the western grower for his white-faced cattle. It was not until after World War II that sizable Angus herds began to be seen in range operations. Since then there has been a steady increase of the breed in the western states.

For many years large numbers of the cattle that go to slaughter in the United States have been raised and finished by the middle western farmer, who traditionally keeps some livestock to supplement the income from his row crops. His herds run from 15 to 25 cows, and infrequently up to 100, and the calf crop is either fed out on the home place or sold to a neighbor to be fed. Such operations have been the stronghold of Angus cattle in the United States.

The American Angus Association, along with the other major beef-breed societies, was slow to orient itself to the new concepts in cattle improvement which evolved after World War II. The strict limitations on artificial insemination that the association prescribed until 1970 undoubtedly retarded the improvement that could have been accomplished with this method. Recently, however, the association has adopted a progressive attitude toward modern procedures.

The Angus Herd Improvement Records program consists of three parts: herd classification, production measure, and carcass evaluation. Herd classification, an attempt at a mathematical approach to conformation scoring, was offered to association members in 1948. It was an obvious attempt to retain a position of importance for the old concept of judging an animal's quality by its appearance. Production measure, which can consist of performance data only or a combination of that with progeny information, was offered to members in 1962. Also in

1962 a program for carcass evaluation for sire progeny groups was set up which was designed to use the USDA yield grade system.

The Angus Herd Improvement Records program was the first such program offered by any beef-breed society in the United States. This program looks toward breed improvement through evaluation of the economic factors involved in beef production.

As another first, the program is open to commercial producers. This is a major breakthrough for purebred beef organizations, which traditionally have given only lip service to the commercial cattleman. The basic data involved in production records—birth date, weaning and yearling weights and dates, and sire and dam identification—are furnished on forms supplied by the association. The data are then processed on the association computers. The program is free to the breeder, and if effectively handled the detailed information which will eventually be accumulated can be a valuable tool in breed-improvement procedures.

For many years following their introduction, Angus cattle claimed a steady 10 per cent of the United States beef-cattle population. Around 1940 the breed began to attract more attention from western cattlemen, and its numbers have increased steadily from that time. Today Angus make up well over 40 per cent of the national beef herd. This popularity appears destined to continue and could well be enhanced by the current interest in crossbreeding. The Angus cow has been a favorite for evaluating the crossbred progeny of the newly introduced beef breeds.

The American Aberdeen-Angus Breeders' Association was organized in 1883. In 1957, the name was changed to American Angus Association (St. Joseph, Missouri).

Red Angus.—Black and Red Angus cattle belong to the same breed, the difference in color having no more significance than the red, roan, or white colors in the Shorthorn cattle. The Aberdeen-Angus Cattle Society in Scotland has always registered Red Angus. The American Aberdeen-Angus Breeders' Association followed the same practice until 1917, when the color line was drawn and the red animals were dropped from its records.

In the formative days of the Angus breed, some farmers in northeast Scotland began selecting for breeding purposes black polled cattle. Unknown to their owners, some of these blacks carried a red gene that was transmitted to their progeny according to Mendelian principles. The mating of two black animals, both carrying a red gene and a black gene, will produce, on the average, a pure-red offspring (two red genes) one time in four; two black offspring will carry both a red and a black

Red Angus bull, seven years old, 2,000 pounds. Beckton Stock Farm, Sheridan, Wyoming.

gene; and the fourth will be a pure black. Pure reds mated to each other will always produce red offspring, and pure blacks will always produce black offspring. A pure red mated to a red-carrying black produces two pure reds and two blacks which carry red genes. (Black and red carriers are always black in color because the black gene is dominant.)

There has been considerable conjecture about the source of the red gene in the Angus cattle. The "history" of all the British breeds, before the days of the herdbooks, is largely romance. Whether the red gene of the Angus came from a distant English Longhorn ancestor or through some other type of early cattle which displayed a more solid red color can be little more than a matter of opinion.

Because the Angus cattle were originally selected for black color, the occasional red calf was often disposed of at birth. A critical breeder might even eliminate the dam, and perhaps the sire as well. This practice decreased the carriers of the red gene in the Angus population in Scotland, as well as in the United States, and the frequency of red births declined until it is now considered to be 1 in 500 among the black cattle. In 1890, a few years after the American Angus Herd Book was started, there were 22 red entries out of a total of 2,700 for that year, or 1 red in 123. This indicates the extent to which the red gene has been bred out of the black Angus population in the United States over three-quarters of a century.

The Red Angus should be a medium red, solid over the body, with only minor white markings behind the navel. The interest in Red Angus began at the time when selection for the compressed type of beef animal was rampant in the United States. The Reds, to a considerable extent, were developed to escape this fallacy. There are no small-type Red Angus similar to the small-type Blacks. The large-type cattle of both colors are comparable in size, as well as in other characteristics.

Occasionally an Angus breeder either in the United States or in Britain with a knowledge of the elementary genetics involved obtained red individuals from herds in which they were dropped and proceeded to raise and interbreed such reds. A red mated to a red always breeds true, and thus a red herd, just as pure Angus as any black herd, could rather quickly be established.

In Britain there was no particular challenge in establishing a line of Red Angus. If of pedigreed parents, there was no discrimination against the Reds, and such individuals were acceptable for recording. The action of the American association, however, prohibiting the registration of red cattle, laid the cornerstone for the Red Angus movement.

In the years following World War II, a number of United States breeders began picking up good red Angus and red-carrier Blacks and established Red Angus herds. By 1954 this movement had reached the point where the breeders wanted a permanent record of their stock. The result was the Red Angus Association of America (Denton, Texas).

Today the most significant difference between Red Angus and Black Angus is in the breed societies that sponsor them, the cattle differing mainly in the color of their hair. It is possible, however, that significant genetic differences may eventually be developed of much more importance than hair color. The breeding and selection policies of the two Angus organizations will determine the extent to which this will occur.

New in the field and unhampered by custom and tradition, the Red Angus Association has adopted breed-improvement policies which could prove a revelation to the industry. Since the beginning of the association performance testing and official inspection have been required for registration. A Red Angus pedigree includes the animal's official performance record. The use of artificial insemination has always been encouraged by the association, and no arbitrary restrictions have been imposed on breeders using this technique. There is no limit on the number of calves that can be registered by a bull used artificially, nor is there any requirement, as in some associations, that the owner of the cow bred by artificial insemination must also have an interest in the bull. Sires used artificially, however, must be blood-typed if their calves are to be registered. This rule provides an authentic check on parentage.

The Red Angus Association also offers an open herdbook program. A Red Angus–sired crossbred heifer which meets the required performance and inspection standards may be recorded. Continued upgrading to $^{15}/_{16}$ Red Angus permits such progeny to be certified. The progeny of Black Angus sires on Red Angus dams can be qualified for registration. Geneticists affirm that such a policy can raise the ceiling of a breed's potential.

The largest concentrations of Red Angus are found in Oklahoma and Texas, although there are herds throughout many states.

Charolais.—Although there were prize-winning herds of Charolais cattle in Cuba in 1914 and there had been importations to Mexico before 1910, the introduction of the breed to the North American continent for which continuity can be confirmed occurred in 1929. In that year 2 bulls, 5 bull calves, and 10 heifers were shipped to Mexico from France. Two subsequent importations increased the total number of Charolais brought into Mexico from France to 37 head—8 bulls and 29 cows. All these cattle were held by Jean Pugibet, a cigarette manufacturer. It was from this group that the Mexican herd of Charolais developed; there were no further importations from France to North America until those brought to Eleuthera, an island in the Bahamas, in 1963 by a group of American investors.

In 1936 the King Ranch of Texas imported two bulls from Mexico. During the next few years a number of other bulls were brought to the United States and became the foundation of the Charolais breed here. A continuous upgrading program, starting with cows of the British breeds, eventually produced the American strain. The USDA restrictions against the importation of cattle from countries subject to foot-and-mouth and other devastating bovine diseases prevented any further introduction of Charolais to the United States until the mid-1960's.

At that time widespread interest in the Charolais, among other factors, led to the establishment by the Canada Department of Agriculture of extensive quarantine and entry facilities for the admission of cattle from certain restricted countries.

Charolais bull, thirty months old, 1,800 pounds. A A Ranch, Richmond, Texas.

A good many Charolais were imported directly from France to Canada under these regulations, and some even found their way from Canada to the United States with USDA approval. By September, 1967, however, the Canadian government had stopped such transshipment.

At the same time 12 bulls and 12 females were moved from the Bahamas to the United States by way of Canada's quarantine station at St. John, Nova Scotia. These were the offspring of the French Charolais imported to Eleuthera.

Such large quantities of Charolais semen have been imported from Canada in recent years that the American gene pool would seem adequate. Breeders, however, will continue to import in their quest for perfection and promotion.

The Charolais is one of the larger breeds of cattle; mature bulls weigh up to 2,500 pounds, and cows 1,400 to 2,000 pounds. The horns are rather small, considering the size of the breed. Those on the bull are thick and short, extending laterally from the head; on the cow the horns are thinner and longer and extend upward. In the United

States even purebred animals are frequently dehorned, and a polled strain has become quite popular. The hair color is an off-white to a very light cream. Bone structure is heavy, and there is a rather prominent dewlap. The body, particularly the rump, displays an excellent beef conformation.

In recent years the Charolais has moved ahead of the Shorthorn to occupy third place numerically among the United States beef breeds. This rapid increase stems from the open herdbook policy of the American International Charolais Association, as well as from the traits of the breed itself. The female progeny of a registered Charolais bull bred to a cow of any of the recognized beef breeds may be recorded with the association. The continued breeding of registered bulls to such progeny, if duly recorded, can lead to a "purebred equivalent" in the fifth generation. Thus a $^{31}/_{32}$ Charolais can be registered in the association herdbook.

Much of the current popularity of the Charolais comes from the fast-gaining feeders that result from crossing Charolais bulls on cows of the other beef breeds. This practice has a duel effect: a more vigorous calf because of the genetic heterosis and a larger calf resulting from the size of the Charolais. This big-framed crossbred gains faster in the feed lot and produces an acceptable, though later-maturing, carcass. Such crosses can at times cause difficult calving. This obstacle is one the cowman must weigh against the heavier birth weight and more rapid gains.

Straight-bred and high-percentage cross steers must be carried to relatively heavy weights to produce a carcass that will make USDA choice grade. First crosses of Charolais on the British breeds, however, produce excellent carcasses which have been winning consistently in competitions across the country. Both feeders and packers discriminate against a straight Charolais steer on the basis that the meat is coarser and will not grade well.

Initially the Charolais cattle imported from Mexico were crossed with the Zebu (Brahman) cattle in South Texas to produce a better beef animal. Thus the first breed organization in the United States involved with Charolais was the American Charbray Breeders Association, organized in 1949. Their main interest was in the Charolais-Brahman cross. The Charbray Association joined the Charolais breeders in 1951, but separated from them in 1958. There were several organizational changes over the years until 1967, when the American International Charolais Association (Houston, Texas) became the official organ of both the Charolais and Charbray breeders in the United States.

Mature Devon bull, 1,800 pounds. Mary Ellen O'Connor Estate, Victoria, Texas.

Devon.—Among the first English cattle to reach America were the red cattle of Devonshire. It was from Plymouth in Devonshire that the Pilgrims sailed in 1620 to establish the Plymouth Colony in Massachusetts. Since the new settlement continued to be supplied from the port of Plymouth for a number of years, the surrounding countryside was the natural source of the cattle that were brought over for draft and for milk. The first shipment of cattle from Devonshire arrived in 1623.

Historians of the Devon, endeavoring to trace an ancient lineage for the breed, outdo the other English cattle societies and go all the way back to the Phoenicians as the first breeders. About all that can be established, however, is that by the early 1800's, nearly 200 years after the founding of the Plymouth Colony, two distinct breeds were emerging from the bovine population of Devonshire and Cornwall. Both were typically draft animals, milk and beef being more or less by-products at that time. Eventually a dual-purpose milk-beef type, the South Devon, came out of the south, and more of a beef type, called simply Devon, was differentiated in the north. Both were red, the Devon cattle a darker shade than the South Devon. The South Devon was considerably the larger of the two. The early cattle taken to Plymouth Colony were contemporaries of these indigenous red cattle of southeast England.

Today's Devon has an excellent beef conformation. Neither the English nor the United States breeders succumbed to the compressed fad that intrigued many breeders of purebred beef cattle four decades ago. The Devon cattle are of good size, off the ground but not leggy. In commercial herds mature cows weigh around 1,200 pounds, and bulls 1,800 to 2,000 pounds. The hair color is normally a rich, deep red although lighter shades of red are seen. In their part of England they were formerly referred to as Red Rubies. The horns are rather short, thick, and curved forward on both the male and female. A polled strain has also been developed and is registered by the same breed society as the horned cattle.

There was less outcrossing of the cattle of the early Plymouth Colony than was the case with many other cattle that reached American shores. Although other settlements along the Atlantic Seaboard followed soon after Plymouth was established, the colony remained isolated from the other communities because of conflicting religious and political attitudes. There are, however, records of the red cattle of Plymouth being moved westward to the vicinity of Springfield, Massachusetts, where a sizable herd was established and apparently kept pure. Market cattle of this herd were trailed nearly 100 miles to the Boston slaughterhouse.

With changing times and the Revolutionary War, continuity of the red Plymouth cattle was eventually lost.

The Devons in the United States today derive from importations from England during the first half of the nineteenth century. The earliest importation of Devon cattle for which continuity became established was recorded in 1817, when Messrs. Caton and Patterson, merchants of Baltimore, Maryland, received several head from England. Descendants of this nucleus in the hands of Patterson and subsequent owners thrived until the late 1880's. There are records of about 15 separate shipments following the one in 1817. By this time the Devon and South Devon breeds in Devonshire and Cornwall had been more or less segregated, but the Devon had received more attention from the English breeders of the day, who established a herdbook in 1850. The South Devon herdbook was not started until about 40 years later.

In the northeastern United States, Devon cattle were raised principally on small and medium-sized farms. They were highly regarded as draft animals. A span of Devon oxen was considered the equivalent in plowing of a team of good horses if both were fed grain. On grass or hay alone they would outperform the horses. For beef the Devon compared with the Shorthorn and the Hereford. In some places it was preferred. In the closing years of the nineteenth century the Devon became popular as a dairy breed. In experiment-station work on dairy cattle in the eastern states the Devon was recognized as one of six main dairy breeds, the others being the Ayrshire, Guernsey, Holstein-Friesian, Jersey, and Shorthorn. A favorite measure of a cow's productive ability in that day was the cost of feed to produce a pound of butter. On some such tests the Devon ranked with the Ayrshire and the Holstein-Friesian.

The breed was also a favorite with the prosperous urban dweller who had a farm in the country, although the Devons were by no means limited to such owners. Eventually islands of the breed were established throughout the states, although for some reason not readily apparent they never became widely popular. There are approximately 25,000 to 30,000 Devons in the United States today.

The development of the polled strain of Devon originated from a polled mutant born in the purebred herd of Case and Elling in Concordia, Missouri, in 1915. This bull became the center of a heated controversy among the Devon breeders of the day. Antagonists owning horned cattle claimed that a strayed Angus or Red Poll bull was responsible for the hornless bull. The arguments subsided, however, and W. E. Gird, a California rancher, eventually developed a pure strain of polls from the stock he purchased from Case and Elling. The preference for polled cattle spread among English cattle growers, and a polled Devon bull and five cows were exported from the United States in 1957 to found the polled strain in England.

The American Devon Cattle Club, succeeded in 1971 by the Devon Cattle Association, Inc. (Goldendale, Washington), has displayed a progressive attitude toward the adoption of modern techniques. The association favors the use of artificial insemination, and semen from Devon bulls is available from some of the bull studs. Performance and progeny records are a part of the program for members. The herdbook is open, and qualified animals $15/16$ Devon may be registered as purebreds.

Many Devon cattle during the first decades of the nineteenth century became absorbed in the Native American cattle, but in the hands of the more meticulous breeders they were often maintained pure. In the southern states, especially Georgia and Florida, they were particularly prominent for many years.

The current enthusiasm for new breeds in the United States invariably

Galloway herd on summer pasture. Bozeman, Montana.

leads to a search for something foreign. The Devon cattle have excellent beef potential and are already at hand with no importation problems involving either cattle or semen. It is difficult to understand why they have not been more widely used.

Galloway.—The chronicle of the Galloway breed in the United States dates from 1870, although there have been herds in Canada since 1853 and some cattle were probably taken across the border in small numbers about that time. The breed attracted considerable attention, and there were a number of importations after 1870. Most of the promotional effort of the Galloway, however, came from Canada. The first herdbook was formed there and accepted United States cattle for registration. Registrations were continued by the North American Galloway Breeders Association, which was formed in 1882. This organization was founded in Chicago by breeders in both Canada and the United States, but became domiciled in Canada. The American breeders broke away a number of years later and formed their own organization.

In the meantime the Aberdeen-Angus breed, which had reached the United States in 1873, had gained wide popularity. The Galloways spread through the northern states lying west of the Mississippi River but numerically were considerably fewer than the Angus.

A rough, curly hair coat characterizes the Galloway. It is otherwise very similar to the Angus, both being strongly polled. Typical of some other black breeds, individuals that are carriers of the red gene appear and occasionally produce red calves. Dun and white are also acceptable colors in some of the breed societies.

The Galloways have a well-earned reputation for hardiness in inclement weather. On northern mountain ranches they calve in the spring with few losses. There are breeders of Galloways throughout the western states, but their herds are small in number.

The breed has never enjoyed the popularity its excellence as a beef animal warrants. Lack of co-ordinated promotional effort by two different breed societies may have contributed to the lack of interest in the Galloway. This situation paved the way for the formation of a third Galloway organization in 1970, Galloway Performance International. This association requires that each animal's performance record be entered on its registration certificate. There are few restrictions on the use of registered Galloway bulls either naturally or through artificial insemination. Full registration in the organization is recognized for the fourth generation. Five color variations are recognized, Black, Dun, Red, White, and Belted.

In 1971 there were three Galloway breed organizations in the United States: Galloway Cattle Society of America (Hennepin, Illinois), American Galloway Breeders Association (Rapid City, South Dakota), and Galloway Performance International (Eureka, Kansas).

Belted Galloway.—The Belted Galloway is the same breed as the Galloway but carries the gene for the band of white hair around the middle that was introduced at some time in the unrecorded past. The white belt in all probability developed from the same recessive gene as that which was responsible for the Dutch Belted cattle. The origin of this gene has been traced to the Tyrol region of the Alps; from there carriers apparently were taken to the Netherlands and then to the British Isles.

The black part of the coat of the Belted Galloway is occasionally replaced by a red-dun color, the same as occurs in the straight Galloway.

The Belted animals in North America trace to an importation in 1948 which was distributed in both Canada and the United States. Although the number of the breed in the United States is small, the owners are distributed throughout the country.

The Belted Galloway Society (Summitville, Ohio) was founded July 1, 1951.

Hereford.—Under this heading only the horned Hereford is discussed.

Belted Galloway bull. White Belt Dairy Farm, Port Mayaca, Florida.

The polled and horned Herefords are the same breed, but in recognition of the Polled Association they are usually treated separately.

Antecedents of the cattle which eventually coalesced in the Hereford breed in England were probably brought to Massachusetts and Virginia during the seventeenth century in some of the cattle shipments of that time. Such arrivals were eventually absorbed in the Native American cattle and left no trace.

The editor of the first Hereford herdbook, L. F. Allen, recorded that Henry Clay imported the first of the breed to Kentucky in 1816 or 1817. A few other importations were also recorded in the early years of the nineteenth century, none of which have left any trace.

In 1840, William H. Sotham, a farmer and cattle buyer of New York State, imported 22 head of Herefords from England. These cattle were used largely as draft animals and attracted considerable attention. A number of small herds were soon established, one in Maine and several

Cattle of North America

in other New England states, but the number of purebred Herefords did not increase rapidly until after the Civil War. Importations from England then began to multiply, and by the end of 1886 nearly 4,000 head had reached the United States. Most of these arrived after 1880.

The well-known red-and-white color pattern of the Hereford is one of the most uniform of any cattle breed. This was not always the case, however. Before 1800 and for some years after, the Herefords in England were described as "mottle"-faced and "spotted" red and white over the body. Today, as the result of long, persistent color selection, the white face predominates, and a white crest, dewlap, and underline are required markings. Not quite as rigidly adhered to are white flanks, white on the legs below the knee, and a white switch. Hair color over the rest of the body is red and may vary from a dark shade to a yellow-red. The yellow-red—a "mellow yellow"—was in high demand a few years ago. The horns are of medium size, curved outward and forward. An excellent beef conformation is typical.

When the compact craze struck in the 1940's, the Herefords lost size, as did the other beef breeds. This situation was eventually complicated with dwarfism. The latter problem has been overcome by extreme measures used to weed out the carriers of the dwarf genes. Some high-reputation bulls were eliminated in the process, however. Today the short, compact animal has practically disappeared. Breeders have emphasized stretch and put more daylight under their cattle.

On farms from the eastern seaboard to Ohio the early Herefords were used both for draft and for meat. With the close of the Civil War a demand developed for more and better beef. As the Longhorn cattle were trailed to the railheads in Kansas, cows became available for breeding in the West. At first Shorthorn bulls were favored, but as the Herefords became more widely distributed they were also used by the western stockman in his upgrading program. The use of Hereford bulls continued to increase, and by the turn of the century there were as many upgraded Hereford as Shorthorn herds in the western states.

In the early 1870's, Colorado followed Kansas in the swing to Herefords. From there they moved to Wyoming. They reached Texas in 1874 and by 1880 were found throughout the high-plains country of West Texas. The Hereford then became the range cow of the western cattle country and still maintains this position.

By the mid-1920's the number of Herefords in the country exceeded the number of Shorthorns. This trend continued at an increased rate until the 1960's, when the ratio of Herefords (including Polled Herefords) to Shorthorns was 25 to 1. The major competitor of the Hereford

Hereford cows on calving ground. Smith Valley, Nevada.

is now the Angus. However, in 1965 the Herefords and Polled Herefords accounted for slightly over one-half the total beef-cattle population of the United States. They are found in every state of the Union, including Hawaii and Alaska. The breed adapts well to any part of the United States except the deep South.

The American Hereford Association took a reluctant approach to the use of performance and progeny records in its breed-improvement program. In 1927, the "Register of Merit" was adopted in an attempt to recognize outstanding breeding stock, both bulls and cows. Accumulation of a sufficient number of show ribbons by its progeny was required in order for an animal to receive the "Register of Merit" award.

In 1962 the association set up a feed-test and carcass-evaluation program, co-operating with commercial feed lots and packers to obtain the necessary data. This program was designed primarily for the evaluation of purebred herd sires. Two years later the program was combined with the Register of Merit into a Total Performance Records program. In 1967 further modification of this program was announced to become effective in 1970.

In this latest program a sire must qualify as formerly on the show ring winnings of his progeny and, in addition, must meet certain progeny standards. These standards require satisfactory records on at least eight progeny, including weight per day of age, calculated per cent of retail yield, and marbling score.

The requirement for carcass data on eight progeny is a forward step in breed improvement, but the retention of a requirement for show-ring performance can hardly be considered progressive. The carcass-data requirements are unrealistic. They permit a carcass of poor cutability (one with only 4.4 yield, grade, or 50 per cent cutability) to qualify a sire as long as it meets a minimal weight-per-day-of-age requirement and a very poor marbling score.

The Total Performance Records program, the latest program of the association, was set up in 1969. It provides for "in-herd improvement" and the selection of desirable sires; it is open to both purebred breeders and commercial producers. Preferably, calves are to be identified with their dams at birth, and weaning weights must be obtained. Yearling weights are optional. A supplementary program for commercial breeders does not require identification of calves with their dams. The data are computer-processed by the association, and a copy of the results is furnished the co-operator.

Only in recent years have calves sired by artificial insemination been accepted for registration, and those within strict limitations. There has been little use of this technique in either purebred or commercial Hereford herds, except in the case of a few highly publicized sires.

The American Hereford Cattle Breeders Association was organized in 1881. The name was changed in 1934 to American Hereford Association (Kansas City, Missouri).

Polled Hereford.—The polled strain of Shorthorn cattle that was developed during the last quarter of the nineteenth century attracted considerable attention among cattlemen of the day. The advantage of hornless cattle was apparent, particularly under confined conditions. There was also the instinctive desire of cattlemen for something new. This concept struck the Hereford breeders in the early 1890's, first in Canada and soon afterward in the United States.

In Canada a breeder named Mossom Boyd crossed purebred Angus bulls on purebred Hereford cows and then bred the hornless male progeny back to Hereford cows. A few Hereford breeders in the United States made similar use of the polled influence at about the same time. In addition to Angus, these breeders used Red Poll and polled Short-

Polled Hereford bull, seven year old, 2,100 pounds. American Breeders Service, De Forest, Wisconsin.

horn bulls. In 1900 a breed society was established for the registration of polled cattle which were predominately Hereford but carried the influence of a polled breed other than Hereford. Such cattle were termed Single Standard Polled Herefords.

It is unfortunate that the breed jealousies of the Hereford owners of the time dictated the abandonment of the Single Standard Polled Hereford. The relatively minor influence of the Angus, Red Poll, and Shorthorn breeds which were used as dehorners could hardly have adversely affected the usefulness of the Hereford as a beef animal.

Again following in the footsteps of the Shorthorn breeders, Warren Gammon, a lawyer and cattle breeder of Des Moines, Iowa, embarked on a program in 1901 of developing a strain of Polled Hereford which would be the straight progeny of naturally hornless purebred Hereford parents. He solicited practically all the purebred Hereford breeders in the United States at that time in an effort to obtain any naturally hornless offspring that might appear in their herds. Four bulls and seven cows were the result of his search and comprised what Hereford history calls the foundation herd of Polled Herefords. From this nucleus he began to develop a line of Polled Hereford cattle that traced to 100 per cent Hereford ancestors. Such hornless progeny were called Double Standard.

Other Hereford growers took up this type of breeding. Selection at first was almost entirely for the polled characteristic, since the primary goal was to build up numbers of purebred cattle that were hornless. Conformation and other desirable characteristics were relegated to a secondary role. Occasionally, as purebred mutations from horned parents were discovered, they were added to the herds of the Polled Hereford breeders. Even with such additions the genetic pool was limited, and the quality of the stock, judged by the conformation standards of the day, did not compare with that of the horned Herefords.

At this point Gammon and others started utilizing the best-quality horned stock they could acquire to improve their polled strain. Polled bulls were bred to horned cows and horned bulls to polled cows. All were registered cattle, and, for the most part, only the polled progeny were retained for breeding. Once the polled base was established, the continued use of the horned Hereford in the breeding is evidenced by the fact that up to 1913, out of 2,250 registrations in the American Polled Hereford Association, only 418 were out of polled dams; the other mothers had horns but, bred to polled bulls, gave birth to polled calves.

Actually, no breed distinction should be made between the horned and the Polled Hereford. The only essential difference between the two is that one has horns and the other does not. In addition to the horned and polled variance, there may be some basis to the claim made by the Polled Hereford Association in regard to size. Since their cattle were seldom selected for the small compact type, the Polled breeders contend that they never had to breed back to a more productive size.

Polled Herefords have enjoyed a phenomenal growth during the past 20 years and now account for well over one-third of the total Hereford cattle in the United States. They are found mostly in the hands of small and medium-sized growers in the corn belt and in the south. The Polled Hereford is not as prominent on the western range as the horned Hereford. Texas has a larger proportion of Polled animals in its Hereford herd than the other western states, but even there the horned Hereford is a sizable majority. Polled Herefords have been exported from the United States to England, the birthplace of the Hereford, and established there in large numbers.

Highland cows on winter feed ground. Dubois, Wyoming.

The breed association has had a varied existence under several names. The American Polled Hereford Cattle Club was established in 1900 for the registration of Single Standard Polled Herefords. With the advent of the Double Standard cattle, the name of the club was changed in 1907 to American Polled Hereford Breeders Association. A new herdbook was established for the registration of Double Standard Herefords. Finally, in 1947, the name of the organization was modified to American Polled Hereford Association (Kansas City, Missouri). Only animals whose ancestry could be traced to the American Hereford Association record of horned cattle were admitted to registry. The polled offspring of one horned parent and one polled parent may be registered today.

Highland (Scotch Highland).—The Highland breed of Scotland, sometimes called West Highland, has come to be known as the Scotch Highland in the United States. The Highland cattle appear to have descended more or less unchanged from an old landrace indigenous to western and central Scotland. The early owners of Highland cattle were content to maintain them without change because of their exceptional hardiness.

Cattle of North America

The breed has probably been but little affected by artificial selection.

There are reports of a herd of Scotch Highlands being brought to Colorado and established there in the mid-1800's. Another early importation was three carloads of bulls and heifers which were unloaded at Moorcroft, Wyoming, and trailed to the Powder River country. Apparently these introductions of the Highland eventually ran out. The present-day population of the breed traces to an importation to Montana in 1922 by Walter Hill and also to several other introductions around this same time to the eastern United States.

The Highland is a small breed, although there is considerable variation in size, depending on the level of nutrition. Mature cows usually weigh 900 to 1,000 pounds and bulls 1,200 to 1,500 pounds. The most noted characteristic is the long shaggy hair coat, with the hair on the forehead usually covering the eyes. A layer of fine, thick hair underlies the shaggy outer coat. The color varies from an off-white or light grey to yellow and brown. The horns are large, spreading outward and usually upward. The breed is inclined to a rather nervous disposition and is alert and quick in movement when disturbed. This characteristic can be modified to a large extent by careful handling. Highland cattle have proved to be extremely hardy in severe northern climates. They have a reputation for calving in the open in snow and cold with little loss from exposure. Herds in the Dakotas and a few in Alaska where the winters are much more extreme than in Scotland have displayed this hardiness.

In Scotland the breed has a reputation for maturing slowly, cows often not calving until three or four years old. This characteristic appears to have been overcome to a considerable extent in the United States. Young bulls on performance test reach weights of just under 1,000 pounds at one year of age with gains of well over 3 pounds a day while on feed.

The Highland breed has spread across the northern United States from coast to coast and is also found in Canada. The number is small, however, interest in the breed being kept alive mainly by a few loyal and enthusiastic breeders. There has been some crossing in recent years with the more common beef breeds, particularly the Hereford. The Highland-Hereford cross produces a high-quality, high-yielding carcass.

The American Scotch Highland Breed Association (Edgemont, South Dakota) was organized in 1948.

Indo-Brazil.—Few of the breeders of Zebu cattle in the United States made an effort to perpetuate any particular strain of Zebu cattle after

Indo-Brazil herd in South Texas. Coquat Ranch, Cotulla, Texas.

the formation of the American Brahman Breeders Association in 1924. The objective of the Brahman breeders was the development of a Zebu type with better beef conformation. The dissident faction, which is mentioned in the section on the American Brahman below, consolidated their position in 1946 and backed the Indo-Brazil breed as the type to be maintained pure in the United States. This led to the organization of the Pan American Zebu Association (PAZA) in that year. The objective of the association was the "registration, propagation and promotion of the Indo-Brazil breed of Zebu cattle."

The only true Indo-Brazil cattle in the United States at that time were 9 bulls included in the 1946 importation of 19 Brazilian Zebu bulls from Mexico. The other Zebu cattle in the United States were indeterminate mixtures of Zebu strains, principally Guzerat, Gir, and Nellore.

In essence, the objective of the PAZA breeders was to upgrade the Zebu cow of mixed Zebu antecedents by the use of the nine Indo-Brazil bulls and their progeny. Some PAZA breeders, however, had females that were predominantly of Guzerat and Gir breeding. They were used as foundation females in the revival of the Indo-Brazil in the United States, since the Guzerat and Gir are the major components of the Indo-Brazil breed as developed in Brazil.

The Indo-Brazil breed as seen in South Texas today may be red, or red-and-white spotted, although the predominant color is medium grey. The ears are long and pendulous and often have a twisted appearance at the base and again at the ends, a typical Gir characteristic. Both

Red Poll bull, four years old, 1,800 pounds. Herman Ebers Farm, Seward, Nebraska.

male and female tend to stand somewhat higher than either the American Brahman or the parent breeds.

The Indo-Brazil breeders in their Pan American Zebu Association (Cotulla, Texas) are still carrying on in spite of the strong position of the much larger American Brahman Breeders Association. Sizable exportations of breeding stock continue to be made to Central and South America. It seems inevitable, however, that the two breed organizations will eventually merge, possibly by maintaining two herdbooks—one for the American Brahman and one for the Indo-Brazil.

Red Poll.—Suffolk and Norfolk counties were among the areas in England from which cattle were brought to America by the colonists in the early years of the seventeenth century and for some time thereafter. Because of the areas of their origin, these cattle were undoubtedly from among the progenitors of the Red Poll breed as it came to be recognized

Red Poll cow, five years old, 1,250 pounds. Herman Ebers Farm, Seward, Nebraska.

in England nearly 200 years later. At that time the red, polled cattle of Suffolk and adjoining Norfolk counties which had been held separately in each county were combined and called the Red Poll. As with all the introductions of cattle to North America, every trace of such forerunners of the Red Poll breed was lost during colonial days, when little attention was given to cattle breeding.

G. F. Taber, of New York State, imported the first recognized Red Poll cattle to the United States in 1873, one year before the first herdbook was published in England. They became popular with the farmers in the northeastern part of the country for draft, milk, and meat. There were sizable importations of Red Poll cattle to the United States until 1900, when they dropped to practically nothing. In all, 350 head of the breed had reached the United States by 1967. The Red Poll was also a much-sought-after breed on some of the Caribbean islands under English influence. There were sizable importations there and along the eastern coast of Central America.

The Red Poll is a solid dark red with only minor white undermarkings permitted. Many herds now show a distinct beef-type conformation, though the dual-purpose appearance lingers. Red Poll steers stand well off the ground and often present a less wasty appearance than some individuals of the beef breeds. The cattle are strongly polled. Mature cows average 1,250 pounds in a good herd, and bulls in breeding condition should weigh around 2,000 pounds.

Historically, the Red Poll has been the small farmer's breed. After the horse displaced the draft ox, the principal emphasis on the Red Poll was as a dairy animal although it acquired a dual-purpose status.

The growth in numbers of the breed was never phenomenal. Representatives of the Red Poll spread gradually over most of the United States except in New England and the western range states. Over the years the breed has had its ups and downs. Much of the time the enthusiastic promotion of the modern successful breed society was lacking.

The Red Poll Advance Registry program was established in 1908. This recognition was granted to cows whose butterfat production met established standards and to bulls with two daughters meeting the same standards. The program accomplished little in the way of either breed or herd improvement since it applied only to individuals invariably selected from the few top-producing cows in a herd. No program or standards were established for herd improvement. The program is still in effect, although practically no use is now made of it.

A decade ago, when the beef-cattle breed societies of the United States began employing more realistic methods for improving cattle, the policies of the Red Poll Cattle Club underwent a radical change. First, emphasis was placed on beef production rather than milk, and, second, the new techniques for evaluating and improving the productive ability of beef animals were recognized and recommended to the membership. The Red Poll Club was among the first of the beef-breed associations in the United States to introduce modern record systems to its members.

The Gain Register, inaugurated in 1960, is a recording of the pre-weaning rate of gain of all the calves in a breeder's herd that are produced in one year. During its first six years, over 10 per cent of the Red Poll calves registered in the United States were in the program. The Gain Register is designed for in-herd improvement of an entire herd. The grower furnishes the basic data to the Red Poll Cattle Club, where it is processed and returned to the breeder.

United States

The Carcass Register, inaugurated in 1963, was designed to give the purebred breeder carcass data which, in conjunction with the Gain Register, would implement the selection of more productive beef cattle. Both the Red Poll Gain Register and the Carcass Register are offered only to the purebred breeder.

During the past few years the Red Polls in the United States have crossed the line from a dual-purpose to a beef breed. There are only a few Red Poll dairy herds left in the country, although a cow in a farm herd may be milked for household use. Most growers of the breed now either feed out their calf crop to slaughter weights or sell their calves at weaning to other feeders.

The breed society was organized in 1883 as The Red Poll Cattle Club of America (Lincoln, Nebraska), preceding by five years the British society.

Shorthorn.—Representatives of the forerunners of the Shorthorn breed, which were raised in the northeastern counties of England, were probably brought to the Atlantic Coast early in the seventeenth century when the colonists were receiving the nucleus of their cattle stock. The Shorthorn was among the first cattle types to be segregated and distinguished by a breed name in England. It was the first recognized breed to reach North America. Shorthorn, however, was not the only name by which these cattle were known in their homeland. The development of the British breeds frequently involved the interbreeding of cattle which had been grown and selected for certain characteristics in adjoining counties. The county of origin was thus often used to designate the Shorthorn cattle. Yorkshire and Durham counties were heavy contributors to the Shorthorn breed. The name Durham eventually came to be generally applied to Shorthorn cattle in the United States and they were so called until after the turn of the century.

For many years the Shorthorns were considered the largest beef animals in the United States. With the introduction of the Charolais and other continental cattle and the development of larger types in other breeds, that is no longer true. Shorthorn cows in good breeding condition weigh from 1,200 to 1,400 pounds and bulls up to 2,000 pounds. Color ranges from red through roan to white, with red-and-white patterns also permitted, though they are considered less desirable. A dark red is currently the preferred color, but any shade is acceptable for registration. The typical conformation is blocky and rectangular.

The Shorthorn emerged in England as a recognized breed during the last quarter of the eighteenth century. Some of these early Shorthorn cattle, though probably of the milking type, were shipped to Virginia in 1783. There is record of another shipment to New York in 1791. Progress of these early arrivals in the United States was not recorded, although the progeny of the 1783 shipment to Virginia are said to have been moved to Kentucky, where they became established.

The real founding of the Shorthorn breed in the United States came with importations into New York and Kentucky in 1817. Western Pennsylvania, Kentucky, and Ohio were just being settled at this time, and the demand for cattle resulted in sizable importations from England for the next 30 years. By 1850 the Shorthorn population of the United States was well established. The commissioner of agriculture in his report for 1866 stated that there were 6,000 "thoroughbred" Shorthorn cattle in the United States and that 2,000 of them were bulls. Imports of purebred stock were continued in limited numbers, however, primarily as promotional efforts of individual breeders and for sales programs of cattle dealers.

Sizable numbers of Shorthorn cattle, both purebred and grade, had been trailed to Oregon with the early settlers in the late 1840's. They did exceptionally well in their new homeland, and the increase was phenomenal. During the last half of the nineteenth century the breed spread across the United States into nearly every area where cattle were raised, except in the Southeast. As they reached Illinois, Indiana, Missouri, and Kansas, purebred herds were established which served to stock the continuing westward movement of cattle.

In the years following the Civil War the historic cattle drives from Texas brought Longhorn steers to the Kansas railheads for shipment east. As the cattle boom expanded into the country east of the Rocky Mountains, the Longhorn cows stocked the ranges from Colorado through Wyoming to Montana and the Dakotas. This movement was well established by 1870, and the market for Shorthorn bulls to upgrade the Longhorns was insatiable. Bulls and in some instances cows to establish breeding herds for the production of purebred bulls were taken back from Oregon to the mountain-range country. The first such movement was to Wyoming in 1879.

Before 1880 the Shorthorn cattle brought to the United States came from England, where the development of the breed had leaned toward a dairy, or at least, a dual-purpose cow. A number of factors tracing to the origin of the breed now seemed to converge to the Shorthorn's detriment in the United States.

Importations of Hereford and Angus cattle, breeds which developed later than the Shorthorn in England, had been increasing, and these breeds, particularly the Hereford, gained an early foothold in the corn belt of the United States. The commercial cattleman of the day, even then susceptible to the glamour of a new breed, began to notice these white-faced cattle. As Hereford bulls became available for upgrading the growing herds of Longhorn cows in the western states, the Shorthorn bubble burst.

Prices for purebred Shorthorn stock had risen to grossly inflated values. In 1873 a cow of highly publicized breeding sold for over $40,000—an unrealistic value to place on a cow in any period, but incredible at a time when fat steers were bringing only $10 a hundred pounds. Twenty-three years later a good purebred Shorthorn herd at a dispersal sale in Kansas averaged only $205 a head. Although the sale took place in a period of generally depressed prices, it underscores the exorbitant prices that had been paid during the boom.

As the commercial breeder began to turn from the Shorthorn, the Hereford ascendancy started in the open-range country. This movement did not, however, occur suddenly. After the first decline of the Shorthorn around 1880, the popularity of the breed was revived for a period by importations of the "Scotch" Shorthorn stock.

While early development of the Shorthorn in England had resulted in the fixing of a distinctly dual-purpose type, Scottish breeders selected for a more pronounced beef-type conformation. The Scotch cattle had been brought to the United States before the Shorthorn breeders found themselves in trouble, but this line of breeding had not become popular. The position was reversed when the older type of Shorthorn went out of favor and the Scotch Shorthorn saved the day for the breed in several decades. The few herds of Scotch breeding in the United States were rapidly expanded and augmented by sizable importations from abroad. Today the beef Shorthorns in the United States are essentially of the Scotch strain.

The Shorthorn suffered equally with the two other major beef breeds when the compact fad was the fashion. Shorthorn breeders had their share of dwarfs, but when the cause was recognized, they began to search out and eliminate the bulls that were responsible.

Once the dominant beef breed on the western range and still widely distributed over the United States, both the commercial and purebred Shorthorn cattle are now seen mostly in small or moderate-sized herds. In numbers the Shorthorn has dropped to the fifth most numerous breed in the United States. By 1960 the breed was estimated at less than 4 per cent of the national herd.

The American Shorthorn Breeders Association (Omaha, Nebraska)

Cattle of North America

was established in 1882. The first herdbook had been compiled privately in 1846 and was acquired by the association when it was organized. The American Shorthorn Breeders Association is considered the parent organization for all Shorthorn cattle, registering both the polled and the horned cattle, but the polled strain has a separate promotional organization.

Polled Shorthorn.—During the last decades of the nineteenth century when the Scotch Shorthorns were replacing the earlier English Shorthorn imports to the United States, the Polled Shorthorn was in the initial stage of development. This strain originated in the United States and is considered separately, although the only practical difference between the two types is the presence or absence of horns. They have the same variation in color—from red to roan to white—and have the same beef-type conformation.

The first Polled Shorthorn cattle in the United States were the result of crossing Shorthorn bulls on cows of a polled breed. These were commonly known as Polled Durhams until after the turn of the century. Continued crossing and rigid selection for progeny without horns developed in a few generations a strain of nearly pure Shorthorn cattle which was substantially polled. This development had reached the stage in 1889 where a number of the breeders of these animals felt justified in establishing a herdbook. The original requirement for registration was three-fourths purebred Shorthorn lineage. This was later changed to $^{31}/_{32}$. All cattle of such breeding that could not be traced to 100 per cent Shorthorn ancestry were known as Single Standard Shorthorns. They were ineligible for registration in the American Shorthorn Herd Book.

As happens occasionally in most breeds of horned cattle, mutations of polled individuals appeared from time to time in the herds of purebred breeders. Interest in polled cattle had been aroused by the promoters of the Single Standard Polled Shorthorns, and a number of purebred breeders sought out these mutations, interbred them, and established small herds of Polled Shorthorns that were the progeny of registered parents. They also were called Polled Durhams.

A cow which became widely known in polled circles as one of the founders of the Polled Shorthorns was Oakwood Gwynne IV, owned by Colonel McCormick Reeve, of Minneapolis, Minnesota. Oakwood Gwynne produced many polled progeny, among which were twin calves born in 1881. As the small herds of pedigreed Polled Shorthorns, which came to be known as Double Standard, increased, the breeders of the Single Standard cattle were pushed aside, and their stock event-

Polled Shorthorn bull. American Breeders Service, DeForest, Wisconsin.

ually disappeared. It has been hinted from time to time that some of the better Single Standard individuals may have surreptitiously crossed over to the registered side. Actually, because more latitude in selection was possible, the Single Standard Shorthorn was probably a better meat animal than the Double Standard animal, which had to be selected rigidly for the polled characteristic. It took time to build up a polled base large enough that Horned Shorthorns could again be used for upgrading.

The first purebred Polled herds were of the early Shorthorn type and carried none of the Scotch influence. Though hornless, they were inferior conformation-wise to the favored type of Horned Shorthorn in the early 1890's. An Indiana breeder, J. H. Miller, set out to correct this defect. He purchased several Scotch Shorthorn bulls with excellent conformation and bred them to his Polled cows. Selection of the hornless animals from such matings led to the segregation of a greatly improved herd of Polled Shorthorns.

Other Polled Shorthorn breeders followed similar programs, and, by thus incorporating the best of the horned Shorthorns into the polled strain, they soon eliminated any difference in quality between the two

Mature Sussex bull, 1,800 pounds. Lambert Estate, Refugio, Texas.

strains. These Polled Shorthorns are now entered in the same herdbook as the Horned Shorthorns, with an X placed before their registration.

The Polled Shorthorn is found throughout the United States in the same general areas as the Horned strain. Since it was developed after the western range was stocked and after the Horned Shorthorn had received its major setback, the Polled Shorthorn was never used to any great extent as a range animal. Over one-third of the Shorthorn cattle in the United States are now Polled.

For breed promotion there is a separate association, the American Polled Shorthorn Society. Registrations are handled by the American Shorthorn Breeders Association (Omaha, Nebraska).

Sussex.—The Sussex cattle are said to have been introduced into the United States in 1884 on the farm of Overton Lea in Tennessee. Descendants of the Lea herd found their way to Texas. In the early 1890's this nucleus formed the small purebred herd on the ranch of T. D. Wood near Refugio. There is apparently no record of their origin. Later,

Mature Sussex cow, 1,250 pounds. Lambert Estate, Refugio, Texas.

during a period of drought, all the cattle were sold to a farmer in Tennessee. At that time all trace of the Sussex as a breed disappeared from the area. Small herds of Sussex cattle were known to exist in Tennessee, Oklahoma, Indiana, and Texas in the first decades of the twentieth century, but they also eventually disappeared.

There is record of a small importation of Sussex animals to South Texas in 1947. Lawrence Wood, a grandson of the earlier owner of the Sussex herd in Texas, made a number of importations from England. In all, 44 females and 14 bulls were taken to the state over a period of eight years.

The Sussex, originally selected as a draft animal in Sussex County in southern England, was later developed into an excellent beef type. The color is a bright, deep red, solid over the whole body. The horns of the bull are rather short, extending laterally from the head. On the cow the horns are larger and curve either forward or upward. As seen in South Texas, the Sussex is a somewhat smaller animal than in Eng-

Cattle of North America

land, mature cows weighing around 1,200 pounds and bulls 1,600 to 1,800 pounds.

In 1971 there were four Sussex breeders in South Texas. They were primarily producing purebred bulls for crossing on Brahman cows. This cross has produced a new breed called the Sabre, discussed in the section "American Developed Breeds."

The number of purebred Sussex in the United States is probably less than 200 head. The excellent fleshing qualities of the breed and the current demand for new breeds for crossing purposes might lead to a new demand for this old English breed. Before the turn of the century, in the early fat stock shows in Chicago, many prizes were won by Sussex representatives.

A polled strain of Sussex has been developed in England by selection from the progeny of a Red Angus bull. They may be registered if they are $15/16$ pure Sussex. Some of the Sussex in the United States herds carry this polled strain.

The Sussex Cattle Association of America (Refugio, Texas) was organized in 1966. The English Society registers United States entries in a separate section of its herdbook.

Introduced Breeds—Dairy Type

Ayrshire.—The earliest mention of the importation of Ayrshire cattle to the United States from their native Scotland is of a few head shipped to Windsor, Connecticut, in 1822. No record of these remains, and the assumption is that they soon disappeared into the mainstream of Native American cattle. The Massachusetts Society for the Promotion of Agriculture made several importations from 1837 to 1858, a total in all of 6 bulls and 18 cows, some of which were in calf. There were also other small importations of the breed by interested farmers during this period. The major importations, however, took place between 1865 and 1900.

The Ayrshire is red and white, the white frequently covering much of the body as the breed is seen in the United States. The red may be any shade; occasionally even a very dark mahogany is seen. The physical characteristics of the Ayrshire appear not to have changed significantly since the breed was developed in the County of Ayr in Scotland early in the nineteenth century. Conformation is of the dairy type, but the cattle are larger and somewhat better muscled than the Jerseys or Guernseys. Ayrshire cows have long been noted for their excellent udders. The horns on the cow are large, upturned, with a twist toward the ends; those of the bull are less prominent. Both cows and bulls,

Ayrshire dairy herd.

however, are now usually dehorned. Polled mutants appear occasionally, but there has been no attempt to select a polled strain. Mature cows in good milking condition weigh 1,200 to 1,350 pounds; bulls on pasture, 1,800 to 2,000 pounds. Ayrshire cattle are a hardy breed and particularly well adapted to cold climates. Introduced into Finland at the middle of the nineteenth century, the Ayrshire has made a remarkable record and is now the most popular dairy animal in that country.

The early Ayrshire owners were situated in the northeastern United States, where the rocky hillsides became a home to which the breed was well adapted. From this section the Ayrshire spread westward, finally reaching most states of the Union.

For a period after the Civil War the Ayrshire and Jersey were the two most prominent dairy breeds in the United States. The Ayrshire was favored by some dairymen because the larger cow brought a higher price when she went for slaughter. Ayrshire herds have increased modestly in size along with the trend to large farm operations. The 50-to-75-cow unit is typical. The Ayrshire continues to be the breed of the family farmer. Changing times—the increased productivity of the national dairy-cow herd without a corresponding increase in the demand for milk and the swing to large Holstein-Friesian herds—have contributed to a decrease in the Ayrshire population in recent years. It was estimated at 90,000 head of purebred cattle in 1970, or a total of approximately 120,000, including commercial herds.

The Ayrshire Breeders Association(Brandon, Vermont) was organized in 1875, when it took over a private herdbook and from then on handled the registration of all purebred animals of the breed. As was the case with all the breed organizations formed around that time, the initial work of the association was largely the recording of both domestic and imported animals and the transfer of titles. Within ten years, however, the association had entered the field of production testing.

In 1884 the Ayrshire Association offered prizes for home tests for the production of milk, churned butter, and soft cheese. The competition applied to the herd total of these products, as well as the quantities produced by the best cow. The program created general interest in production testing among the Ayrshire breeders and led to the establishment of the Advance Registry test in 1902. This registry gave official recognition to the individual production record of a high-producing cow and served to keep interest focused on production records. Next it was realized that the herd production was what really counted, but it took time to get a herd-testing program formulated and approved. The big step forward came in 1925, when the herd test was inaugurated.

The Ayrshire was the first breed society in the United States to inaugurate this type of test. It led to sire evaluation based on dam-daughter comparisons. Artificial insemination was employed in the late 1930's; the best of the proven sires were used as widely as possible. Through the various steps of production testing and herd improvement down to the Dairy Herd Improvement Registry, the Ayrshire has been among the most progressive organizations in the dairy industry.

Brown Swiss.—The Swiss Brown cattle which became segregated in the valleys of Switzerland and the adjoining mountainous terrain of Austria retained much of the character of the landrace from which they derived. They responded to selection first as a draft animal; then as a dual-purpose, milk-meat type; and finally in America as a heavier milk producer. Throughout these changes the "Big Brown Cow" retained her well-recognized identity. In North America and throughout the Western Hemisphere the breed is called Brown Swiss instead of Swiss Brown as in continental Europe.

The first Brown Swiss cattle brought to North America were imported in 1869 by Henry M. Clark to his farm in Belmont, Massachusetts. This shipment consisted of one bull and seven cows which had been purchased in Arth, in the Schwyz Valley of Switzerland. Subsequent importations, all from Switzerland, brought the total number of Brown Swiss introduced to the United States of which there is definite record to 155 head—25 bulls and 130 cows.

Development of the United States Brown Swiss cattle proceeded along dairy lines. This selection for milk production has effected a conformation that is quite different from the breed as seen today in Germany and Austria, which, although a dual-purpose type, has a good beef conformation. The Swiss Brown in Switzerland is intermediate between these types.

The Brown Swiss was the last of the dairy breeds to enter the United States and arrived at the time when the practice of selecting dairy animals on the basis of their production records was beginning to attract attention. The early owners of the Brown Swiss naturally followed this trend of the times in the management of their cattle.

The Brown Swiss is one of the larger dairy breeds. Bulls in breeding condition weigh upwards of 2,000 pounds, and cows from 1,300 to 1,500 pounds. The color is nearly solid brown, varying in individuals from a medium dark shade to a light, somewhat greyish color. The underline is generally lighter but should not show white. Horns are rather small, curved forward, and shorter on the bull. The breed is exceptionally hardy and has a reputation for a long productive life.

In the United States the Brown Swiss has always been the farmer's cow and was not subject to the ups and downs of promotional activities, exorbitant prices, and selection for extremes in fashionable colors and size.

After its establishment in New England the breed spread over the eastern states and then became prominent in the dairy land of Wisconsin and Minnesota. Today it is found in all 50 states and has acquired a reputation for milk production and adaptability throughout the Western Hemisphere. It is exported to many countries in South and Central America, to Mexico, and to Canada.

The Brown Swiss Cattle Breeders Association of America (Beloit, Wisconsin) was organized in 1880. After adopting a scale of points for selection of animals for registration, it followed with production testing and has been exceptionally progressive in the many steps it has taken for breed improvement. Particularly noteworthy are two recent innovations: Identity Enrollment and the Beef Registry.

Under the Identity Enrollment plan, the female progeny of a grade

Cattle of North America

Brown Swiss cow may become eligible for full registration in the herdbook if the following three steps are completed:

1. Identification: Each cow whose progeny are to become eligible under the program must be characteristic Brown Swiss color and type. These females, which may be of any age, must be inspected and identified by a representative of the Brown Swiss Cattle Breeders Association.

2. Identity Enrollment: Female offspring of an *identified* cow, sired by a registered Brown Swiss bull, are eligible for registration in the Identity Enrollment Herd Book. These females must be production-tested and classified to determine performance.

3. Performance: When a cow in the Identity Enrollment Herd Book reaches a level of production with a 305d 2× record above the current DHIR average for her age and is classified "good plus" or better, her *female* offspring will be eligible to enter the Official Brown Swiss Herd Book. The prefix C.I.E. must be used in the name of the offspring of the "above average" Identity Enrollment cows entering the Official Herd Book. (Such offspring to qualify for registration must have passed an above-herd-average production test and have been classified as "good plus" or better.)

Limiting the gene pool of a breed to the descendants of whatever animals happened to be on hand when the first herdbook was started has been the generally accepted practice of most breed organizations. The Brown Swiss Association has taken a forward step in permitting new genetic material to be utilized in the manner described above.

In 1971, Brown Swiss Beef International, Inc. (Beloit, Wisconsin), was founded for "the specific purpose of recording and registering the offspring of Brown Swiss bulls used in beef cattle crossbreeding and upgrading programs." To this end the new organization established two recording systems, the Brown Swiss Beef Purebred Registry and the Brown Swiss Beef Cross Record.

The Brown Swiss Beef Purebred Registry will record the male and female offspring of registered or Identity Enrollment Brown Swiss cows, females that are ⅞ Brown Swiss, and bulls that are $^{15}/_{16}$ Brown Swiss. If this move strikes a responsive chord among Brown Swiss breeders who develop an interest in the beef possibilities of the breed, it could well lead to a successful selection for a beef-type Brown Swiss that will be competitive with the established beef breeds.

The Brown Swiss Beef Cross Record registers Brown Swiss percentage females that are from ½ to less than ⅞ Brown Swiss. The founda-

Brown Swiss meat-type sire: American Breeders Service, De Forest, Wisconsin.

tion cows used in this program may be of any of the established beef or dairy breeds or crosses thereof.

In 1971 Brown Swiss semen was imported direct from Switzerland, affording United States breeders the opportunity to introduce better fleshing characteristics into their Brown Swiss herds.

Dexter.—The Dexter breed originated in Ireland and was taken to England in the 1880's. Dexter cattle were imported to the United States in 1912.

The Dexter is characterized by its small size; a mature bull weighs 800 to 900 pounds and a cow around 650 pounds. The color is solid black with only minor white markings on the underline. The breed carries a recessive red gene and occasionally an all-red calf is dropped. The horns are rather small and grow outward. The Dexter is usually

Young Dexter bull, 650 pounds. Black Hawk Ridge, Wisconsin.

considered a dairy or dual-purpose type. Its conformation is smooth, well rounded, and blocky. The legs are noticeably short in proportion to body size. The small Dexter cow is a surprisingly good milk producer, good representatives giving 5,000 pounds of 4.5 per cent butterfat.

Both in its homeland and in the United States the Dexter produces an abnormally high percentage of the bulldog-type dwarf calves. This tendency probably resulted from the early selection in Ireland for a small, compact conformation, the same selection criteria that resulted in the wave of dwarfism in the British beef breeds in the United States 30 years ago. Breeding practices are now aimed at eliminating the dwarf carriers.

While the Dexter cattle in the United States must be considered in the hobby class, they are spread across the country from New England to California. The herds are small, often only numbering a few head. Milk is usually taken regularly by the owners.

Mature Dexter cow, 600 pounds. Black Hawk Ridge, Wisconsin.

The American Dexter Cattle Association (Decorah, Iowa) was organized in 1914, and a large proportion of the cattle are registered.

Dutch Belted Cattle.—The Dutch Belted cattle are known in Holland as the Lakenfeld. Their history in the United States goes back as far as most of the other dairy breeds. Dutch Belted cattle were first imported in 1838 by P. H. Haight, a United States consul to the Netherlands. Two years later the renowned showman P. T. Barnum imported a few for exhibition in his circus, eventually placing them on his farm in New York State as breeding stock. Continuity of the Dutch Belted in the United States is said to have begun with the Barnum herd. Subsequent importations resulted in the spread of isolated herds throughout the United States. Good Dutch Belted cows will compare favorably in production with other dairy cattle, but, because of the small numbers, the breed is not of much commercial importance.

The distinctive mark of the Dutch Belted is the broad white band

Dutch Belted cow herd. White Belt Dairy Farm, Port Mayaca, Florida.

completely encircling the body, the rest of which should be solid black. The white belt should be sharply defined, beginning behind the shoulder and extending nearly to the hips. The horns are upturned and moderately spread, long and thin. The Dutch Belted cow is a medium-sized dairy type which has been selected for milk production in the United States as long as the other dairy breeds. The cow should weigh from 1,150 to 1,300 pounds, and bulls weigh up to 2,000 pounds.

The white belt is one of the most unusual color markings to be found on cattle. As far as is known, it is confined to two breeds, the Dutch Belted and the Belted Galloway. Both are black with the white band around the middle. The white band is said to come from a recessive gene which was first found in the cattle of the Tyrol region of the Alps. From there cattle bearing the gene must have traveled across Germany to the Netherlands in the prebreed days. There by selective breeding the Dutch Belted cattle were developed. The theory is that carriers of the gene then reached Scotland, and the Belted Galloway developed.

The Dutch Belted Cattle Association of America (now situated in Miami, Florida) was incorporated in New Jersey in 1909. At that time most of the breed was to be found in New Jersey and New York. One of the foremost breeders of the Dutch Belted cattle, John G. Dupuis, Jr., of Miami, Florida, maintains a bank of semen from high-record bulls as a service to members of the association.

Guernsey cow, large type. Elwyn Fowler Farm, Southampton, Massachusetts.

Guernsey.—The Channel Islands, the small group of islands in the English Channel, include Jersey and Guernsey and the two very small islands Sark and Alderney. The Channel Islands were often given as the source of the early cattle brought to the United States from either Jersey or Guernsey. Sometimes the name of the specific island was mentioned; sometimes it was not. The official histories of the two breeds quote from the same letter, written by Reuben Haines, of Germantown, Pennsylvania, to authenticate the early presence of their breed in the United States. Haines wrote on October 20, 1818: ". . . with this you will receive a pound of butter made from the Alderney cow imported in 1815 by Maurice and William Wurts, and now in my possession." The assumption was that the cow was of the breed covered by the history.

The Guernsey as seen in the United States today is tan or fawn color with distinctive white markings, usually more prominent on the underside. The horns are thin and small and curved forward. Most cows and many bulls are now dehorned as calves. The size of the Guernsey has been increased in recent years until the cow in a modern herd weighs between 1,100 and 1,200 pounds, and bulls in breeding condition weigh upwards of 1,700 pounds.

Cattle of North America

In the early years of the nineteenth century there was very little difference between the cattle on the islands of Jersey and Guernsey. In 1789 steps were taken against the importation of cattle from the mainland to the Channel Islands, but there appear to have been little or no restrictions on the movement of cattle from one island to another. There were exhibitions in which cattle from both islands were shown together; and if a particular cow or bull took a farmer's fancy, he did not hesitate to buy it on one island and take it to his home on the other. During the first half of the nineteenth century the dairymen of the two islands began breeding for different types of milkcows—on Jersey, for a smaller, more finely boned animal; and on Guernsey, for a larger type. As late as 1896, however, individuals could be found on either island which met all the breed requirements of the other.

Guernsey owners have gone to great length to authenticate the colorful story pertaining to the establishment of their breed in the United States. This bit of history, substantiated by the affidavits of early farmers with direct knowledge of the events, is outlined as follows.

Around 1830, a Captain Prince of Boston called at the island of Guernsey and picked up two young cows and a bull. He took them back to the United States and presented them to his brother, who had a small farm on Cow Island in Lake Winnepesaukee, New Hampshire. One of the cows apparently died without issue, but the other cow and the bull founded a substantial herd.

Another farmer, Joseph Barnard, subsequently obtained some of the progeny of what came to be known as the Prince Cattle, and they eventually passed to his son, who was responsible for consolidating the history of this strain. On its completion this record was called the New Hampshire Guernsey Herd Book. The primary evidence that the Prince cattle were from Guernsey and not another Channel Island, is a reference to the elder Barnard's journal, which referred to his cattle as "Alderneys, sometimes called Guernseys." The culmination of this story was an act of the New Hampshire Legislature in 1933 changing the name of Cow Island to Guernsey Island in America.

The next noteworthy importation of Guernsey cattle occurred in 1840. It included three cows in calf which terminated their pilgrimage at the estate of Nicholas Biddle, in Bucks County, Pennsylvania. Scattered shipments from Guernsey continued over the next 35 years. Throughout this period there was probably little physical difference between the cattle on Jersey and those on Guernsey, but, in spite of the occasional transfer of an animal from one island to the other, the two breeds from that time on essentially trace to their respective island.

United States

The imports of both breeds to the United States began to increase in the last quarter of the nineteenth century and reached a peak in 1913–14. There were only minor importations of either breed after 1930. In all, almost 13,000 Guernseys were brought to the United States during the period of active importation. That is about five times the current number of cows on the home island of Guernsey.

The American Guernsey Cattle Club (Peterborough, New Hampshire) was organized in 1877. The sole objective of the founders was the establishment of an accurate and reliable procedure for recording pedigrees. The consuming interest in their cattle, however, soon led the founders of the club to wider fields—milk testing, advanced registry, herd improvement, and promotion.

For the most part Guernseys are still held on family-type farms in strong dairying areas. There are good representatives in Pennsylvania, New York, and Wisconsin and in the Northwest, in Oregon and Washington. The largest Guernsey herd in the world, 3,100 head, is maintained in Arrey, New Mexico.

The trend in recent years has been toward a larger cow; this trend is particularly noticeable in the herds in Washington and Oregon. In 1938, as with the other dairy breeds, artificial insemination came into use and spread rapidly. This technique made possible some of the rapid changes that were effected in the breed. The height of representative cows is said to have increased five inches during the past 15 years, and they now weigh from 1,100 to 1,200 pounds. In the past decade average milk production of recorded registered cows has been increased from 9,200 pounds to 10,000 pounds (two milkings a day for 305 days, the accepted dairy standard). Following the recent trend for less fat in the American diet, much less attention is now devoted to the high butterfat content of Guernsey milk.

To promote the breed and its product in the years after World War I, the Guernsey Cattle Club publicized "Golden Guernsey" milk. In 1934 Golden Guernsey, Inc., was formed as a subsidiary of the American Guernsey Cattle Club. This organization was effective in spite of—or possibly even because of—the publicity gained by its opposition to federal milk-marketing orders. In the years just before World War II, it was highly successful in selling Golden Guernsey milk as a premium product. This promotion is given credit in some circles for advancing the Guernsey to second place in popularity among the United States dairy breeds.

The adaptation of Golden Guernsey to today's demand for low-fat foods is unique in dairy-marketing publicity. Three low-fat milks are offered under franchise: Guernsey Royal Milk, 3.7 per cent butterfat; Guernzgold, 2.0 per cent; and Gurnzskim, 0.5 per cent.

Guernsey cows were entering and winning experiment-station and fair contests as far back as 1888. The tests by which these competitions were rated were usually based on the cost of producing a pound of churned butter. Nothing explicit was stated about what made up the cost other than feed, but presumably the bases were comparable for the different breeds. Cows entered in these contests were cared for and milked on the ground, their feed measured, and the butter churned there. At the 1893 World's Fair in Chicago the Guernsey cows on test, with the cost of their butter calculated at 13.2 cents a pound, barely nosed out the Jerseys, with a cost of 13.3 cents a pound. Both were ahead of the Shorthorn, the only other breed in the contest, at 15.85 cents a pound. No Holstein-Friesian cows were entered, the association having objected to the test conditions stipulated by the fair authorities.

The first effort of the Guernsey Cattle Club to establish production testing was made in 1894, when a procedure was begun whereby a farmer could milk a cow out one day a week and send the milk to a laboratory for the determination of butterfat content. In 1898 this method of testing was superseded by the Home Butter Test. This test required the recording of the amount of all milk produced by a cow over a period of one year, starting January 1. A monthly sample was taken for the determination of butterfat content, the test for which was made at the nearest agriculture experiment station, and a check on the weights of the milk recorded was made at least twice a year by an official of the Guernsey Cattle Club. This program was well received by Guernsey owners, who soon demanded an Advanced Registry program.

The first attempt at Advanced Registry was made in 1901. The requirements involved a combination of the point-system rating and the results of a milk and butterfat test. The point-score requirement was soon abandoned, and thereafter acceptance to the Advanced Registry was conditioned only on the Home Butter Test for cows, and, for bulls, the Home Butter Test of their daughters.

A Herd Improvement Test similar to that initiated by the Ayrshire Society in 1925 was added to the Guernsey program in 1930. Obtaining the milk production of all the individual cows in a herd led to the bull index. This index was the evaluation of a bull's ability to transmit to his daughters a milk-producing ability exceeding that of their mothers.

The Guernsey cow has maintained herself in second position among the United States dairy cows. The leveling off of the demand for milk and milk products in this country in recent years, the increase in the

A high-producing Holstein-Friesian herd in Idaho.

productivity of the average cow of all breeds, and the phenomenal increase in the popularity of the Holstein are factors that have resulted in a decline in the actual number of Guernsey cows. But it seems certain that as long as milk is produced in the United States the Guernsey cow will furnish a share of it.

Holstein-Friesian.—The Dutch colonists who landed on Manhattan Island in 1621 brought with them the black-and-white cattle common in the Netherlands at the time. They were the ancestors of the cattle from which the Friesian breed of the Netherlands was developed in the latter half of the nineteenth century. There are various references to black-and-white cattle imported to Manhattan Island in 1625. There are also records of black-and-white cattle, undoubtedly of Netherlands origin, taken from Denmark to New England in 1633. As late as 1725 the Dutch West Indies Company was still taking cattle from Holland to the settlements around Manhattan. No record remains of these early forerunners of the Holland Friesian in the American colonies. They interbred with the various types of British cattle which had been brought to the colonies during those days. All trace of these types was lost; in the early colonial days cattle were merely cattle. They were worked,

Cattle of North America

milked when possible, and butchered for food when no longer useful, but practically no attention was given to their propagation.

The Holstein-Friesian is the largest and most popular of the United States dairy breeds. A good cow in milk weighs from 1,400 to 1,600 pounds, and a bull in average breeding condition weighs 2,200 to 2,500 pounds. The color is the familiar black-and-white pattern, or since 1971 even red and white. There are breed fancies in color, though, with restriction against black hairs in the switch, black from hoof to knee, and others. Conformation is large, rangy, and long-legged.

The official history of the Holstein-Friesian Association of America credits Winthrop W. Chenery, of Belmont, New York, with establishing the first herd of Dutch Black and White cattle in the United States in 1852. In that year he purchased from a Dutch captain a cow which had served as a ship's store for milk on the voyage from The Netherlands. The cow's milk production was so unusual that Chenery imported several cows and at least two bulls from Holland, and his herd grew to sizable proportions. In 1872, when the first herdbook was published, the progeny of Chenery's herd was found in 12 states, including even California. In those days the black and whites were known only as "Dutch cattle."

The Chenery herd was apparently of mediocre quality according to the standards of the time, for its progeny soon faded from the herdbook. Their lasting contribution was the interest they generated in "Dutch cattle," although there were no further significant importations of their kind until after the Civil War. Then, in the 1870's, breeders and dealers began to import the black and whites in considerable numbers. The high production of these cows and the wide publicity Chenery had given them led to their ready acceptance. Imports increased rapidly from the early 1870's, and around 10,000 head had been brought into the United States by 1905.

During this period the name Holstein-Friesian, largely a misnomer, evolved in the United States in a curious manner. Chenery always referred to his cattle as "Dutch," with no other qualification. In 1864 he wrote a report on his operations for the commissioner of agriculture in which he consistently referred to his animals as Dutch cattle. When the article was published in the *USDA Annual Report for 1864*, it was entitled "Holstein Cattle, by Winthrop W. Chenery," and three illustrations were captioned "Holstein," though the term Dutch was used throughout the text. There is no record of why or by whom the change from Dutch to Holstein was made. Chenery blamed the USDA, but because of the publicity which the article had gained for his cattle, he went along with the new nomenclature and thereafter called his cattle Holstein. Holstein, a province of Germany, had no significant connection with the black-and-white Dutch cattle brought to America.

Two competing breed societies sponsored the Holstein cattle as they became established in the United States. The first of these, the Association of the Breeders of Thoroughbred Holstein Cattle, was formed in 1871 through the efforts of Chenery and his associates. The first volume of the Holstein Herd Book was published the following year.

At the same time Thomas E. Whiting of Concord, Massachusetts, another breeder of the Holstein cattle became involved in a bitter controversy with Chenery, stemming from Chenery's refusal to accept some of Whiting's cattle for registry in the Holstein Herd Book. The result was the organization of the Dutch Friesian Association, which was formed in 1877 by Whiting's associates after his death. The establishment of the Dutch Friesian Herd Book followed at once, and there were then two herdbooks for the black-and-white cattle.

After the demise of Chenery and Whiting, the two breed societies got along amicably for a few years, each registering cattle recorded in the other's herdbook and permitting breeders to register simultaneously in both herdbooks. In 1880, however, the feud broke out anew, beginning with a controversy at a state fair over breed names. Antagonisms were bitter until 1885, when the two associations joined in a new organization, the Holstein-Friesian Association of America. This organization prospered beyond the wildest dreams of any of its founders.

One unturned stone, however, lay in the path of the new organization immediately after the unification. The merger agreement provided that imports from the Netherlands would not be accepted for registration after March 18, 1885. A number of cattle traders had Friesian cattle either in quarantine in the United States or on the high seas. Failing in their efforts to have such animals accepted for registration, they organized the American Branch of the North Holland Herd Book. This registry was maintained until 1892, when the Holstein-Friesian Association agreed to accept for registration qualified animals whose parents were registered in the American Branch of the North Holland Herd Book.

The present Holstein-Friesian Association of America (Brattleboro, Vermont) is not only the largest of the dairy-breed organizations but also one of the most progressive. Evidence of this is the recent decision to accept red-and-white Holsteins for registry. For years the red-and-white cattle were ineligible for acceptance in the Herd Book, despite the fact that red-and-white Friesian cattle have always been registered in Friesland in the Netherlands. It has also been recognized for years that the red-and-white progeny of black-and-white parents are simply the result of the expression of the recessive red gene which has no effect on the genetic potential of the offspring, though in many instances it has helped identify some outstanding producers.

For some years now red-and-white Holsteins have been selected and bred together, and a pure red-and-white strain has been developed. After the Red and White Dairy Cattle Association was formed, the Holstein-Friesian Association reassessed its position. As a result, since January 1, 1971, red-and-white offspring of registered Holstein-Friesian parents, the latter either red and white or black and white, have been accepted for registry in a separate herdbook maintained by the Holstein-Friesian Association of America.

Around 1880 the concept was gaining acceptance in some circles that a concrete measure of a cow's productive ability, such as pounds of milk or butter produced in a given period, could be a more useful criterion for the selection of breeding stock than hair color and conformation. Soloman Hoxie, an advanced dairy farmer and a major factor in the Dutch Friesian Association of the 1880's, is credited with originating this idea. Because of his connection with the Dutch Friesian Association, he was able to insist, when the merger with the Holstein Breeders Association was effected, that one of the binding agreements should be "that the new Association establish a thorough system of advanced registration."

This was in 1885, and the Holstein-Friesian Association thus came to be the first cattle organization in the United States to utilize records of production for systematic breed improvement. A program almost identical to that adopted by the Holstein-Friesian Association at this time had been introduced by Hoxie and used by some members of the Dutch Friesian Association before it joined with the Holstein group.

The growth of the advance registry concept was slow until the Babcock Test for Butterfat came into general use. Before that, milk from individual cows might be weighed for a day or a week, in very exceptional cases for a year. The butterfat content of milk was determined by separating the cream and churning it to obtain the actual butter. Conformation was put on a point-score basis in an attempt to reduce it to mathematical measurement. While any genetic advancement from these procedures was questionable and though the procedures were not widely applied, they created an interest, focused attention on breed improvement, and established a receptive state of mind among breeders when better test measurements of productive ability were devised.

A few herd owners took the necessary steps to enter their prize cows in the Advance Registry, but the growth of the program was slow. The Babcock Test for Butterfat was accepted instead of the churn test in 1892. The following year Hoxie agreed to drop type inspection and measurement as a requirement for entry to Advance Registry, and the cow's production then became the only requirement. Neither of these steps, however, did a great deal to encourage participation in the program. In 1894 in an attempt to revive interest in Advance Registry, the association offered prizes for butterfat production on a seven-day test made under the supervision of an association inspector or an experiment-station representative. This was the turning point. Participation in Advance Registry increased, and though it was eventually discarded, it bridged the gap until other, more sophisticated programs were devised.

Advance Registry went through various modifications: establishing classes for 305 or 365 days' milking as the seven-day test went out, recognizing the volume of milk production as well as fat, taking into account the number of milkings per day; but it still told the story only on a breeder's highest-producing cows. As a more realistic measure of breed performance, the Herd Improvement Test was authorized by the association and came into use in 1928. Conformation re-entered the picture a year later with Herd Classification for type. This led to sire-recognition programs with bronze, silver, and gold medals, the dam-daughter relationship of milk production, the herd mate comparison, and finally, computer-processed data and all the modern procedures for increasing productivity by the use of production records.

The Holstein-Friesian Association has approved the use of artificial insemination from the time of its introduction. The technique is now used extensively in both purebred and commercial herds.

Accounting for 80 per cent of the national dairy herd, the Holstein-Friesian blankets the United States, with concentrations around the larger towns and cities. There is also a sizable export market for Holstein-Friesian breeding stock, chiefly in Central and South America. In some instances the cattle have even been shipped to European countries to improve the milk production of the parent Friesian breed. Foreign marketing is now promoted by a subsidiary corporation of the association, The Holstein-Friesian Services, Inc.

Holstein-Friesian herds are usually larger than those of the other dairy breeds in the United States. Even a family operation often has 100 cows or more. The Holstein-Friesian is the largest in size of the dairy breeds and has always commanded a high salvage value. In recent years this trait has provided a definite advantage for Holstein steers. Fed out to reasonable slaughter weights they have found a very satisfactory market.

The two attempts to market Holstein milk under a breed label were perhaps premature. In the early 1920's "White Nectar" was promoted in national advertising. Again in 1940 a try was made at "Holstein Holsome." Neither program achieved any reasonable measure of success, and both were abandoned. At a later date the low fat content of Holstein milk might have been capitalized on. By that time, however, the breed had most of the market anyway.

To the Holstein-Friesian Association of America and its predecessors must go much of the credit for the widespread use of production testing and artificial insemination. Without these tools the Friesian could never have become the premier milk animal of the world.

Jersey.—In the early years of the nineteenth century there was little if any distinction made between the two breeds on the Channel Islands. Both were rather indiscriminately referred to as Alderneys, Jerseys, or Guernseys. Regardless of this confusion in nomenclature, however, it seems probable that in the years 1815 to 1817 the first cattle from Jersey were taken to Delaware. There are also records of later shipments in 1830 and shortly thereafter, although no trace of these first Jersey cattle remains.

The typical Jersey color as seen the world over is the familiar light tan or fawn, although a much darker shade is permissible. The shoulders and neck of the bulls are often nearly black. White marks, usually confined to the underside of the body, are allowed. Horns are rather fine and curled forward on the cow. Dehorning is permitted and is now common. Cows in milk of the preferred type now weigh around 1,000 pounds, but there are still many herds with 800- to 900-pound animals. Bulls weigh up to 1,500 pounds, sometimes more.

The actual history of the Jersey in America begins with small shipments which were landed in Connecticut and Massachusetts in 1850. There were sizable imports of Jersey cattle both from Jersey and from England in the years that followed. By the time the American Jersey Cattle Club (Columbus, Ohio) was formed in 1868, the breed had gained a wide reputation in the United States for rich milk. After the formation of the club there was an increase in the flow of Jersey cattle to the United States, nearly 500 being imported between 1868 and 1870.

Mature Jersey bull, 1,600 pounds. Knolle Farms, Sandia, Texas.

The purpose behind the organization of the American Jersey Cattle Club was to authenticate the pedigrees of individuals offered for registration. This it did in a fashion that adequately fulfilled the tenets of its constitution that "only such animals shall be admitted to the herd book as are proven, on proof satisfactory to the Executive Committee, to be either imported from the Island of Jersey or descended only from such imported animals." Still, it is highly probable that some cattle from Guernsey were registered in the Jersey herdbook and vice versa because of the confusion in names, the similarity of the cattle, and their transfer from one island to the other.

The popularity of the Jersey cow spread rapidly over the United States, prompted by the general appetite for cream and the acknowledged richness of Jersey milk. The Jersey herds were usually small and were held on family farms. This is still true today in spite of the trend to larger farm and dairy operations. A few of the largest dairies in the country, however, have Jersey cattle. One herd in Florida has nearly 15,000 head of all ages with 7,000 cows in milk. For years the largest concentration of Jersey cattle was owned by a Texas family, which had 5,000 cows.

The breed is especially popular in Oregon, where it accounts for more than half the dairy herd in that state. The fact that the climate there more nearly approaches that of the island home of the Jersey than do other parts of the United States is probably only coincidental, for the Jersey cow has proved that she is well adapted to most parts of the country.

Jersey owners were among the first dairymen to employ testing procedures to determine the productivity of their animals. The first full-scale test of a Jersey cow was recorded in 1853 for the cow Flora 113, which produced 511 pounds of butter in 350 days. Breeders then started testing their favorite cows, principally for butter production, which meant actually recovering the churned butter from cream. *Country Gentlemen* magazine began publishing a list of the yields of high-producing cows in 1882, and in 1885 the Jersey club employed an official tester to verify churn tests in an attempt to quiet the furor which had arisen over the testing practices.

This confusion on testing procedures continued for many years and led to much controversy among Jersey breeders. When the Babcock Test for Butterfat was introduced in 1894, the Jersey club refused to authorize its use, and three years later agreed to it only as a means of confirming the churn test. Happily, the struggles of the club concentrated attention on the desirability of some form of production testing until a satisfactory method evolved. The Registry of Merit, adopted in 1902 to give recognition to high-producing individual cows, eventually led to a sound herd-improvement program and a reliable method of evaluating sires.

With the adoption of the Herd Improvement Registry in 1928, the Jersey Cattle Club entered on the modern phase of improving production. All cows in a herd are required to be tested and recorded under uniform procedures and competent supervision and a meaningful base established for breed improvement. In 1932 the club adopted a type classification applicable to herds which is essentially an improved system of conformation scoring. The inspectors are appointed by the board of directors of the club. A breeder requesting classification of his herd agrees to the cancellation of registration of any animals classified as "poor" and to cull bull calves from cows classified no better than "fair."

There has been a trend in all the major United States dairy breeds to increase the size of the mature cow and to pay less attention to the

Herd of large type Jersey cows. Coastal Plains Experiment Station, Tifton, Georgia.

butterfat content of her milk. Jersey breeders have made considerable progress along these lines.

The beautiful little Jersey cow of a few decades ago not only is now larger in size but also is a less bountiful cream producer. Twenty years ago selection for the small, smooth Jersey had brought the average cow down to around 750 pounds when she was in good milking condition. Today in a modern herd she will reach 1,000 pounds or more. There has also been a decided trend toward a decrease in fat content and an increase in the volume of milk she produces. These changes demonstrate both the versatility of the breed and the effectiveness of the selection procedures which have been applied.

Two trade-name projects have been sponsored by the American Jersey Cattle Club. The trademark "Jersey Creamline Products" was registered in 1927, and approved milk distributors were licensed to use this label. The project met with only mediocre success and was abandoned in 1953. A second program emerged in 1957 with "All Jersey" as the trademark. This program is more in keeping with the times, in that various low-fat products, such as buttermilk and 2 per cent milk, are promoted. Its ultimate success is still to be determined.

Jersey breeders have not utilized artificial insemination as extensively as some of the other dairy breeds. Probably 50 per cent of the cows in the registered herd are still bred naturally, although the proportion in commercial herds is much lower.

Milking Shorthorn bull, four years old, 1,800 pounds. Sam Beadleston Farm, Springfield, Missouri.

Milking Shorthorn.—The Milking Shorthorn (including the polled Milking Shorthorn), the Shorthorn, and the Polled Shorthorn are one breed of cattle, although each is represented in the United States by a separate breed association. Because ability as a draft animal was a major factor in the selection of cattle during the first half of the nineteenth century, the Shorthorns were not differentiated as either a beef or a milk type until around 1900. The first 70 years of Shorthorn development in England resulted in an animal that tended toward the dairy type but not to the extremes of the specialized dairy breeds. During the 1840's some Scottish breeders made considerable progress in selecting for a beef-type animal from the English Shorthorn. When this type reached the United States, it became exceedingly popular among the Shorthorn breeders, who had little interest in milk production. It was at this point that the milk type initially diverged from the beef type in the eyes of the United States breeder, although there was no

Mature Milking Shorthorn cow, 1,500 pounds. Sam Beadleston Farm, Springfield, Missouri.

clear-cut distinction between beef and milk Shorthorns for many years thereafter.

The Milking Shorthorn Club was organized in 1912 by a group of dairymen who were members of the American Shorthorn Breeders Association but also wanted a place in the sun for their kind of cattle.

The Milking Shorthorn has the same color pattern as the beef type—red, red and white, roan, and white. The two types are similar in appearance if the beef animal is from a herd which escaped selection for the compact conformation a few decades ago. The Milking Shorthorn breeders never followed this selection practice, which engrossed the beef Shorthorn men in the 1930's and 1940's. As a result the Milking Shorthorn is today a larger animal by some 200 pounds than the beef-type Shorthorn. Cows in a good herd weigh 1,400 to 1,600 pounds; and bulls, if raised under conditions where they can attain their full growth, weigh

Cattle of North America

over 2,000 pounds. The Milking Shorthorn stands higher and does not have the rounded appearance of the beef type. The horns of the two types are similar, but many Milking Shorthorn cows are now dehorned. There is also a polled strain that is carried in the same herdbook as the horned cattle. The polled Milking Shorthorns trace to the same twin heifer calves that were the foundation of the polled beef Shorthorn.

The Milking Shorthorn is the only breed in the United States now considered as a dual-purpose type. Good herds have an average production of 12,000 pounds per lactation. The salvage of culled cows, as well as the price brought by steer calves, is greater than for any of the other five recognized dairy breeds.

While the breed is scattered widely over the United States, the corn belt has always been the stronghold of the Milking Shorthorns. Herds are usually small in number, probably because the owners are often engaged in a general farming operation in addition to dairying. Milk production, however, has remained the prime objective of most Milking Shorthorn breeders and they have not swung completely to a beef type as have the Red Poll breeders. Artificial insemination has not been widely utilized in the breed. This may explain the relatively limited milk production of the breed, which is lower than might be expected from such large cows.

Though most of the dairy breeds in the United States have been declining in total numbers for some years, Milking Shorthorn registrations seemed to be increasing in 1970. The current enthusiasm for crossbreeding in beef production has focused considerable attention on the utilization of the dairy breeds to increase milk production. The Milking Shorthorn can make a contribution here and can supply a large frame and beef-type conformation as well.

The American Milking Shorthorn Society maintains an open herdbook. By following authorized procedures, females with ⅞ purebred Shorthorn breeding are eligible for full registration.

The Milking Shorthorn Club for a time worked within the framework of the American Shorthorn Breeders Association which registered both beef and dairy Shorthorns. The conflicting interests of the beef-cattle breeders and the dairy members led to wide disagreement. In 1948 the owners of the dual-purpose Shorthorns withdrew from the parent organization and formed the American Milking Shorthorn Society (Springfield, Missouri).

Red and White Dairy Cattle (Red and White Holstein).—Initially red-and-white Holstein-Friesian cattle in the United States were the chance

Young Red and White bull. American Breeders Service, DeForest, Wisconsin.

offspring of black-and-white Holstein-Friesian parents. As with other black breeds, some Holstein-Friesians carry a black gene and a recessive red gene instead of the more usual two black genes. (The Mendelian principle governing the inheritance of this color pattern is explained under "Red Angus.")

In the Netherlands, the home of the Friesian cattle from which the Holstein-Friesian was derived, red-and-white Friesians have always existed alongside the black-and-white animals, although the black pattern was strongly preferred by most Dutch farmers.

There were probably two reasons for the preference for the black-and-white pattern. The red-and-white lowland breed in the Netherlands, the Meuse-Rhine-Yssel, was probably being selected out of the common cattle population around the time the Friesian was being developed. It was natural for a chance red-colored offspring of black-and-white parents to be frowned upon, for such occurrence cast suspicion on the authenticity of its parentage.

Somewhat later, heavy exportations of Dutch Black and White Cattle were being made from the Netherlands to the United States. Their importers were already exhibiting strong breed jealousies and wanted no red skeletons in their barns. Eventually the largest of the Dutch herdbooks, The Netherlands Cattle Herd Book Society, discontinued the registration of red-and-white Friesians. There are relatively few registered red-and-white Friesians in the Netherlands today, although the Friesian Cattle Herd Book, which handles all registrations in the province of Friesland, continues to register them.

The red-and-white Holstein-Friesians in the United States have derived from specific selection. In 1946, Larry Moore, a Wisconsin mink breeder who had been very successful in developing a number of mutations in his stock, became interested in the color pattern of the red-and-white Holstein-Friesians and, out of curiosity, visited the herd of Henry du Pont in Delaware. There he explained his technique in developing new colors in mink, which technique was based on combinations of recessive genes.

Impressed by Moore's command of practical genetics, the du Ponts presented him with a red-and-white Holstein-Friesian bull calf from a high-producing mother and a sire with an exceptional line of high-producing daughters. Larry Moore King, as the calf was named, became the foundation sire of the American Red and White Dairy cattle.

Realizing the potentials of the red-gene-carrying cattle, Moore proceeded to comb the country for red Holstein-Friesians from registered herds and discovered the red-gene-carrying black-and-white animals by examination of countless pedigrees. Over the years he developed a select herd, pure for red and white, all descendants of registered Holstein-Friesian parents. These were called Red and White cattle to avoid the term Holstein or Holstein-Friesian.

These selected Red and White cattle were as pure Holstein-Friesians as any registered representative of the breed. They had all the general characteristics of the black-and-white animals. The large angular frame, strong udder, horns, and general dairy conformation were unchanged. The high-milk-producing ability was also characteristic of the Red and White cattle, and the claim has even been advanced that they are superior to the black and whites.

Holstein-Friesian breeders in the United States have always attempted to sweep the reds under the carpet. The red calf was usually disposed of at birth. Often its sire and its dam, unless she was a particularly heavy milker, were also discarded. Thus there was possibly some unintentional

United States

selection over the years for red-gene-carrying females which were exceptionally productive.

In 1962 a group of Milking Shorthorn breeders saw the potential in the Red and White cattle for improving milk-producing ability. The USDA Experiment Station at Waseca, Minnesota, had crossed Red and White bulls on Milking Shorthorn cows and obtained significantly improved milk production in the progeny. These developments led to the establishment by the Milking Shorthorn breed society of the Appendix Registry for Red and White Holstein crosses on Milking Shorthorns. The abandonment of this appendix led to the founding of the Red and White Dairy Cattle Association (Elgin, Illinois) early in 1964.

The registration policy adopted by the association was exceedingly broad in scope. The design was to provide for the registration of both purebred Red and White Holstein-Friesians and the crosses of purebred Red and White Holstein-Friesians with other dairy breeds.

The recording procedure involves the use of symbols to show breeding and percentages of blood, indicated as follows:

F100 (100 per cent Friesian)
F50MS (50 per cent Friesian–50 per cent Milking Shorthorn)
F50A (50 per cent Friesian–50 per cent Ayrshire)
F50G (50 per cent Friesian–50 per cent Guernsey)
F50D (50 per cent Friesian–50 per cent Red Dane)

Second-generation offspring are designated as follows:

F75MS (75 per cent Friesian–25 per cent Milking Shorthorn)
F75A (75 per cent Friesian–25 per cent Ayrshire)

Starting January 1, 1971, the Holstein-Friesian Association of America began the registration of purebred Red and White Holstein-Friesians in a separate section of its herdbook. Some Red and White Cattle breeders now have their cattle recorded in both the Holstein-Friesian Association and the Red and White Cattle Association herdbooks.

In addition to registration the association handles the sale of semen of approved Red and White bulls.

American Developed Breeds

Under this heading are presented the breeds which have been developed in the United States by crossing recognized breeds.

American Brahman.—The American Brahman was developed in the United States from three East Indian breeds of Zebu cattle. During the formative years there was also a strong influence of the major British beef breeds. Some of the foundation Zebu stock was shipped directly from India to the United States, though most of the individuals to which the American Brahmans trace were Zebu cattle which had been raised in Brazil, exported to Mexico, and taken across the Río Grande to the United States. The Brazilian cattlemen of the late nineteenth and early twentieth centuries were good selective breeders and were free from government restrictions when importing from India. They learned early that the Zebu was well adapted to the tropical climate.

The Zebu breeds which entered into the composition of the American Brahman were mostly draft animals, but by selection for what the American breeder considers a beef-type conformation, a deep, broad body has been obtained. The hair color may be either a light to medium grey or a medium to dark red, but either should be solid. Darker shades of the base color are usually evident on the neck, shoulders, and rump of the male. The horns are rather short and thick, often upturned. While both male and female carry a pronounced hump, that in the male is much larger. The ears are long and droopy. The hide is thick and extremely loose, gaining full expression in the pendulous sheath and dewlap. The resulting high ratio of skin surface to body size is undoubtedly a factor in the high heat tolerance of the Zebu as is the presence of sweat pores over much of the skin surface. Bulls weigh from 1,600 to 2,200 pounds in breeding condition and cows from 1,000 to 1,400 pounds.

The first Zebu cattle in the United States are said to have been brought to South Carolina from Egypt in 1835. The next introduction was also in South Carolina in 1849, when two Indian Zebu bulls were obtained by Dr. James Bolton Davis. Any trace of the progeny of these Zebu importations was lost during the Civil War.

In 1854 the British government presented two bulls to Richard Barrow, a Louisiana planter, as a reward for his services in sugar-cane culture. The offspring of these bulls, which had been bred to Native cows of the area, came to be known as Barrow cattle, and were noted for being well-doing. These crosses appear to have been the seed from which sprang the wide interest in Zebu breeding among cattlemen along the Gulf Coast. There were no planned matings in the methods followed. Any animal with discernible Zebu influence was recognized as being well acclimated to the harsh environment and any humped bull that could achieve this effect on his progeny was considered a desirable sire. Occasionally small additions to Zebu breeding stock were made through the acquisition of an exhibition bull from a circus, but it was not until after the turn of the century that the major importations were made.

American Brahman bull, three years old. J. D. Hudgins, Inc., Hungerford, Texas.

In 1906, with the assistance of the USDA Bureau of Animal Industry, arrangements were made to import a shipment of Zebus direct from India. A total of 51 animals, only 3 of which were females of breeding age, were shipped. Only 33 survived the quarantine procedures and were released to the two Texans who had masterminded the importation. Of those, 16 went to Thomas O'Connor and 17 to Abel Borden, manager of the Pierce Estate. These two herd nuclei had a significant influence on the subsequent development of the American Brahman.

Over the next 37 years there were several significant introductions of Zebu cattle to the United States breeding herd. These started with a few circus bulls that were obtained by a Texas breeder in 1910. In 1924, 90 bulls which had been shipped from Brazil to Mexico were taken across the border at Eagle Pass, Texas. These were followed a year later by 120 bulls and 18 heifers, also of Brazilian origin. These 1924 and 1925 importations were widely distributed throughout South Texas and constituted the real foundation of the Brahman breed. The final arrivals were 19 head of Brazilian-bred bulls, which entered by way of Mexico in 1946. Since that date import restrictions relating to the control of foot-and-mouth disease have prevented any further introduction of Zebus to the United States.

Fewer than 300 individuals of the various Zebu breeds entered into the composition of the American Brahman. The ratio of bulls to females was over 10 to 1. There is no authentic record of the numbers of the different breeds that were involved, but the Guzerat, Gir, and Nellore breeds are known to have been represented, and probably also the Krishna Valley. The Brazilian-developed Indo-Brazil breed, which is predominantly a Guzerat-Gir cross of unknown proportions, was also represented in the shipments from Brazil by way of Mexico. The 1946 importation in particular is said to have included a number of Indo-Brazil individuals.

The early crossing of Zebu bulls on cows that were predominantly of the British beef breeds produced progeny that were highly resistant to insects and disease and tolerant to heat. These were the attributes which made the Zebu attractive to the Gulf Coast cattleman. The cows available in the area at the time of the infusion of this Zebu blood were mainly commercial Herefords and Shorthorns upgraded from the Longhorn. The indiscriminate breeding pattern led to such a dilution of the Zebu influence that the more knowledgeable breeders began re-emphasizing the Zebu characteristics in their selection to maintain a line available for crossing with the British types. A certain amount of hybrid vigor was a valued by-product of these programs.

At the same time an effort was made to improve beef conformation. The Zebu imports had descended from cattle selected largely for draft purposes. Such selection had produced animals with a pronounced slope to the rump, the antithesis of good beef conformation. Careful selection on the part of the Gulf Coast breeders eventually put a better-rounded rump on the regenerated Zebu, together with greater width and depth of body, a less wasty dewlap, and a less pendulous sheath.

The efforts of the widely scattered Brahman breeders were unco-ordinated until the American Brahman Breeders Association (Houston, Texas) was organized in 1924. One of the first moves of the association was to establish much-needed standards of excellence. For the next 20 years some progress was made toward a distinctive beef-type Zebu.

United States

Factions divided the Brahman breeders in the mid-1940's. One group wanted to concentrate on breeding the British type of beef animal, insofar as that was obtainable on a Zebu frame. The other faction desired to retain more Zebu characteristics. This controversy led in 1946 to the formation of the Pan American Zebu Association (see the section "Indo-Brazil"), which settled on the Brazilian-developed Indo-Brazil as the Zebu breed which should be perpetuated. Eventually many of the members of the new Pan American group returned to the American Brahman Breeders Association.

The American Brahman breed remained a typical Zebu albeit with a much better beef-type conformation than any of the Indian Zebu breeds from which it derived. The degree of influence which now remains of the British breeds is probably negligible.

Zebu cattle have made a major contribution to the beef-cattle industry of the United States, as well as to many tropical and subtropical countries throughout the world. Their heat tolerance and insect and disease resistance in the days before the elimination of the fever tick greatly improved cattle production in the Gulf Coast areas of the United States. Even with the tick eliminated, cattle carrying the Zebu influence are more efficient in such areas than straightbred *Bos taurus*.

The American Brahman is receiving renewed attention in the crossbreeding programs which are currently generating such widespread interest. This breed, however, as well as other Zebu types, has been utilized for several decades for crossbreeding in commercial herds by southern cattlemen.

Barzona.—The Barzona breed (the name is a combination of Bard and Arizona) was developed during the past 25 years in the arid mountainous region of central Arizona. E. S. Humphrey, a cattleman with an analytical bent and some South Texas experience, was managing the 330,000-acre ranch property in central Arizona belonging to F. N. Bard, of Chicago. Both Humphrey and Bard embarked on a project to develop a more productive beef animal for the dry, hot Arizona climate. The work was begun after the end of World War II, with Humphrey directing the breeding and selection work. The foundation herd is still maintained on the Bard Ranch in Skull Valley, west of Prescott.

Hereford, Africander, Angus, and Santa Gertrudis cattle were utilized in developing the Barzona. Color is nearly solid red, varying in shade from a light cherry red to a near mahogany. There is very little white except on the underline, although an occasional animal may have a few minor white markings around the head. Selection for color played no

Barzona bull, three years old, 1,600 pounds. Bard Kirkland Ranch, Kirkland, Arizona.

part in the development of the breed, and there is no color requirement today. Although dehorning is now generally practiced on all breeding stock, the horns, when allowed to grow, are somewhat longer than those on the Hereford but still do not approach the typical long Africander type of horn. Around 7 per cent of the calves are naturally polled. The face is long and nearly rectangular in outline and decidedly reminiscent of the Africander. Although noticeable on some males, the hump is never pronounced and is usually not discernible on the cow. Mature cows weigh 1,100 to 1,250 pounds, and bulls weigh up to 1,800 pounds.

The objective behind the development of the Barzona was the production of an animal which would grow and reproduce satisfactorily in

Mature Barzona cow, 1,100 pounds. Bard Kirkland Ranch, Kirkland, Arizona.

the dry, mountainous country of central Arizona. Summers there are hot with practically no precipitation; in winter temperatures well below freezing are experienced. The plant growth induced by such spring rains as occur soon dries up when the rains cease. There are long drought periods which not infrequently extend from one year to the next. Browse is more prevalent than grass on the thin rocky soils of the hillsides. The Barzona have proved remarkably hardy in this environment and make excellent use of the browse and prickly-pear cactus, which is practically the only feed available during drought periods and which they attack even before it is burned.

Humphrey was familiar with the Africander cattle which the King Ranch had imported from South Africa in 1931. The Kenedy Pasture Company had acquired three bulls and six females from the neighboring King Ranch. These were of the original imported stock, and the Kenedys proceeded to breed pure Africander cattle from this nucleus for a number of years.

In 1945 a number of Africander bulls were purchased from the

Kenedy Pasture Company, brought to the Bard Ranch and bred to a herd of mountain-raised purebred Hereford cows. Some good-producing Hereford cows from a grade herd were also bred to these bulls. After the Africander-Hereford cross had been established, two-thirds of the heifers were bred to Santa Gertrudis bulls and the remaining one-third to Angus bulls. The Santa Gertrudis bulls were purchased from the King Ranch. Some of the Angus bulls utilized in this program were obtained from the Paleface Ranch in Texas, where the Red Brangus were developed. The Paleface bulls were known to include an unusually high percentage of red-gene carriers. This probably had some bearing on the nearly solid red color which has been characteristic of the Barzona.

The Bard herd was closed in 1960, some time after the introduction of the Santa Gertrudis and Angus bulls. Meticulous records were maintained on weaning weights, pasture gains after weaning, beef-type conformation, and dams' productivity. Adaptability to the harsh environment was a major factor in selection. Cows either calved on their own and raised a calf or were culled. These selection methods made it imperative that the cow be able to raise her calf while grazing a high proportion of browse and that the calf be able to put on satisfactory postweaning gains.

Because selection was based on the individual's record of performance, there is no way that the proportions of the various breeds entering into the Barzona can even be approximated. The Santa Gertrudis is itself a recently developed breed embodying about three-eighths Zebu and five-eighths Shorthorn. Taking this into consideration, one might say that five established breeds were blended into the Barzona (if the Zebu element is considered as one breed). Mathematically the makeup of these five base stocks would have been as follows if selection had not altered the picture: Africander, 25 per cent; Hereford, 25 per cent; Shorthorn, 20.8 per cent; Angus, 16.7 per cent; and Zebu, 12.5 per cent.

Tests conducted by some of the experiment stations in the southwest have shown Barzona steers to have a better feed conversion than the breeds with which they were compared. This would seem to indicate that an animal with the ability to utilize inferior nutrients can be expected to make superior use of good feed when it has the opportunity to do so.

A marked degree of uniformity is evident in the Bard Ranch Barzona cattle today. This is particularly true of the replacement females.

Greatest demand for the Barzona is for bulls to be used on Hereford herds in the area where the breed was developed. Some of the southern experiment stations have obtained representatives of the breed for

Replacement heifer calves, eleven months old. Bard Kirkland Ranch, Kirkland, Arizona.

experimental feeding and breeding work. One of the large artificial-insemination studs recently put a Barzona bull in service.

Cows are now recognized as purebred which are the third-generation offspring of the mating of registered Barzona bulls and cows of any of the recognized beef breeds. A bull must be of the fourth generation of Barzona breeding to be eligible for registration.

The Barzona Breed Association of America (Prescott, Arizona) was organized in 1968.

Beefmaster.—Unique selection criteria have been combined in a simulated natural selection along with rigid artificial selection for the characteristics desired in the Beefmaster. Such factors as hair color, size of horn, and shape of head, which have always been major considerations to the purebred breeder, have been ignored in the development of the Beefmaster.

Forty years of such selective practices have resulted in a large, hardy type with a good beef conformation. Under range conditions the average cow weighs around 1,400 pounds with individuals up to 1,600 pounds. The average bull weighs 2,200 pounds and individuals up to

Mature Beefmaster bull. Lasater Ranch, Matheson, Colorado.

2,600 pounds. The color is usually a solid, or nearly solid, medium red. Reds and whites and an occasional brindle animal are seen. There is no color requirement. The female appears to be nonhumped; the mature bull carries a modest hump. The Beefmaster is normally horned, but polled animals occur.

In 1908, Edward C. Lasater, a South Texas rancher, began to use Zebu bulls on his cow herds, as was common practice in that day. He was maintaining two large herds, one of grade Herefords and the other of grade Shorthorns. Lasater was also a Jersey breeder, at one time having the largest Jersey herd in the United States. The three Zebu breeds used on the Hereford and Shorthorn herds were Gir, Nellore, and Guzerat, the same breeds that entered into the composition of the American Brahman. There were no planned matings in this upgrading program, and the proportionate influence of the three Zebu breeds is not known. The most promising females of these and subsequent breedings were selected and continuously bred back to Zebu sires. By 1940 a nearly straight Zebu herd had been developed whose genetic composition closely resembled that of the American Brahman.

During these same years, Lasater was also developing a registered Hereford herd which was selected primarily for three characteristics: heavy milking ability, large size, and brockle faces. Each animal kept in the herd had a patch of red hair around the eyes as protection against pinkeye and cancer eye.

Lasater's son Tom took over the ranching operations in 1930 after his father's death. He later moved the herds to eastern Colorado and began to interbreed the Lasater Zebu herd and the Lasater registered Hereford herd. He also crossed Shorthorn bulls on cows of the Zebu herd. The Hereford-Zebu and Shorthorn-Zebu crosses were then interbred, and the progeny of this mating became the Beefmaster. Since the entire breeding program was carried out under range conditions, the proportionate influence of the parent breeds is not precisely known. Considering the time and numbers involved, the blend probably consists of slightly less than one-half Zebu and a little more than one-fourth each of Hereford and Shorthorn.

The foundation herd, which has been closed since 1930, has averaged less than 1,000 brood cows, and breeding has been in multiple sire groups segregated by age of cow. The cows are run the year round in the open at an elevation of 5,800 feet, are subjected to a 65-day winter breeding season, and calve in the open. No assistance is given at parturition except in the event of an abnormal presentation of a young heifer. Such practices, rigidly adhered to, obviously carry a considerable degree of natural selection into the management program. Superimposed on these are certain artificial-selection practices. These include culling cows open on pregnancy test or with nonfunctioning udders and culling calves that do not meet weaning and yearling weights and conformation standards. Individuals of either sex displaying an unruly disposition are culled. The six-point standard of selection covers disposition, fertility, weight, conformation, hardiness, and milk production.

Bulls for the breeding herd must meet even higher standards. They must fall in the upper brackets of yearling weights, meet fixed conformation standards, and prove sound of limb and body.

The Beefmaster enjoys particular favor in South Texas and is spreading over portions of the Rocky Mountain area. Probably the best measure of its acceptance by both purebred and commercial breeders is the mail-order method of selling that Tom Lasater inaugurated in 1949. By that time demand for his bulls exceeded the number he had to offer. This sales procedure is as follows:

Buyers must submit a written contract for the number desired by November 30 of the year preceding that in which delivery is to be

Mature Victoria bull, 1,600 pounds. J. F. Welder Heirs, Vidauri, Texas.

made. A drawing determines the order in which the buyers make their selection. Contracts received after November 30 follow in the order in which they are received. On the delivery date in August the purchasers select in the order of their priorities the bulls for which they contracted, from a group of about 100. The price for the first bull sold is established at the time the contracts are submitted to prospective purchasers, and the price of each selection after the first drops $5 a head. Thus, if the price on the first bull is set at $1,000, the second goes for $995, and if 100 bulls are sold, the price for the last one is $505.

Beefmaster Breeders Universal (San Antonio, Texas) was organized in 1961. Straight descendants of the Lasater foundation herd are eligible for certification (the association does not offer registration), as are also the progeny of three consecutive top crosses on cows of any breed by bulls of recognized Beefmaster breeding.

Mature Victoria cow, 1,100 pounds. J. F. Welder Heirs, Vidauri, Texas.

A second Beefmaster association was organized in 1922, the Foundation Beefmaster Association (Denver, Colorado).

Braford.—Ever since Zebu cattle were first brought to the United States, bulls of the East Indian breeds have been in demand among southern cattlemen for use on their commercial cows. Many herds in Texas, Louisiana, and Florida had been upgraded to Herefords at the time the practice of breeding to Zebu bulls became widespread, and when the American Brahman emerged as an improved type, it gained popularity wherever the Zebu influence was desired.

In the early 1930's efforts were made in Texas to establish a fixed Brahman-Hereford cross, to which the name Braford was given. Brahman and Hereford cattle continued to be crossed indiscriminately and in various combinations in southwest Texas. Many cattlemen en-

Braford bull, seven years old, 1,800 pounds. Alto Adams Ranch, Fort Pierce, Florida.

deavored to keep one-eighth to one-fourth Brahman influence in their breeding herds. At times attempts were made to establish *inter se* breeding of some of the Brahman-Hereford crosses. Two South Texas ranchers, Claude McCann and J. F. Welder, had a one-fourth Brahman, three-fourths Hereford line of breeding they were maintaining as far back as 1946. The line came to be known locally as Victoria cattle, from the county where the founding cattlemen lived and where the line of breeding is still kept by the heirs of the originators. The Victoria cattle are as much Braford as any Brahman-Hereford cross.

The Braford, as it is usually recognized today, originated from a herd of grade Brahman cows on the Alto Adams Ranch, near Fort Pierce, Florida. Starting in 1937, a herd of Florida Scrub cattle were bred up to an average of seven-eighths Brahman through the use of purebred American Brahman bulls. When Alto Adams, Jr., took over the management of the ranch in 1948, he began breeding the Brahman cow herd to purebred Hereford bulls. This breeding program was continued until 1954, when *inter se* breeding of selected Hereford-Brahman cross bulls and females was started.

Braford cow and calves on improved pasture. Alto Adams Ranch, Fort Pierce, Florida.

Weaning-weight records kept on the male calves were the basis for selecting the highest-ranking bulls that went into the single-sire cow groups. Five lines were eventually established of selected cow groups, from which the range bulls for the main herds were produced.

Selection of replacement females from the main herds was based on type, size, and thriftiness. The highest-ranking individuals went to the single-sire cow groups. While pigmentation around the eyes was a requirement, there was no other color preference. All bulls and females were subject to rigid culling after an annual inspection. Any female not in calf after the breeding season was culled. This program, with minor changes, has been continued to the present day.

By 1971 selection from *inter se* breeding had been in progress for 17 years. Today 90 per cent of the calves dropped are considered true to type. The Braford so constituted is approximately three-eighths Brahman, five-eighths Hereford. It carries much of the Hereford conformation, although it is somewhat more rangy in appearance. External evidence of the Brahman influence is slight. There is little if any hump on the bull and none on the cow. The sheath shows a Brahman influence

453

but is of only moderate size compared to most Zebu crosses. Color is predominantly red, varying in shade and with some white often showing on the body and face. The hair is short. Cows weigh 1,200 to 1,300 pounds and bulls around 1,800 pounds. Good udders are a characteristic of the cow. The Brafords are heat-tolerant in the south Florida climate; they are fertile and remarkably free of eye trouble.

The size of the Adams herd—5,000 cows and 300 bulls—has been a major factor in the success achieved in the development of the Braford. The rigid selection and culling practiced on a population of this size has produced remarkable results. The Braford should probably be considered as in the final stages of becoming an established breed.

The International Braford Association (Fort Pierce, Florida) was organized in 1969.

Brangus.—In the early 1930's the USDA Experiment Station at Jeanerette, Louisiana, began to cross Zebu cattle with the British beef breeds with the objective of developing a better tropical meat animal. This experimental work showed that crossing Zebu bulls on Aberdeen-Angus cows gave a very promising beef type. This line of breeding was soon adopted by a number of cattlemen in the South who used American Brahman for the Zebu element.

The Brangus breed is fixed at five-eighths Angus and three-eighths American Brahman. This percentage requires at least three generations and may be arrived at by any sire-dam combination that will give this result. The herdbook is open, and Brangus of the required parentage of Angus and Brahman (each of which must be registered in the herdbook of its breed) may be registered in the Brangus Herd Book. Newly established breed societies are usually in a hurry to close their herdbooks on the assumption that it will enhance the prestige of the cattle of the founders. By avoiding this pitfall, the Brangus Association has left an opening through which the genetic potential of the breed may be broadened. Most Brangus cattle now being registered, however, are the progeny of Brangus parents. Animals of both sexes must be officially classified before they are accepted for registration.

Breed requirements are for a solid black with only minimum white markings permitted behind the navel. The animal must be polled. The hump on the male is small and solid to the body; it usually is not discernible on the female. Bulls in breeding condition will normally weigh 1,950 to 2,050 pounds, cows, 1,200 to 1,400 pounds. Brangus have an excellent beef-type conformation with exceptionally smooth body lines, a characteristic of the Zebu-Angus cross. This same clean-lined, beef-type conformation is a feature noted on the Jamaica Black, a Zebu-Angus cross very similar to the Brangus, developed on the island of Jamaica.

Interest in the Zebu-Angus cross developed, first, because of its heat tolerance and resistance to tick-borne diseases and, second, because the carcass was of considerably better quality than that of a straight Zebu.

The two breeders most prominent in developing the Brangus were Tom Slick, of the Essar Ranch, in Pandora, Texas, and Frank Buttram, owner of the Clear Creek Ranch, at Welch, Oklahoma. In the early 1940's both ranchers started almost simultaneously to cross the best Angus and American Brahman cattle they could obtain. The three-eighths Brahman, five-eighths Angus cross was the end product of both breeders. Other breeders then followed much the same program as that set by the two pioneers.

The Brangus Association is progressive in its policies. Performance testing has been encouraged, but very few breeders have started weighing their calves. The employment of artificial insemination is permitted with only minor restrictions, although up to 1971 it has been used very little.

The Brangus are concentrated in the South. Texas is first in the number of registered cattle, with Oklahoma second. The breed, however, is now scattered over much of the United States and also has a firm foothold in many Latin-American countries.

The American Brangus Breeders Association was organized in 1949. The name was later changed to the International Brangus Breeders Association, Inc. (Kansas City, Missouri).

Red Brangus.—In 1946, Mike Levy, a cattleman in the hilly country of South Central Texas, was following a program on his Paleface Ranch of crossing purebred Brahman bulls on his black Angus cows. An unusually high percentage of red calves was dropped from this breeding, undoubtedly because the herd contained a considerable number of cows carrying the recessive red gene. Since these red calves appeared to grow out exceptionally well, Levy began saving them for replacements and eventually accumulated an all-red herd of mixed American Brahman–Angus breeding. These red cattle attracted the attention of other Angus breeders in the locality, and soon a sizable nucleus of these American Brahman–Angus crosses, all red in color, was formed. Mating a red bull of this breeding to a red cow, of course, always produces red progeny, and so the number of such red cattle multiplied rapidly.

This was the principal procedure which produced the Red Brangus

Mature Red Brangus bull, 2,200 pounds. Paleface Ranch, Spicewood, Texas.

cattle as they are known today; a small group, however, possibly 15 per cent, of the breed is the result of mating American Brahman bulls to Red Angus cows.

The Red Brangus and the Brangus (black) are identical in external appearance except for the difference in hair color. The Red Brangus, however, is not required to be exactly five-eighths Angus and three-eighths American Brahman as is the Brangus. Officially, the Red Brangus must "show characteristics of both the American Brahman and Angus breeds without preponderance of either." While this may appear a rather loose specification, all individuals for which registration is desired must be approved by an inspector of the breed society.

There are a number of Red Brangus herds scattered through the Gulf Coast states, but the largest concentration is in South Texas. Efforts of Red Brangus breeders to obtain recognition by the International Brangus Breeders Association failed because that organization would not accept the red color and required an exact blend of five-eighths Angus

Mature Red Brangus cow, 1,250 pounds. Paleface Ranch, Spicewood, Texas.

and three-eighths Brahman. As a result the American Red Brangus Association (Austin, Texas) was organized in 1956.

Cattalo.—Since the days of the Longhorn occasional efforts have been made to cross the American Bison (buffalo) with cattle. Cattalo (a name formed from "cattle" and "buffalo") is usually accepted as including any cross of cattle and buffalo. While Cattalo cannot be considered a breed, there have been a number of attempts to interbreed the two bovines, and at least one breeder in the United States is continuing in this effort.

One of the earliest breeders of buffalo and cattle was Charles Goodnight of Texas, who was also one of the first to move the Longhorns out of Texas by organized trail drives. Other cattlemen of those days also produced some of these hybrids, but nothing of commercial importance came of their efforts.

The first-cross male Cattalo is invariably sterile, and the first-cross

Cattalo bull (¾ bison, ¼ Hereford), Jim Burnett Ranch, Luther, Montana.

female has a very low level of fertility. Bulls that are three-fourths buffalo are also usually infertile, although one bull of this percentage has produced semen with 20 per cent live sperm cells. Recent breeding efforts have indicated that about 30 per cent of the first-cross females will become pregnant. Because of this low level of fertility, most Cattaloes have been the F_1 hybrids. Jim Burnett, of Luther, Montana, a

Cattalo cow (½ bison, ½ Hereford), Jim Burnett Ranch, Luther, Montana.

breeder of Cattaloes for the past 15 years, has succeeded in producing a number of three-fourths buffalo—one-fourth Hereford, and one-fourth buffalo—three-fourths Hereford crosses.

The F_1 Cattalo cow usually has a black or nearly black body, although a brindle color sometimes results. The face is invariably white if the cross has been to the Hereford. The pronounced hump of the buffalo disappears, and all that remains is a muscular thickening spread over the shoulders. The cows weigh 1,700 to 1,900 pounds in average condition.

The three-fourths buffalo bull, which results from mating an F_1 cow with a buffalo bull, has the general appearance of the buffalo, although it is a somewhat smaller animal. A nearly black color, solid over most of the body, is more prevalent than the dark brown of the buffalo.

Some of the Canadian experiment stations made an extended effort to establish a breed involving a buffalo-cattle cross, but the project was eventually abandoned in 1964. (See the section "Cattalo" in Part III, "Canada").

Burnett is continuing his breeding program, which is aimed at producing a stable buffalo-Hereford cross. It is problematical at this

Cattalo cow and calf. Cow is ¾ Hereford, ¼ bison; the calf's sire is ¾ bison, ¼ Hereford; the calf is therefore ½ Hereford, ½ bison. Jim Burnett Ranch, Luther, Montana.

time whether anything of commercial importance will come from this endeavor.

Charbray.—The Charbray is a Charolais–American Brahman cross which at one time was defined as three-fourths to seven-eighths Charolais. Recording and registration of the Charbray is handled by the American International Charolais Association. Presently interest in the Charbray in the United States is overshadowed by the great popularity of the Charolais. An animal which is three-fourths or more Charolais approaches so closely the $^{31}/_{32}$ "pure" Charolais that it is worth more as a Charolais prospect than as a registered Charbray.

Many a Gulf Coast cattleman, however, places considerable value on the Charolais which carries an infusion of the Brahman breed. The Charbray, whether specifically identified as such or not, seems destined to play a continuing and important role on the farms and ranches of that region.

The American Charbray Association was organized in 1949. Two

United States

years later this organization became affiliated with the American Charolais Association but separated in 1958. In 1967 the American Charbray Association merged with the reorganized American International Charolais Association, and it is this organization that now registers any Charbray cattle for which registration is requested (see the section "Charolais" above).

Makaweli.—During the latter half of the nineteenth century, when the British breeds were being imported to the Hawaiian Islands, representatives of the Devon and Shorthorn breeds were taken to the island of Kauai. The two breeds were interbred on the ranch of the Robinson family, and a type of cattle known as Makaweli resulted. It is locally considered a breed but was never adopted to any appreciable extent by other cattle raisers.

The Makaweli cattle are still grown by the Robinson family on Kauai and are shipped to the Hawaii Meat Company feed lot on Oahu and processed in their plant. They are large, rangy animals, red in color, and extremely wild. Difficulty is experienced in getting them to eat and drink in dry lot. When a pen is finished for slaughter, one animal is left behind to put with the next lot to be received as an aid to getting them started on feed.

Sabre.—The Sabre is a Sussex-Brahman cross that has recently been developed on the Lambert Ranch at Refugio, Texas. Starting in 1950 good-quality Brahman cows were bred to purebred Sussex bulls. Females of this cross were then bred back to Sussex bulls. The resulting progeny was three-fourths Sussex, one-fourth Brahman. A bull was selected from the calves of this breeding and bred back to Sussex cows, giving a seven-eighths Sussex–one-eighth Brahman cross. This was the base of the Sabre, which is now bred *inter se*. Critical selection has been practiced to obtain a beef-type conformation similar in character to that of the Sussex breed.

The Sabre as seen today on the Lambert Ranch displays uniform characteristics which would warrant its consideration as a breed. It is very similar in general appearance to the Sussex; the Zebu characteristics have practically disappeared. There is no hump on either the bull or the cow, although the bull is thick over the shoulders. The dewlap and sheath also display a tendency toward the pendulous Zebu type, though not as extreme as in the Brahman. The color is a deep red, solid over the whole body except for minor white markings on the underline. Mature cows on the range average 1,000 pounds in weight, bulls around 1,700 pounds.

Mature Sabre bull, 1,750 pounds. Lambert Ranch, Refugio, Texas.

While the Lambert Ranch is the only breeder of the fixed Sabre cross, these and other Sussex-Brahman cross bulls have been used very successfully to produce feeder cattle on neighboring properties. Bull calves weigh 525 pounds at 205 days and heifers 475 pounds.

There is no breed association for the Sabre. The Lambert Ranch is the principal breeder, and the organization most interested in the Sabre is the Sussex Cattle Association of America, Refugio, Texas.

Santa Gertrudis.—In general the breeds of cattle which have been developed in the United States during the present century have been formed from two or more already established breeds. Such parent breeds have in general been arbitrarily selected without experimental work designed to evaluate the final results to be expected. The Santa Gertrudis breed, developed on the King Ranch of Texas, is unique in that many years of painstaking experimentation preceded the final decision about its composition.

The King Ranch, which was established in 1853, eight years after Texas was annexed to the United States, was stocked with Mexican cattle. Early in the 1880's the cattle were divided into two herds, and

United States

an upgrading program started, Shorthorn bulls used on one herd and Hereford bulls on the other. As the influence of the well-acclimated Mexican cattle decreased, the nearly pure British breeds were observed to be less productive, particularly in the size of the calf crop, and at one time consideration was given to the desirability of backcrossing to Mexican bulls as a revitalizing influence.

In the early years of the present century the two herds of the King Ranch consisted of 25,000 cows each—one Shorthorn and the other Hereford. This was at the time when the tick tolerance and heat resistance of the Zebus were gaining recognition along the Gulf Coast. In 1910 the mating of a Zebu-Shorthorn–cross bull with a group of the purebred Shorthorn cows resulted in progeny that were markedly superior to either the Hereford or Shorthorn cattle in the way they maintained themselves during drought periods and in the heavier calves weaned. A program of experimental breeding was initiated and comparisons were made between Brahman-Hereford crossbreds and Brahman-Shorthorn crossbreds. After a period of 10 years the advantages of the Brahman influence on both the Shorthorns and Herefords was clearly demonstrated.

Forty years of hard work had gone into the upgrading of these herds, which were at this stage essentially pure. It must have been a heartbreaking step to take, but the decision was made to expand the Brahman influence to the entire Shorthorn herd.

To initiate the program, 52 bulls of ¾ to ⅞ Zebu breeding (embodying unknown proportions of the Nellore, Krishna Valley, and Guzerat breeds) were divided among two single-sire herds of 50 cows each and 8 multiple-sire herds. In all there was a total of 2,500 purebred Shorthorn cows in the initial program. Experimental results had indicated that a ⅜ Zebu–⅝ Shorthorn combination was the most desirable under conditions pertaining on the ranch. A systematic breeding program was outlined to achieve this relationship and at the same time select for a good beef conformation and a red color.

Because the required number of Zebu bulls of the desired quality could not be obtained, a Zebu breeding herd of the best animals available was established, and a selective inbreeding program was designed to produce the desired type of bulls for the crossing project. Although not connected with the development of the American Brahman breed, this Zebu herd was founded on the same Indian breeds as those which entered into the formation of the Brahman. Pictures of the two lines of cattle show them to have been practically identical.

In the course of the crossbreeding program, a remarkably prepotent

Santa Gertrudis bull and cow with calf. Bull 2,900 pounds, cow 1,800 pounds. 9 Bar Ranch, Houston, Texas.

Young polled Santa Gertrudis bull. King Ranch, Kingsville, Texas.

sire was disclosed by the superiority of his progeny. This was Monkey, a bull that resulted from the chance mating of a bull in one of the single-sire groups with a milk cow that was $1/16$ Zebu and a direct descendant of the Zebu–Shorthorn–cross sire which had first aroused interest in Zebu breeding in 1910. With the discovery of Monkey the breeding program was modified to incorporate close inbreeding and line breeding to him and his descendants. This procedure was carried to the point where the entire purebred Santa Gertrudis herd of the King Ranch traces its descent from this bull.

One of the most noteworthy features of the whole development program was the meticulous identification of all individuals in the purebred herd and of their progeny in card-index records.

The Santa Gertrudis is a large breed of solid cherry-red color with no white permitted other than minor spotting on the underline. No hump

Cattle of North America

is discernible in the female, and in the male it is of moderate size. The horns grow outward from the head and are thick and rather short. Frequently the horns bend downward or forward. The breed shows a marked degree of resistance to ticks and biting insects and carries a high degree of heat tolerance, markedly superior to any of the European breeds. The Santa Gertrudis are excellent beef producers, either off grass or as feeders. To grade choice, however, steers must be carried to relatively heavy weights. Mature cows in good condition average 1,350 pounds and up to a maximum of 1,850 pounds. Bulls average 2,000 to 2,400 pounds and some individuals up to 2,700 pounds.

A polled strain of Santa Gertrudis has been developed from polled mutants which have occurred in the King Ranch purebred herd. These carry all the other characteristics of the horned Santa Gertrudis.

The breed has been exported to many foreign countries, including the Argentines, Brazil, the Philippines, Australia, and many countries in Africa. The King Ranch itself has established large Santa Gertrudis herds in many of these countries. Some of them rival in extent the parent ranch in South Texas. The King Ranch operations in Australia, Brazil, and the Argentine are discussed in the chapters on those countries in Volume I.

The Santa Gertrudis Breeders International (Kingsville, Texas) was founded in 1951.

Senepol (Nelthropp).—The Senepol, an unheralded breed of cattle, has been developed during the past 50 years on St. Croix, the largest of the Virgin Islands. There has been some exportation of these cattle to other Caribbean Islands, but their obvious potential as a good tropical beef animal has attracted little attention other than on the islands of St. Croix and St. Thomas. This is probably due to a lack of promotional efforts by the breeders, who, for the most part, have been content to raise and improve their cattle and sell on the local market.

In 1860 a shipment of 60 heifers and 2 bulls of the West African N'Dama breed was taken to St. Croix by a man named Elliot, owner of the Longford Estate on that island. This was the nucleus from which a number of N'Dama herds were established on St. Croix. They were maintained pure by a few breeders of the time and were known as Senegal cattle, since the initial shipment was made from the French colony of Senegal, West Africa. There appears to be no record of any subsequent importations of N'Dama cattle to St. Croix. It is said that Puerto Rico also received some cattle at the time of the shipments to St. Croix, but they left no trace.

United States

One of the larger breeders of the N'Dama (Senegal) cattle on St. Croix was Bromley Nelthropp, manager of the Granard Estate. In 1889 his herd of pure N'Damas numbered 250 head. He maintained a pure herd of the breed until 1918, when he started a limited crossing program, using a Red Poll bull named Douglas, which had been imported from England to the island of Trinidad. This bull was purchased by Nelthropp and taken to St. Croix for use in his herd. Subsequently two of the best sons of Douglas out of N'Dama cows were also used in the herd. This nucleus was maintained as a closed herd until 1942, at which time a second Red Poll bull and two cows, purchased from a breeder on St. Thomas, were introduced into the herd. After that the herd was again closed for a number of years. The type of cattle which resulted from this crossing of the Red Poll on the N'Dama came to be known as Nelthropp cattle and had a wide acceptance among growers on St. Croix.

Other cattle breeders on the island who had been raising N'Dama cattle began using Nelthropp bulls in their herds. Later Nelthropp introduced selected progeny of some of his neighbors' cattle into his own herd. One of the major suppliers of this outside stock was Carl Lawaetz, who had been a consistent user of the Nelthropp bulls on his pure N'Dama cows.

The cattle which resulted from this interbreeding of the Red Poll and N'Dama showed uniform characteristics and bred true. Later they became known as Senepol rather than Nelthropp cattle. They can be defined as an indeterminate mixture of the N'Dama and Red Poll breeds in which the Red Poll appears to be the stronger factor. The Senepol, however, shows some strong characteristics of the N'Dama. Selection by the early breeders was for a beef-type conformation, a solid, bright-red color, and the polled characteristic.

The pure N'Dama is a small, hardy, tan-colored African breed, usually with prominent horns but having a polled tendency as shown in occasional individuals. It is heat tolerant and unusually fertile under tropical conditions (see Volume II, page 755). The N'Dama is one of the very few African breeds which carries a strong tolerance to trypanosomiasis carried by the tsetse fly. This disease decimates nearly all types of cattle, even most indigenous African breeds. The N'Dama has been shown to respond remarkably well to selection for a beef type in the Congo by a large breeder who employed modern breeding practices (see Volume II, page 729).

The Red Poll (Volume I, page 310), while originally developed in England as a dual-purpose type, has a good beef conformation. In the

Mature Senepol bull, 1,600 pounds. Annaly Farms, St. Croix, Virgin Islands.

Mature N'Dama bull, 1,225 pounds. Jules Van Lancker Farm, Congo, Africa. Note similarity of the N'Dama and Senepol breeds.

Mature Senepol cow and calf. Annaly Farms, St. Croix, Virgin Islands.

United States it is now more of a beef than a dual-purpose type. Representatives of the Red Poll introduced into the Caribbean area and some parts of Central America have displayed an ability to adapt more readily to a tropical environment than most breeds transplanted under similar conditions. It is therefore not surprising that combining the N'Dama and Red Poll into the Senepol resulted in an outstanding tropical beef animal.

There is a marked resemblance between the true Senepol and the N'Dama of West Africa, as shown by the accompanying illustrations. The N'Dama bull is from the selected herd of the Jules Van Lancker Farm in the Congo. The Senepol bull is from the Annaly Farms herd on St. Croix.

The Senepol is a strongly polled breed of medium size, a bright solid red with only very minor white undermarkings on some individuals. Several writers have commented on the fertility of the Senepol as superior to that of many other tropically adapted breeds. Its beef-type conformation is superior to that of the Zebu breeds and also generally to that of the Zebu-nonhumped crosses.

Mature Senepol bulls weigh up to 1,750 pounds, and cows in a good herd average 1,100 pounds. Bull calves, weaned at 8 months of age, weigh up to 550 pounds. The common practice on the island is to sell young bulls for slaughter. These under good management on pangolagrass pasture will weigh 800 to 850 pounds at 12 to 14 months of age.

In 1970 the Annaly Farms on St. Croix was the largest breeder and the principal sponsor of the Senepol. The breed name is registered in the United States, but there is no breed society and no official registry. About 20 years ago there was some minor use of Santa Gertrudis bulls in the Annaly herd, but this modification of the breed was discontinued after a number of years. The current objective is to improve the Senepol by selection within the breed.

The number of Senepol cattle in existence is small. There are probably 2,500 on St. Croix, 1,300 of which are in the Annaly Farms herd. This is one of the best nonhumped tropical beef breeds and is superior in both size and beef-type conformation to the other few nonhumped breeds that are adapted to the tropics. It is to be hoped that some means will be found to assure the perpetuation of the Senepol breed.

New Breeds

The new breeds have been introduced to the United States largely through importations of frozen semen from Canada. The major breeds involved before 1972 were the Limousin, Maine-Anjou, and Simmental. The Murray Grey was introduced by direct importation of semen from Australia. Live animals of the Lincoln Red, South Devon, and Welsh Black breeds were imported either directly from England or by way of Canada. Representatives of the Canadian-developed Hays Converter have also arrived in the United States.

With interest in foreign breeds riding high in 1970, artificial insemination centers took steps to import semen direct from Europe under the USDA procedures discussed in the section "Northern European Cattle" above. By the close of 1971 semen was available for distribution in the United States from the following countries and breeds:

Austria: Gelbvieh
France: Blonde d'Aquitaine
Germany: Fleckvieh (German Simmental)
Italy: Chianina, Marchigiana, and Romagnola
Switzerland: Simmental and Brown Swiss

The donor bulls in Europe are held in strict quarantine conditions supervised by the USDA and, like their counterparts exported to North America, they must never have been vaccinated against foot-and-mouth disease.

Blonde d'Aquitaine.—This breed, also known as the Garonne, was introduced to the United States through the importation of semen direct from France in 1971. The Blonde d'Aquitaine was developed in southwestern France as a draft animal and more recently for beef. It is a large animal with an excellent beef conformation. Cows weigh 1,600 pounds and bulls 2,500. The cows have the reputation of being poor milk producers. The color is fawn, sometimes with a reddish tinge. In the past Limousin bulls were used to a considerable extent and the Blonde d'Aquitaine today probably carries Limousin influence.

Chianina.—This was the first Italian breed to reach North America. Live animals arrived in Canada in 1971, and in the same year semen imported direct from Italy was available in the United States. The breed is discussed in Volume I, page 190.

Gelbvieh.—The Gelbvieh was developed in both Austria and Germany as a dual-purpose milk-beef type from the original draft animal. It has the best beef conformation of any of the breeds in those two countries. In 1971 the Gelbviehs were brought to Canada from Germany as live animals, and in the same year semen of the breed was imported to the United States from Austria. See Volume I, page 86, for the Austrian Gelbvieh, and page 341 for the German Gelbvieh.

Hays Converter.—The Hays Converter was developed in Canada by Senator Harry Hays in a three-breed crossing program which utilized the Hereford, Holstein-Friesian, and Brown Swiss breeds (see the section, "Hays Converter," Part III, "Canada"). The only purebred representatives in the United States are three leased bulls at American Breeders Service, DeForest, Wisconsin.

Limousin.—This is one of the few European breeds which has been selected primarily for beef characteristics following its initial development as a draft animal. It is a particularly hardy type and is usually kept in the open the year round. The home of the Limousin is south-central France, where the winters are raw though free of extremely low temperatures.

The Limousin is a bright reddish tan, the color being solid over practically the entire body except for lighter shades on the extremities. Mature cows weigh from 1,300 to 1,500 pounds and bulls 2,200 to 2,500 pounds. Conformation is a good beef type. In France steer calves have been fed out to weigh 1,200 pounds at 12 to 14 months of age.

The first Limousin bull arrived in Canada in 1967. Up to the present there have been only token representatives of Limousins in the United

Hays Converter bull, five years old, 2,150 pounds. American Breeders Service, DeForest, Wisconsin.

States, bulls brought in from Canada under bond for exhibition and semen collection. Approximately 92,000 cows throughout the United States were bred with Limousin semen from 1968 through 1970. An estimated 150,000 to 200,000 were bred in 1971.

The breed society is the North American Limousin Foundation (Denver, Colorado).

Lincoln Red.—A few Lincoln Red bulls have been imported from Canada in recent years. There is no breed society and the breed has attracted little notice in the United States (see the section "Lincoln Red," Part III, "Canada").

Maine-Anjou.—The Maine-Anjou is reputedly the largest French breed. Its home is northeastern France, where, in the mid-nineteenth century, it was developed as a draft animal by crossing the English Shorthorn on the local Mancelle cow. Subsequently it was selected for a dual-purpose, meat-milk type.

In 1962 the Maine-Anjou and the Armoricaine breed societies in France joined in a crossing program using Red and White Lowland bulls of the Netherlands and Germany to increase milk production. The Armoricaine is a breed that also was developed in Brittany by crossing with the Shorthorn. The Maine-Anjou-Armoricaine federation was dissolved in 1970 to emphasize the Maine-Anjou as a beef breed and thus encourage its acceptance as one of the new breeds being exported to Canada.

Maine-Anjous are predominantly dark red, usually with some white markings. Mature cows weigh 1,650 to 1,900 pounds with some individuals exceeding 2,000 pounds. Bulls weigh 2,600 to 3,000 pounds. The body is exceptionally long, even considering the large size of the breed. A rather coarse appearance is characteristic, and the conformation is not as distinctly beefy as the Charolais, Simmental, and Limousin (see the section "Maine-Anjou," Part III, "Canada").

Maine-Anjou semen imported from Canada was first used in the United States in 1969, and the breed enjoyed immediate popularity. An estimated 100,000 cows were bred artificially to Maine-Anjou bulls in 1971.

The International Maine-Anjou Society is situated in Kansas City, Missouri.

Marchigiana.—Marchigiana semen was imported to the United States from Italy in 1971. This is a large draft type later selected for beef conformation. The color is a light to medium grey. Bulls weigh upward of 2,500 pounds and cows around 1,400 pounds. There has been considerable admixture with the Chianina and possibly with the Romagnola breeds.

Murray Grey.—The Murray Grey goes back to the progeny of a very light roan Shorthorn cow, bred to an Angus bull in 1905 in Victoria, Australia. This calf was a silver-grey color, and subsequent matings of the same cow to other Angus bulls also produced silver-grey calves. For years the Sutherland family, owners of Thologolong, the property where the grey calves were born, continued to breed Angus bulls to such grey cows as were produced. The offspring were nearly all the same grey color. No attempt was made at that time to establish a breed.

Around 1940 there was a change in ownership of the grey cattle and the Angus herd in which they had been produced. Grey bulls were then bred to Angus cows, and the grey color continued to persist. *Inter se* breeding was started, and other cattlemen in the area began to use what had come to be called Murray Grey bulls.

The Murray Grey is a medium-sized, strongly polled animal of the

Cattle of North America

silver-grey color mentioned. Cows weigh 1,050 to 1,500 pounds, and bulls weigh 1,700 to 2,200 pounds.

Before 1972 artificial insemination was the only means of bringing the Murray Grey to either the United States or Canada, the frozen semen coming from Australia. Movement of live cattle to both countries was banned because of the incidence of bovine pleuropneumonia in Australia. The USDA approved importation of cattle from Australia under rigid quarantine procedures in 1972, and two bull and three heifer calves were imported in that year. Murray Grey semen was first used in the United States in 1969, and the first crossbred calves were dropped in 1970. An estimated 20,000 cows had been bred to Murray Grey bulls by the end of 1971.

The registration organization in the United States is the American Murray Grey Association, Inc. (Shelbyville, Kentucky).

Romagnola.—The Romagnola arrived in the United States in 1971 through semen importations direct from Italy (see Volume I, page 192).

Simmental.—The Simmental is one of the oldest and most widely distributed of all the European breeds. The breed originated in western Switzerland as a draft and dairy animal, and during the nineteenth century it spread into nearly all the countries of Europe.

When introduced to France, the breed was used in upgrading local cattle to three present-day French breeds—the Pie Rouge de l'Est, the Abondance, and the Montbéliard, all of which are now basically Simmental. The German Fleckvieh was derived from the Swiss Simmental after many generations of selection in Germany. The Austrian Fleckvieh was the result of an upgrading process using Swiss Simmental bulls on local types of cattle. Both the German and Austrian representatives of the breed are now better dual-purpose milk-beef animals than the Swiss type. The use of artificial insemination was rigidly restricted in Switzerland, and better selection practices for both milk and beef were employed in Germany and Austria. The breed is popular in the Soviet Union, as well as the other countries of Eastern Europe.

The Simmental was among the nineteenth-century introductions to American soil, but since it made no lasting impression at the time, it is treated as one of the new breeds.

The *Fifteenth Annual Report* of the Bureau of Animal Industry places the first introduction of the Simmental to North America in 1886, when a small herd was established in Texas. The following year an Illinois farmer made another importation. In 1898 a herd of purebred Simmentals in New Jersey provided bulls for experimental crossing on

Jersey cows. This experiment apparently produced little in the way of significant results, since no reference to it is to be found in the literature of the time.

The American Simmental Herd Book Association was organized in 1896, when 10 bulls and 15 cows were registered. The enthusiasm of these late-nineteenth-century Simmental importers failed to match that of the present-day promoters, and the breed soon disappeared from the scene and was forgotten until reintroduced into Canada in 1966.

The Simmental is a large dual-purpose breed. Mature cows weigh 1,400 to 1,800 pounds and bulls 2,300 to 2,600 pounds. The face is white and the body is usually pied with prominent cream to red markings on a white background, although a nearly solid red over most of the body is sometimes seen.

Currently the Simmental is the most popular of the new breeds in the United States. In 1971 an estimated 150,000 cows were bred with Simmental semen, and there were probably over 200,000 crossbred Simmentals throughout the country.

The importation of Canadian semen continues to be the principal source of Simmental genes, although there were two bulls in service in the United States in 1971. The breed had been represented exclusively by the French Pie Rouge and the Swiss strain until 1971. In that year there were importations of semen directly to the United States from Switzerland. The Fleckvieh (German Simmental) was also introduced to the United States through the frozen semen route in 1971.

The breed association makes no restrictions about the use of artificial insemination, there are no color restrictions, and performance records are mandatory for registration. Cattle can be bred up from cows of any breed, and ⅞ females and $^{15}/_{16}$ males are recognized as purebreds.

The American Simmental Association (Bozeman, Montana) was incorporated in 1968.

South Devon.—The South Devon breed was late reaching North America, although it had been successfully imported into other parts of the world, particularly into southern Africa. Three head of South Devon cattle were brought to the United States in 1936 by the USDA at the suggestion of Henry Wallace, then secretary of agriculture. It is also recorded that in 1947 two South Devon bulls were imported to New York State by Thomas E. Milliman. Nothing developed from either of these introductions, and the South Devon is therefore placed in the new-breed category.

In April, 1969, after four years of endeavor, Arthur N. Palmer suc-

Mature South Devon bull, 2,500 pounds. Big Beef Hybrids, Inc., Stillwater, Minnesota.

ceeded in bringing 44 purebred bulls and 40 cows of the breed to the United States. A few months later a second shipment of 96 head arrived in Canada, where 20 head were retained and the remainder went to the farm of Big Beef Hybrids International at Stillwater, Minnesota. In 1971 this company merged with a number of other organizations promoting the new breeds, and the name was changed to Big Beef Hybrids, Inc. (BBH).

The South Devon is the largest of the British breeds, although earlier in the century it was exceeded by the Shorthorn. Mature cows weigh 1,400 to 1,700 pounds, and bulls range from 2,200 to 3,000 pounds. The color is a bright, yellowish red, generally solid but sometimes slightly mottled. Horns are small considering the large size of the breed, and there is very little tendency to be polled, although efforts have recently been made in England to select a polled strain.

First developed as a draft animal and then for milk production, the breed was never selected for the small compact conformation in vogue a number of years ago. Although now usually considered a dual-purpose

Mature South Devon cow, 1,750 pounds. Big Beef Hybrids, Inc., Stillwater, Minnesota.

type, the South Devon in England has been selected for beef by some owners and for milk production by others. It responds well to both types of selection.

The approach of BBH in introducing the South Devon has been somewhat different from that of the organizations promoting the other new breeds. The latter have based their upgrading programs on cattle already established in the United States. BBH seeks to produce a superior first-cross female which can be bred to a bull of a third breed to produce premium feeder cattle. The first step is the selection by progeny test of superior South Devon sires. Such progeny-proven sires are then to be bred artificially to selected cow herds for the production of the F_1 females to be used as brood cows in commercial herds.

The program calls for breeding South Devon bulls artificially on contracted cow herds, and from the offspring BBH obtains progeny records on the South Devon sires. Progeny records are also to be obtained on offspring of the South Devon crossbred females. Superior sires selected by these records will then be used for the production of outstanding females. It is estimated that five years will be required to complete this first cycle of the program.

In 1971, South Devon semen was offered to the public for the first time. This digression from the basic program was made in an effort to hold interest in the breed until the crossbred females can be placed on the market.

BBH introduced the French mini straw—a recent innovation from the older ampule method of storing semen—to their co-operators in the South Devon progeny-testing program.

The South Devon Breed Society in the United States is headquartered at Albert Lea, Minnesota.

Welsh Black.—The Welsh Black has not attracted much attention in the United States, although the first importations of the breed to Canada originated in this country. A small herd of Welsh Black is held in Reno, Nevada.

There is no breed society in the United States, but the Canadian Welsh Black Cattle Society is active in Canada (see the section "Welsh Black," Part III, "Canada").

Other Breeds

Africander.—In 1931, 16 bulls and 11 cows of the Africander breed were imported from South Africa by the King Ranch in Texas for experimental breeding during the time the Santa Gertrudis breed was being developed. It was thought that they might be useful in fixing the deep-red color that was desired in the new breed, but a satisfactory color was fixed from such red genes as the Shorthorn and Zebu contributed, and the Africander did not enter into the development of the Santa Gertrudis.

The Kenedy Ranch at Sonita, neighbor of the King Ranch, obtained some of the Africander cattle, as did a few other breeders in the area. Later the USDA Iberia Livestock Experiment Station at Jeanerette, Louisiana, used two of the bulls for crossing experiments on British beef breeds.

The incorporation of the Africander into the Barzona is the only known contribution the breed has made to United States cattle (see the section "Barzona" above). Descendants of the original herd of Africander are still extant on the King Ranch. They have been maintained as a separate herd and in 1971 included 20 cows of breeding age.

British White Cattle.—A representative group of the British White (Park) cattle was shipped to the Bronx Zoo in New York City during the early days of World War II. This was one of many steps which the

Africander bull from the original herd imported by the King Ranch. Iberia Livestock Station, Jeanerette, Louisiana. National Archives Photograph.

British government took to ensure continuity of the British heritage in event of a German invasion of England. The zoo, lacking facilities for handling even a small herd of cattle, accepted the offer of the King Ranch to provide for them. The breed has been maintained in complete isolation and is probably the only example of the British White cattle in existence outside the British Isles. In 1971 there were seven cows of breeding age in the King Ranch herd (see Volume I, page 288).

Canadian (French Canadian).—French Canadian Cattle, as they were called until the name was changed to Canadian Cattle in 1930, were taken by the Jesuits from Canada to their missions in the Mississippi Valley toward the end of the seventeenth century. These cattle also accompanied the Acadians on their emigration to Louisiana in 1765. There they were used as dairy animals for over a century. Eventually

British White Cattle herd. King Ranch, Kingsville, Texas.

these isolated populations of the breed were absorbed in the mainstream of Native American cattle.

French Canadian Cattle were also imported into the northeastern states from Canada in the early years of the twentieth century, and a herdbook was established for their registration but was subsequently discontinued. Representatives of the breed were exhibited at the Pan-American Exposition in Buffalo, New York, in 1901 and placed first as the most economical producers of butter in a competition with other dairy breeds (see the section "Canadian Cattle," Part III, "Canada").

Longhorn (English).—The English Longhorn was declining in popularity in England in the opening years of the nineteenth century, when the first Shorthorn representatives were arriving in the United States. The Shorthorn had only recently gained ascendency over the Longhorn in England, and some loyal Longhorn breeders managed to get representatives of their breed into the United States as early as 1817. There were a few other importations later in the century. The numbers were small, and there are only minor references to their presence in writings

United States

of the time. The Longhorns were soon absorbed in the Native American cattle (see Volume I, page 307).

N'Dama.—Individual breeders on St. Croix, one of the Virgin Islands group, introduced N'Dama cattle from Senegal in West Africa around 1865, when the island was a Danish possession. These N'Dama entered into the composition of the Senepol breed (see the section "Senepol" above), which exists today on St. Croix. Some N'Dama were shipped at about the same time to Puerto Rico but left no trace.

The N'Dama as a breed has disappeared from the Caribbean area.

Normandy.—A few Normandy cattle were taken from northern France to Illinois in 1885. Other importations followed; a number of head reached New York in 1886 and others went to Massachusetts in 1887. Another shipment was distributed in New York, New Jersey, and Vermont in 1895. Photographs of these Normandy cattle show them to be the same breed as the Normande (Normandy) of France, as described in Volume I, page 139.

An American organization of Normandy breeders was formed at the time the cattle were being imported. A herdbook was started but was never published. After a time the breed was apparently absorbed in the upgrading of the Native cattle and disappeared from the scene in the United States.

Red Danish.—The Beltsville, Maryland, Agricultural Research Center of the USDA imported a herd of about 20 Red Danish cattle in 1934. They were used in experimental work and crossbreeding, and in 1939 and in a few subsequent years a number of the Red Danish bulls were sent to Michigan. County bull associations were organized to foster the breed in the state, and upgrading to Red Danish became popular.

The American Red Danish Cattle Association was organized in 1948, and females of ⅞ and bulls of $^{15}/_{16}$ Red Danish breeding were declared eligible for registration. Production records of female ancestors were required.

The Dairy Science Department of Michigan State University investigated the lethal defects resulting from inbreeding in farm herds in the adjacent area. Two lethal defects, paralyzed hindquarters and ankylosis, frequently occurred in the second cross to Red Danish sires, and both calf mortality and cow sterility increased as the number of crosses increased from one to three. It is not clear to what extent these difficulties affected the popularity of the breed, but it has practically disappeared from the United States. The Dairy Science Department investigated

Cattle of North America

these abnormalities but denies that its work with the Red Danish was abandoned because of them.

The last Red Danish cattle imported to the United States were four bulls that went to Beltsville in 1958. They originated in Sweden. The only known herd remaining in the country today is the one moved a few years ago from Beltsville to the Meat Animal Research Center at Clay Center, Nebraska.

The American Red Danish Cattle Association was disbanded in 1968.

Red Sindhi.—The Red Sindhi is one of the better milk-producing Zebu breeds. At some of the government experiment stations in India it has undergone extensive selection for milk production.

Two bulls and two cows of the Red Sindhi breed were imported in 1946 by the Iberia Livestock Experiment Station, Jeanerette, Louisiana, from the Allahablad Agricultural Institute in India. They were used in crossbreeding experiments with Holstein-Friesian, Jersey, and Brown Swiss cows in an effort to breed a better-producing animal for a hot climate. The Red Sindhi crosses, however, showed a lower milk production than that of the straight European breeds in the Louisiana environment. Work carried out at the USDA Agricultural Research Center at Beltsville, Maryland, and at the Georgia Coastal Plains Experiment Station at Tifton, Georgia, confirmed the Jeanerette results. The Red Sindhi were subsequently disposed of, and the breed has left no significant trace in the United States.

Siberian Cattle.—When the United States took possession of Alaska in 1867, all the major Russian settlements were stocked with Siberian cattle. Sitka, then known as New Archangel, was a thriving seaport, and Kodiak had been growing since its establishment in 1792. When the Russian settlers first moved to Kodiak Island and the Kenai Peninsula, they took their herds with them. They were the only cattle in Alaska until some years after the United States took possession.

The Siberian cattle must have been of an old landrace that resulted from natural selection in eastern Siberia. Descendants of these early Alaskan cattle are described by C. C. Georgeson in a bulletin of the Alaska Agricultural Experiment Stations in 1929 entitled *Brief History of Cattle Breeding in Alaska*:

"The animals were small, slim in all proportions, and had a narrow head with thin, upright horns. The average weight of the mature cow was about 500 pounds. In color the stock was brown, or dark red, and occasionally the body was mottled. The milk yield was low and had a fat content of

United States

about 3 per cent. The cattle had deteriorated not only in general conformation, but also in milking qualities and in suitability for beef production.

This description fits well with that of Mongolian cattle, as given in Volume II, page 963. The Siberian cattle that reached Alaska may have been of similar origin.

The early Russian settlers were primarily interested in the fur trade and in shipping. They gave little attention to their cattle, which had to survive the winters on such forage as they could obtain from grasses standing from the luxuriant summer growth. Often this forage lay under heavy snow cover even in the more equable climate of the Kenai Peninsula and Kodiak Island. The only protection available to cattle in the winter was the brush and timber. Under such conditions and because no attention was given to breeding practices, there was deterioration in the quality of the cattle. An exceptionally hardy strain, however, survived. These free-ranging cattle were utilized by the Russians for both beef and milk.

The Siberians were the principal bovine inhabitants of Alaska until the early years of the twentieth century, and when the experiment station was opened at Kenai in 1898, the Siberians were the first used in its livestock programs. A small dairy herd was developed from them at the station in 1902. The accompanying illustration was published in 1906 under the caption "A So-Called Native Cow." There appears to be no record of any cattle having been previously brought from the outside to this part of Alaska. The white markings on the face, underline, and legs are somewhat similar to those of the Mongolian bull shown in Volume II, page 963. This circumstantial evidence would indicate that the cow pictured was a descendant of the Siberian cattle.

There had been a few introductions of cattle from the United States before this time to other parts of Alaska, but they left little impression on the cattle population. In the late 1880's a San Francisco firm had planted a few head of cattle on uninhabited Chirikof Island, 125 miles south of Kodiak Island. Also about 1900 a Seattle packing company shipped some cattle to Kodiak in an attempt to establish a herd for the production of beef. This effort failed when the herd was unable to survive the winters.

In 1906, in an effort to develop a more productive type of cattle, two Galloway bulls and seven cows were shipped from Missouri to the Kenai and Kodiak experiment stations. Then followed the introduction of more Galloways, Holstein-Friesians, Milking Shorthorns, Guernseys, and Jerseys and later Herefords and Angus. Eventually the early Si-

Cow in the Kenai Experiment Station. From Annual Report of Alaska Agricultural Experiment Station for 1906.

berian cattle disappeared, being both supplanted and bred out by the more glamorous British breeds. During this period cattle husbandry was almost entirely the province of the experiment stations, and a cow did not necessarily have to pay her way. Whether the British breeds gave Alaska a more productive type of cattle for her environment than could have been developed from the Siberian cow will never be known.

Yak.—During the period 1919 to 1930, nine head of Himalayan Yak—three bulls and six cows—were taken from Wainwright, Alberta in Canada to the Alaska Experiment Station at Fairbanks. They were the descendants of animals which had been imported from England to Canada in previous years. The Yaks were taken to Alaska for crossing with the British breeds in an endeavor to develop a type of cattle that could satisfactorily withstand the Alaskan environment.

The small Yak herd was maintained for 11 years at the Fairbanks station, where the elevation is 420 feet. They did not reproduce satisfactorily, although a number of crosses on Galloway cows were obtained. Two Yak cows did not breed at all, and one bull and one cow died.

United States

There was a marked deterioration in the condition of all the animals. In August, 1930, the Yak herd, including the Yak-Galloway crosses, was moved to the high-plateau area near Lignite, 100 miles south of Fairbanks, where the elevation ranges from 1,200 to 2,500 feet. Following this change in environment, all the animals showed a marked improvement. They became more active, the bloom of the hair coat improved, and they put on weight. The following year they bred satisfactorily. The effect of this change following the movement to a higher elevation is pertinent. In their native Himalayan highlands, it is said that Yaks will not survive at elevations below 8,000 feet. Experience with Yaks in Canada also had shown that they did not reproduce satisfactorily at low elevations. The hybrid females grown at Fairbanks produced offspring, though the hybrid males were sterile. This reproductive ability of the female and the sterility of the male is the same condition that pertains in Nepal, where the crossing of Yaks and the small hill cattle is common practice. (see Volume II, page 907).

The work with the Yaks and Yak hybrids was discontinued in 1933 before any conclusive results were obtained. Both the Yaks and the crosses had demonstrated their ability to survive in satisfactory condition the cold winters of inland Alaska. The animals had only the remaining grass from the summer growth, often under snow cover, to carry them through the winter. The major difficulty in raising cattle in the Alaskan environment is the problem of providing winter feed. Although the grass growth is ample, climatic conditions are such that it is frequently impossible to put up even poor-quality hay. Frequent short rains, cloudy weather, and temperatures too low for drying make a nightmare of haying. The preparation of silage in most locations is economically unfeasible. A beef animal which could carry through the winter on its own would be a boon to the country. The Yak hybrid conceivably could meet this requirement.

In 1971 a few Yaks were maintained on farms in the United States. They had probably originated from exhibition animals.

MANAGEMENT PRACTICES
THE COLONIAL PERIOD

The first English settlers in America stepped into as rugged an environment as any future pioneers were to encounter as they moved westward across the continent. Continuous exposure to the elements was a strange experience for many, and there were those who, unfamiliar with an ax, had to start with the standing tree to build a shelter for their families. A dugout, half below the ground surface, was usually the first housing.

For sustenance they often became hunters, gleaners, and fishermen. Such pursuits were usually rewarding, for the coastal areas had both game and fish, and the woods furnished berries and nuts. When the first crop was finally harvested, the situation improved. Providing the bare necessities for survival, however, required most of the settlers' time during the first decade of their lives in the new land.

Under such conditions cattle received very little attention when they were brought over, usually some time after a settlement had been established. The Jamestown Colony in Virginia is reported to have consumed whatever livestock it had to survive the first few winters. The Pilgrims at Plymouth in Massachusetts had no cattle until 1624, three years after they landed. The settlement which grew around the trading post that the Dutch started in Manhattan in 1613 is reported to have obtained its first cattle in 1625.

The early communities grew up in timbered areas along the coast. There were coarse grasses along the stream beds and sedges in the swamps, but little grass in the woods. Back from the shore there were prairie areas enclosed by trees, but it was several decades before the settlers moved inland to these better grass areas. The common practice throughout the early years in the colonies was to turn cattle loose in the timber to glean such forage as they could. Browse was obtained from the leaves of brush and trees. Young calves were penned around the dwellings to encourage the return of the cows at nightfall.

It was many years before hay was put up in any quantity. In emergencies when supplemental feed was essential to get livestock through the winter, sedges and any standing dry grass were cut and brought in. Crops for human consumption naturally came first, and the ground had to be cleared before they could be sown. Violence from the Indians had to be contended with, and theft continued to be a problem long after they had been subdued. It was years before the settlers even thought of barns for their stock.

The cattle which the Spaniards had introduced to Mexico and Florida were not available to the first settlers in New England and along the eastern seaboard. Their livestock had to come from Britain or the Continent. Because of less favorable weather, transportation across the North Atlantic was a more hazardous undertaking than the route of the Spanish galleons to the Indies. The small sailing vessels were not equipped to handle cattle, and at times losses were severe. One movement of 200 cattle to the Massachusetts Bay Colony was mentioned as losing 70 head.

The early cattle importations were usually made by the companies who held royal charters for the development of the new country or by joint-stock companies that were organized to buy, transport, and sell livestock to the colonists. Payment for cattle was sometimes in tobacco. The development company, commonly referred to as the "Company" in its own area, often furnished cattle and other livestock to the settlers on a communal basis as part of the settlement understanding.

Under such conditions the cattle population increased slowly, even though in many areas the slaughter of female cattle was frowned on. The shortage in Virginia was such that in 1619 a law was enacted prohibiting the slaughter of any cattle without the consent of the governor. Importations, principally from England, were encouraged at that time. Such measures were effective, and a year later Virginia reported having 500 head of cattle.

As the number of cattle increased, the practice of running them in "common" herds was started. This means of caring for cattle was adopted when communities grew to the point that the common pasturelands became overcrowded. The selection of bulls to be used on the common herd became a community responsibility. A common procedure was the appointment of a committee, which selected what it considered the two or three most desirable bull calves; the others were castrated. Final selection was then made of the bull to be retained when the animals neared breeding age. The bull so selected was kept by a townsman for a fee paid by the town authorities and was available for all cattle owners to use.

Importations of cattle were in demand all along the coast to supply new settlements and increase the stock in the established ones. The large "yellow cattle" of Denmark, to which there is frequent reference, were taken to New Hampshire in 1629. In 1638 a few cattle were shipped from Sweden to the newly established Swedish colony in Delaware. Most of the stock for the expanding settlements, however, came from neighboring communities. Some years later a number of shipments of Spanish cattle from Barbados in the West Indies reached Maryland.

Intercolony exchanges of cattle also began at this time. Boston and Plymouth were trading with Manhattan and Virginia. New England farmers liked the large Dutch cattle which they obtained from Manhattan better than their English cattle. It was during this period that the unplanned mixing of the various types of European cattle began.

By 1650 the colonies had become practically self-supporting in cattle. Local shortages could be met by trade among the colonies. Note had already been taken in some of the northern colonies that the home-raised cattle were more hardy than those which had recently arrived from Britain. In the four decades that cattle had now been bred in the New England climate, natural selection apparently was having some effect.

The first of the "fence laws" was enacted in Virginia in 1632. These laws required landowners to enclose their ground or "else to plant uppon theire owne perill."

Annual fairs at which cattle were shown were being held in New Amsterdam in 1641. Cattle had reached a position of considerable economic importance throughout the colonies by the mid-1600's.

In New England, where the community-type settlement predominated, the "cowherd" had become an institution. The cowherd was hired by the village authorities. In the early morning he would gather the milk cows and draft cattle of his patrons, blowing a horn to announce that he was ready to collect them. He drove the cattle to the common grazing ground and returned with them at night to the village green. The arrangements with the cowherd were often formalized in writing, his compensation being in terms of so much produce per head tended. Corn, wheat, or butter were common mediums of exchange by which he was paid. In the fall the stubble fields were grazed in the same manner as the commons. The usual pasture season was from May through October in southern New England and a month to six weeks shorter in the north.

After the middle of the seventeenth century the practice grew of providing some form of shelter for cattle in the winter, although such facilities were of a rudimentary type. Some hay and straw were stored for winter feed, and a bit of grain was fed the milk cow. The average size of New England farms in 1675 was around 40 acres, and by that time the practice of planting improved pastures was taking hold. The seed, usually bluegrass and white clover, was imported from England. For specialized markets farmers began to stall fatten steers on grain. In 1670 winter-fattened cattle (descendants of the Devonshire cattle) were trailed from the Springfield area in Massachusetts to the Boston slaughterhouse. The tradition of the town bull grew, and more attention was given to his selection.

Large farms under more prosperous conditions developed south of New England, particularly in New Jersey and Pennsylvania. Dairying increased rapidly in Pennsylvania. During the latter half of the seventeenth century slaves and indentured labor were employed as far north as New Jersey on large plantations of several thousand acres. There, as in New England, cattle were the main source of draft power; horses were used mostly for riding and carriage work.

Through the South cattle numbers had also increased. Virginia, which previously had found it necessary to prohibit the export of cattle, was shipping live cattle and salted beef to Barbados by 1657. Except for the expansion into Pennsylvania and New York, the area of settlement and cultivation was confined to the coastal plains in the middle and southern colonies, and it was here also that the cattle were raised.

During the 1670's population pressure began to be felt along the coastal plains, particularly in Virginia and North Carolina where these areas were closely held by large plantations. Settlers seeking their own establishments moved inland toward the Piedmont, which stretches from Virginia to Georgia. This foothill and low-plateau region lies between the Appalachian Mountains and the Fall Line, which separates the Piedmont from the coastal plain. Early attempts were made at raising sheep in this region, but they were more difficult to protect from the ever-present wolves than the cattle. The Piedmont was more cattle country than a farming area, and the early settlers readily adapted to the frontier stockman's life. Open-range stock raising, the first instance of this practice north of Mexico, became the principal occupation of these first Piedmont settlers.

The forest was not as dense there as on the coastal plain, and prairie areas were scattered throughout the region. The native grasses were stronger, of higher protein content, and more nutritious than the sedges and water grasses on the lowlands along the coast. As the herds increased in size, corrals were built for handling stock during calving, branding, and weaning. These enclosures were called "cowpens"; and "cowpenning" became the name by which this method of cattle management was known. The term "cowpen" had been in common use in the colonies to define any enclosure for confining cattle, but in the Piedmont "cowpen" now came to have much the same meaning as "ranch" did in the West 200 years later. The term covered the living quarters, corrals, and often a small pasture or two enclosed by a rail fence for holding stock preparatory to driving it to market.

These operations reached their peak in pre-Revolutionary days. Some of the cowpens were sizable operations. In one particularly good grass area, across the South Carolina line in Georgia, individual herds at one time ranged from 1,500 to 6,000 head. The methods of handling cattle were similar to those on the western plains at a later date. The hardships endured were those of the pioneer family, including harassment by marauding Indians. Cattle were handled by men on horseback. Market herds were driven to the population centers periodically, the largest drives occurring in the fall. Such tidewater towns as Baltimore and Philadelphia were the principal markets.

After the Revolutionary War the rapid westward movement of the population eventually engulfed the cowpen stockmen. At first, as farmers encroached on their range, the cowpen people moved west to new ranges in an endeavor to continue their way of life. But even after they had crossed the Alleghenies, they soon found themselves again hemmed in by farmers. Eventually the butchers in the cities, who had been getting their beef from the Piedmont herds, had to go to the farmers (who had replaced the cowpen people) for their meat supplies. Cattle for slaughter again became the more-or-less salvage product of a general farmer's operation, the same situation which had existed throughout the earlier settlements.

Most of the cowpen people moved westward and then eventually entered other pursuits. The cowpen practice of cattle raising, however, moved into Georgia in the late 1780's. The operations there were not as extensive as they had been in the Carolina Piedmont, but the activity continued until after 1800.

The closing years of the eighteenth century saw the first purebred cattle entering the United States. Not many cattle had been imported to the colonies after 1650. All those which arrived previously were of no particular breed. In 1783 a shipment of the newly developed Shorthorn breed is reported to have been made from England to Virginia. A second shipment of the breed is recorded as having been sent to New York State in 1791. No continuity of these cattle was maintained, and they were undoubtedly absorbed in the Native cattle of the time. The initial step had been taken, however, and an interest in better cattle was to spread throughout the country during the next century.

At the close of the eighteenth century the draft ox and the milk cow comprised nearly the entire cattle population. Raising cattle for slaughter had always been something of a special case, and there was no indication of a beef industry as it exists today.

THE NINETEENTH CENTURY

At the opening of the nineteenth century the Native cattle of the eastern states made up the major portion of the cattle population of the United States. The western movement of cattle had started several decades earlier and was to continue until after the emigrants reached the Northwest and Pacific coast.

Throughout the country cattle continued to be held mainly for draft, but beef was becoming a more substantial element in the national diet and an important article of trade. The milk cow remained largely a household adjunct.

During the new century, three separate populations of Spanish cattle that were previously existent on foreign soil were to be included in the United States—those in Texas, the Southwest, and California; a group in Louisiana; and those in Florida.

The Mexican cattle, later to become the Longhorns, that were acquired with the annexation of Texas in 1845, were the most important of these cattle. Part of the same population was in southern California with a few scattered herds in New Mexico and Arizona, all of which areas were acquired in 1848 with the Mexican Cession.

The first Spanish cattle on United States soil were those the French settlers husbanded in Louisiana which came to the United States with the Louisiana Purchase in 1803. Years later they were absorbed by the Native cattle and left no great influence on the national herd.

The Spanish cattle in Florida entered the United States scene when the territory was purchased from Spain in 1819. They were not absorbed into the mainstream of the national herd until after the opening of the twentieth century.

The United States cattle story of the nineteenth century involves the movements and mixing of the various populations. The principal external influence during this period was the introduction of the purebreds from Europe, which assumed major proportions after 1865.

Eastern United States

The western movement of the agricultural population which was in evidence at the opening of the nineteenth century was the result of farmers looking for better pasture for their livestock, as well as for new land to till. Both New York and Pennsylvania provided better land for raising cattle than the rocky hills of New England and were settled first in the move westward. Draft oxen and milk cows were widely used in Pennsylvania even before the Revolutionary War.

Cattle were moved next to Kentucky and then to Ohio, where cattle raising had entered into many farm operations before the turn of the century. The Ohio farmer soon began to feed corn to his stock and then drive them in the fall to the markets of Philadelphia and New York. Regular drovers' routes were established to eastern cities with provision for caring for man and beast at rest points. For the first two decades of the nineteenth century Ohio was the major cattle-producing state. Starting in 1825 there were overland movements from Ohio to Michigan for stocking purposes. Native cattle comprised almost the entire reservoir which maintained this western flow of cattle.

Cattle of North America

During the first half of the nineteenth century the draft cattle and the limited number of steers grown out for beef were raised on the family farms in the north and, to a lesser extent, on the plantations in the southeast. While a beginning had been made in beef production in the north, the bulk of the country's beef was a product salvaged from draft and dairy stock.

Some preference had been given to certain types for the family milk cow and also to those used for work, but, except for the isolated groups in Louisiana, Florida and the Southwest, the Native cattle comprised almost the entire bovine population. They were extremely useful and adaptable, however, and fulfilled well their functions of drawing the plow and keeping the household supplied with milk. At the close of the first half of the nineteenth century nearly three-fourths of the United States population lived on farms, and cattle raising was fundamental in the rural economy.

The Deep South—Alabama, Mississippi, and Louisiana—was not considered good cattle country in the days of western expansion, but as settlers moved out of the Carolinas and Georgia to the virgin lands along the Gulf of Mexico, they are said to have encountered herds of wild "black Spanish cattle," as well as wild hogs and horses. Neither this stock nor that which the first wave of pioneers brought with them entered very extensively into the agricultural development of the South. Slave labor was used for cotton cultivation, and draft power was provided by mules instead of oxen. Horses were generally used for transportation. Before the Civil War cattle were being driven from Texas for slaughter in New Orleans. While some plantations had small herds of cattle, many planters did not even own a milk cow. This is not to say, however, that the South did not have a sizable cattle population, but for the most part it was in the hands of small farmers. In 1840 the four eastern Gulf Coast states had 2.5 million head, one-sixth of the United States total. By 1860 this population had increased to 3 million but by then comprised only one-eighth of the national herd.

Throughout the first half of the century the western movement of cattle paralleled that of the human population. By 1850 the Illinois farmer had started his barnyard feeding operation and found cattle a good way of marketing the corn. The Chicago packing plants which had started in a small way in 1845 were killing 25,000 head a month in 1852. Homesteaders were settled in areas west of the Mississippi River and in eastern Iowa and Kansas by the mid-1860's. All took Native cattle with them. Horses had practically replaced the draft oxen in the east, and the settlers moving westward took horses as well as ox teams. Milk and beef were becoming important to the expanding economy.

In the days before the Civil War the American farmer became cognizant of the more productive types of cattle which had been developed in England and had read how breeds had been segregated and developed for specific purposes. Thomas Jefferson brought over some stock from France, and Henry Clay was perhaps the first Hereford importer. The first Shorthorn herd in the United States was founded in 1817. Continuity of the Hereford breed was established with an importation from England in 1840. The founding herds of the dairy breeds arrived in this same period, starting with the Guernsey in New Hampshire in 1830. With the possible exception of the Shorthorn, none of these foundation herds generated sufficient numbers to have much effect on the country's cattle population until after the Civil War.

Men of means in the cities, statesmen, and other influential citizens became seriously interested in agriculture, and their interest often centered on cattle. This involvement in the husbandry of cattle where immediate financial return was not the prime consideration followed the pattern of the landed gentry of England in the development of the English breeds.

The desire for better stock increased following the Civil War. When normal shipping was resumed, heavy importations of purebred cattle, at the time often referred to as "thoroughbreds," were soon underway. This began the second major movement of northern European breeds to the United States, most of which came initially from England and then from Scotland and the Continent.

During the post–Civil War years the expansion of the cattle business in the settled parts of the country remained on the family farmstead. The farm had outgrown the subsistence-level operation that the vanguard of pioneers had carved out of the wilderness two centuries earlier. Fluid milk was produced in quantity on farms near the cities. Farther out in the country the milk was churned and the butter traded for staples in the village store. The dairy herd, however, was only one facet of the general farm operation, which usually included grain, hogs, chickens, and possibly fruit.

Toward the close of the century the specialized dairy farm came into existence around the large cities. There milk was the sole farm product, and was sold to a pasteurizing plant, which marketed bottled milk and milk products. In some areas that were well situated for dairying, such as Wisconsin, manufactured milk products led to local dairy-farm concentrations.

Cattle of North America

As horses replaced the oxen for draft on the farm, beef became a primary product instead of the result of salvaging a no longer useful ox or cow. Steers were fed grain as well as hay to produce a heavier and more palatable carcass. For many years this grain feeding continued as an on-the-farm enterprise and paved the way for the later day feed lot. Such feeding operations gradually increased in size, and by 1900 there were farmers with 100 head or more on feed at one time. Most farm feeders of this era ran light hogs with the cattle to utilize the undigested corn in the droppings. The common ratio in the corn belt was one 80-pound pig to two steers on a full feed of corn. The pig needed no additional grain beyond what he could pick up.

The growth of the purebred herds went hand in hand with the expansion in dairy- and beef-cattle husbandry during the last half of the nineteenth century. All the major breeds now common to the United States became established during this period. The increased productivity of the progeny of the Native cow bred to a purebred bull fascinated livestock people over the country. The milk cow was surpassed in production by her daughter, and the nondescript cow that had produced the draft ox brought forth a calf that was heavier and better fleshed than its dam. The influence of hybrid vigor may not have been appreciated, but the benefits of the purebred bull were obvious. There was also the glamour, as there is today, in new animals of foreign origin. These were the factors which enabled the purebred breeder to gain a dominant position.

The farmer who turned to the production of purebreds toward the end of the century enjoyed a seller's market for several decades. There was a strong demand for cows and heifers, as well as bulls, and the livestock shows which were gaining prominence provided both a showcase and a competitive atmosphere for the promotion of their sale. The competition engendered by the show ring in its early decades led to genuine improvement in both beef and dairy animals. The milk cow also had to compete productionwise in the quantity of butter and milk she produced, an incentive that was lacking in beef-cattle raising.

Northwest

Long before the northwestern boundary between the United States and Canada had been established the Hudson's Bay Company had introduced a few head of cattle into Oregon Territory (present-day Oregon, Washington, and Idaho) for provisioning some of its outposts. By 1836 the Hudson's Bay posts were said to have 1,000 head of cattle west of the Rocky Mountains, most of which, however, were in Canada.

The emigrants to the Northwest first settled in the Willamette Valley in western Oregon in the 1830's. The land not needed for cultivated crops was ideal for raising cattle. A herd of around 1,000 head of Native cattle was trailed from St. Louis to the vicinity of Walla Walla, Washington, in 1834 by a band of missionaries. It was the first sizable drive of cattle to the area. In what was called the "Great Cattle Drive" to Oregon, 729 head of Mexican cattle left the San Francisco Bay area in 1836 and, after much trouble with the Indians, reached the Willamette Valley with 630 head. Most of the herd had been bought around San Diego for $3 a head by a company which the Oregon settlers had organized to buy and drive cattle up from California.

There were a few subsequent drives of Spanish cattle from California to the Oregon country; one of 600 head in 1841 and another of 1,250 head in 1843 are mentioned. By this time large migrations from the East had started—1843 was the year of "Oregon Fever." The cattle which came with the wagon trains were soon recognized as superior to the "long-horned, light-bodied, half wild Spanish cattle."

In 1845 people were arriving from the East "by the hundreds and cattle by the thousands." Then, according to the records of the time, by 1860 people were coming "by the thousands, cattle by the ten thousands." Such reports may have been exaggerated to some extent, but the Indian agent at Umatilla, on the Columbia River west of the Cascades, recorded 6,449 people, 9,077 oxen, and 6,518 cows passing his station in 1853. The inflow of Spanish cattle from California soon ceased as the Native cattle from the East supplanted them. It seems certain that the limited number of Spanish or Mexican cattle which reached Oregon by the middle of the nineteenth century could have left no significant influence on the cattle population of that area.

The Willamette Valley was the Mecca at the end of the Oregon Trail. Here were the first concentrations of cattle. Rainfall was ample for good growth of the native grasses, and the winters were moderate. The herds established served to stock much of the Northwest in later years. By 1857 cattle were being trailed back to California from Oregon in considerable numbers, a movement which started with the gold rush of 1849. By 1871 cattle were being driven eastward, both to market and for breeding stock in eastern Oregon and Idaho. Later these drives reached farther eastward and helped establish range herds in Wyoming and Montana.

As early as 1885 purebred cattle from the East were driven to Oregon and employed in upgrading the Native cattle. This process followed the same pattern as in the western plains, with the introduction first of Shorthorn bulls and later of Herefords.

During the early days in the Oregon country the lack of markets was a major handicap to the cattle industry. The number required for home consumption was small, and the long distances made it impractical to trail cattle to the population centers east of the Great Plains. The good market for breeding cows in Wyoming, Montana, and Nevada disappeared after a few years as the ranges became stocked. A reliable outlet did not develop until the railroads became established.

The Mormons arrived at the site of Salt Lake City in 1847, two years before the gold rush to California. They were more oriented to the horse, and their initial emigrations included twice as many horses and mules as oxen and cows. The trails to the Northwest and to California forked in Mormon territory, and the Mormons became adept cattle traders as well as horse dealers. Oxen and horses that had gone lame on the trail had to be replaced at once, and the incapacitated animals could be acquired cheaply, brought back to a useful condition in a short time, and sold at a profit. Such seems to have been the extent of the Mormons' role in the early range operations in the West.

Until 1883, when the Northern Pacific closed the last transcontinental rail link, Oregon was largely isolated from the rest of the United States. For years cattle had been a drug on the market and yet continued to increase in number. Many ranchers and farmers began raising horses for which there was a demand to meet the needs for transportation and railroad construction. From the early 1880's, with the opening of rail transportation to the East, until the end of the century was a period of rapid expansion in the cattle industry of the Northwest.

The eastern market absorbed cattle so rapidly that in 1888 there was again movement of cattle from northern California to Oregon. These cattle were longhorn-type animals, mostly four-to-five-year-old steers for slaughter, and therefore had no influence on the breeding stock.

As the demand for cattle continued to increase, the cattlemen moved into the desert and semiarid regions of central Oregon. There shorter grass, periodic droughts, and harsh winters were experienced. At first cattle were wintered without hay or other supplemental feed. Heavy losses soon forced the cattlemen to put up hay, and this led to irrigation in many instances.

Large ranches came to control the range areas, some so extensive that roundups were continuous through the summer. Cows were gathered over a wide area, the calves were branded, and the roundup crew then moved on to the next station. Three-to-five-year-old steers were taken off in the fall for sale. The cow herds were numbered in the thousands and received very little attention.

Toward the end of the century dairy herds were established around the centers of population. Native and Shorthorn cows were principally used for milking, and there was not much interest in the purebred dairy breeds until the end of the century.

Southwest and Western Plains

As the Spaniards moved from the West Indies to Mexico during the sixteenth century and up to what is now Texas, they took their cattle with them. The husbandry practices imported from the homeland were modified to the extent demanded by the new environment. The cattle were handled by men on horseback, and all the accouterments of the American cowboy—the chaps (*chaparejos*), the lariat (*la reata*), the boots, and even the western saddle (introduced to Spain by the Moors) —originated with the conquistador and his retainers. Cattle were hot-iron-branded with the mark of the mission or the coat of arms of the *don ranchero*.

In 1598, nearly 60 years after Coronado had explored as far north as Colorado, Juan de Oñate took a reported 7,000 head of cattle across the Río Grande and led his expedition into what is now New Mexico, Arizona, and southern California. A few missions were founded, but because of the hostility of the Indians it was more than a century before they were permanently established. During the eighteenth century ranches, as well as missions, were developed in the area south of San Antonio in Texas and in scattered locations across New Mexico and Arizona to southern California. Some cattle were taken by ship from the peninsula of Lower California to San Diego, but for the most part even that area was stocked from the herds of northern Mexico.

The Spanish missions were the principal cattle owners until Texas seceded from Mexico, and Indian converts were trained in the handling of cattle. Most of the ranches established east of California had only meager facilities and small herds. In contrast, California ranches rivaled the missions in the size of their operations.

Following Mexico's independence from Spain in 1821 settlers from the southern states began to migrate to the sparsely populated region that today is Texas. They became the cattlemen of the Southwest. The movement from the more settled parts of the South accelerated after the Republic of Texas emerged in 1836.

After the annexation of Texas in 1845 and the purchase by the United States of California and the adjoining areas from Mexico in

1848, the old Spanish missions soon were abandoned, and cattle raising became the province of the rancher. The methods of husbandry continued much as they had in the past. Cattle roamed the open range with little attention from man, except for branding, slaughter, and moving from one area to another as necessary for forage or water or to counter an excess of predators. In California the major products of the cattle business were hides and tallow. The same was true in Texas, where there was probably less of a market for such cattle products than in California. Wild cattle continued to multiply throughout the Southwest.

When the future Texas cattlemen began arriving from the southern states, the cowboy replaced the Mexican vaquero. Occasionally markets for live cattle opened. Cattle were trailed to New Orleans in 1842, while Texas was still under the Mexican flag. In 1846 a drive of 1,000 head was made to Ohio, the first recorded movement of cattle from Texas to a northern state. During the early days of the gold rush to California, there were a few drives to that state. The first drive of Texas cattle to Chicago was made in 1856. Such movements of cattle for slaughter were minor, and the cattle in South Texas multiplied rapidly. Numbers in those days, however, were often wildly exaggerated. One report estimated that in 1860 there were more than 5 million head in the Neuces Valley alone, mostly in a wild or semiwild state. The USDA estimate of the cattle population of 1860 of all of Texas, however, was only 3.5 million, which probably was considerably underestimated.

With the onset of the Civil War the Texas cattle movements to the North, which had begun to show considerable promise, came to an abrupt halt. After the war the drives to the North began again and supplied the expanding eastern markets. It took time for shipping facilities at the rail terminals in Kansas to be built, and the experience essential to driving a few thousand head of semiwild cattle across an unknown land menaced by hostile Indians as well as white outlaws, had to be acquired. Before long, however, cattle were moving north at a faster rate than they had before the war. Soon the southwest chaparral was depleted of wild cattle, and range animals from the large ranches took their places in the northern drives. Recognized cattle trails came to be known, and drives to the Kansas railheads became routine.

Branding the owner's mark on his cattle was a practice that had been introduced to Texas with the first Spanish cattle trailed north from Mexico. The wild cattle which had broken off from the mission herds and multiplied rapidly belonged to whoever applied the first hot iron to them. Calves were marked with the brand of their mothers at the roundups. Many were missed, however, and ownership of slick (unbranded) cattle was established by the first man to rope and apply his brand to them. Recognition of a brand as legal evidence of ownership of cattle became general throughout the western-range country and has continued to this day.

The boom days of the western cattleman began in the early 1870's. The spring drives of steers to Kansas markets were matched by movements of breeding cows to Colorado, Wyoming, and Nebraska and into Montana and the Dakotas. Ranching operations expanded rapidly over this immense region, some under the direction of Texans but most under that of newcomers. Many of them had the sound financial backing of British and other European investors. Initially much the same management practices as had developed in Texas were employed, though they were eventually modified to meet the requirements of the colder climate.

The drives which drained the surplus cattle from Texas began in 1866 and continued for 20 years. Volumes have been written romancing the trailing of the Texas Longhorns to the wild-west towns of Kansas, but to the men engaged in the operation it was mostly hard work at times involving some nasty episodes.

Many a rider drowned at a flooded ford, and there was the constant danger of a stampede, which could start seemingly for no reason at all. A blowing tumbleweed or a lone coyote's howl might send 2,000 head of crazed cattle on a dead run until exhaustion overcame them. The stampede induced by a revolver shot or even a whistle was a favorite device of cattle thieves and Indians to scatter the herd for easy picking. The night herder was often trampled under the hoofs of the stampeding beasts, and a shallow grave was all his comrades could provide for him.

The most serious problem on the northern drive was often the antagonism of the Missouri and Kansas farmers. After the first few drives had passed through a settled area, armed farmers often blocked further movements. The Texas cattle carried the fever tick, which was transmitted to northern herds. With no built-in resistance against the disease, the death loss among the northern cattle ran extremely high. A Missouri farmer may never have seen the herd which caused the outbreak among his cattle, but he and his neighbors were out with their shotguns when the first trail herd appeared on the horizon the next year.

The economics of the cattle drives of the 1870's were based on a price of $10 for a steer in Texas, compared with $25 at the railhead in Kansas. The starting point was any location in Texas where a herd of 2,000 to 2,500 head could be gathered. They could be the branded property of one or more owners, newly branded animals acquired without a bill of sale, or, in the days of the early drives, completely wild

Two-span ox cart in Laredo, Texas. 1904. National Archives Photograph.

cattle which had been held for a period in the company of domesticated animals to condition them for the trip north. The terminus was usually the closest railhead in Kansas. In 1867 it was Abilene; from there it moved westward to Wichita, Ellsworth, Ellis, Dodge City, Caldwell, and finally, in 1880, Hunnewell.

In the closing years of the 1880's the drives dwindled to almost nothing. The huge surplus in Texas had disappeared; the steers had gone to the eastern slaughterhouses, the cows to stock the northern ranges. Estimates of the total number of cattle moved are only guesses, but the figure often used is 10 million over the 20-year period. After that time the Texas cattleman began to settle down, upgrade his Longhorns, and raise cattle for a more normal market.

As the cattleman became established on the plains north of Texas, two major changes in management occurred: the upgrading of the Longhorn and the practice of winter feeding. In the late 1870's bulls of the British breeds were brought in to upgrade the Longhorn. In many areas there was a two-step development—the Shorthorn (called Durham at the time) supplanted the Longhorn, and later the Hereford bred out the Shorthorn.

United States

In the Texas cattle country there was usually grass for both summer and winter feed if the range was not overstocked. Even if grass was short, there was "pickin' " sufficient to carry a cow through the winter. As the first breeding herds moved north, cattle were brought through the winter on whatever standing dry growth remained at the end of summer. In some areas, with the range conservatively stocked, cattle could be wintered in this manner and show up in fair condition in the spring.

In 1881, however, the cattlemen on the northern plains received a warning that went unheeded. Cattle were unable to graze because of heavy snow, and there were serious death losses. Then came the winter of 1886–87, which was far worse. Many individual losses were estimated at one-half to two-thirds of an entire herd. Large cattle companies as well as individual owners were wiped out. Provision for winter feeding became standard practice thereafter.

Throughout the northern-range country (from the Rockies to the eastern edge of the Great Plains and north to the Canadian border) three factors determine the necessity of winter feeding: elevation, wind, and natural protection from storms. At higher elevations snow is often much deeper and temperatures lower, adverse conditions for cattle survival. Prevailing winds can keep a range open most of the year so that cattle can graze in a winter with heavy snowfall. Neither a low elevation nor strong prevailing winds, however, will permit cattle to survive a severe winter storm unless good natural protection is available. All that is needed is a creek bottom with a good growth of trees, a cover of brush, or a sheltered area in the lea of a line of hills. Even today in the range-cattle country, man-made shelter for the breeding herd is practically unknown. Given the proper combination of the essential conditions mentioned, cattle can survive a severe winter satisfactorily without winter feed or, at most, with only a small supplement of "cake" (ground grain compressed to a hard cube which can be fed on the ground with little loss).

The free grass that grew on the public domain was the base for the rapid expansion of the range operations in the last decades of the nineteenth century. With railroads across the country, the Indians on reservations, and land along the watercourses taken up by ranchers, cattle raising became a more settled operation. The collapse of the cattle boom in 1885 foretold the end of the big Texas drives to the North and hastened the day of the settled stockmen in the West. A nucleus of deeded land became essential to the rancher, although grazing on government land at certain times of the year was still necessary for a

Indians receiving an allotment of partially upgraded cattle in New Mexico. Late 1890's. National Archives Photograph.

successful operation. The barbed-wire fence, long the curse of the early cattlemen, became all-important for controlling the rancher's cattle under the changing conditions of land ownership.

Management practices of western ranches in the last years of the century varied to some extent with climate and local conditions but still followed a rather common pattern. The herds of several owners were frequently intermingled during the summer breeding season on a common open range. This summer range was in high country where the rainfall was sufficient for a good growth of grass, and typically was on government land. The owner of a large "spread," a rancher with several thousand cows, however, often managed to hold a particular

Longhorn herd in process of being upgraded. 1890. Kansas. National Archives Photograph.

range for his own use, sometimes excluding others without due process of law. Seasonal roundups were held for branding calves in the spring and again for weaning in the fall, at which time the bulls were taken out.

Where the climate demanded, hay was put up in the summer and fed out during the winter months. This hay was from native grasses, grown on unirrigated land where annual rainfall was upward of 15 inches. In areas of lower rainfall irrigation was necessary to raise hay.

Before the days of the baler, hay was stacked in the open and enclosed in a crib or fence as protection from cattle or game. In the winter it was forked from the stack onto a hayrack drawn by horses and was scattered on the ground in a wide circle so that it could be picked up by the cattle without undue wastage. In the high country a hayrack on sled runners was used when the snow became too deep for wagons.

In the spring cows calved in the open with a minimum of attention. When the grass started to grow, they were trailed to the summer range, and the bulls were turned in.

The cattle were usually sold in the fall and trailed to a rail point for shipment to Kansas City, Chicago, or Omaha. In the early days the cattle so trailed were three-to-four-year-old steers and whatever bulls were no longer useful. As prices gradually increased, some ranchers shipped old cows, but it was usually considered a better gamble to keep an old granny in hope of another calf rather than to accept the two or three cents a pound she would bring on the market. It was not until well after the turn of the century that younger cattle were shipped; and just before World War I yearling steers began passing through the central markets to be fattened in corn-belt feed lots.

Louisiana

When the vast region known as the Louisiana Purchase became a part of the United States in 1803, the southern extremity that touched the Gulf of Mexico (later the state of Louisiana) was the only area with cattle. These were the Spanish type, genetically of the same heritage as the Texas Longhorn and the Florida Scrub cattle. These three isolated populations not only were effectively separated from each other by wide wilderness areas but were even more remote from the Native cattle in the eastern United States. The Louisiana cattle never obtained an identity of their own. They were later absorbed into the main body of Native cattle as the latter moved westward and southward.

The history of cattle in Louisiana traces back to the colonial period, but this background is given here since the area only became a part of the United States in the opening years of the nineteenth century.

The early French settlers had difficulty obtaining foundation stock. The first fort was built in lower Louisiana in 1700, and a community was established in 1714. Efforts were made to import cattle as well as other livestock from Cuba and Santo Domingo; it was doubted that any cattle could survive the ordeal of a drive from South Texas or Florida. The Spanish authorities in Cuba, however, were reluctant to release cattle to the French in Louisiana because of the strong Spanish resentment toward French occupation of the area. Eventually a nucleus of Spanish cattle was obtained, a few head at a time, by shipment either from Cuba or from Florida.

The numbers grew slowly. In 1723 killing any kind of cattle was prohibited by law, as a means of increasing the population, and even as late as 1737 notice of intended slaughter had to be given the authorities. Good husbandry for the times and the cumulative effect of these measures was that southwestern Louisiana prairies were well stocked with Spanish cattle by the middle of the eighteenth century.

The Acadians managed to bring their French Canadian cattle with them in their forced migration to Louisiana in 1755. These cattle were the Brittany strains which the French had taken to Canada in the seventeenth century and maintained there.

In southwestern Louisiana east of the Sabine River there were extensive prairie areas which could support stock, although the grasses were not of the best quality. The Spanish cattle were not a pampered type and soon became acclimated there, reaching sizable numbers before the soil was turned for cotton and rice.

The Spanish method of handling cattle on the open range prevailed. Herding was done on horseback. The first brand book was dated 1739. French terms were given to the various words in the Spanish cattleman's vocabulary. A "vacher" was the French equivalent of vaquero; the establishment running cattle was a "vacherie," which might be defined as a Louisiana hacienda. Herds as large as 3,000 head were accumulated. They required very little care, and because they were well acclimated to an equally harsh environment before reaching the prairies, they did not suffer unduly from heat and insects.

The influence of the northern states began to be felt even before Louisiana was admitted to the Union in 1812. The grasslands in the southwestern part of the state had already felt the plow. Cotton, rice, and sugar cane displaced the herds of Spanish cattle which had gradually merged with the Native cattle that by this time had arrived from the North. The French Canadian cattle of the Acadians also disappeared in the Native cattle. Commercial agriculture with little livestock dominated the economy through most of the nineteenth century and well into the twentieth.

Florida

For several years before Florida became United States territory in 1819, there was considerable activity in obtaining land grants for cattle-raising operations in that area. The Florida Scrub was the only type of cattle known. Cattle were run under open-range conditions; they were well acclimated to the heat and parasites and required little attention. Bulls were usually left with the cows the year round, and calves were permitted to wean themselves. Because the Scrub cow usually calved only every other year, calf crops averaged 30 to 35 per cent. There were exceptional operations, however, where 70 per cent calf crops were said to have been obtained.

On the face of it an operation with such low yields would hardly be considered profitable, but the cost of maintaining cattle in Florida during the nineteenth century was practically nil. An Englishman who visited Florida to investigate stock-raising prospects, wrote home that "it costs less to raise a cow in Florida than a chicken in England." Pasture could often be had for payment of the taxes, which were nominal. (As late as the 1930's, pasture was renting for $1 a head a year, the lessee maintaining the fences.) It is obvious that under such conditions ranching could be profitable even with the low calf crop and the small Florida Scrub cattle.

Early attempts were made to bring better cattle to Florida after the Indians were finally subdued in the 1830's. Considerable numbers of Native cattle were taken into the region from Alabama, Georgia, and the Carolinas as conditions became more settled, but these either failed to survive or were soon absorbed in the Scrub population. After the British purebreds began to enter the other states, occasional attempts were made to introduce them to Florida. These efforts, too, were discouraging. One cattleman in northern Florida shipped in a carload of Hereford bulls from Texas and turned them loose on the range that spring. Only three head survived till fall. While this is probably an extreme example, the outcome was not unlike what usually resulted when cattle from outside were taken to Florida.

It was not until the early 1900's that a few purebred herds became established. In general, the British breeds did not make much headway until after the fever tick was eliminated in 1941.

Another factor which delayed the improvement of Florida's cattle was the open-range grazing policy. Fences did not come into general use until the 1930's.

The Zebu was late in finding acceptance in Florida. Four Zebu bulls purchased in Texas by J. S. Turner were taken to Cedar Key in 1880, but nothing of importance developed from them. Little mention is made of the Zebu breeds in writings on early Florida cattle. By 1930, however,

United States

American Brahman bulls were being crossed with the Florida Scrubs and soon became popular.

In the early years of Florida's cattle industry, Cuba was a good market. During the 1840's, 30,000 head were shipped there annually, but as the state became more settled and the winter tourist trade increased, the beef produced locally was far from sufficient to supply the home market. It is only in latter years that production has reached the point that feeder cattle can be shipped out of the state.

THE TWENTIETH CENTURY

Beef-Cattle Husbandry

Population density, climate, and topography are the basic factors which determine the various patterns followed in the raising of beef cattle.

In the accompanying map conterminous United States is divided into six regions. The figures shown represent the dairy-cow population, the beef-cow population, and the percentage of the total national beef-cow population in a particular area. The following table summarizes this information.

UNITED STATES COW POPULATION, 1972

District	Number of Dairy Cows	Number of Beef Cows	Per Cent of Total U.S. Beef Cows
I. Northeast	5,927,000	1,958,000	5.0
II. Southeast	1,452,000	7,778,000	20.2
III. North Central	2,789,000	13,419,000	34.7
IV. South Central	499,000	7,689,000	19.9
V. Mountains and Plains	489,000	5,859,000	15.2
VI. Pacific Coast	1,111,000	1,931,000	5.0
Total 48 states	12,267,000	38,634,000	100.0
Alaska	2,000	2,500	
Hawaii	13,000	89,000	
Puerto Rico	91,000*	70,000*	
Total outside	106,000	161,500	
Grand total	12,373,000	38,795,100	

*Estimated.
Source: USDA, Statistical Reporting Service.

The precipitation and summer and winter temperatures are the elements of climate that have a major influence on cattle husbandry. Elevation is often a determining factor in temperature. The six cattle-raising regions are outlined on the accompanying rainfall, temperature,

ALASKA
2.5 / 2

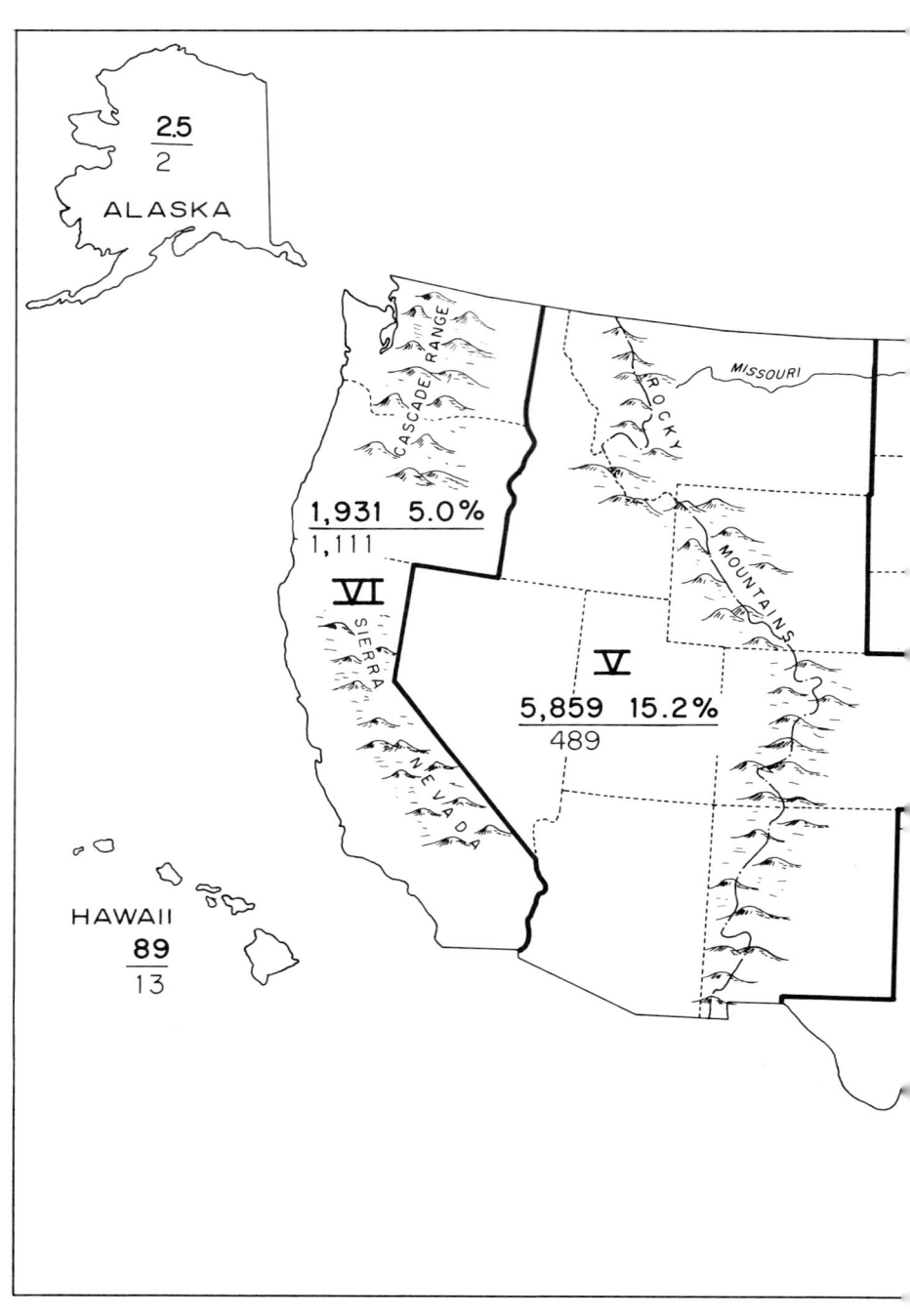

1,931 5.0%
1,111
VI

V
5,859 15.2%
489

HAWAII
89 / 13

CASCADE RANGE
ROCKY MOUNTAINS
MISSOURI
SIERRA NEVADA

The six major cattle-raising regions of the United States, drawn to coincide with areas in which USDA cattle-population estimates are available. The figures in boldface type are the numbers of beef cows in thousands. The figures in lighter

Key:

Number of Beef Cows in thousands.	Per cent of total Beef Cows in conterminous U.S.

Number of Dairy Cows in thousands.

I

$$\frac{1,958 \quad 5.0\%}{5,927}$$

I

III

$$\frac{3,419 \quad 34.7\%}{2,789}$$

MISSISSIPPI RIVER

II

$$\frac{7,778 \quad 20.2\%}{1,452}$$

IV

$$\frac{89 \quad 19.9\%}{99}$$

Zebu Line →

PUERTO RICO

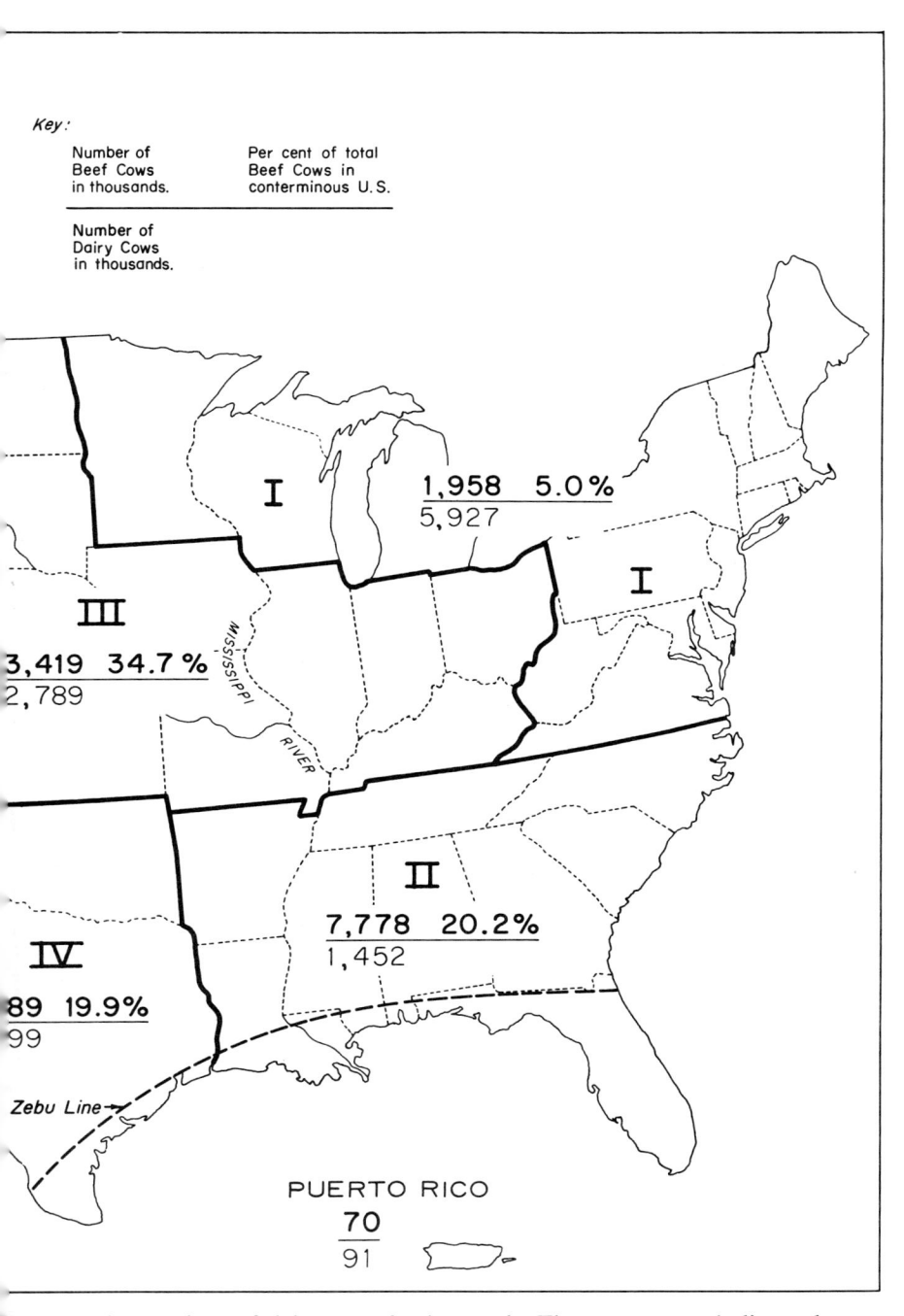

$$\frac{70}{91}$$

type are the numbers of dairy cows in thousands. The percentages indicate the per cent of total beef cows in the conterminous United States.

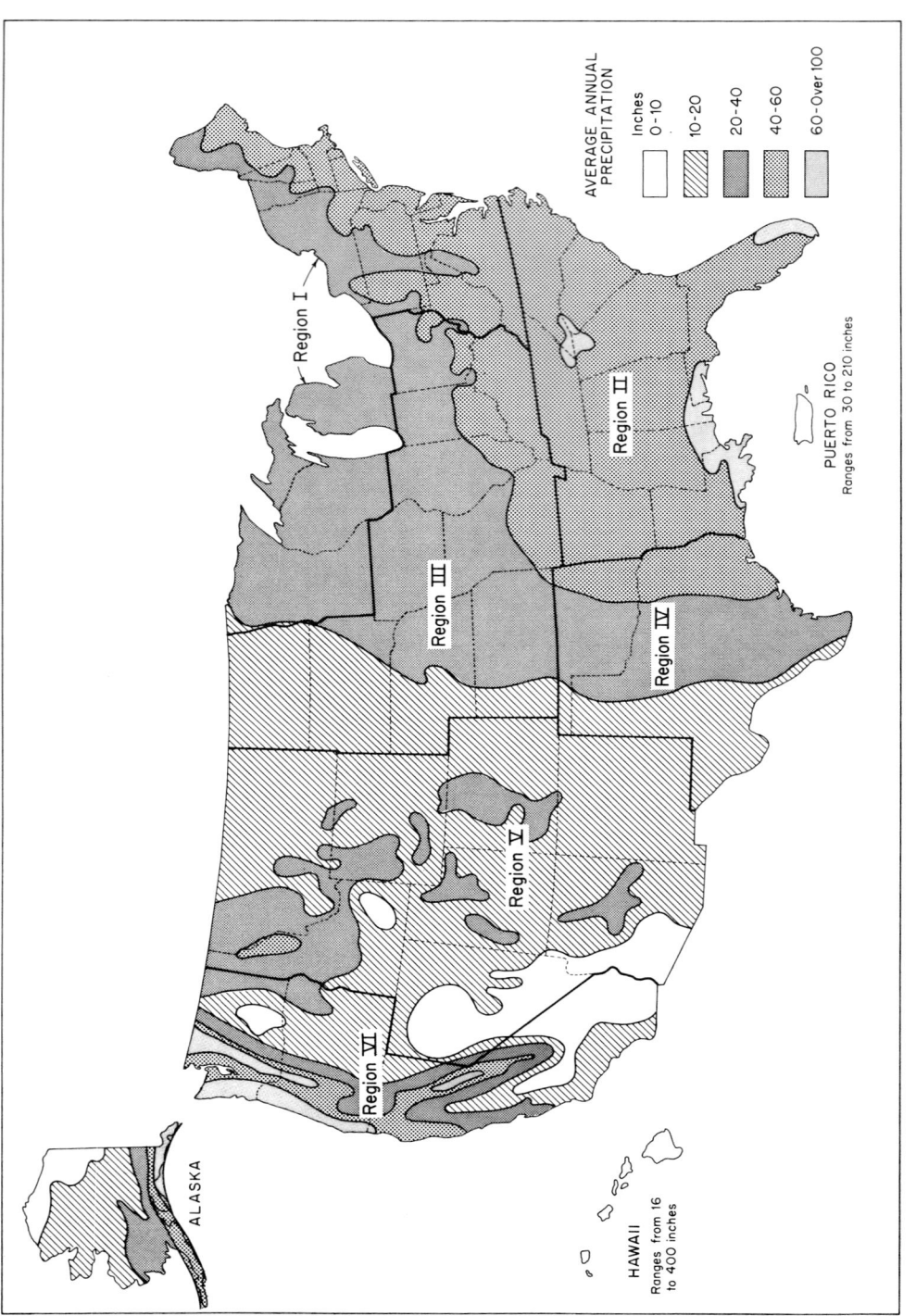

Precipitation map of the United States (major cattle-raising regions indicated).

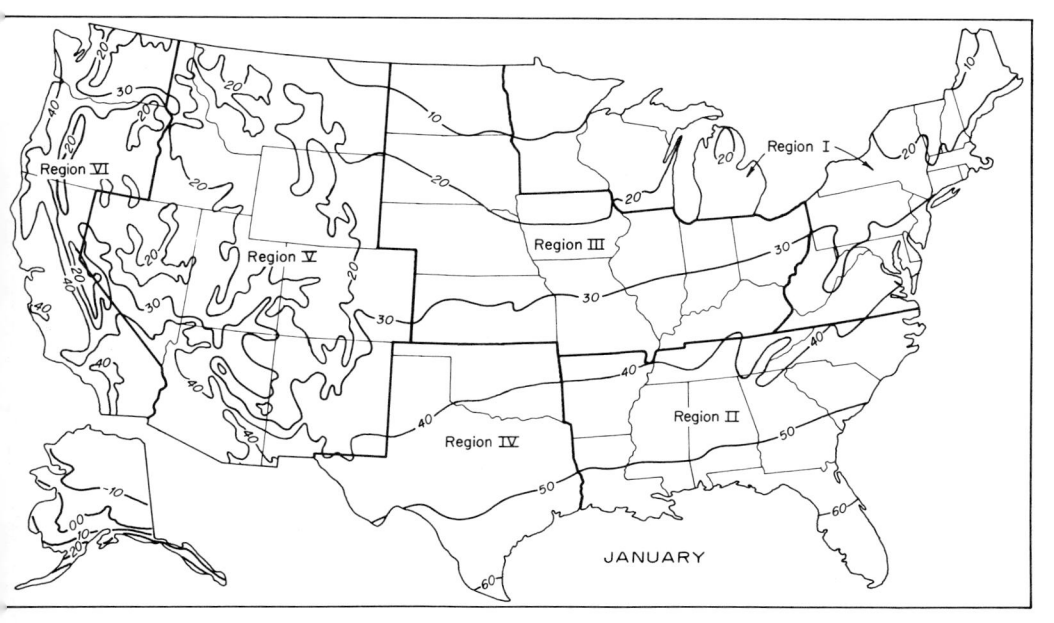

Mean temperature charts of the United States, January and July (major cattle-raising regions indicated). Based on maps of U.S. Geological Survey.

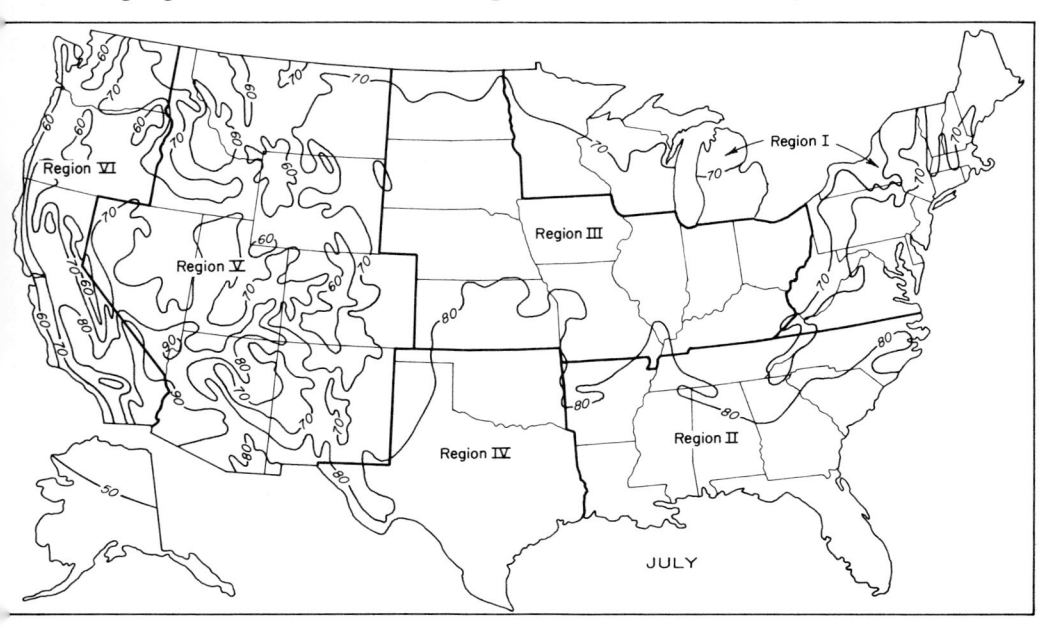

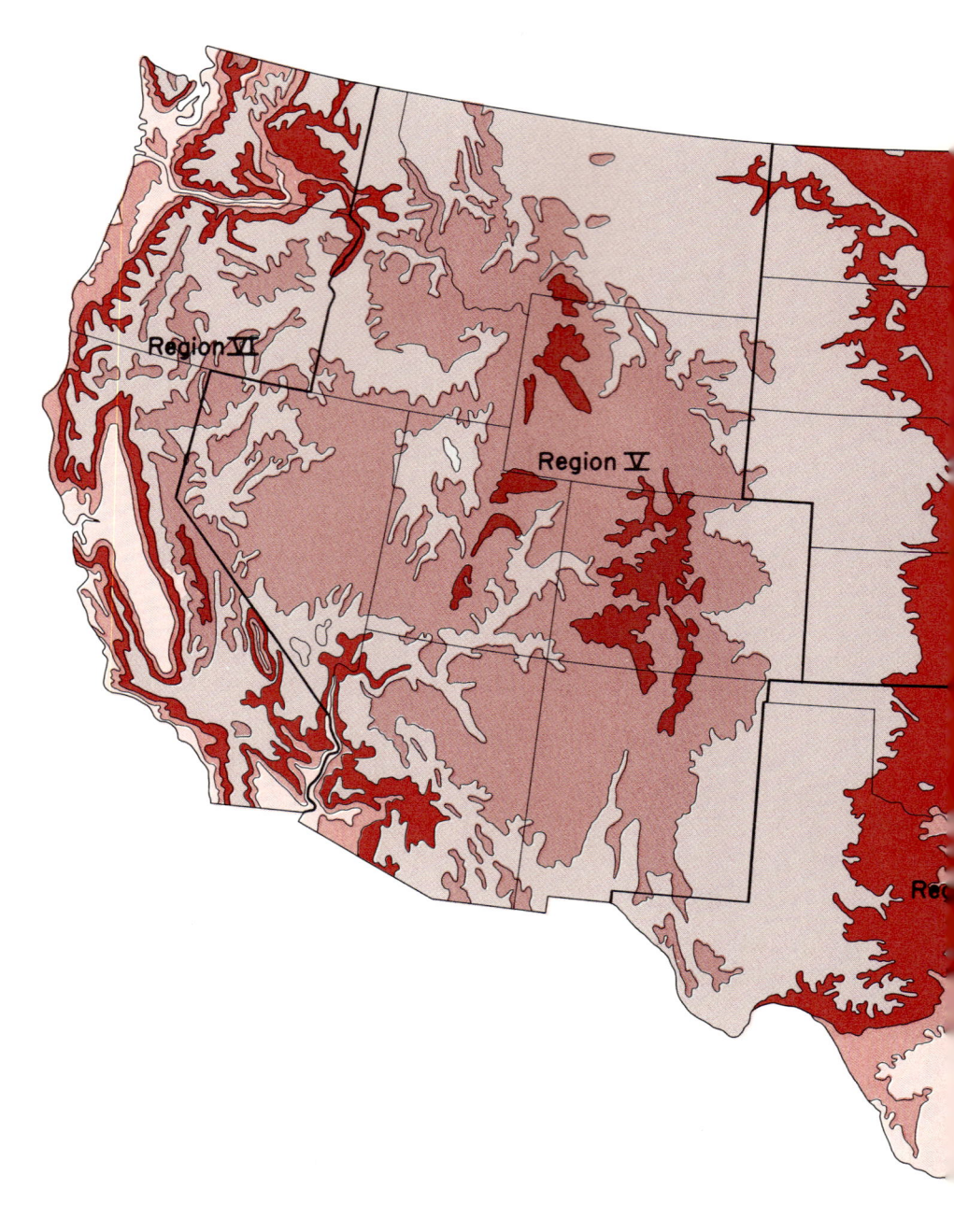

Physiographic map of the United States (major cattle-raising regions indicated). Based on map of U.S. Geological Survey.

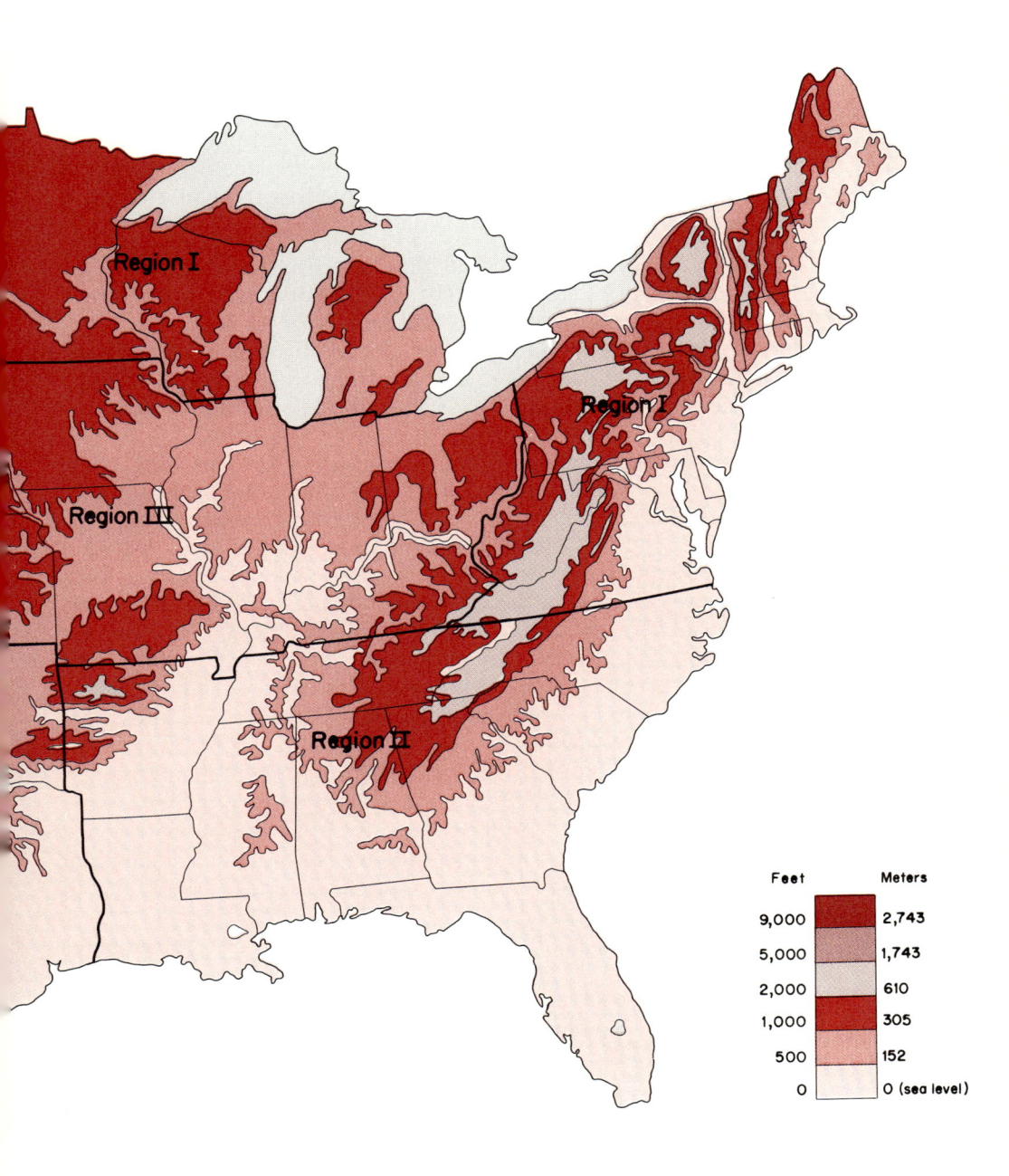

Cattle of North America

and elevation maps. They give a hint of how physiography and climate influence the handling of cattle.

The last two states to enter the Union—Alaska and Hawaii—and Puerto Rico, because of their isolated locations, cannot be indicated in the regions in which the other states have been placed. A section at the end of "Management Practices" is devoted to the cattle of these three political divisions.

I. *Northeast Region.*—Throughout the New England states and the Great Lakes area and south to North Carolina dairy cattle predominate. Beef herds are typically one- or two-bull units of 15 to 40 mother cows. Both calves and yearlings are marketed, depending on local conditions. Most of the beef cattle of this region are Herefords, although in some areas Angus herds predominate, as in Virginia. The Shorthorn representation is small.

There is scattered evidence of crossing with Charolais, but, while crossbreeding accounts for 10 to 20 per cent of the total beef calf crop, there is no evidence of any trend to a definite crossbreeding system.

The small beef herds which predominate are either incidental to a general farming operation or are held by an owner who works in town and runs a few cattle as the easiest way to obtain some return from his land. There are a few purebred herds of excellent type and performance in the region and many mediocre ones. Some large herds are owned by business or professional men, but, in spite of their size, these operations can usually be put in the hobby classification. There is very little large-scale commercial production.

The beef herds are often run on small pastures to utilize land unsuited for cultivation. Row-crop residues are also used. Hay, often with some grain, is fed to young stock in the winter, for five to six months in the northern part of the region and for three to five months in the south. Cattle are usually given some winter protection, such as free access to a shed or barn.

There is some finishing of steers in barnyard feed lots to slaughter weights of around 1,000 pounds, but the usual marketings are either calves or yearlings. Most sales are made at local auction barns.

II. *Southeast Region.*—Before the days when cotton dominated the agriculture of the South, half of the cattle in the United States were grazed in this region. There were more cattle in Georgia and in the Carolinas in the 1850's than there were in those states a century later when the expansion of the cattle industry in the South began.

As the world demand for cotton declined, the least productive cotton land reverted to, or was developed as, pasture and stocked with cattle. After the control of the fever tick in 1941 and the screwworm in 1966, all of the Southeast became potentially good cattle country. The mild climate and good rainfall, modern practices for insect and parasite control, and better management in handling the cow herd probably made beef production more profitable there than in any other region of the United States. In 1970 the Southeast had 30 per cent more cows of beef type than the traditional cattle country of the Mountain and Plains States.

The northern European breeds can be maintained satisfactorily north of the "Zebu" line, shown in Figure 10. Below this line such breeds require increased care, and their productive performance is generally surpassed either by Zebu breeds or by crossbred cattle carrying some Zebu influence.

North of the Zebu line herds are small in the East, 20-to-40-head size, but grow larger toward the West. In Arkansas herds run from 20 to 80 head. Only a few number more than 100 head. In Louisiana herds are larger, many with several hundred, some even a few thousand, head.

More cattle are marketed as calves every year though the small growers are still inclined to carry their calves through the winter and sell them as yearlings.

From the Southeast Region, especially the Gulf Coast states, the calves and light yearlings move westward to be grown out on wheat fields and other pasture before going to the feed lot. This pattern provides a good market for the young stock from the area and enables the seasonable forage in the West to be used to the best advantage.

Beef herds have been established on farms where cotton remains the major crop but where the less productive land has been converted to pasture. There are also properties running several hundred head of brood cows where all the formerly cultivated land has been converted to pasture and cattle are now the only farm enterprise. Good management is often evidenced in these larger operations by rotation of pastures, control of breeding seasons, and rigid parasite and insect control. In many small herds, however, cattle are allowed to shift for themselves, and bulls are run with the cow herd the year long.

Cattle are grazed throughout the year in this region. This may be on either improved pastures of coastal or common Bermuda grass, Dallisgrass, lespedeza, ladino clover, or other good semitropical forage crops, or merely the regrowth of wire grass or other less productive plants which return after the cultivation of cotton is stopped. In good areas of heavy soils one and one-quarter acres of improved pasture will support one animal unit for a year, while native grasses require up to six acres per animal unit.

The practice of putting up hay or silage for winter feed varies with the local climate. In the northern part of the region hay or silage usually supplements the pastures for two or three months during the winter. Farther south, little forage is stored, but a pound or two of grain per day is fed during the few winter months that the plant growth is inhibited. Salt-controlled feeds are used to a limited extent.

Local breed preferences vary, but in general the straight British breeds predominate north of the Zebu line. Below this line the beef herds take on a more nondescript appearance, and the Zebu influence is prominent.

Throughout most of Florida, the southern third of Louisiana, and the extreme southern parts of Mississippi and Alabama, most cattle show Zebu influence or are of a Zebu-type breed, generally the American Brahman, Santa Gertrudis, and Brangus. In Florida most of the herds are an American Brahman–Hereford mixture of no fixed proportion, running from ⅛ to ¼ Brahman. Some of these present a definitely nondescript appearance while others are uniform and might be called Brafords. The cattle raised in Florida are generally smaller than those of similar quality and breeding in areas farther north. Even with the tolerance a degree of Zebu breeding affords to heat stress, insects, and parasites, the harsh Florida climate still takes its toll of any bovine other than a Zebu.

North of the Zebu line crossbreeding is seen to a varying degree. The crossbred cattle are usually Angus-Hereford or Charolais crosses on the straight British breeds. The Charolais cross is prominent in the small herds in South Carolina. Some scattered upgrading with Simmental and Limousin is also seen. Over all, probably 25 per cent of the beef-cattle production north of the Zebu line is today the result of crossbreeding. Few of the crossbred cows are more than the F_1, and there is little evidence of planned crossbreeding programs.

III. *North Central Region.*—The area from Ohio westward to Kansas and Nebraska and north through the Dakotas was the cradle of the beef-cattle industry of the United States. In this subregion east of the Mississippi River the Native cattle were upgraded to the British breeds. It was here that the production, marketing, and slaughtering of cattle for beef became a major industry during the last half of the nineteenth century.

The modern feed lot had its inception in the barnyard feeding operations of the middle western farmers, who fed out home-raised steers to weights of at least 1,400 pounds and sent them to the Chicago stockyards. This region has produced over one-third of the total number of cattle marketed for beef in the United States since the beginning of the century. In addition to the beef cattle raised there, large numbers of feeder cattle are shipped in for fattening.

With the exception of the extreme western part, the North Central Region has the good soils, moderate climate, and normally adequate rainfall, that make it one of the best cattle-raising areas of the world. The excellent farming conditions and the demands on the soil for cultivated crops limit the expansion of breeding herds. From Ohio westward to eastern Kansas and Nebraska cattle usually play a minor role to the main farm operation. Most beef-cattle herds number no more than 40 to 50 brood cows with many in the 15 to 20 range.

In general, the agricultural units become larger toward the west and herds increase in size. In the Sand Hills of Nebraska a typical ranch consists of 6,400 acres and runs 400 to 500 head of cattle. Large operations are also found in the Flint Hills of Kansas and through the Dakotas to the Canadian border. These and many smaller units are often exclusively devoted to the cattle business, and the cattle provide the principal source of income.

In recent years there has been a movement of large-scale beef operations into areas of the corn belt where the terrain and soil are poorly adapted to the cultivation of row crops. During war years and other periods of heavy demand for grain much marginal land was cultivated and cropped. Government conservation and crop-control programs have resulted in the restoration of large acreages of such land to pastures in a pattern identical to that of the conversion of marginal cotton lands in the southern states. This has permitted the development of sizable beef-cattle operations. In Missouri some producers are running upwards of 1,100 head of brood cows.

Beef herds in the North Central Region are pastured from spring until fall and are fed hay or silage during the winter. Cows are bred to calve in the spring, and probably more than half the growers now sell calves in the fall. The remainder, however, continue to sell yearling steers, which have been wintered on crop residues, wheat pasture, and grass and then taken to the feed lots during the summer or fall. In the eastern part of the region barnyard protection is often provided in the winter. West of the Mississippi River cattle are usually in the open the year round, during the winter running in pastures which afford natural protection of brush or river bottoms.

Throughout the general farming areas better management has followed the trend in recent years to larger farms. Better utilization of crop residues, especially cornstalks, and the conversion of marginal crop land to pasture allowed for an increase in cattle numbers. The corn combine has been a heavy contributor to wastage of the cornstalk, one acre of which with a little protein meal will winter a cow. Methods of salvaging this feed are now being adopted, and the corn belt stands in an excellent position to increase its cow numbers.

The dairy herds have increased in size, but the number of herds has declined. The total number of dairy cows was reduced by 5.5 million head from 1951 to 1971 which helped to make room for the increase of 9.5 million head of beef cows that occurred over the same period.

Hereford and Angus cattle are the major breeds. In some areas a strong preference for one or the other breed prevails. In parts of Missouri there is a large number of Angus herds. Throughout much of the Dakotas there is a large majority of Herefords. Considerable evidence of crossbreeding is seen scattered through the region, though there is less in the small herds in the east.

In 1971 the cattle in the sales barns of the western part of the region were running 25 per cent crossbreds. Most of these were Angus-Hereford, followed by Charolais crosses on either Angus or Hereford. Herds large or small with a definite crossbreeding plan are practically nonexistent outside the experiment farms of the land-grant colleges.

Those states of the region lying east of the Mississippi River were the home of the purebred breeder until after World War I. Since that time the purebred industry has spread westward and is now general throughout the country. The Middle West, however, is still a stronghold of the purebred breeder.

IV. *South Central Region.*—Oklahoma and Texas, the two states which comprise this region, produce one-fifth of the beef cattle of the country. Before World War II these states had nearly 3 million cows of the beef breeds, which comprised one-fourth of the national herd. By 1972 the number had increased to nearly 8 million head but was then only one-fifth of the beef-cow population.

The cattle country of Texas lies west of the 97th meridian, which passes between Fort Worth and Dallas. Known as the High Plains and the Southern Plains, it is the traditional range country of the Southwest. There are good grasslands over much of Oklahoma, often in land too rough for cultivation, and in the western part of the state are large areas in wheat which provide winter pasture for feeder cattle. Except in the extreme western part of Texas, good grass can be expected in normal years, and cattle can survive a winter without hay.

Below San Antonio is South Texas, the area where the Longhorns were first consolidated into herds and western ranching began. Summer temperatures are higher, and insects and parasites take a heavier toll of cattle. Before the fever tick and screwworm were controlled, the Longhorn and then the Zebu were the only cattle that could adapt well to South Texas. The extreme western part of Texas directly south of New Mexico, where semiarid conditions prevail, is also hard country for cattle.

In eastern parts of Oklahoma and Texas rainfall is more abundant, and cultivated crops are the major farm enterprise. There cattle are usually held in small herds and supplement the farm income to a minor extent. Where the terrain is too rough for cultivation, as in the Ouachita Mountains of eastern Oklahoma, there are small ranch-type operations, the largest of which seldom has more than 100 brood cows.

Throughout the region the size of the ranch unit increases to the west and south. Rainfall decreases, more land is needed to support a cow, and more cows are necessary for a profitable operation. The cattle empires of the early twentieth century which spread over the vast Southern Plains with only a few exceptions have been broken up into smaller holdings. Ranches in South Texas have also become smaller, although there are many operations running 1,000 to 2,000 brood cows on units of 15,000 to 25,000 acres. The fabulous King Ranch, still in the 1-million-acre range, stands as a monument to the immense cattle enterprises of the past.

Income from oil royalties made it possible for many of the large Texas ranches to weather the lean years and hold their spreads together. Through the region as a whole, however, the herds now probably average less than 400 cows.

Today commercial cattle south of the Zebu line can be placed in three categories: first, the Zebu, or breeds in which Zebu influence has been incorporated, such as the American Brahman and Santa Gertrudis; second, the undefined crosses of the Zebu types and British beef breeds, predominantly American Brahman-Hereford crosses that have been continued for several generations under controlled conditions; third, a nondescript type of cattle showing some Brahman but considerably more Hereford influence, of inferior quality, the result of indiscriminate crossing.

There are many excellent purebred herds. These are mostly American

Brahman, Santa Gertrudis, Charolais, and Hereford. Good herds of the European breeds are usually maintained under carefully controlled conditions to minimize the effects of heat, parasites, and insects.

North of the Zebu line the British breeds are more important, with the Hereford predominant. The number of Charolais and Charolais-Hereford crosses, however, increases every year. There are very few Shorthorn herds left. In Oklahoma also there are many Angus herds, commercial as well as purebred.

In the entire South Central Region, even in the more mountainous areas, cows can be brought through the winter without supplemental feed following a growing season of normal precipitation. A little hay or grain cake, however, is often fed in bad weather to help produce a larger calf crop and healthier calves.

The entire region is subject to periodic droughts. When they occur, losses run high, herds are drastically reduced in size, and marginal operations are liquidated.

The eastern part of the Southern Plains was one of the first areas where the value of performance-tested bulls was realized. In the years following World War II the commercial cattlemen there began to use bulls that had been through the early test stations. They also weighed their calf crops at weaning. These modern methods of herd improvement are now common throughout the Southern Plains.

V. *Mountain and Plains Region.*—This region includes three major areas: the Western Plains, which lie between the Great Plains and the Rocky Mountains; the Mountain Parks and Valleys; and the Great Basin, which lies west of the Rockies and east of the Sierra Nevada and Cascade Mountains.

WESTERN PLAINS. It was here that the colorful expansion of range herds occurred during the last quarter of the nineteenth century. This trend continued well into the twentieth century, and in the years between the two world wars the mountain states produced over one-fourth of the nation's beef cattle. Although the cow herd in this region doubled between 1941 and 1971, this increase was sufficient to supply only one-sixth of the national beef slaughter cattle. The more rapid expansion of the industry in the Southeast and Central regions made up the deficit thus created.

The Hereford gained a majority in the herds of this region in the late 1920's. The Longhorn cattle had disappeared by that time, and the Shorthorn was reduced to a very spotty representation a few years later.

Breeding stock that was predominantly Shorthorn was moved from the Oregon country into Idaho and as far east as central Wyoming even before the Texas drovers arrived with their Longhorns. These Oregon cattle were also bred up to Herefords within a few generations.

After World War II a small representation of Angus was established in the region. They increased rapidly after 1950 and in some localities became a majority. Over the region as a whole, however, the Hereford continues to predominate.

The Angus herds were moving into the range country at the time when the practice of breeding yearling heifers to calve as two-year-old cows was starting. The old-time cattleman had always adhered to the theory that a cow should not be bred until she was two years old. In an effort to increase production during World War II some producers began to breed yearling heifers (a previous start at this breeding practice had begun in the days of World War I but apparently soon died out). Calving difficulties increased because of the immaturity of the two-year-old, but the extension of the productive life of the cow more than compensated for the trouble in helping her calve. Breeders soon learned that an Angus bull on a Hereford heifer not only reduced calving difficulty but also produced an exceptionally vigorous and good-gaining calf. The cattle feeder liked the white-faced black calves or yearlings and often paid a little more for them.

Many loyal Hereford breeders naturally refused to have a black bull on the place, but Angus-Hereford crosses became popular among more dollar-conscious cattlemen. When interest was renewed in crossbreeding in the mid-1960's, there was a marked increase in the number of white-faced black calves. In 1971 the large majority of crossbred cattle in the western feed lots were Angus-Hereford hybrids.

Among the Hereford breeders who used Angus bulls on their yearling heifers, the usual practice was to sell both steer and heifer calves as feeders. Under the influence of the new crossbreeding philosophy, however, some producers are now retaining such heifers with the object of embarking on a further crossbreeding program.

Throughout the mountain states there are growing numbers of Charolais herds. They have resulted almost entirely from an upgrading program on Hereford or Angus cows. Both artificial insemination and natural breeding have been used in the process.

The number of cattle in western feed lots that show some Charolais breeding is increasing each year, although actual numbers of such cattle are still small. Many Charolais breeders find an active demand for low-percentage Charolais bulls and are still keeping their heifers as part of an upgrading program.

Cattle of North America

The Taylor Act of 1934 is usually referred to as marking the end of the open-range method of running cattle. Although it definitely ended the free-grass days, most cattle were already run under fence by that time. The use of leased lands, both federal and state, is still essential to many cattle operations in the western states. Utilization of such lands decreases from year to year as limitations on the grazing are increased.

The theory of multiple use of public lands followed the demand for recreational use. The trend would seem, however, to be to one of single use for recreational purposes only. After the turn of the century most cattle operations in the mountainous areas of the western states relied on public lands for their summer pasture. Now many ranchers have to depend entirely on their fee-held land to graze their cattle.

On the Western Plains from Montana to New Mexico in a normal year cattle can be carried through the winter without supplement feed. Some old-time ranchers still operate on this basis, feeding only a little cake during an unusually severe winter. Most operations, however, provide some feed regularly through the winter months. In the southern part of the region the feeding period may last only a month or two. Farther north it may extend to four or five months, and even to six months at high elevations. Hay is put up where irrigation water is available or in some areas where rainfall is sufficient to raise a fair crop on unirrigated land. Where hay cannot be raised, grain cake is fed at the rate of two or three pounds a day.

Cows are now usually bred to calve in the early spring in the south and in the late spring in the north. The herd is run on the short grasslands during the summer. The calves may be sold at weaning in the fall, or, if grass growth has been sufficient, they may be carried over and sold as yearlings. In good grass years many ranchers purchase calves, carry them through the winter, and graze them until the following fall, when they sell them as feeders.

The entire plains area east of the mountains is subject to drought periods. These times demand rapid adjustment on the rancher's part. At such times calves are sold early, rigid culling is effected to eliminate older cows, and, if the drought is unusually severe, the breeding herd may be even further reduced or shipped to another area for wintering.

MOUNTAIN PARKS AND VALLEYS. Cattle ranching in the mountain region did not begin until after the plains to the east were stocked with the cattle brought from Texas in the drives of the 1870's and 1880's. Ranches were then established in the foothills, along the river and creek bottoms, and in the high mountain parks. Since winter feeding

Angus herd on summer pasture on a Wyoming mountain ranch.

was a necessity at these higher elevations, the size of an operation depended on the amount of hay land it included. Cattle were pastured on the mountain slopes and parks during the summer and in the fall were brought down to the meadows around the ranch headquarters to be wintered on the hay which had been put up during the summer. In favorable locations many operations were developed in the 150,000-acre range running several thousand brood cows. Steers were usually marketed when two or three years old. Long drives to the nearest railhead were made in the fall to get the cattle to market. Old cows were seldom shipped.

Today's operations follow much the same general pattern, although the ranches are much reduced in size, the cattle are moved out by truck, and the two-year-old steer is gone. A herd of 1,000 cows is now a rather large unit, though a limited number of cattlemen run herds of several times this number. Improved meadows, more efficient irrigation methods, modern hay-harvesting machinery, and all-round better management have greatly increased the productive capacity of the mountain ranches.

Some ranches continue to carry their calves over to sell as yearlings, but most now sell calves in the fall, keeping only the heifers needed for replacements. The rancher who is dependent upon the government land for summer grazing is slowly being crowded back to his own land, and one method of retrenchment is to ship calves instead of yearlings.

THE GREAT BASIN. Ranching operations in the mountainous areas scattered through the Great Basin are similar to those in the Rocky Mountains. In the lowlands, other than the desert and arid areas, of this intermountain region conditions are generally similar to those on the plains east of the Rockies. There is less rainfall and less plant growth, and more land is required to maintain a cow and calf.

VI. *Pacific Coast Region.*—California, Oregon, and Washington have about 5 per cent of the national beef-cow herd, in contrast to 12 per cent of the human population.

The early cattle in California were the Spanish type which had been driven up from Mexico in the eighteenth century. The Native American cattle, many of which had a strong Durham (Shorthorn) background, were the foundation of the early Oregon herds. After the settlements had been established and the Indians forced onto reservations, many cattle were trailed to Oregon Territory from California. They were of Spanish ancestry but were so greatly outnumbered by the Native American cattle that they had little genetic effect on the total cattle population.

Upgrading to the Hereford became popular early in the twentieth century over all the region, and before long the beef herds of the Pacific Coast states were largely white-faced, a condition which has obtained to the present.

The best cattle-raising area in California is the central valley lying between the Sierra Nevada and the coastal ranges. Only a few ranches remain today with as many as 1,000 head of brood cows. Most operations are in the 300-head range on ranches of four or five sections.

Cows are normally bred from January to April and calve in the fall as the rainy season starts. Ranches in the foothills area make considerable use of mountain grazing lands on the public domain that are leased from the federal or state government. In dry years some feeding of hay or cake is necessary during the summer months, but in years with 14 to 20 inches of rain there is good grass throughout the year.

Population pressure is a serious threat to ranching on much of the

Commercial Hereford herd on salt controlled self-feeder. Empire Ranch, southern Arizona.

better grassland in California. In many locations, property taxes are $7 to $8 an acre, which at 10 to 12 acres per animal unit puts a tax burden of close to $100 per head on every calf before it is sold. As a result, moderate-sized ranches are being broken up into building sites of a few acres.

Today only a few herds west of the Cascades exceed 1,000 brood cows. The representative herd has 200 to 350 head. Hay is fed as is required to bring them through the winter in good condition, and calves are sold in the fall. Some of the older cattlemen still keep steer calves to sell as yearlings, but this practice is dying out. Hay is raised on irrigated meadow and is usually baled for storage.

Crossbreeding has become general in the Northwest. At times some of the feed lots in the Columbia River basin have 50 per cent of their pens filled with crossbred cattle, mostly Angus-Hereford, with some Charolais crosses on either Angus or Hereford.

Cattle of North America

In California crossbreeding has been less general, although feeder cattle of mixed breeding are seen throughout the state. Most producers appear to be proceeding on a trial basis, crossbreeding only a few females at a time.

In many feed-lot areas across the region, a significant number of Holstein-Friesian steers are seen on feed. In southern California the feeding of this type of feeder has increased rapidly in recent years as the number of dairy cows approaches the number of beef cows in the state.

Dairy-Cattle Husbandry

Modern dairy facilities range from the hillside farm in Vermont with its 25 to 50 cows to the factory-type enterprise milking several thousand Holstein-Friesians on the outskirts of a major city. Mechanization is complete in the larger plant, and even the smaller unit has a milking machine, perhaps a milking parlor, and a refrigerated tank, which receives the milk directly from the milking machine. From the tank the milk is transferred directly to the tank truck of the hauler. Hand milking and the 10-gallon milk can set by the mailbox have disappeared.

Throughout the country, however, the family farm geared solely to "making milk" is still an important factor in the dairy industry. One man and his wife with modern equipment can now milk 100 cows or even more quite handily. Milking is still a twice-a-day chore seven days a week, but even the small dairyman can usually arrange a vacation, with cows giving 12,000 pounds of milk at a price frequently hitting six cents a pound or better.

Feeding practices vary with the size of the operation, the climate, and the terrain. Units up to 200 head in the hilly parts of Wisconsin or New England pasture the milk herd during the growing season. On land better adapted to cultivation larger herds are usually kept in a loafing yard, possibly with an adjoining small exercise pasture, and the cows are taken to the barn or milk parlor for milking. Some of the industrial-type dairies in southern California handle their cattle in a manner similar to a beef feed lot.

Silage or haylage is almost universally the basis of the ration, along with some hay. Grain is fed while the cow is milked, and more is added to the roughage if needed. While cows in full lactation may be fed at one level and those in the later state of production at another, the practice of feeding each cow in line with her production is no longer as widespread as it once was. Many commercial dairies have found that a less than maximum ration is conducive to a more moderate functioning of the

Loafing yard of a 5,000-cow dairy. Corona Farms, Corona, California.

mammary glands and results in better health and a longer productive life for the cow.

Calves are usually taken from the cows at birth and in no event are permitted to suckle for more than three days. Disposition of the calves varies widely. Large dairies often farm out at a few days of age the heifers desired for replacements. These are returned as yearlings or bred heifers. Bull calves or unwanted heifer calves may be sold to farmers, who veal them at about 200 pounds or take them to 400 or 500 pounds weight for the feed lot.

The Knolle Farms in South Texas raise all their replacement heifers. This operation has the largest concentration of Jersey cattle in the world, with 5,000 cows in milk and a total of 10,000 head of registered cattle. Other dairies buy all their replacements and sell all their calves when a few days old.

Individual calf raising pens. Corona Farms, Corona, California.

The smaller units of the family type usually raise their own replacement heifers and may keep a few good bull prospects. Male calves with no particular recommendations and unwanted heifers are grown out for veal. In some areas the bull calves are steered and raised as feeders.

Calves are commonly raised in individual pens for the first three months on milk or milk replacer with high-quality hay and grain supplement always before them. The pens are in a barn or a separate building and have slatted floors which provide for waste disposal.

At Colorado State University, Fort Collins, Colorado, where the winters are moderately cold, calves are raised in the open. Each calf is chained to an individual doghouse-type shelter which it may enter or leave at will. Fifty such houses are placed on a small lot near the dairy barn and moved to fresh ground for each new occupant. At about three

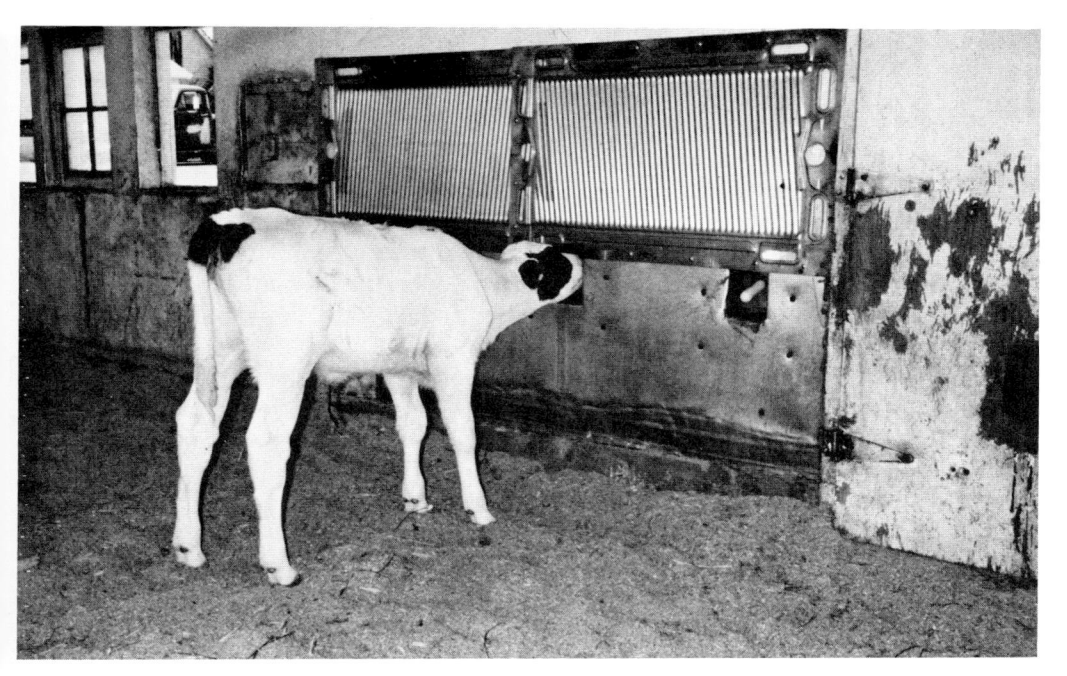

Automatic calf feeder. Jim Stewart Farm, Nampa, Idaho.

months of age the heifer calves are grouped in pens according to age and given access to an exercise yard.

While young stock are often pastured on small dairy farms, most replacement heifers are raised in dry lot. Hay, silage, and three or four pounds of grain make up the common growing ration.

The heifers are usually bred at 12 to 14 months of age. Currently 47 per cent of the cows in the national dairy herd are bred artificially. Large, privately owned bull studs and co-operative centers supply most of the semen and employ full-time technicians to do the inseminating. Most large dairies keep bulls for natural service of difficult breeders, and a surprising number of small herds also include a bull.

Automatic calf feeders are used to some extent in moderate-sized dairies. These are operated electrically and have been designed to replace the old method of hand feeding from a nipple pail. A ration of the proper amount and proportion is mixed from powdered milk replacer and water and is delivered to a container with protruding nipple. The device eliminates overfeeding by supplying the ration only at definite intervals. A one-nipple unit serviced by a sanitation-minded individual with considerable mechanical ability will feed 15 to 20 calves a day.

The calf nursery has become a fixture in recent years in some areas of heavy dairy concentrations. In southern California a typical nursery with a capacity of 2,500 head of calves finishes 10,000 calves a year. The calves are usually maintained in small individual pens and are carried to 200 to 250 pounds in 8 to 10 weeks. Some nurseries employ a mechanical-type feeder (described under "Dairy Cattle Husbandry," Part III, "Canada"), but most feeding is done with nipple pails delivered to each calf pen.

Initially these nurseries were set up for the production of veal, but there is now a large demand from feed lots for these young dairy steers.

The purebred dairy herd and the purebred breeder occupied a dominant position in the dairy industry until the advent of artificial insemination. The dairy breed societies in general immediately accepted this technique, and the rapid progress that followed in the development of high-yielding dairy cows is now history. The purebred breeder suffered somewhat in the process, and his herd is now primarily an efficient producer of quality milk.

Formerly the purebred man derived his income largely from the sale of young stock, particularly bulls, for breeding purposes. The milk check of the registered herd owner is now 70 to 75 per cent of his total income; another 15 per cent comes from the salvage sale of discarded animals for beef. He is fortunate to realize 10 per cent of his gross from the sale of breeding stock.

Breed Improvement

Much of the effort devoted to breed improvement in the mid-nineteenth century was based exclusively on visual appraisal. Breeders of both beef and dairy stock selected for characteristics that they felt, with little or no evidence to back them up, led to more efficient production. While some dairymen took notice of the milk in the pail and the amount of butter that could be churned from a three-day production of cream—and some beef breeders were astute enough to select the kind of conformation that indicated good muscling—color patterns, shape of horn, and other superficial marks of bovine beauty had an exaggerated importance in selection.

Business and professional men from the cities played a large role in the development of the breeds. A place in the country and a fine herd of cattle was the goal of many a tycoon, who then proceeded to import

Compact Hereford bull on left, standard-type bull on right. 1951. Colorado State University, Fort Collins, Colorado. National Archives Photograph.

large numbers of the British and Continental purebreds which otherwise would not have reached American shores. The large western ranching operations established during the last decades of the nineteenth century were widely publicized, attracting investments of outside capital which led to a much more rapid stocking of the ranges than would have occurred if the cattle industry had had to generate its own capital. Many purebred activities over the years have been similarly financed by outside interests.

Cattle for Beef.—The process of upgrading the United States beef herd to the British breeds proceeded rapidly during the last third of the nineteenth century and continued during the first decades of the twentieth century. Various new elements then began to enter the picture.

The benefits derived from crossing the Zebu on the British breeds attracted serious attention in Texas and Louisiana, but there was difficulty finding Zebu bulls. Eventually the requirements were filled through importations from India and later from Mexico and from Brazil by way of Mexico.

Then a start was made in developing new breeds which incorporated the Zebu influence. First was the American Brahman, which is 100 per cent Zebu; then came the Santa Gertrudis, Brangus, and Braford, which combined British and Zebu breeds. All these breeds are more productive in the Gulf Coast environment than are the straight British breeds.

The beef-breed societies contributed little during this period toward the improvement of their cattle. The livestock shows served a useful purpose until the 1920's when they became mere bovine beauty contests, which led down the path to dwarfism.

In the 1930's some of the experiment stations and some commercial cattlemen began to weigh calves at weaning and to identify them with their sires and dams. At the same time many purebred breeders were encountering serious troubles because of their negative selection practices. Bull calves whose prize-winning mothers could not give enough milk to support them were raised on heavy-milking dairy cows. Short-legged, compact animals of both sexes were selected for breeding, and the good, growthy kind were culled.

Then, beginning in the 1950's, a few producers of commercial cattle, along with a mere handful of purebred breeders, began to employ records as a guide to breed improvement, and performance testing came into play.

A chronology of the development of performance testing follows:

1936: First progeny testing of beef-breed sires, United States Range Livestock Experiment Station, Miles City, Montana
1941: Performance testing of yearling bulls, Balmorhea Experiment Station of Texas A&M College
1947: Determination of heritability of rate of gain and related characteristics in beef cattle sires, Balmorhea Experiment Station of Texas A&M College
1953: First performance-tested bull sale, Texas Agriculture Experiment Station, McGregor, Texas
1954: First requirement by a breed society that individual production records be furnished for registration, adopted by Red Angus Association of America
1955: Virginia established the first state Beef Cattle Improvement Association
1955: Organization of American Beef Cattle Performance Registry Association, forerunner of Performance Registry International
1959: American Angus Association, first purebred organization to recognize performance testing in its breed-improvement program
1961: First organized judging contest involving both live-animal and car-

United States

cass grading, University of Washington Junior Livestock Show, Spokane, Washington
1962: First offer by a breed association to process performance records of either purebred or commercial cattle, American Angus Association
1968: Formation of the Beef Improvement Federation

A vast amount of experimental and test data pertaining to performance records have been accumulated by the agricultural experiment stations and private breeders. Among the most important have been the heritability estimates for various traits. These have been summarized by the USDA Agricultural Research Service as follows:

Trait		Approximate Average Heritability*
Calving interval	Low	0 to 15
Birth weight	Medium	35 to 40
Weaning weight	Medium	25 to 30
Weaning conformation score	Medium	25 to 30
Maternal ability of cows	Medium	20 to 40
Steers or bulls fed in dry lot from weaning to final age of 12–15 months		
Feedlot gain	High	45 to 60
Efficiency of feed-lot gain	High	40 to 50
Final weight off feed	High	50 to 60
Slaughter grade	Medium to high	35 to 40
Carcass grade	Medium to high	35 to 45
Area rib-eye per cwt. carcass weight	Medium to high	30 to 50
Fat thickness over rib per cwt. carcass weight	Medium to high	25 to 45
Tenderness of lean	High	40 to 70
Summer pasture gain of yearling cattle	Medium	25 to 30
18-month weight of pastured cattle	High	45 to 55
Cancer-eye susceptibility	Medium	20 to 40
Mature cow weight	High	50 to 70

*Summarized from many published sources. Wider ranges indicate characters for which fewer estimates have been made and for which probable average heritability is less precisely known.
Source: E. J. Warwick, *Effective Performance Recording in Beef Cattle.*

These percentages indicate the degree to which the parents can be expected to pass on a given trait to their offspring. Inversely, they indicate the degree of the trait that is determined by climate, type of feed, and other environmental factors. Fortunately they show that the breeder has considerable opportunity to improve his stock. Many of these traits have been markedly improved within a few generations when performance records were used in selection.

533

By the 1960's commercial cattlemen had begun to use performance-tested bulls. Such bulls had acceptable weaning weights and made satisfactory gains on postweaning feed tests, standardized at 140 days. Weaning weights are standardized at 205 days of age by adjusting for the actual age at which the calf is weighed.

The next step was progeny testing, in which the sire was evaluated by the weaning weight and feed-lot performance of his sons.

Replacement heifers were also selected by their weaning weights, and by their mothers' records as producers of superior calves.

These procedures involve the permanent identification of each member of the cow herd and include recording the birth date of every calf born, identifying it with its mother, and weighing it at weaning. The labor and equipment involved in these procedures have retarded the universal adoption of the practice.

The work of the pioneers in performance testing had little direct impact on the industry. Their efforts gained no more than grudging comment from the breed societies, in which circles the lament was heard that the work of the master breeders would be lost forever if weight became a major criterion in selection.

Because they were unable to purchase purebred sires with performance records, some commercial cattlemen began raising their own bulls. Those used in their herds were carefully selected individuals with above-average weaning weights and outstanding gains on feed test. Their calf crops at weaning clearly reflected the superiority of these sires.

A few purebred breeders who had not followed the popular trend also began keeping performance records. It was these breeders, along with a few commercial cattlemen, who saved the day for the beef breeds in the United States.

The first Beef Cattle Improvement Association was founded in Virginia in 1955, and began testing yearling bulls for their daily rate of gain. Thirteen years later 34 states had at least one bull-testing station, and the Beef Improvement Federation was formed to co-ordinate the efforts of the state organizations. The Beef Improvement Federation has become closely affiliated with the Agricultural Research Service of the USDA and can someday be expected to furnish the kind of sire evaluation for the beef industry that the Dairy Herd Improvement Association has provided for the dairy industry for the past 45 years.

Another organization which has played an important part in performance testing is Performance Registry International. Formed in 1955, this organization now offers a computerized service for performance-test data to all breeders.

Gradually the purebred breed societies came to recognize the fact that performance records were essential to their continued existence. To a varying degree they incorporated performance-test procedures into their programs, though some still emphasize show winnings for superior recognition. A few of the new and smaller associations have made performance records a requirement for registration.

As performance testing gained recognition, attention was also directed toward carcass evaluation, or yield grade (as discussed in the section "Grading," below). Cutability is largely determined by the relationship of fat to muscle in the carcass and is a trait of fairly high heritability. Selection is complicated, however, by the fact that it involves slaughter and inspection of the progeny of the animal whose qualifications as a breeder is being determined, but genetic improvement of carcass traits will undoubtedly increase in the near future.

There is some evidence that gainability is directly correlated to muscling. Such a relationship could lead to rapid improvement in both traits.

The essentials of most performance-testing programs can be summarized as follows:

Individual Performance Records:
1. Weaning weights on all calves adjusted to 205 days of age and corrected for the sex of the calf and age of dam
2. Weight at 365 days
3. Daily gain of bulls on 140-day feed test. (Such feed tests are not recommended for heifers)

Progeny Records:
1. The above data on all calves of specific bulls and cows
2. Carcass data on randomly selected progeny. Such data include:
   a. Hot carcass weight, both actual and as percentage of liveweight
   b. Fat thickness over twelfth rib
   c. Area of rib eye at twelfth rib
   d. Percentage of kidney and pelvic fat
   e. Quality grade (prime, choice, and so on)
   f. Yield grade (designated by numerals 1 through 5 to indicate the cutability or amount of red meat in the carcass: 1 indicates an extremely lean carcass; 5, an extremely fat carcass)

These measures of carcass evaluation are still in the process of refinement.

Cattle of North America

Dairy Cattle.—During the 1830's the Frenchman Gueonan advanced his "escutcheon theory" for the selection of productive dairy cattle, and for many years it was accepted as the standard. The escutcheon is the pattern made by the hair which grows upward on the thighs and udder of a cow. It supposedly indicated her ability to produce milk. While the theory has long since been abandoned, the "scale of points" which followed it has had much to do with the selection of dairy cattle down to the present day. Although progress was slow and unpredictable, there is no denying that this standard aided in the development of a more productive cow. While the dairy breeds still have the Unified Score Card, it has been relegated to second place in selection programs.

Long before the turn of the century the breeders and their breed societies had become production-minded. In 1880 Soloman Hoxie, founder of the Holstein-Friesian Society, advanced the theory that a cow should have a superior production record to be rated superior. At the Chicago World's Fair in 1893 the principal dairy breeds competed on the basis of the cost of producing a pound of butter.

Production Testing: By the last decade of the nineteenth century all the five major dairy-breed societies had taken initial steps in production testing. These steps took the form of churned butter tests, milk production for short periods, and Register of Merit programs in which a cow met a minimum standard for milk or fat production. All these were forward steps, but only short ones since only individual performance was emphasized. Then, during the first quarter of the twentieth century the herd-improvement registries were adopted whereby the average herd production was recognized as the standard. This has proved a far more effective selection tool than the individual cow test.

The Ayrshire Association established a herd test in 1925 which required reports on the production of all registered females of milking age in a herd. The following year the USDA set up the Dairy Herd Improvement Section, which was designed to co-ordinate dairy-testing procedures on a nation-wide basis, and within a few years all the dairy-breed associations had adopted herd testing. This development signified the coming of age of the United States dairy industry.

Co-operative associations now periodically weigh the milk from individual cows and obtain the butterfat content. The first such association was founded in Newaygo County, Michigan, in 1905, and after the USDA became involved in the program many similar organizations sprang up. They eventually grew into the Dairy Herd Improvement Association (DHIA), which is probably the most effective co-operative agricultural organization in the country.

United States

The grass-roots unit of the Dairy Herd Improvement Association is usually a group of 25 to 30 dairy farmers in a specific area. These local associations, of which there are currently over 1,250 in the United States, hire one or two milk testers, who proceed from farm to farm on a monthly basis, the date of their visit usually unannounced. They weigh the evening's and morning's milk from each cow and take a sample for butterfat determination. Data on the breeding of each cow, her age, date of entering and leaving the herd, and other pertinent facts are recorded by the tester and forwarded to the National DHIA Coordinating Group (a bureau of the USDA) for electronic processing. Printout sheets are furnished the co-operating herd owner every month. The annual cost to the co-operating dairyman varies with the locality and the concentration of herds but is in the range of $7.50 to $10.00 a cow.

In recent years the DHIA plans have been enlarged to include certain unofficial testing programs. This move reflected an effort to obtain wider participation by providing records at lower cost. Two programs are offered, the "Owner-Sampler" and the "Weigh-a-Day-a-Month," or WADAM. In the Owner-Sampler program the dairyman weighs and samples the milk, and one of the official testers picks up the weight records and samples. The cost is approximately 60 per cent of that for the official plan. Under the WADAM program the dairyman weighs the milk, uses the milk check for butterfat percentage, and mails in all data. The cost is about 10 per cent that of the official program. The National Co-operative Dairy Herd Improvement Program also recognizes a number of miscellaneous programs. A few state and private testing plans are also available. Only about 115,000 cows are under test in these programs. The following table summarizes the extent of dairy testing programs in 1970.

NUMBER OF HERDS AND COWS IN UNITED STATES TESTING PLANS*

	Number of Herds	Number of Cows	Per Cent of Total Dairy Cows
Official DHI Program			
Purebred herds	3,814	235,829	1.9
Commercial Herds	30,494	1,866,182	14.8
Total official DHIA	34,308	2,102,011	16.7
Owner-sampler	23,281	839,343	6.6
WADAM	703	47,493	0.4
Total	58,292	2,988,847	23.7

*National Cooperative Dairy Herd Improvement Program, January 1, 1970. The above table is based on 12,689,000 cows in the National Dairy Herd as of January 1, 1970.
Source: Agricultural Research Service, USDA.

Cattle of North America

The best dairy cows in the United States have now arrived at a state of mammalian perfection that may be close to the point of diminishing economic return. If the top cow in many herds today is fed a ration that enables her to reach her maximum production, there is difficulty in getting her bred back to start another lactation, she is much more susceptible to the common cow ailments, and her useful life is likely to be reduced to less than three years. The prima donna holding a world record of over 300,000 pounds of milk and five and one-half tons of butter in a lifetime represents more of a tribute to skillful handling by her keepers than to her individual productive ability. While it in no way detracts from the superb performance, it must be realized that such a cow may have had unheralded sisters who, given equal opportunity, would have been worthy competitors for the world record.

During the 50 years from 1920 to 1970 the production of the DHIA-tested cows more than doubled—from slightly over 6,000 pounds to 12,300 pounds per 305-day lactation. But this average production for 1970 applied to only one-fourth of the national herd; the other three-fourths were at the 7,000-pound level. This is the level at which improved production is most important. If the average production of the entire national dairy herd could be brought up to 12,300 pounds, only 8 million cows would be required to supply the present market, compared with the 12 million now in production.

The accompanying chart shows the advances made in the United States dairy industry through the application of production records and artificial insemination, the two major tools for achieving genetic improvement.

The first DHIA organization was started in 1906. By 1925 records were being reported on only 42,300 cows, but from that year on enrollment increased rapidly, rising to nearly 2 million by 1970. The production of reported cows rose from 7,000 pounds in 1925 to 8,000 pounds in 1935.

Artificial insemination was introduced in 1938, and as its use increased during the next 12 years, the annual production per cow rose to 9,000 pounds. The introduction of frozen semen in the early 1950's made outstanding sires available to more and more herds, and production per cow climbed from 9,200 to 12,200 pounds in the following 15 years. Artificial insemination contributed to this increase, but the wider use of progeny-proven sires was an even more important factor.

The total milk production of the country has steadily increased and is now sufficient to saturate the market, even though the national dairy herd has been steadily decreasing in number. Along with this trend

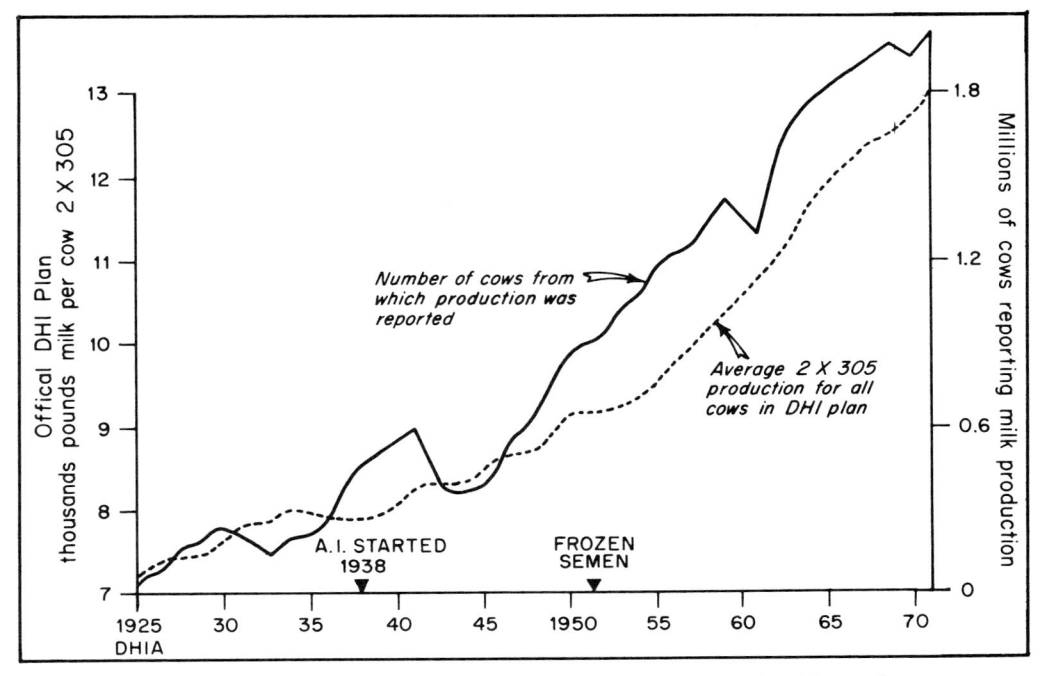

Increased milk production per cow and number of cows for which milk production was reported. Based on data from USDA.

there has been an increase in the size of the average herd. Over the past 50 years the DHIA herds have grown from an average of 16 cows to 62 cows.

In a well-managed herd replacements are now selected from the progeny of the best-producing cows and sired by a proven bull. Even the technique for proving the bull has been refined. Formerly it was considered sufficient if his daughters had better milk production records than their mothers. Today an unselected group of his daughters must outproduce the daughters of his contemporaries.

Crossbreeding

Crossbreeding in simplest terms is merely a matter of mating bulls of one breed to cows of another. It can be either haphazard or planned. In the former case it produces mongrelization, a situation which has made all crossbreeding suspect to large numbers of cattlemen. Well-planned crossing, however, can be a means to achieve highly desirable

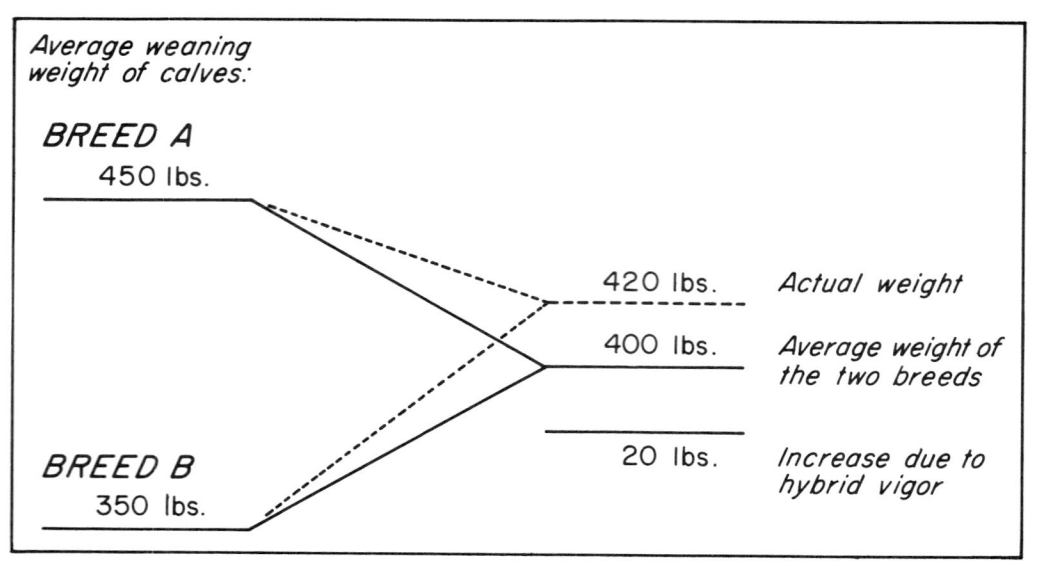

```
Average weaning
weight of calves:

BREED A
  450 lbs.
_____
                  ╲╲╲
                     ╲╲╲
                        ╲╲╲  420 lbs.   Actual weight
                        ╱╱╱  ─ ─ ─ ─
                     ╱╱╱     400 lbs.   Average weight of
                  ╱╱╱                   the two breeds
               ╱╱╱
BREED B     ╱╱╱              ─────────
  350 lbs.                    20 lbs.   Increase due to
_____                              hybrid vigor
```

Effect of hybrid vigor on weaning weight.

ends. The upgrading of the western range herds was an outstanding example. The Brangus, the Barzona, and the Santa Gertrudis are among the breeds that have been developed through controlled crosses.

It has been well established that crossing brings with it distinct advantages in health and general thriftiness, known as heterosis, or hybrid vigor. It is this factor that has largely contributed to the success of the Brahman crosses in the South and has led to the popularity of the Hereford-Angus crosses in the feed lots. The crossbred cow usually proves herself a better mother than her straightbred sister. The crossbred calf is hardier and a faster gainer than his straightbred pasture mates.

This phenomenon has proved highly attractive to commercial cattlemen, as well as to research and extension workers. It has become the focal point of much discussion and to practical activity during the past decade. Among the crossbreeding plans advocated by the various experiment stations rotational crossing and the use of the terminal cross are two programs which usually receive the most attention.

In rotational crossing, cows of breed A are mated to bulls of breed B. The resulting F_1 heifers may then either be bred back to bulls of breed A or bred to bulls of a third breed. A high degree of heterosis is main-

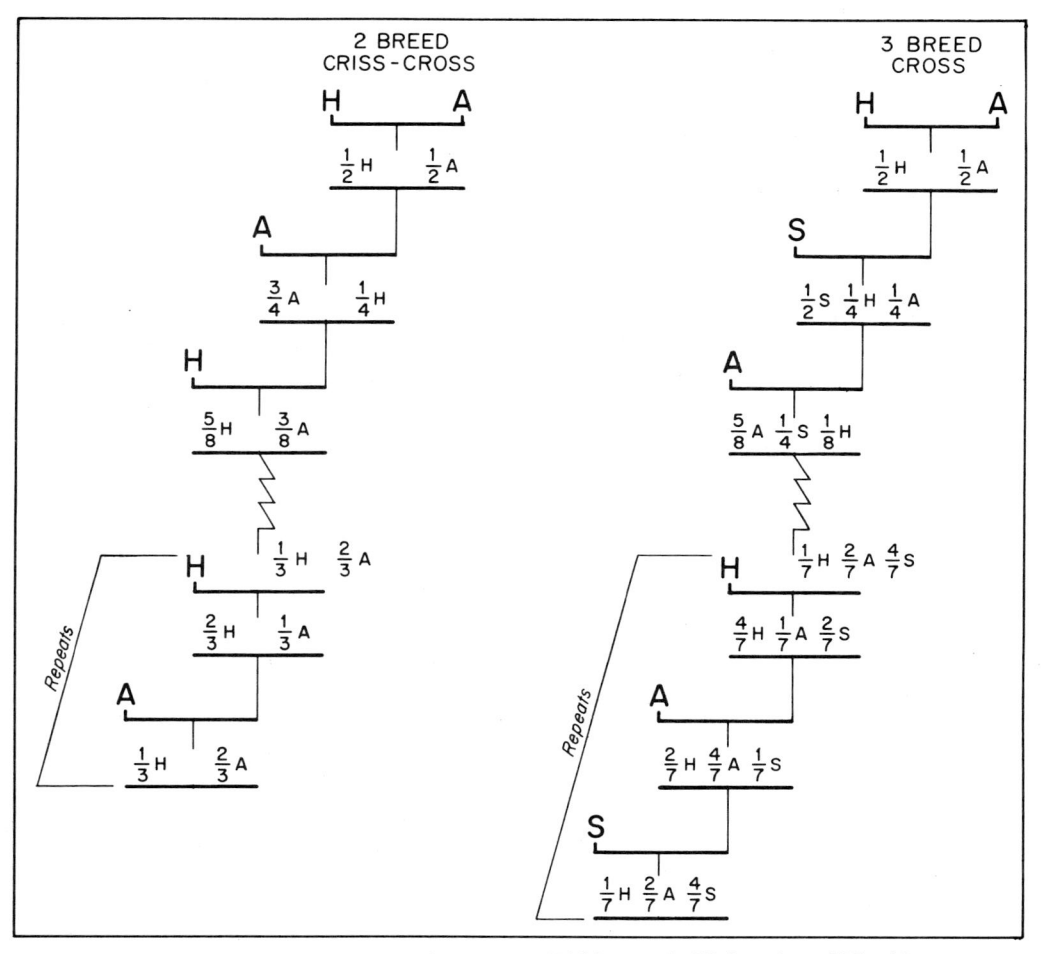

Rotational crossing schemes. Based on R. W. Kidder et al., University of Florida *Bulletin* 673, 1964.

tained throughout these crosses, the steer calves make profitable feeders, and there is a good market for the heifers either as feeders or as replacements.

The terminal cross is based upon the concept that some breeds are "maternal" breeds and that others are "sire" breeds. Thus an Angus bull might be bred to Brown Swiss cows to produce F_1 females combining both excellent beef characteristics and outstanding milking and maternal

541

traits. These females would then be bred to a Charolais bull. This would be considered the terminal cross, since all the progeny—female as well as male—would express the ultimate in gainability and carcass traits, and all would be fed out and sent to slaughter.

Theoretically, these plans have great merit. Practically, they call for management skills of above-average level. They demand the resources to maintain two or more breeding herds and the ability to manage those herds, or at least to be able to maintain replacements through judicious purchase arrangements.

The recent introduction of the continental European breeds has stimulated interest in crossbreeding, and the next few years will undoubtedly see many fascinating as well as many unsuccessful attempts to weave this stock into better breeding plans.

Whatever the merits of crossbreeding, it is obvious that it demands increasingly better genetic material in the purebred base stock.

Artificial Insemination

The organized use of artificial insemination began in the United States in 1938. Enos J. Perry, an extension professor at Rutgers University in New Jersey, had observed the use of the practice in Denmark and interested a number of New Jersey dairymen in trying this method of breeding. The following year an estimated 646 herds used artificial insemination to a varying extent, and 7,359 cows were reported to have been so bred. The use of artificial insemination, which was confined solely to dairy cows, increased rather slowly until after World War II but then advanced rapidly. The frozen-semen technique which had been developed in England in 1952 came into use in the United States in 1955 and was instrumental in achieving a more widespread use of artificial insemination since it eliminated the difficulties involved in the transportation of fresh semen and permitted storage of the semen for an indefinite time. The number of dairy cows in the national herd began to decrease in 1946, but the number of cows bred artificially continued to increase. In 1971 over 47 per cent of the national dairy herd, 6.7 million cows, were bred in this manner.

Some commercial beef herds and a very few purebred herds were bred artificially in the late 1950's. Armour and Company started its Beef Cattle Improvement Program in 1958. The goal was to prove superior sires by determining feed-lot gains and carcass characteristics of their progeny. Commercial herds were bred artificially to selected bulls at no cost to the producer other than the labor involved in detection. Although the program was abandoned a few years later, it was instrumental in familiarizing a number of commercial cattlemen with the management problems involved in breeding beef-cattle herds artificially. At the same time a few of the larger artificial-insemination centers began offering breeding service by bulls of the major beef breeds.

Because of the opposition of some segments of the beef industry, progress was slow for the next decade. When they did not ban the use of artificial insemination completely, the breed associations employed registration requirements which made it impractical. Many practicing veterinarians in the cattle country discouraged the use of artificial insemination for beef cattle, claiming that the genital tract would be so injured that future conception would be endangered. How the beef cow differed from the dairy cow in this regard was not explained.

It took the progressive beef-cattle breeder a number of years to learn that, while there is no physiological difference between a beef cow and a dairy cow, there is a difference in the way artificial breeding has to be practiced if it is successful in a beef herd. Heat detection is the basic factor involved in this method of breeding. The dairy cow is under close observation two times a day, and her estrus period is readily detected by her handler. The beef cow on pasture and uncontrolled is much more difficult to detect in heat. An experienced observer, riding from sunrise to sunset, is essential to determine the cows that should be bred. The lesson also had to be learned that the cow in heat should not be run into the ranch chute but must be handled gently in a corral adapted to artificial breeding.

For these reasons artificial insemination of beef cattle progressed slowly. The USDA did not even report the number of beef cows bred artificially until 1963, when 235,000 were recorded so bred. Artificial insemination of beef cattle, however, can be considered as starting in 1959, and in 1969, after 10 years, 924,000 cows, or 2.6 per cent of the national beef herd, were being bred in this manner. Considering the handicaps under which the artificial breeding of beef cattle started, this indicates considerable progress. (In 1949, after 10 years of artificial insemination in dairy cattle, 2,091,000 cows or 8.7 per cent of the national dairy herd, were so bred.) By 1971 around 1.5 million cows of the beef breeds, or 4 per cent of the national herd, were bred artificially.

The rectal technique of inseminating is used exclusively. Until recently semen has been packaged and stored exclusively in ampules. Some centers are now experimenting with straws, but their use is limited. Liquid nitrogen is now generally used as the refrigerant for semen storage.

The National Association of Animal Breeders was organized in 1947 and has been a major factor in advancing the use of artificial insemination in both dairy and beef herds. It is an association of individuals and organizations concerned with all the various aspects of artificial breeding. There is a sizable foreign representation from countries in the Western Hemisphere. The association's monthly publication, the *A.I. Digest*, has a wide distribution in the industry. Annual conferences emphasizing this method of breeding beef cattle have been held since 1967, and they are widely recognized as a medium for breed improvement, as well as for the spread of information on artificial insemination.

Free-enterprise organizations, either breeder co-operatives or private concerns, produce all the semen originating in the United States and handle the importation of semen from foreign countries. The number of bull studs producing semen commercially had increased to 97 by 1950, when 2,619,000 cows of the dairy breeds were bred. Through consolidations and mergers and the abandonment of smaller operations, the number has since decreased to 30 in 1971, when over 8 million cows, both dairy and beef, were bred artificially.

The breed improvement and the resulting increase in average milk production per cow which has resulted from the wide use of artificial insemination in the dairy herds was furthered by the national sire-evaluation summary prepared from the records of the Dairy Herd Improvement Association. It is surprising that the individual dairyman, who may use whatever bulls he chooses, does not always choose bulls that will increase the productivity of his herd. Some estimates indicate that as many as one-third of the cows bred artificially are being bred to bulls that cannot be expected to increase the milk production of their progeny over that of the dam and frequently can be expected to lower it. The only explanation for this failure to select "increaser" bulls is either a lingering preference on the part of some dairymen for type or a false economy in using bulls with lower production records at a lower cost.

Some of the large bull-stud services are now offering semen from bulls of the beef breeds on which there is excellent progeny data, including carcass evaluation. The beef-cattle purebred societies have adopted a more realistic attitude toward artificial insemination although the practice is far from being wholeheartedly accepted. A start has been made toward a national sire-evaluation summary in the *North American Beef Sire Directory*, first published in 1972. Such factors, together with the incentive added by the new breeds and crossbreeding, spell an increasing use of artificial insemination in the beef-cattle industry in the years ahead.

Early twentieth-century feed lot in the corn belt. National Archives Photograph.

The Feed Lot

The Pilgrims' beef came from the draft ox too old for further work. Toward the end of the seventeenth century, however, farmers started feeding grain to slaughter cattle. City markets, such as those in Boston, New York, and Philadelphia, began paying a premium for superior beef.

Cattle of North America

As agricultural lands were developed in western Pennsylvania, Kentucky, and Ohio following the Revolutionary War, farmers soon found that they were growing more grain, especially corn, than could be consumed locally. The marketing of this product was seriously hampered by transportation problems. Distilling was resorted to as a means of getting the corn to market, but excise taxes soon restricted this activity, and cattle came into their own as highly efficient corn movers.

At first the object was merely to move the corn, but eventually the quality of the beef became a matter of concern. The public liked the improved product and demanded more.

Grain feeding became more common in the days following the Civil War, when the movement of purebred cattle from England to the United States was well under way. A ration of corn, it was found, improved a bull's appearance for sale and kept the breeding herd in better condition. Grain was fed, too, to put more weight on an animal or to obtain a superior carcass for a particular customer. Such practices inevitably led to the regular feeding of butcher cattle.

By the opening of the twentieth century many corn-belt farmers were feeding grain to cattle as part of a diversified farm operation. The number of cattle varied from 2 to 3 steers, calved by the farmstead milk cows, to 50 or even 100 steers brought to the farm from the western range country by way of a central market.

The operation was largely seasonable; the cattle were bought in the fall, fattened, and sold the following spring or summer. The barnyard was the site of the project, and the grain and hay were fed by hand.

As farm enterprises increased in size, cattle feeding became more specialized. The successful farmer-feeder even bought corn from his neighbors to increase the size of his operation, and by the 1930's it was not uncommon to find several hundred head in the individual lot.

By that time the better feeders were using a protein supplement, such as linseed or cottonseed meal, with their corn, and most had hogs following the steers to salvage the undigested corn in the manure. The small farmer, however, feeding 30 head or fewer, accounted for well over half the total slaughter cattle sold.

The commercial feed lot, as distinct from a farm operation, made its appearance in the 1920's. The first units, usually in the 1,000-head bracket, were commonly located in a corn-growing area near a country sales barn. Feed lots also sprang up in the vicinity of central markets, like Chicago, and river markets, which in turn served neighboring packing plants.

Concurrent with the development of the feed lot came the stocker

Feed lot 100,000-head capacity. Monfort Packing Company, Greeley, Colorado.

program. The stocker is a feeder calf or yearling. Since it is often desirable to get more growth on these stockers before putting them on feed, they are usually wintered on some form of pasture. It may be a stalk field in the corn belt, a stand of winter wheat in Kansas or Oklahoma, or a mature grass pasture where they will possibly be fed some silage or hay. Gaining one-half to one pound daily through the winter, the steer acquires a bigger frame and can be fed out more economically than if it has been started on full feed in the fall. Payment for winter pasture and the feeding of stocker cattle is usually on a per-head basis for a certain number of months, or at a certain price per pound of gain, though various other pricing methods are also used.

With growthier cattle coming into the picture, more calves now go direct from the producer to the feeder, and the practice of roughing them through the winter is declining.

Following World War II the wider use of new farming techniques—improved methods of irrigation, more productive plants, better understanding of the nutritional requirements of cattle—paved the way for

Feed bunk arrangement. Stanley Brown Feedlot, Santa Monica, California.

larger feed lots in the grain-growing regions. This trend was stimulated as the major packers closed their plants in Chicago and Omaha and other long-established centers and along with newer firms moved closer to the supply of cattle. In some instances the newly located plants became integrated with feed-lot operations.

Colorado, Kansas, the Texas and Oklahoma panhandles, Arizona, and California became major cattle-feeding regions. The numbers fed in the traditional cattle-feeding states, Iowa, Illinois, and Nebraska, continued to increase but at a slower rate, while the very large feed lots were built in the West.

Currently the two largest lots in the country are those of the Monfort Company in Greeley, Colorado. One of the lots has a one-time capacity of 110,000 head; the other, 100,000 head. Combined, the two lots finish over 500,000 head of 1,150-to-1,200-pound steers annually for slaughter in the Monfort Packing Plant, also at Greeley.

An even larger cattle-feeding complex stretches from Lubbock north to Garden City and Dodge City, Kansas. It embodies over 100 units with a total one-time capacity of 1.7 million head and can finish 4 million head annually. Other large cattle-feeding centers have developed in

Feed car for delivering feed to bunks. Stanley Brown Feedlot, Santa Monica, California.

irrigated grain areas in the Imperial and San Joaquin valleys in California and in southern Arizona and in eastern Nebraska.

A typical western feed lot has a capacity of 25,000 to 35,000 head of cattle at one time and handles two to two and one-half times this number in a year. Construction costs, including land, are around $50 per head of one-time capacity, or $1.5 million for a 30,000-head lot finishing 60,000 to 75,000 head annually. Such an operation is completely mechanized, and highly qualified personnel handle all phases of the work. The feed preparation and handling requires no manual labor. The purchased grain is dumped in an elevator, and the ration is blended, loaded into trucks, and metered into feed bunks which are never allowed to become empty. One large feed lot in California distributes feed to the bunks from a railroad car hauled by a locomotive.

Least-cost rations in some lots are formulated by electronic computers, and mixing is controlled automatically. Antibiotics, vitamins, and other additives prescribed at only a few parts per thousand are blended in with the utmost precision. Manure is handled by heavy earthmoving equipment.

The steer in a large feed lot enjoys a more nutritious and better balanced diet than the supermarket patron who eventually consumes the beef. The rations are formulated by a livestock nutritionist and are transcribed on computer punch cards for each pen of cattle. The cards activate the automatic hopper scales and mixing mechanism so that the correct quantities of the various ingredients are delivered to the loading hopper. Trucks deliver the feed to pens for which it was formulated, and the computer charges each pen with the quantity of feed delivered. In another system an automatic scale on the truck weighs and records the quantity of feed delivered to each pen.

The total quantity and cost of feed are computed for each pen from the time the cattle are delivered until they go to the packer. Such items as death loss, veterinary expense, cost of the cattle, sale price, and initial and final weights are entered into the final accounting for each lot. The summary of this information gives the average daily gain, the quantity and cost of the feed consumed per pound of gain, and the profit or loss on each pen of cattle.

A full-time veterinarian supervises the animal health program. Cattle are routinely dipped to eliminate external parasites and are vaccinated against diseases suspected in the area. Sick animals are treated in a hospital section.

Many of the older feed lots in the corn belt have installed facilities similar to the western lots.

Corn is the basic feed grain in the middle western and plains states. Milo is fed in the Southwest, and barley in the Pacific Coast states. Many feed lots now steam-cook and roll the grain immediately before feeding in order to obtain better utilization. Where corn is grown, corn silage is usually fed, the silage being stored in large pits in the fall. The roughage element of the ration depends on local availability. Alfalfa hay is generally used with or without silage, and, where conditions permit, green-chop alfalfa is fed. Many feed lots are situated adjacent to sugar factories or breweries to utilize the by-products at a minimum handling cost. In recent years in wheat-growing areas this grain has been used to an increasing extent in feed-lot rations, replacing other feed grains to as much as 30 per cent.

Until late in 1972 stilbestrol (diethylstilbestrol) was included in almost all rations. This synthetic compound, carrying the properties of the female hormone estrogen, was fed at a rate of no more than 10 milligrams a head a day, the maximum intake permitted by the Food and Drug Administration. Federal regulations further required that the drug be withdrawn from the ration a minimum of seven days before slaughter. The use of stilbestrol increases feed efficiency some 10 per cent.

United States

Although stilbestrol (DES) was used in cattle-feeding operations for nearly two decades, no serious objection to incorporation of the drug in cattle feed was voiced until 1971. Because mice fed high concentrations of DES had developed cancerous tissue, a furor arose over the use of the drug as a feed additive. Difficulty in enforcing the seven-day withdrawal requirement and improved analytical methods for detecting minute quantities of DES added fuel to the fire. At the end of July, 1972, the Food and Drug Administration issued a statement banning the use of DES in animal feeds after January 1, 1973.

Urea is another important item in cattle feed. A nitrogen compound manufactured from ammonia, it has been found to replace satisfactorily one-third of the plant protein necessary for a balanced fattening ration. It is widely used because the cost is considerably less than the vegetable protein it replaces.

There is a primitive side to the large feed lot that is in distinct contrast to the computer-controlled and automated feed-handling system. This disadvantage is the problem of mud. Very few lots are paved. Although the sites take advantage of the best drainage available, the pens are often a sea of mud and manure in wet weather, relieved only by islands of relatively dry manure, which serve as a bedding ground for the cattle. Most large lots are unprotected by anything but a windbreak. Various kinds of shade are provided in some lots.

This poor housekeeping has its utilitarian side, however. There has been considerable experimentation with enclosed feeding areas, and slatted floors have been used to provide for liquefied manure collection and removal. Results have generally shown, however, that the saving in feed derived from such facilities is usually insufficient to compensate for the cost of the construction.

With the expansion of cattle-feeding operations before World War II, the large packers began to feed on their own account. Initially the object was to ensure a ready reserve of cattle to keep the plants operating on a reasonable level, as well as to obtain some leverage on cattle prices. In recent years the packers have fed around 6 per cent of the total slaughter. Another 4 per cent has been fed by industry-controlled feeders with a close relationship with the packer.

The Monfort Packing Company is the only major packer which has completely integrated feeding and slaughter operations. The company's average kill is 1,500 head a day.

The farmer-feeder in the 1,000-to-4,000-head bracket, is holding his own in competition with the large commercial feed lots and is even increasing his volume in a modest way. His is usually a family-type

Nursery raised Holstein-Friesian calves starting on feed. 300 pounds average weight. El Toro feed lot, Heber, California.

operation with only a few hired men. Most of the corn that is fed is home-grown. Both the farming and the feeding operations are well mechanized, and the owner is a shrewd buyer of cattle.

The small farmer-feeder is also alive and healthy. He is typically a middle western farmer primarily involved in row crops who continues to feed out a dozen or so steers in his barnyard. The number of such farmers has been decreasing but not nearly as rapidly as had been anticipated. On January 1, 1968, the average number of cattle on feed in the four major corn-producing states was only 30 head per feed lot. Since there are feeders in this area handling from 1,000 head up, an average of 30 indicates that there are many farmers still feeding a few head of cattle.

The wide fluctuations in the prices of both feeder cattle and fat cattle make feeding for slaughter a high-risk business. As a result increasing numbers of cattle are being fed on contract or on a custom basis. This

Nearly finished Holstein-Friesian steers. 900 pounds average weight. El Toro feed lot, Heber, California.

practice gives the feed-lot operator an assured profit and places all the risk on the owner of the cattle. This owner may be a cattle producer, a packer, a large-volume retail beef outlet, or an investor looking for a speculative vehicle or a tax shelter.

There are various methods of allocating the charges in a custom arrangement. The most common practice is to charge a fixed fee for labor and use of the facilities. To this are added the price of the feed and the cost of any veterinary services. Another, less common, method is to charge a fixed price per pound of gain.

Most of the large western feed lots now operate on a custom basis. In the corn belt the practice is much less common. Many of the feeders in the latter area have been established for years and have developed buying patterns for both cattle and feed which tend to minimize possible losses in periods of adverse prices. Their risk factor is also less because they handle fewer cattle.

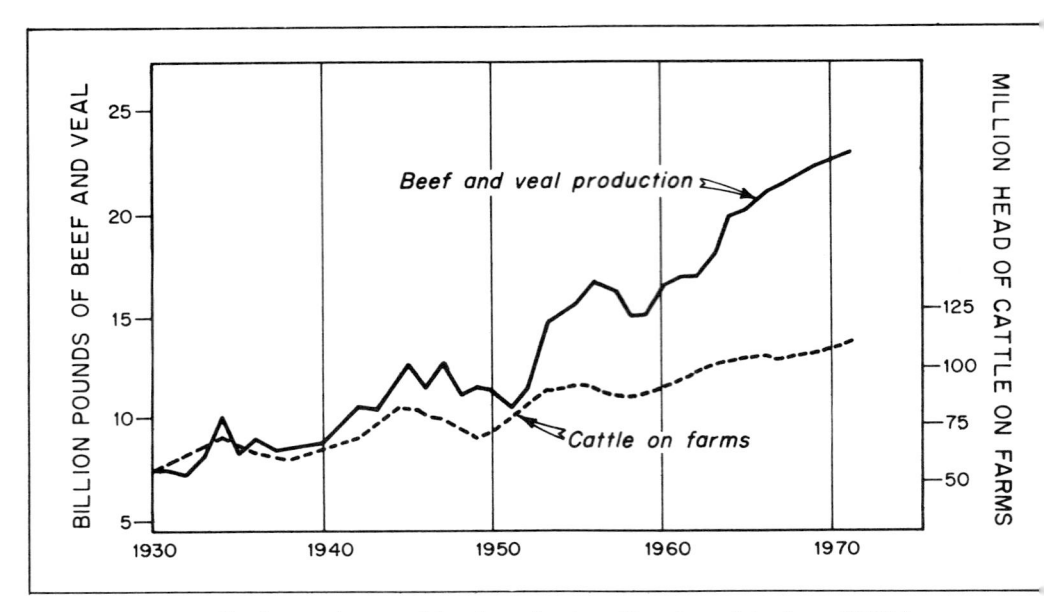

Cattle numbers and beef production. Based on data from USDA.

The great majority of feed-lot cattle are of the beef breeds, either straightbred or crosses. The numbers of young dairy stock seen in the lots is growing steadily. The large dairies of southern California produce more calves than the declining veal market can absorb, and this factor, along with an increasing demand for feeders, has resulted in a new use for the dairy calf.

The calf nurseries (see "Dairy Cattle Husbandry" above) are now raising Holstein steer calves to about 300 pounds. They then go into the feed lots and are finished at 950 to 1,100 pounds. These carcasses grade from 20 to 60 per cent choice with the remainder high-good to good. In 1970 an estimated 150,000 Holstein steers were fed to slaughter weights in California, a significant number in a state which annually produces only 375,000 beef-type steers.

Few outside the business realize the dominant position which the feed lot has achieved in the United States. Of the total weight of the choice beef consumed by the American public 45 per cent is put on in the feed lot. The total number of cattle fed in the commercial lots has more than doubled during the last decade. More than half of these cattle are handled in some 2,000 large feed lots, patterned along industrial lines. This trend appears certain to continue.

Outlying States and Puerto Rico

Alaska

After the United States purchased Alaska in 1867, the representatives of the War Department took over a frontier economy that had been under development by the Russians for more than a century. The Danish navigator Vitus Bering, in the employ of the Russians, had landed on the mainland in 1741, but the Russians established trading posts on Kodiak Island before moving across to the Kenai Peninsula and down the coast. Sitka was founded in 1799 and was a thriving port when San Francisco was still a mission village.

Russian America, as Alaska was known in that day, was looked upon mainly as a source of furs, but, in establishing the settlements, a nucleus of Siberian cattle was usually included as a provisioning measure. These small, rugged cattle are described in the section "Other Breeds" above. Although there may have been instances when the odd animal or so was brought in from "outside" (the Alaskans' term for the Forty-Eight States), the Siberian cattle remained dominant until 1900.

The United States Agricultural Experiment Station established at Sitka in 1898 purchased a number of Siberian cattle locally. The following year the station at Kenai was founded and later obtained a Siberian cow and calf. Both stations began investigations on Siberian cattle for milk production. The cow at Kenai proved to be an exceptional milker, producing an average of 29 pounds a day for the first three months of one lactation period, but no systematic effort was made to develop a milking strain of Siberian cattle by selection within the breed.

At the opening of the twentieth century many white settlements had small dairy herds of Siberian cattle, some with as many as 25 head. They were maintained on pasture in the summer and on hay and grass silage in the winter. There was one herd of 60 head on Kodiak Island.

In an attempt to develop a more productive type of cattle, the Kenai Experiment Station imported one bull and five cows of the Galloway breed in 1906. Most of these animals were later transferred to the station on Kodiak Island, and the herd was subsequently increased by other Galloway importations.

The eruption of Mount Katmai volcano in 1912 covered Kodiak with 18 inches of ash, and for safety the cow herd was shipped to the state of Washington. Two years later the herd, which had been reduced by an outbreak of blackleg, was returned to Kodiak, and the effort continued toward the production of a milking strain of Galloways.

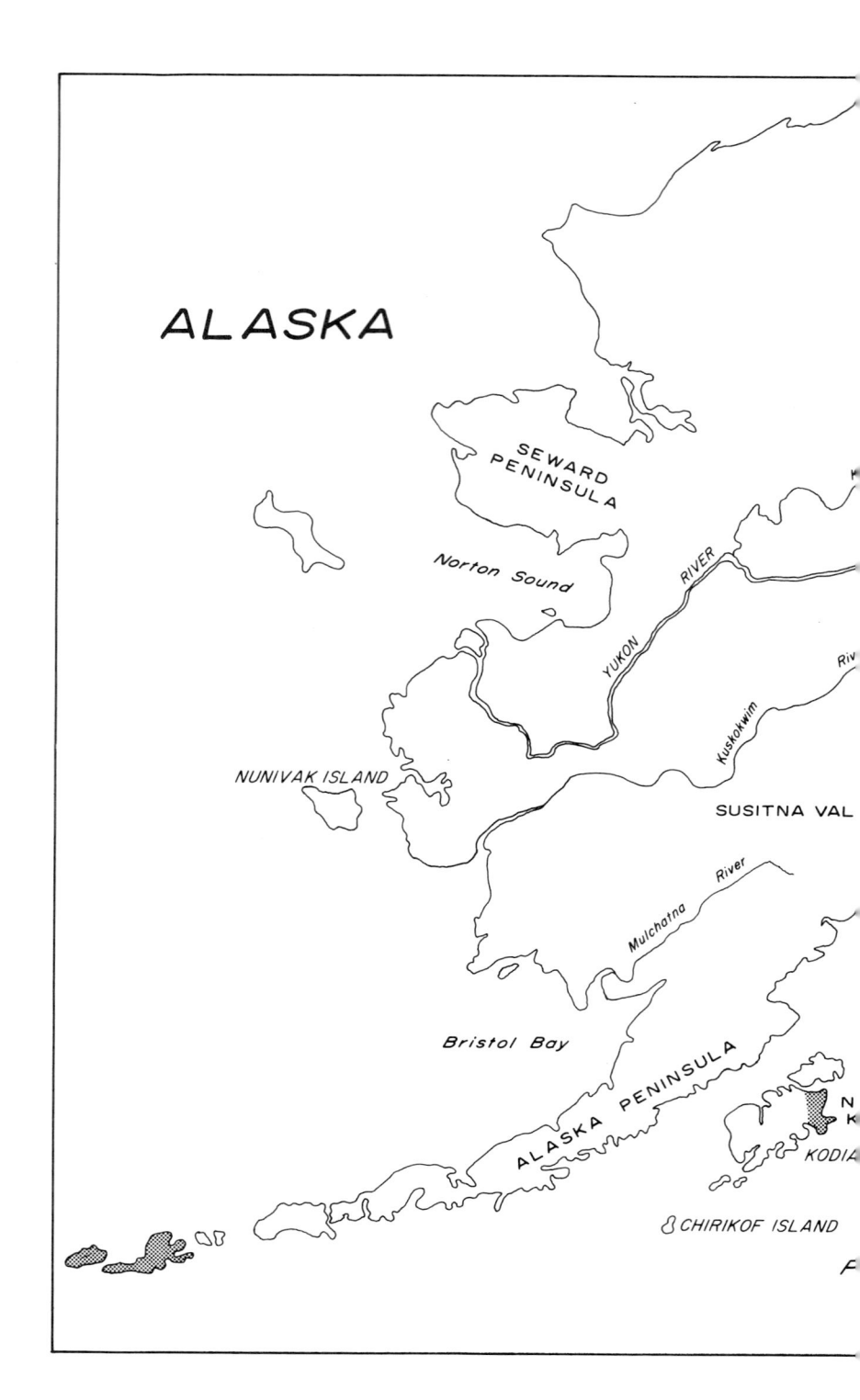

ALASKA

SEWARD PENINSULA

Norton Sound

YUKON RIVER

Kuskokwim River

NUNIVAK ISLAND

SUSITNA VAL

Mulchatna River

Bristol Bay

ALASKA PENINSULA

KODIA

CHIRIKOF ISLAND

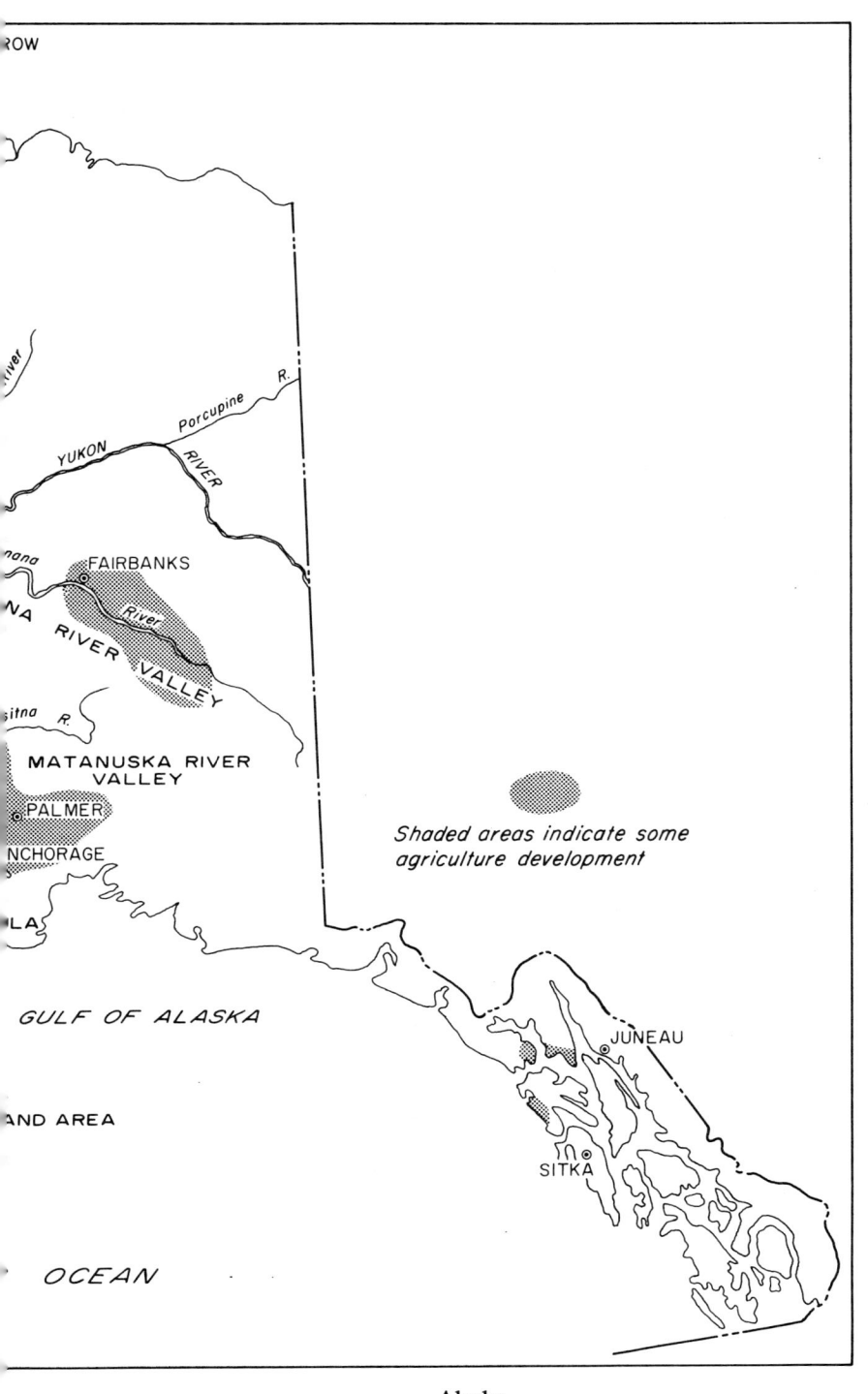

Alaska.

Cattle of North America

The station management became dissatisfied with the progress made with the Galloways and in 1917 imported one Holstein-Friesian bull and five cows to initiate a program of crossing Galloway and Holstein-Friesians. The goal was to combine the hardiness of the Galloway, which had done well in the Alaskan climate, and the milking ability of the Holstein. In 1919, in another attempt at genetic improvement, a Yak bull and cow were obtained from Canada, which were to be crossed with the Galloways. In 1920 a small herd of Milking Shorthorns was sent to the Matanuska Experiment Station, which had been established in 1917. A number of Red Danish cattle were sent to Alaska in 1948 from the USDA Research Center at Beltsville, Maryland. Many years of experimental work with these individual breeds and their crosses produced no more startling result than the usual conclusion that the Holstein-Friesian outproduces the other dairy breeds.

Winter forage presented the most serious problem with all kinds of cattle. The curing of hay was always difficult because of late rains that fell during the harvesting season. Grass silage was also difficult to hold through the winter since it froze if the moisture content was too high. Better management and modern machinery have now minimized these problems, and dairy herds can be profitably maintained if approved practices are followed.

Parts of Alaska are nearly as good cattle country as is to be found in some northern areas "outside." The following table presents a comparison of the growing seasons and annual precipitation rates in areas where cattle are raised in Alaska and in what are considered good livestock areas in some of the western states.

	Last Frost	First Frost	Days Growing Season	Annual Precipitation, Inches
Alaska				
Anchorage	May 25	Sept. 12	110	14.3
Fairbanks	May 29	Aug. 26	89	11.8
Sitka	May 10	Oct. 16	159	87
Kodiak	May 5	Oct. 12	160	61
Colorado				
North Park	June 29	Aug. 28	69	11.6
North Dakota				
Bismarck	May 10	Sept. 27	140	15.4
Wyoming				
Saratoga	June 6	Sept. 11	97	10.7

United States

The climate of southern Alaska is similar to that of the southern parts of Scandinavia and Finland, where productive cattle husbandry was developed by people who had lived off the land for many generations. There agriculture began at a subsistence level. Either the farmers learned to take care of themselves and their stock through the severe winters or both perished. There was no alternative for survival, and management practices evolved under which man and beast have both fared well.

These same agricultural processes appear to have been evolving during the Russian occupancy of Alaska; but when the United States took over, the new inhabitants brought with them what they needed, and there was no incentive to learn how to live off the land. The Russian element disappeared, and the newcomers, whose major interest was gold or trade, shipped in their routine living necessities.

The suitability of some areas for livestock husbandry is demonstrated by an event that occurred in the late 1880's. A company operating a fleet of whaling vessels in the Bering Sea and Arctic Ocean endeavored to establish a readily available fresh-meat supply in the area. Chirikof Island, 90 miles southwest of Kodiak, was chosen as a site, and one Shorthorn bull, one Jersey bull, two Jersey cows, and one Holstein-Friesian cow were released there and left on their own. There is no explanation recorded for the selection of this variety of breeds. The whaling industry fell on hard times a few years later, and the company pulled out and abandoned the cattle.

Fifty years later, in 1944, a party of government officials visited Chirikof in connection with a contemplated lease. In one area they counted 500 head of cattle. This was undoubtedly only a partial count, for large segments of the 100-square-mile island were not covered. There are no predatory animals on Chirikof, and with only the elements to contend with the cattle had multiplied and flourished.

The Alaskan cattle population now totals around 9,000 head, including 1,700 dairy cows and 2,500 beef cows. The development of beef strains, incidentally, was left to the individual rancher after the failure of the Galloway-Yak crossbreeding experiment mentioned earlier.

The state produces only 35 per cent of its fresh-milk requirement and probably less than 15 per cent of its beef requirement. Reindeer meat, produced and distributed mostly on the Seward Peninsula and Nunivak Island, provides more than 70 per cent of the beef produced. The reindeer herds are maintained by the Eskimos.

The lack of transportation for both beef and milk is a major handicap to expansion of either industry. The frontier atmosphere is another drawback. With no backlog of an established agricultural community

A dairy herd in the Matanuska Valley being fed green chop.

to draw on, there is little interest in any branch of the livestock industry. The newcomer to Alaska is not looking for a farm job but seeks the high-paid work he heard of before leaving home. Such work, he often finds, is nonexistent.

Of the 28 commercial Alaskan dairies 26 are situated in Matanuska Valley. The average unit has 300 acres of land and 40 to 50 cows, although one unit is building up to a 165-head, free-stall operation. Cows are normally pastured during the summer and kept in a loose housing shed and in the winter in a tight barn or in an enclosed loafing yard with free stalls. The use of free stalls and a loafing area throughout the year is growing. Under such conditions green chop is fed in the summer and hay and silage in the winter.

The Holstein-Friesian is the dominant breed. The average milk production of the Alaskan cow was 10,300 pounds in 1970. The high cost of feed leads the dairyman to cull his low producers quickly. With milk at 11 cents a pound, dairying can be a profitable operation under careful management.

Artificial insemination was introduced in 1948 and is now used by most dairies. The Matanuska Valley Breeders Association maintains a local bull stud and also brings in semen from outside bull studs. The major milk-processing plant is at Anchorage.

A Hereford herd being crossed with Angus. Kodiak Cattle Company, Kodiak Island.

Beef cattle are held for the most part on Kodiak Island, although there are small herds on some of the other islands and in southeastern Alaska. The Northeastern Kodiak Island Area is at present the cattle country of Alaska. The area includes 307,000 acres, of which only 100,000 are suitable for grazing. Except for 34,000 acres in a military reservation and a few small privately owned plots, the rest of the island, 1,920,000 acres, is in the Kodiak National Wild Life Refuge.

The Northeastern Kodiak Island Area is divided into 13 units, some of which are poorly adapted to stock raising because of location and terrain. The units with the better grazing lands are leased for a 20-year term to seven ranchers. These ranchers run 2,000 mother cows. The Hereford is the principal breed, and Angus make up the remainder, except for a few recently introduced Charolais bulls. Angus-Hereford crosses are being produced on some ranches for slaughter cattle.

Much of the terrain of Kodiak Island is rough and mountainous. Bluejoint and bunch grasses are the principal forages on the higher ground, while sedges are common in the low areas. All plant growth is luxurious during the long summer days, and in a normal year ample forage is left for winter pasture. Cattle are in the open the year round, but most ranches put up some hay to be fed for two to six weeks during the raw spring weather. That is the period when the Kodiak bear, emerging hungry from hibernation, is most dangerous to cattle.

The slaughtering facilities on Kodiak are limited, and there is no adequate transportation to Anchorage, the nearest population center, 300 miles up Cook Inlet. One of the ranches has a small slaughter house and kills two-year-old steers that have been fed grain for six weeks. A large plant with good facilities and adequate capacity for all the cattle produced in the area is under construction.

The available area of Kodiak Island is probably stocked to its practical carrying capacity; yet there are other areas in Alaska where cattle could be grown to advantage. As noted above, the environment in several locations is comparable to mountainous parts of the western states and the Northern Plains that have long been recognized as good cattle country. Lower temperatures are encountered in certain areas but are no more severe than those in the Peace River country in Canada, where cattle are raised successfully. In the Tanana Valley at Fairbanks there are 89 frost-free days, compared to the 69 days between killing frosts in the long-established cattle country of North Park, Colorado.

Except in a small way in a few favored locations the cattle industry of Alaska has failed to progress. The essential local know-how is lacking. Someday, when the need becomes more acute, Alaska will easily become self-supporting in both milk and beef production.

Hawaii

The Polynesians who went to the Sandwich Islands (now the state of Hawaii) sometime before 1000 A.D. took with them pigs, dogs, and fowl. The first cattle were introduced by Captain George Vancouver in 1793, 15 years after the discovery of the islands by Captain James Cook. Vancouver landed five cows and a bull at Kealakekua Bay on the Island of Hawaii. They were Mexican cattle which had been picked up in California, and they were reportedly presented to King Kamehameha by the captain. When the first missionaries arrived in 1820, cattle had migrated well inland to the uninhabited slopes of Mauna Kea and Mauna Loa and lived in a completely wild state.

For about 30 years after these feral bovine inhabitants had established themselves, there was little or no slaughter, and the natural increase was so large that they became a nuisance in the cultivated areas. Initially slaughter was permitted only at the direction of the king, but later the *kapu* (taboo) on cattle was relaxed, and they were killed for their hides and tallow. These products were exported to the United States, and

Hawaii.

Young wild cattle collected for rodeo work on Parker Ranch.

for many years were among the main sources of income in the islands. Most of the carcasses were wasted; the native Hawaiians had not developed a taste for beef, preferring fish.

Later Mexican cowhands were taken to Hawaii from California to help in the handling of the wild cattle. The preferred method was to rope an animal and leave it tied to a tree until it became starved to the point where it could be led or driven to slaughter.

Following the early missionary days the European population of the Hawaiian Islands increased, and with the growth of the sugar-cane plantations in the 1840's small dairies were established by Portuguese immigrants. The dairies were stocked first with imported Shorthorn cattle, known as Durhams, and then with dairy-type Shorthorns. Later other European dairy breeds were imported. Butter for export to California was the principal product.

In the early 1850's interest began to develop in beef herds, and small importations of Hereford, Angus, and Devons were made. Considerable salted beef was exported to California in the gold-rush days.

The first ranching operations had been established for raising horses and mules for use on the cane plantations. Goats and sheep were also introduced in considerable numbers and, like the early cattle, became feral in many areas on the island of Hawaii. As small railroads replaced the mule and horse, beef-cattle ranching expanded. The husbandry of

United States

goats and sheep also declined, but even today many large ranches on the island harbor sizable numbers of wild sheep and some wild goats.

In the early 1880's swamp buffaloes were taken to the island of Oahu from China for use in the rice fields. They followed the arrival of Chinese laborers who had been brought in for agricultural work. Considerable rice was grown for a number of years, but its cultivation was eventually abandoned for the more profitable sugar cane, and the buffaloes disappeared.

By the closing years of the nineteenth century the cattle industry was firmly established. Dairies, horse and mule ranches, and sheep operations were converted to cattle ranches. The largest of these, which at one time comprised 300,000 acres, was the much publicized Parker Ranch.

In 1809, as a youth of 19, John Parker abandoned ship on Hawaii when the vessel on which he had sailed for the Orient stopped at the island. He became a faithful retainer of King Kamehameha I, and was rewarded with title to lands which formed the nucleus of the Parker Ranch. The original holdings were increased by marriages with the royal family, but the operation did not keep pace with the times, and at the close of the century it was beginning to disintegrate. Alfred W. Carter, a loyal trustee appointed by one of Parker's descendants, began the development of the property which brought it to the dominant position it maintained for many years.

After the turn of the century other ranches were developed, small in comparison with the Parker spread but large by mainland standards, varying in size from 30,000 to 60,000 acres and running several thousand mother cows. The Hereford became the dominant breed through the importation of purebred stock and the upgrading of the wild Mexican cattle on a large scale. There was also some representation of the Shorthorn and Angus breeds. As these herds became established, the wild cattle were killed off to make room for them and to prevent interbreeding.

Today the Hawaiian cattle industry is in the hands of medium to large operators, generally running 1,000 to 4,000 mother cows. The Parker Ranch with 18,000 cows is in a class by itself. Hawaii Island, the "Big Island," has most of the cattle. Maui is next, and there are a few ranches on Kauai and Molokai. The cattle on Oahu are mostly in dairies. With 40 head per square mile in density of cattle population Hawaii ranks with such states on the mainland as Texas, California, and the western range states.

Large ranches often trace back to the Ahupuaa land grants, which the island kings gave to their relatives and faithful followers. These

Purebred Polled Hereford cows on pasture. Kukaiau Ranch, Hawaii.

grants extended from the seacoast to the last traces of vegetation on the mountains to enable a chieftain to supply all his needs: fish from the sea, taro (an edible tuber) from the lowlands, vines for his nets from the rain forest, and the Koa trees of the high slopes, from which canoes were fashioned.

In recent years a number of ranches have been organized by consolidating cane plantations closed down by various economic factors, not the least of which have been the demands of the labor unions. Lands acquired and held for appreciation in value are often put in cattle to obtain some current return on the investment. In general, land values are much higher in relation to the return for cattle than they are on the mainland. Much of the cattle land sells for $50 to $500 an acre, even up to $5,000 for sizable parcels in the path of development projects.

The terrain of the cattle country varies from rolling to hilly land that was formerly in cane to rough lava flows of sharp ridges and mounds on which vegetation has taken hold only in recent years.

Rainfall varies widely with location and in extreme cases may range from an average of 15 to 120 inches in different areas on a single property of 50,000 acres. Extended drought periods are frequent. Many ranches include large areas of rain forest and dense brush that are interspersed with open spots where grass and edible plants have taken hold.

Some of the pastures utilizing land formed by more recent lava flows are extremely rough. Ridges and huge hammocks of lava form pockets which would hold a fair-sized house. Since gathering cattle on such terrain is practically impossible by driving, the water trap is resorted to. Such a trap consists of a strong enclosure built around the stock tank which is the only source of water in a pasture. The narrow entrance to this enclosure is so constructed that an animal can force its way in to drink but cannot get out. In this manner all the animals in such pastures can be collected in a day or two and then moved by cowboys and trained dogs.

Cattle on the large ranches are ordinarily handled much as they are in the western-range country on the mainland. Men do the work on horseback, and their skill with horse and rope in the old days equaled the best that the West could produce. In 1908, Ikua Purdy, a Hawaiian cowboy, won the steer-roping event in the Cheyenne, Wyoming, rodeo.

Most of the vegetation, particularly the forages, on the islands comes from imported plants. The Kikuyu grass of East Africa is now widespread on pastureland. Pangolagrass was introduced from the southeastern mainland, and there is considerable guinea grass at lower elevations. Some napiergrass, or elephant grass, is grown for green chop by the dairies. The Ekoa bush is a favored forage legume that is highly prized since most of the other edible plants are low in protein. If not well grazed, though, it grows too high for the cattle to reach the leaves. Various clovers have been introduced but are hard to establish on the phosphorus-deficient soils.

Because of the volcanic origin of the islands, springs are almost unknown. The lava formation holds no underground water. Stockwater comes principally from the surface. Runoff is collected in areas of heavy rainfall and run by gravity or pumped to storage tanks or reservoirs. From there it is distributed by gravity pipeline systems to stock tanks in all the pastures on a ranch. Maintenance of the water facilities is the full-time job of at least one man on even a moderate-size ranch.

Many of the large ranches are very well managed. Purebred herds supply bulls for the commercial herds. Artificial insemination is practiced in some purebred herds with the use of semen from the progeny-proven bulls of mainland bull studs. Performance records are maintained and

Cattle of North America

utilized in selecting breeding stock. The practice is growing of pregnancy testing and culling all open females.

Three breeding seasons, each of 100 days, distributed evenly throughout the year are sometimes employed to provide a steady output of feeder cattle. A common ratio is 1 bull to 17 cows in the commercial herds, but on rough pastures this ratio is sometimes increased to 1 to 10. In single-sire purebred herds one bull is run with 40 to 50 cows the year round.

Calves are branded at 3 to 4 months and weaned at 8 to 10 months. Even at this late age most weaned calves undergo a stress period of 4 to 6 months, during which they practically stand still. While this phenomenon has been attributed to no definite cause, it is thought to be due to a nutritional deficiency. Deficiency of trace minerals has also been considered as a possible cause, even though, in some investigational work, it has persisted in spite of elaborate mineral-supplement programs.

High moisture content of the forage grasses (up to as much as 85 per cent at times in some of the wetter areas) and the low nutritive value of pastures also contribute to a calf's poor condition after weaning. Supplemental feeding helps, but even then the condition of the calves is still far from what could be expected in a more temperate climate.

Internal parasites plague young cattle the year round. Frost occurs only above elevations of 5,000 feet and continues for only two months. At lower elevations there is no break in the life cycle of parasites and insects. The usual practice is to treat calves for worms at weaning and again at shipping time. Some ranches also worm before weaning, when calves are four months old.

Calfhood vaccination for blackleg is universal. Many ranches spray with an insecticide whenever cattle are worked.

Anaplasmosis, brucellosis, and tuberculosis have been practically eliminated. Island breeders are particularly conscious of anaplasmosis. All animals imported must show a negative test before shipping and two negative tests at 30-day intervals after arrival before they are allowed free movement in the islands. The usual interstate requirements for brucellosis and tuberculosis are enforced for importations.

Cattle entering the feed lots from most ranches are 16-to-20-month yearlings, both steers and heifers, in the 550-to-650-pound range. The culled cows and bulls go direct to the packer sometimes after a little pasture-fattening on the ranch. Very few feeders are preconditioned on the ranch.

Until recent years slaughter cattle were 2½-to-3-year-old steers, killed off grass. As late as 1960 only 10 per cent of the cattle passing through the island packing plants were grain-fed. In 1971, 50 per cent were grain-fed.

Slaughter cattle remain the property of the rancher through the feed lot and packing plant, settlement being on the dressed-carcass weight.

Beyond the ranch the Hawaiian cattle industry has been dominated by the Hawaii Meat Company, of which 75 per cent is owned by the Parker Ranch interests. This company operates the only large feed lot in the state and the major packing plant. Both are located in Honolulu. The feed lot has a capacity of 16,000 head and feeds 32,000 head annually, or over one-half of the 60,000 killed in the state. The packing plant of the Hawaii Meat Company slaughters 30,000 head annually, including culled cows as well as fed cattle.

In recent years other ranching interests have been making inroads into the near monopoly enjoyed by the Parker Ranch interests. On Oahu, Kahua Beef Sales Company, which was organized by the Kahua Ranch and in which a number of large ranches hold an interest, annually slaughters 12,000 head of the shareholders' cattle. Most of these are fed in the Hawaii Meat Company feed lot. There are several small packing plants on Maui and Hawaii which process cattle raised on those islands for the local trade. Some of these are supplied with fed cattle from small feed lots; others process grass-fattened 2½-to-3-year-old steers and culled cows. The 800-head feed lot of the Ulupalakua Ranch on Maui handles up to 2,000 head annually.

Transportation is the major obstacle to feed lot operations in the state. Feeder cattle must be barged from the other islands to the one large feed lot on Oahu. The trip involves a loss in weight of 10 to 12 per cent, as well as the stress involved in the movement. All concentrates employed in feeding, as well as some of the roughage, must be shipped from the West Coast. Pinebran, a product produced at the canneries by grinding and dehydrating pineapple peelings, is a fair roughage, but most of it is bought by the dairies, which will pay a higher price than the feed lots. Other than that, very little roughage is available except what comes from the mainland. The frequency of labor-union strikes in recent years, both on the islands and on the mainland, also adds to the problems in feed-lot operations. The charge per pound of gain was around 30 cents in 1971 at the Hawaii Meat Company feed lot, based on the weight of feeders as received off the barge and slaughter animals going to the packing plant with no shrink allowed. Under the usual United States weighing conditions this would increase to 35 cents per pound.

Beef grown and processed on the islands supplies 40 per cent of the state's requirement; the remainder is divided about equally between shipments from the West Coast and those from Australia and New Zealand. The beef from the mainland is mostly of choice grade, and this factor determines the price level in Hawaii, where choice-grade carcasses run 4 cents a pound higher than the going West Coast price; this differential represents the cost of transportation and handling.

Hawaiian dairies are geared to the production of fluid milk, since nearly all dairy products are shipped in from the mainland. The largest dairy is the Meadow Gold Farms on Oahu, which milks 1,500 cows. All heifer calves are raised for replacements. They are grown out on pasture in age groups under a strict routine for control of internal parasites. The milking herd is maintained in dry lot, but cows calve on pasture. Four lactations are considered normal for a cow, and milk production averages 13,000 pounds. The herd is nearly all Holstein-Friesian, only a few Guernsey cows now remaining from the days when a previous manager favored this breed. Bull calves were formerly disposed of at birth, but recently a market has been developed for them in California, and calves a few days old are now shipped by air to West Coast nurseries, where they are grown out for veal or feeders.

Transportation and labor problems hamper dairy operations. Practically all concentrates must be imported. Much of the roughage is also imported from California in the form of hay pellets or cubes. Some pineapple tops are chopped and fed but are low in nutrients and are not readily available even when pineapples are harvested.

There are 27 dairies on Oahu, a number of which milk several hundred cows. These dairies adequately supply the local demand for fluid milk. Currently the farmer receives 23 cents a quart. The other islands have dairy herds which completely meet their needs. The Halaekaka Ranch and Dairy on Maui milks 500 Holstein-Friesian cows, which are run on pasture throughout the year and average 12,000 pounds of milk per lactation. This operation also includes 3,000 Hereford cows producing yearling feeders.

There is some thought in agricultural circles that Hawaii is turning from sugar to cattle. Some years ago marginal cane lands were put into grass and used for raising cattle. Now even more productive areas are finding sugar production unprofitable. In recent years considerable experimental work has been devoted to raising corn and milo, both for grain and for silage, on land formerly in cane. Some of this development now appears to be nearing the stage of practical application. Grain or forage production on the high-value land may be an economic possibility with the improved seeds and fertilizing techniques that are avail-

United States

able and when two or even three crops a year can be produced. During intervals of the past century considerable barley and wheat were raised in Hawaii and shipped to California, evidence that the basic limitations on grain production are economic.

Whatever the future may hold for harvested feeds, it seems certain that Hawaii will maintain at least its current cattle population. There are at present 1,250,000 acres of land in pasture or in areas which could be converted to pasture. Much of the latter land is now unused because of dense vegetation, and grazing would be its only possible agricultural use. Under current economic conditions the cost of clearing and seeding obviates the use of such land for cattle. There is, however, a possible potential to increase cattle numbers if land values should ever revert from a highly speculative to a realistic level.

Puerto Rico

The Commonwealth of Puerto Rico is a nearly rectangular island which lies 1,200 miles southeast of the southern tip of Florida. It is the eastern extremity of the Greater Antilles and marks the division between the Atlantic Ocean on its north coast and the Caribbean Sea on the south coast. Puerto Rico, 3,440 square miles in area, is less than half the size of New Jersey. The climate is warm and pleasant, with an average temperature of 76.5 degrees. Rainfall varies widely in different parts of the island, generally averaging 69 inches annually on the northern plains and 40 inches or less in the south. In the mountainous center precipitation exceeds 90 inches and increases to 180 inches in a small rain-forest area near the eastern coast. There are no pronounced rainy seasons, although from May through December rainfall is generally heavier than during the remainder of the year.

Puerto Rico, along with Cuba, was the last remaining vestige of Spain's dominion in the New World. After the Spanish-American War Spain lost control of both islands in 1898. Cuba subsequently became independent, but Puerto Rico elected to maintain a close political relationship with the United States. The Puerto Rican is a citizen of the United States. The island's status as a commonwealth is unique in American history. Neither a state (in the sense of the fifty states) nor a possession, Puerto Rico is in a highly advantageous economic position. Although personal income taxes imposed by the commonwealth are on a par with federal income taxes in the United States, corporate taxes are materially less, and a fantastic degree of tax exemption has been granted to new industry to entice it to the island. This has resulted in such rapid economic growth in recent years that Puerto Rico now exceeds any other Caribbean or Latin-American area in its economy and is nearly on a par with some of the southern states of the United States.

The people are predominantly of mixed Spanish and Negro descent, although there remains a highly influential core with predominantly Spanish ancestry. The population is nearly 2.75 million and is highly literate. The population density is nearly 800 per square mile, higher than any state except New Jersey. Exodus from the agricultural areas to the cities has been at a high rate, but the rapid growth of industry in the past 20 years has avoided excessive unemployment. Heavy and continuous emigration to the United States has also been a major factor in this respect. There are now said to be as many Puerto Ricans living in or near New York City as in the capital city of San Juan, which has a population now approaching 1 million.

The agricultural pattern of Puerto Rico has followed several stages from the early settlement days, when hides, cassava, and root crops were the principal products. Then came the cultivation of ginger, followed by coffee and, a little later, cacao. Tobacco was the next major crop, and was followed by sugar, which was the backbone of the economy for over a century. Following the rapid industrial expansion that began after World War II, sugar production declined. In recent years mills have been sold for dismantling and exported for reassembling in such sugar-growing countries as the Philippines, Venezuela, and the Dominican Republic. During 1969 sugar production in Puerto Rico declined 25 per cent. Dairy farming is replacing cane growing in some areas. While the economics of such a transposition is hardly profitable, the fact that there are fewer problems with labor and an excellent market for milk seems to satisfy the landowner.

The Criollo descendants of the cattle taken to Puerto Rico by the early Spanish settlers were, after the middle of the nineteenth century, crossed with Zebu cattle for better draft animals. This was the principal function of cattle until well after the turn of the twentieth century. When mechanization replaced the draft oxen on the large plantations, cattle husbandry turned to raising beef cattle and, to a lesser extent, to dairy cattle. There were minor importations of some of the British beef breeds, but they had no noticeable influence on the dominant Zebu-Criollo cross.

The original Criollo cattle have practically disappeared, and only a few animals of the Zebu-Criollo draft type are to be seen.

Later, when the industrial expansion began to draw the rural population to the cities, more attention was given to both the genetics and the management of cattle. This followed naturally to keep production

Mature bull, 950 pounds, showing strong Criollo influence.

levels up with less labor. The major emphasis was then on milk production, and northern-type dairy breeds were imported. The mixed Criollo cattle were upgraded principally to the Holstein-Friesian, although there are some dairies with representatives of Jerseys, Brown Swiss, and Ayrshires. The cattle of the small landowner increased in number and in a minor way followed the same upgrading trend.

In 1970 there were 205,000 cows in the commercial dairy herds. A large number of these are held by the owners of the large plantations formerly in sugar production. There are dairies milking over 500 cows, but more typical are the 100-to-200-head herds. In most of these operations cows are kept in loafing yards between the morning and evening milkings. Shade is provided, and green chop is brought to them. At night they are turned to pangolagrass pastures. There are some operators who maintain the milking herd continuously in small enclosures and all feed is brought to them. Prepared feed, imported from the United States, is widely used as a supplement. In some instances it is the only feed used, since the preparations contain sufficient ground roughage to make a balanced ration. Some dairy concentrates are mixed locally but embody imported grains. The dry cows and young stock over seven to nine months of age are usually kept on pasture but are given some supplement as well.

Artificial insemination is employed in breeding most dairy cattle. Some herds, however, include a bull or two for shy breeders or for emergencies. The government artificial-insemination center at Dorado breeds around 36,000 cows annually. Other breeding services raise this total to 50,000 head. Many of the larger dairies import semen direct from United States breeding centers.

Bull calves are commonly sold for slaughter when a day or two old at a standard price of $25 (1970). Heifer calves are grown out either for replacements or for sale as producers. From birth to 3 or 4 months, they are generally in individual, slatted-floor pens. They are then segregated into age groups and kept on slats until around 8 months of age, at which time they are put on pasture. Heifers are bred when 16 to 18 months old.

Most dairies make use of the herringbone gate and walk-through-type milking parlor. Bulk storage tanks are in general use, and the milk is transported to the pasteurizing plants by tank truck except from the small farm producers. Sanitation is good.

The practice of keeping production records on individual cows is common. Average production per cow is around 6,000 pounds per lactation but in some herds reaches 10,000 pounds.

The small farmer maintaining a dairy herd usually milks 30 to 50 cows, often by hand. His operation, insofar as possible, parallels that of the larger dairies. Milk production per head will generally range from 3,500 to 6,000 pounds per lactation.

The beef-cattle herds of Puerto Rico are predominantly upgraded American Brahman, varying in size from a few hundred head to at least one herd of 2,500 head. These are run on pangolagrass or guinea grass pastures, which are commonly cross-fenced to provide good grazing control. Bulls are usually with the cow herd the year round. The average calf crop on well-managed properties is around 65 per cent. Steers are sold at 18 to 24 months of age at weights ranging from 800 to 1,000 pounds.

For the past several years there has been an increasing use of Charolais bulls in the beef herds. This practice has not been of long enough duration to give evidence of upgrading. It is thus too early to determine

Young United States feeder bulls growing out on pasture in southeastern Puerto Rico.

whether the Charolais in a near-pure state will prove as productive as the cattle carrying some Zebu influence. Because the Charolais seems to be more tolerant of heat stress than many other European breeds, it is possible that the upgraded Charolais will at least equal, if not surpass, the production of the present Zebu-type cattle. They may be helped by the fact that the Puerto Rican climate is not as tropical as that of some of the other Caribbean islands.

A recent innovation has been the importation of 6-to-9-month-old feeder calves from Florida, Mississippi and Louisiana to grow out on grass. Calves weighing 400 pounds are said to gain up to 2 pounds a day on good pangola pasture to reach a weight of 1,000 pounds at 18 to 24 months. Such gain can be obtained only in the 60-to-80-inch rainfall areas. Some 8,000 head of feeders were handled in this manner in 1969. With a 1,000-pound steer selling off the grass for around $260, apparently this can be a profitable operation.

There is no organized system of marketing cattle in Puerto Rico. Buyers visit the larger operators and buy directly from them. Prices are based on an arbitrary dressing percentage of 40 per cent for bulls or steers and 38 per cent for cows. These percentages are used in calculating the number of arrobas for which the owner is to be paid (an arroba equals 25 pounds of dressed meat). Thus a 1,000-pound bull is credited with a dressed weight of 16 arrobas, or 400 pounds; a cow of the same weight, 15.2 arrobas, or 380 pounds. In early 1970, $16.50 per arroba was paid for bulls, $14 for cows. This was equivalent to 26 cents per pound liveweight for bulls and 21 cents for cows.

There is no consistent relationship between price and quality. Some recognition of quality exists, however, for a young animal in good flesh brings a somewhat higher price than the prices mentioned above, and a thin cull correspondingly less. The lack of an organized market and of a consistent recognition of quality grade in the price paid for an animal are the most serious handicaps in the beef-cattle industry in Puerto Rico.

The dairyman sells milk to the pasteurizing plant for .095 cent a pound. It retails at 14 cents, delivered to the door.

There is a modern abattoir in San Juan conforming to average United States standards. It was constructed optimistically to handle a much larger number of cattle than are presently available. The major towns have small municipal plants with only elementary facilities. Sanitation and inspection, however, are improving.

Puerto Rico is free of foot-and-mouth disease, is certified free of tuberculosis, and is a modified brucellosis-free area. Cattle have suffered little from the common cattle diseases. As the result of a federal-state program of fever-tick eradication, the country is now considered free of piroplasmosis.

Brucellosis is controlled in the large dairy herds by annual test and slaughter of the occasional reactor. A number of years ago vaccination of heifer calves with Strain 19 was common, but the practice was generally discontinued as the incidence of brucellosis declined. Pasteurizing plants employ the ring test to determine whether the herds from which they buy milk are free of brucellosis. There is thought to be considerable brucellosis in the beef-cattle population, particularly among the cattle of the small owners. There is now a federal-state control program.

Except for the cattle belonging to the small farmers, herds are sprayed regularly. The interval between treatments varies with the tick and horn-fly population in the particular area. Where systematic spraying has been practiced for years, treatment at monthly or even longer intervals gives satisfactory control. In some areas, however, spraying every week is considered necessary.

Worming of calves once or twice before weaning is a general practice in well-managed dairy and beef herds. The small dairyman milking 20 or more cows is also beginning to follow the health-protection practices of the large milk producer, although not as systematically. Small farmers, with only a few head, however, do practically nothing for the cattle's health.

There is considerable effort on the part of government to encourage both milk and beef producers. The farmer is subsidized to the extent of 50 per cent of the cost of developing improved pastures. Breeding stock can be imported free of duty, and transportation charges on imported animals are paid by the government. The government artificial-insemination center trains farmers as well as technicians in the technique of breeding. Semen is furnished to individual technicians at 55 cents an ampule with the stipulation that their charge per cow may not exceed $4.50. This price covers three inseminations if necessary. Organizations sponsoring the use of artificial insemination and the keeping of dairy-herd-improvement records have been very successful.

The position of cattle in the economy has undergone marked changes since the rapid industrialization which occurred after World War II. The draft oxen disappeared as better-producing beef and dairy cattle replaced them. Parallel with this genetic improvement, husbandry practices have advanced rapidly during the past 25 years. Native-grass pastures and marginal cane lands on the most progressive farms were converted to the much more productive pangolagrass, cross-fencing was adopted, and better provisions were made for stockwater. The large dairies feed an adequate ration even though it must be imported. They have modern milk parlors, refrigerated storage tanks, and tank-truck transportation to the pasteurizing plants.

In spite of such advancements as these in the cattle industry, Puerto Rico continues to import 60 per cent of its milk products and beef. Improvement in the economic level of the people is continuing, and industrialization is providing employment for the growing population. The national consumption of beef and milk therefore appears certain to increase.

If all the cane land were immediately converted to grass and the growing of other animal feeds, Puerto Rico might approach a position of self-sufficiency in beef and milk. The economics of such conversion, however, would not support a total change from cane to pasture. Sugar still produces a greater return from the land than do cattle. The difficulty in obtaining labor to work in the cane fields has been a basic reason for putting some cane land into grass. Also, the older plantation owner with an ample holding has come to prefer the more simple routine of a modern dairy operation as opposed to the labor problems and weather hazards involved in growing sugar cane. His materially lower net return from dairying is compensated for to a considerable extent by the increase

Cattle of North America

in the value of his land. The price of all land in Puerto Rico continues to rise phenomenally.

Cane cutting is still largely a hand-labor endeavor. Mechanical cutting equipment is being used to an increasing extent on fields level enough to permit its use, but as yet such equipment has not been adapted to uneven terrain. Much of the cane land is of this type and will probably eventually be put into grass. If this conversion is carried to the limit, beef and milk production will be materially increased. By the time such a procedure is completed, however, population growth and increased consumption levels could be expected to catch up with the additional production.

The possibility exists for increased production through better pasture management, but no sudden improvement can be anticipated in this direction. The average Puerto Rican farmer cannot be expected to move any more rapidly in such matters than his counterpart in the rest of the world. While the size and productivity of the island's beef and dairy herds will probably increase, the importation of beef and milk products will remain essential to the Puerto Rican economy in the foreseeable future.

MARKETING

BEEF

In colonial days the town butcher purchased his slaughter animals directly from nearby farmers. As the population increased, markets were held in the larger towns on established days, and eventually the cattle dealer entered the picture. He bought from the farmer and sold to the butcher, who had become more specialized in his trade and killed for retail stores. To supply the large cities that grew up on the eastern seaboard, cattle were bought in the areas where they were raised and driven to the large markets or direct to the slaughterhouses, and the role of the dealer became ever more important.

The industrial and railroad expansion of the nineteenth century brought about the centralization of stockyards and slaughterhouses. The yards attracted trade from a wide area, and commission houses which received cattle on consignment from the producers and negotiated sales with the packers or other buyers replaced most of the country traders.

Scales came into general use and the practice of selling cattle by weight rather than by the head became widespread.

The major stockyards were invariably surrounded by a large complex of packing plants, as in Chicago, Kansas City, and Omaha, and became known as terminal markets. They dominated the cattle trade until the middle of the twentieth century, and prices historically were based on the Chicago market after allowance for freight differentials.

By the 1940's the marketing pattern had begun to change, and during the next 30 years transactions at the terminal markets declined until they represented no more than one-fifth of the cattle sold nationally. A number of elements were responsible for this change. They included the decline of the railroads and the increase in truck transportation, the decentralization of the packing plants, loss of confidence in the commission firms and in the private-treaty method of selling, the increase in direct buying and the operation of buying stations by packers, and the increasing popularity of the auction market.

In 1971 the Chicago Union Stockyards, at one time the largest of all the terminals, closed its gates for good, following the action taken by the Denver yards the year before. The question now seems to be how much longer the remaining terminals in Kansas City, Omaha, and a few other cities can survive.

The growth in truck transportation which has taken place since World War II has done much to change the pattern of the cattle industry across the country. In the past cattle movements traditionally followed the railroads from producer to feed lot, from feed lot to packer, and finally from packer to retail warehouse. As the railroads were typically east–west oriented, so were the routes that cattle followed from ranch or farm to retail outlet. North–south lines were either nonexistent or provided poor service. The stock truck for live cattle and the refrigerated truck for dressed beef now move freely in any direction. Except for very long hauls, cattle are now seldom moved by railroad. Because of the greater availability of transportation, cattle in small lots can be grown to advantage in remote areas where getting them to market was formerly a serious problem.

The accompanying chart indicates the general flow of cattle from the producer to the packing plant (1970 numbers).

Grading

There was some recognition of the importance of quality in beef as far back as pre-Revolutionary days. The carriage trade in Boston was particularly discriminating. To supply this market, the Boston butchers paid a premium for the grass-fattened pre-Devon-type cattle that were raised in western Massachusetts. Philadelphia and New York were also known as markets for high-quality beef. There were not, however, any

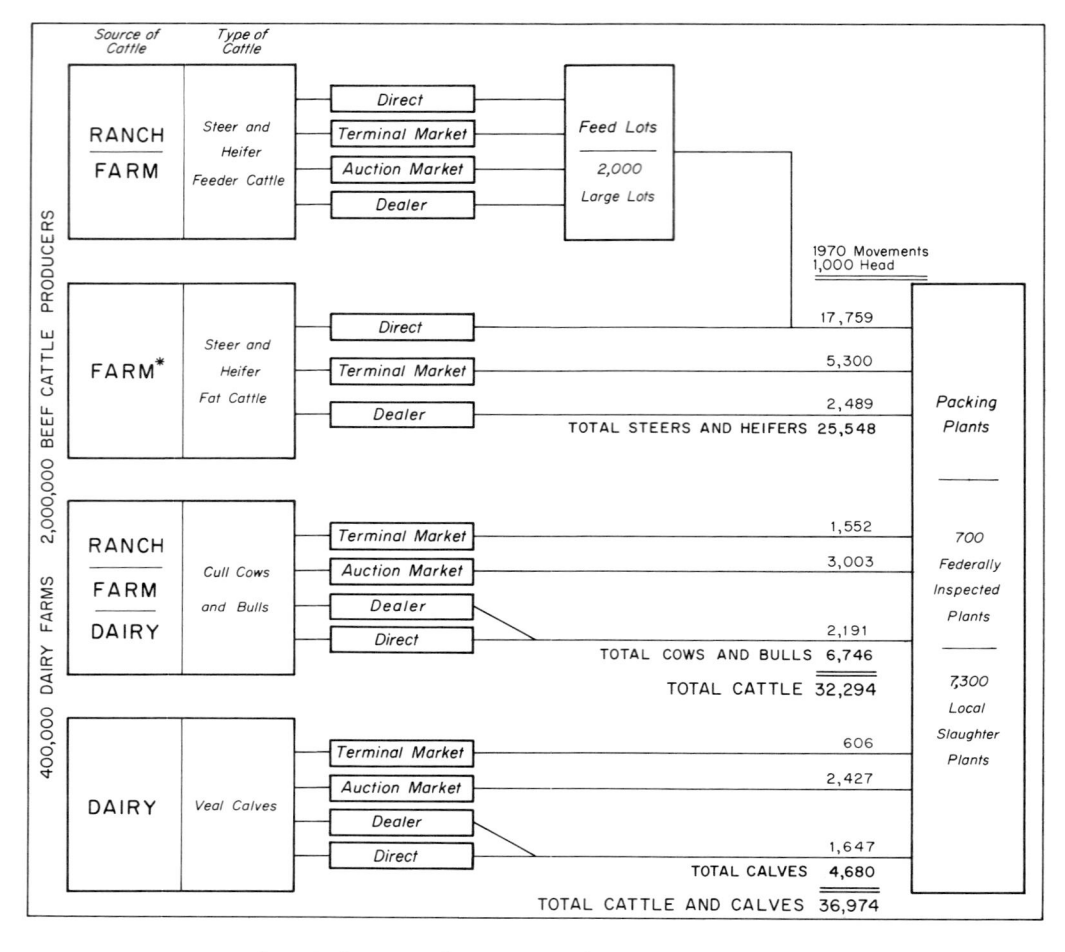

The flow of cattle from producer to packing plant. Based on data from *Packers and Stockyard Résumé*, USDA.

*There are an estimated 206,000 small feed lots—those with less than 1,000-head capacity—on farms.

nationally recognized standards of beef quality until the first USDA grades were established in 1926. These have been revised over the years and have served the trade well.

Quality Grade.—Quality grades as now employed apply to feeder cattle (generally calves or yearlings), to slaughter cattle, and to dressed carcasses. Eight quality grades have been established, but prime, choice, good, standard, and utility account for almost all the beef sold. The three grades of minor importance are commercial, which applies mostly to aged cows in thin condition, and cutter and canner, which apply to aged cows in particular but also to cattle of any age in very thin condition.

An outline of the relationships among the five principal grades for feeder cattle, slaughter cattle, and dressed carcasses is given in the table on the following page.

Historically the quality grades of beef as defined by the USDA have applied only to steer and heifer carcasses. Bull carcasses, whether from an old or young animal, are required to be classified as "bull." Many comparisons of fed steer and young, fed bull carcasses have shown that steer and young bull beef are comparable in quality and palatability, although a slight preference for the steer beef has usually been mentioned. Such preference could well be due to the taster's familiarity with steer beef and the fact that his unfamiliarity with the flavor of bull beef has prejudiced him against it. With the exception of Great Britain, beef from young bulls rather than from steers supplies most of the quality beef market in European countries. The advantages pertaining to the production of beef from young bulls, such as higher feed efficiency and less fat, was officially recognized by the USDA in early 1972. Revisions in quality grade standards were proposed by the Agricultural Marketing Service of the USDA in March of that year which would distinguish between the beef from young and old bulls. The quality grades heretofore applicable only to steer and heifer carcasses would be made applicable to young bull carcasses, which would, however, have to carry the designation "bullock." While no grading standards for young bulls were established immediately, it now appears certain that they will be forthcoming. When this occurs, there will probably be no immediate major movement to feed bulls for slaughter, but over the years the practice will undoubtedly increase as the advantages of higher feed efficiency and less fat in the carcass become generally recognized. The banning of DES as a feed additive could also hasten the advent of bull feeding. See the following pages for tables and illustrations of the different grades of feeder and slaughter steers, of rib-eye sections, and of yield grades.

USDA Grading System
Feeder Cattle (Calves, Yearling Steers, or Heifers)

PRIME: Top-quality stock, well muscled, growthy, with a smooth finish. Usually shows character of a straightbred beef breed.

CHOICE: Very close to prime in many respects but without the eye appeal of the animal put in prime grade. A fair per cent will feed out to prime slaughter animals.

GOOD: Displays a lack of good beef conformation—narrow body, unduly long-legged, small for age. Small percentage only will feed out to choice.

STANDARD: Small for age, angular, narrow. Poor conformation. Usually shows some dairy or mixed breeding. Usually short-fed and finish standard.

UTILITY: Thin, rough cattle, small for age and in poor condition. Mostly put on short feed and finish in utility grade.

Slaughter Cattle

PRIME: Young steer or heifer usually under 3 years of age (42 months allowed maximum age), very smooth cover, well-muscled finish. Can result from a well-fed choice steer.

CHOICE: Young steer or heifer; similar in appearance to a prime grade except not quite as fat.

GOOD: Shows a thin fat cover and an inferior beef conformation. Narrow rump and shallow body.

STANDARD: Rangy, thin young steer or heifer. Angular and long-legged.

UTILITY: Thin and angular. Little fat cover, and what there is, very patchy.

Carcass

PRIME: Young steer or heifer. Smooth, fat cover; rib eye shows abundant marbling. Well muscled throughout. Usually obtained by 10 to 12 months' feeding of a choice or prime feeder. Heavy fat cover, usually ¾ inch or over in thickness.

CHOICE: Young steer or heifer. Usually differs from prime in having less fat cover, although this is not required. Good marbling in rib eye but definitely less than in prime.

GOOD: Rib eye shows light marbling and is small for the size of the carcass. Fat cover is thin. Conformation is narrow.

STANDARD: Rib eye shows very little marbling and fat cover is thin. Rib eye small and narrow.

UTILITY: Narrow rib eye with only trace of marbling. Fat patchy; poor general conformation.

USDA Grades of Young Feeder Steers
These photographs illustrate the quality grades of young cattle destined for the feed lot. Photographs courtesy Consumer and Marketing Service, USDA.

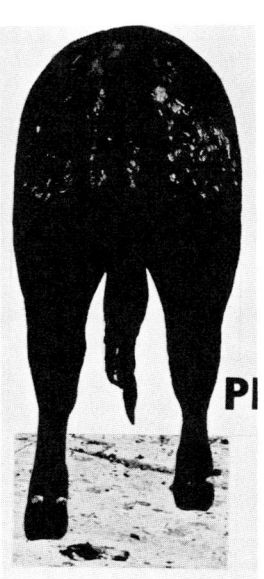

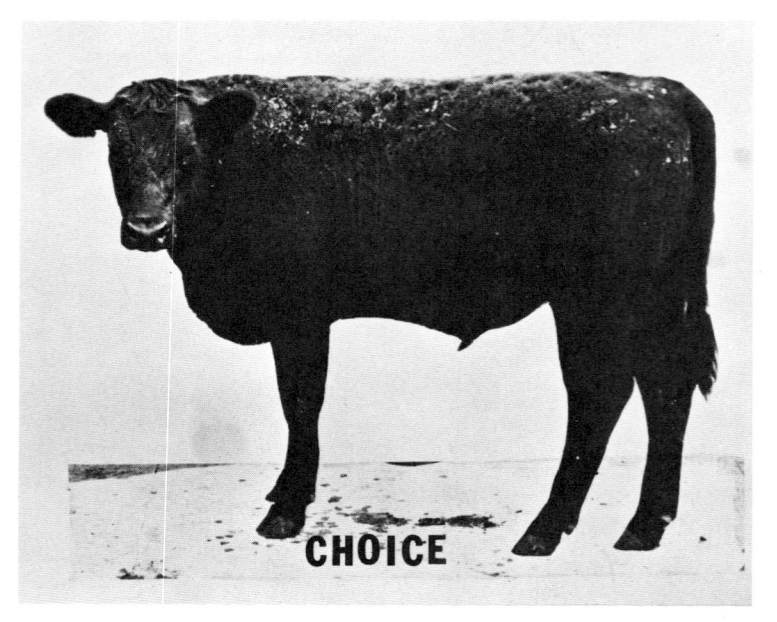

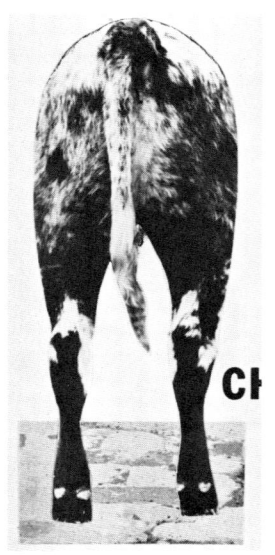

CHOICE

GOOD

STANDARD

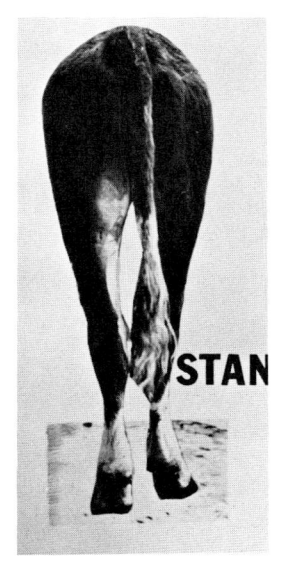

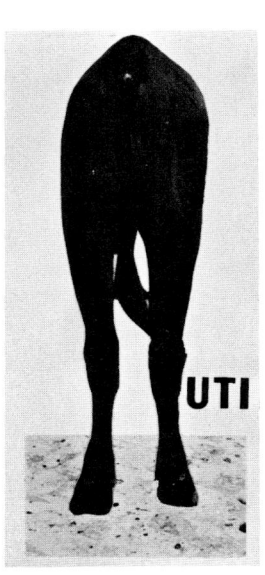

UTILITY

Cattle of North America

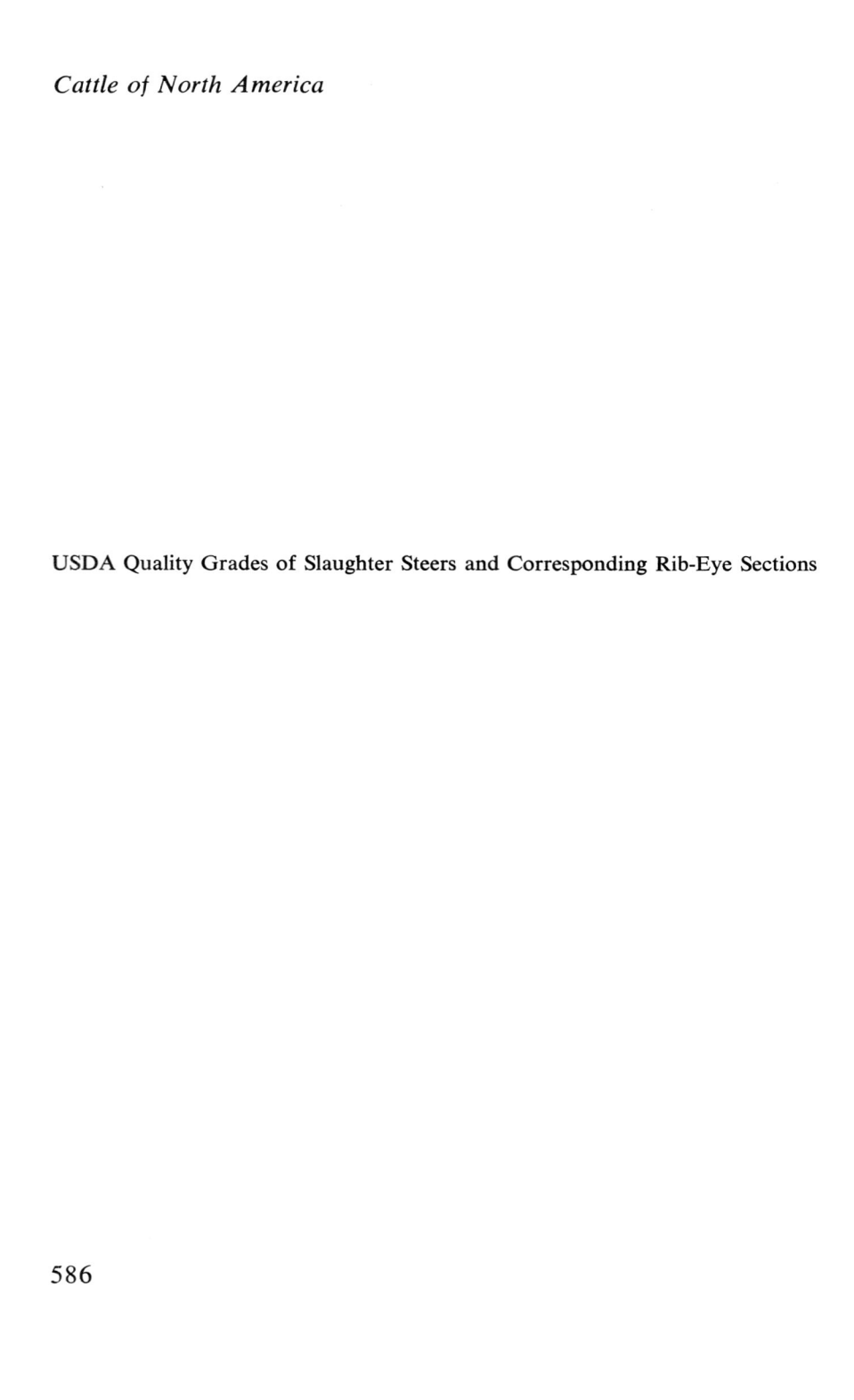

USDA Quality Grades of Slaughter Steers and Corresponding Rib-Eye Sections

Prime grade slaughter steer. Drawing from Agricultural Marketing Service, USDA.

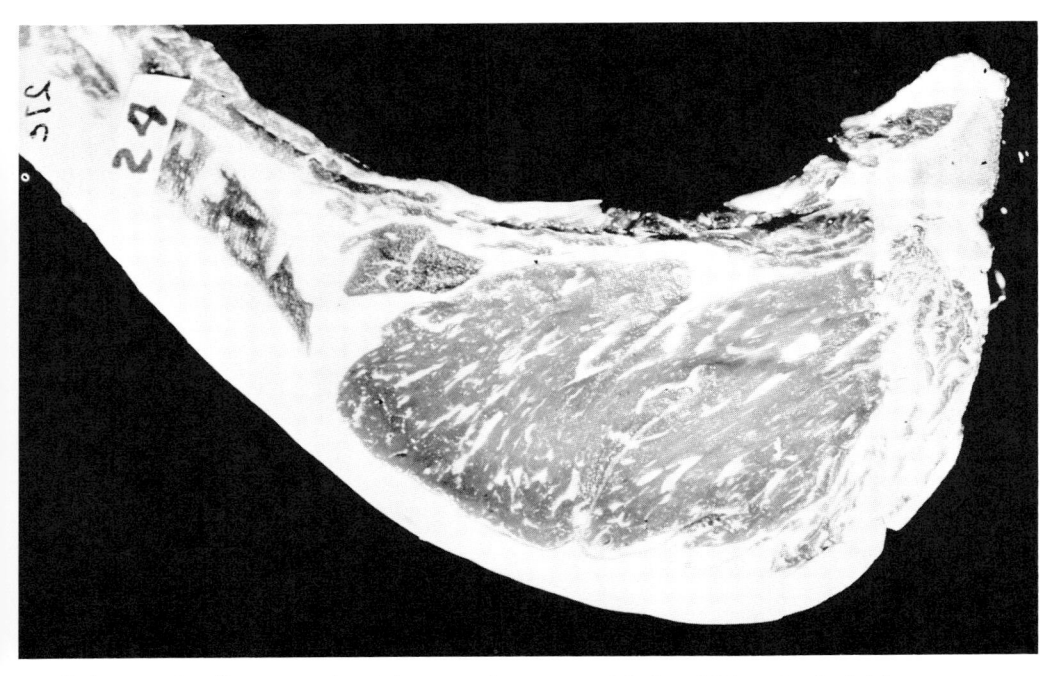

Prime grade rib-eye section. Photograph courtesy National Live Stock & Meat Board.

Choice grade slaughter steer. Drawing from Agricultural Marketing Service, USDA.

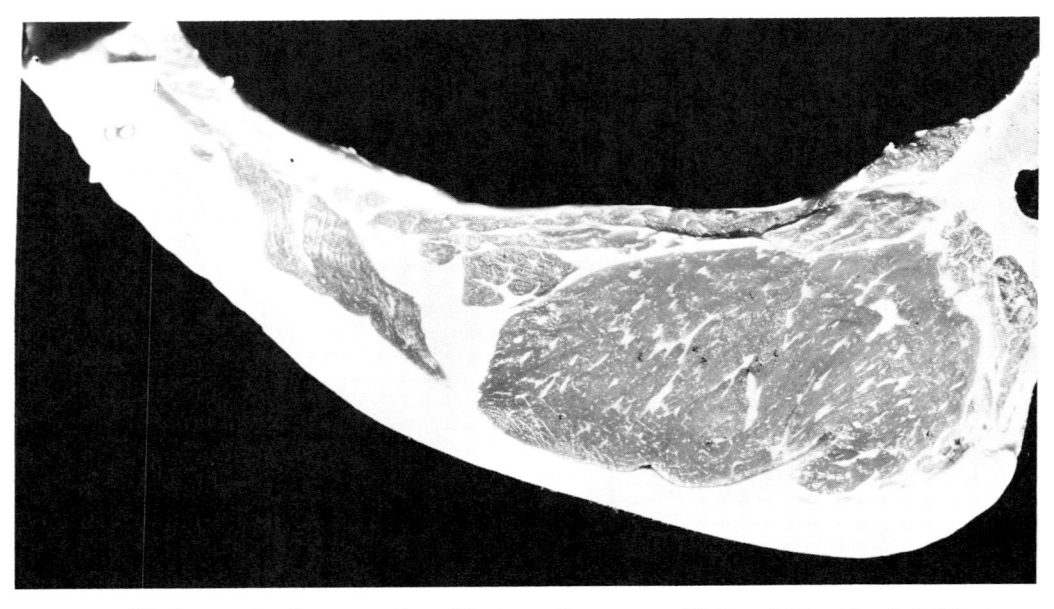

Choice grade rib-eye section. Photograph courtesy National Live Stock & Meat Board.

Good grade slaughter steer. Drawing from Agricultural Marketing Service, USDA.

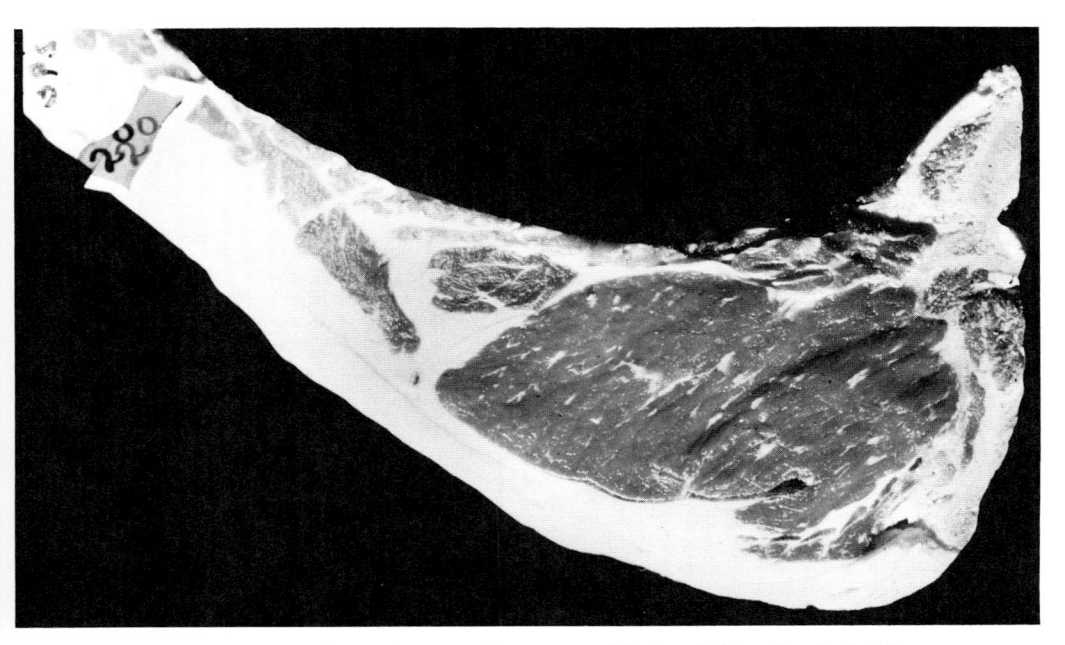

Good grade rib-eye section. Photograph courtesy National Live Stock & Meat Board.

Standard grade slaughter steer. Drawing from Agricultural Marketing Service, USDA.

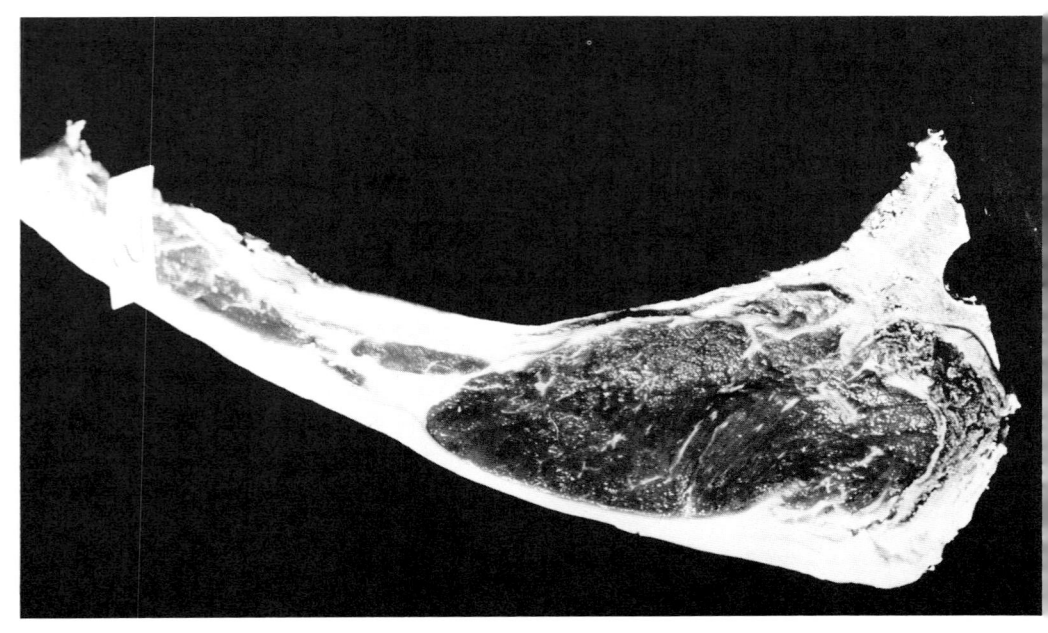

Standard grade rib-eye section. Photograph courtesy National Live Stock & Meat Board.

Utility grade slaughter steer. Drawing from Agricultural Marketing Service, USDA.

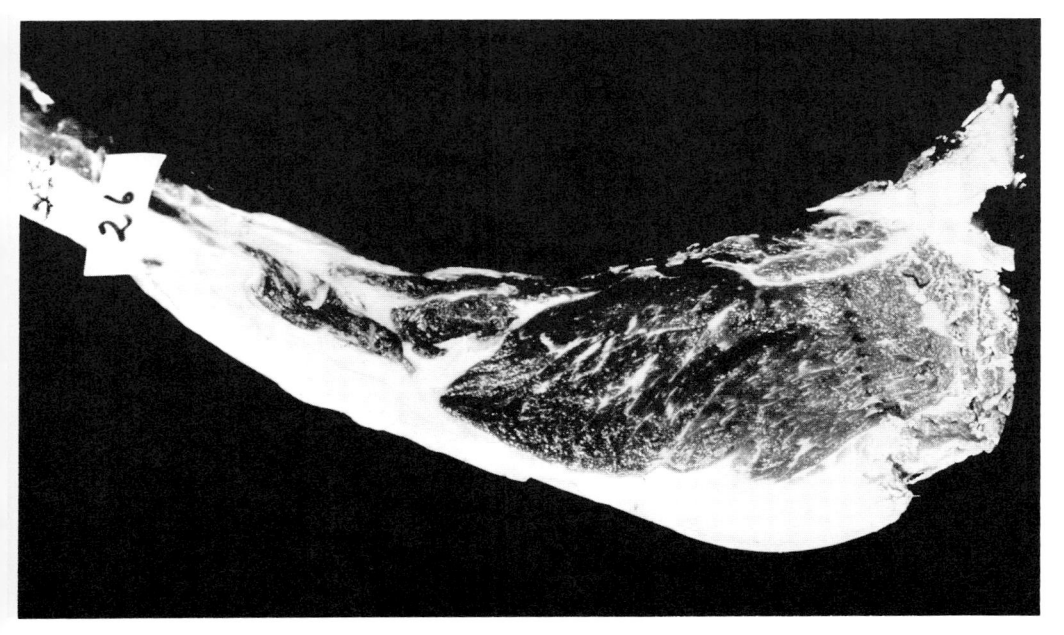

Utility grade rib-eye section. Photograph courtesy National Live Stock & Meat Board.

USDA Yield Grades of Slaughter Steers and Corresponding Rib-Eye Sections. The following illustrations of yield grades are for animals of USDA choice quality grade.

Yield grade No. 1, choice quality grade steer. Drawing from Agricultural Marketing Service, USDA.

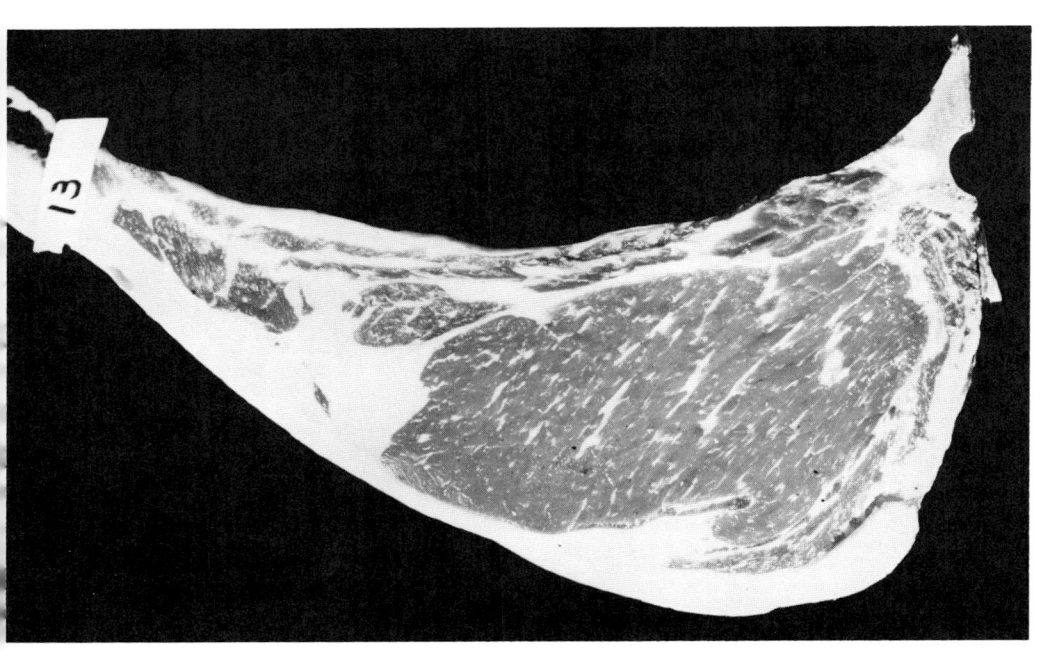

Yield grade No. 1, rib-eye section. Cutability 52.8–54.6. Photograph courtesy National Live Stock & Meat Board.

Yield grade No. 2, choice quality grade steer. Drawing from Agricultural Marketing Service, USDA.

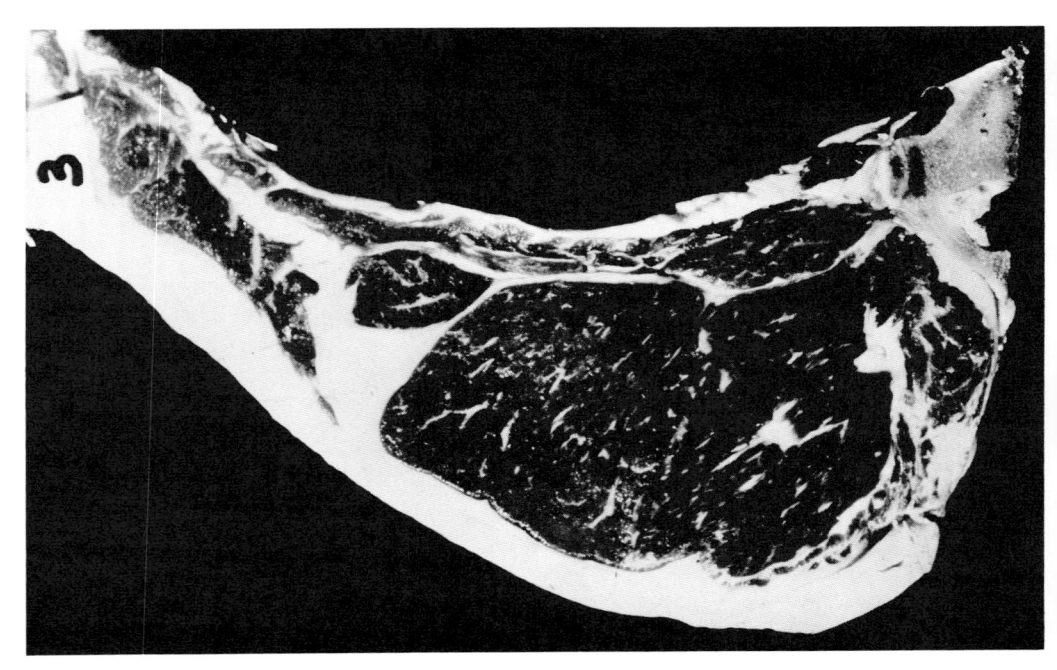

Yield grade No. 2, rib-eye section. Cutability 50.5–52.3. Photograph courtesy National Live Stock & Meat Board.

Yield grade No. 3, choice quality grade steer. Drawing from Agricultural Marketing Service, USDA.

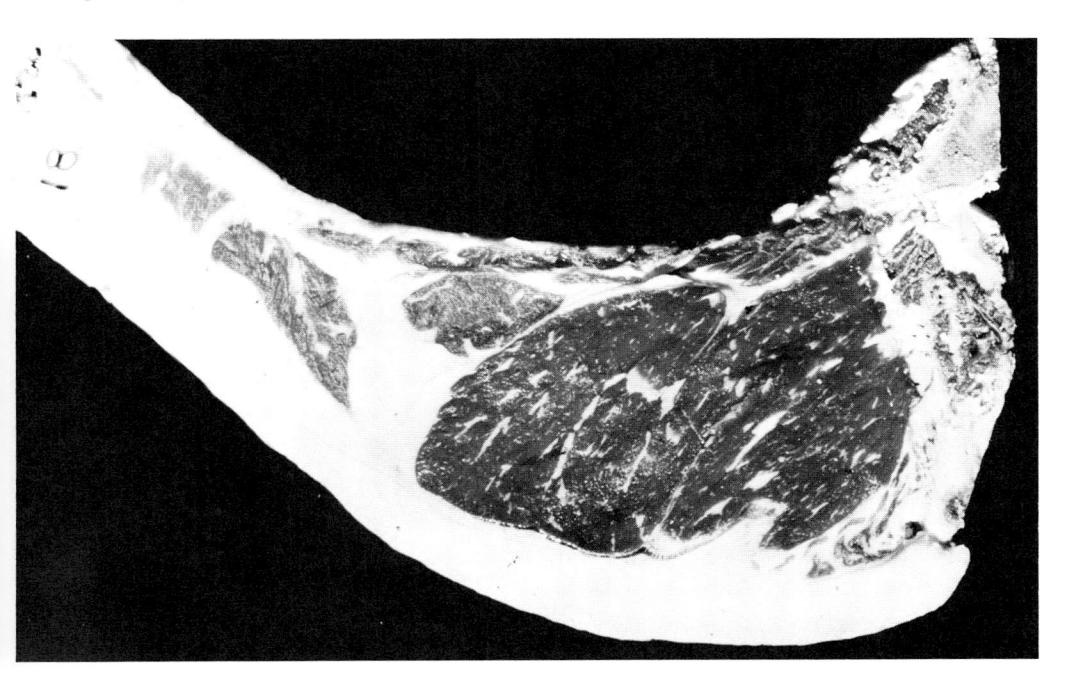

Yield grade No. 3, rib-eye section. Cutability 48.2–50.0. Photograph courtesy National Live Stock & Meat Board.

Yield grade No. 4, choice quality grade steer. Drawing from Agricultural Marketing Service, USDA.

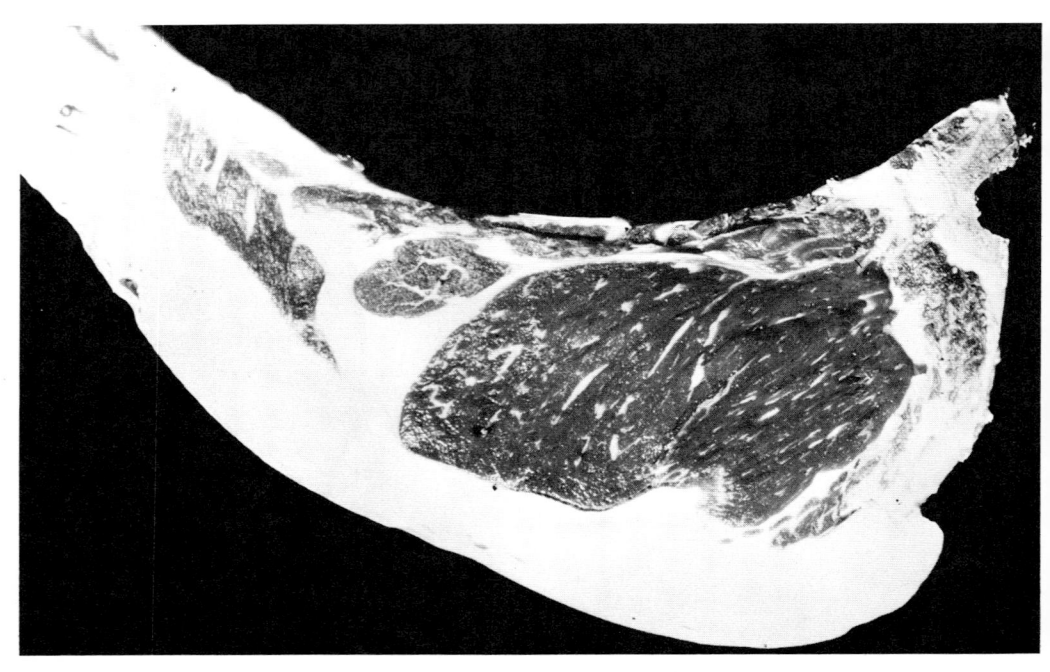

Yield grade No. 4, rib-eye section. Cutability 45.7–47.7. Photograph courtesy National Live Stock & Meat Board.

Yield grade No. 5, choice quality grade steer. Drawing from Agricultural Marketing Service, USDA.

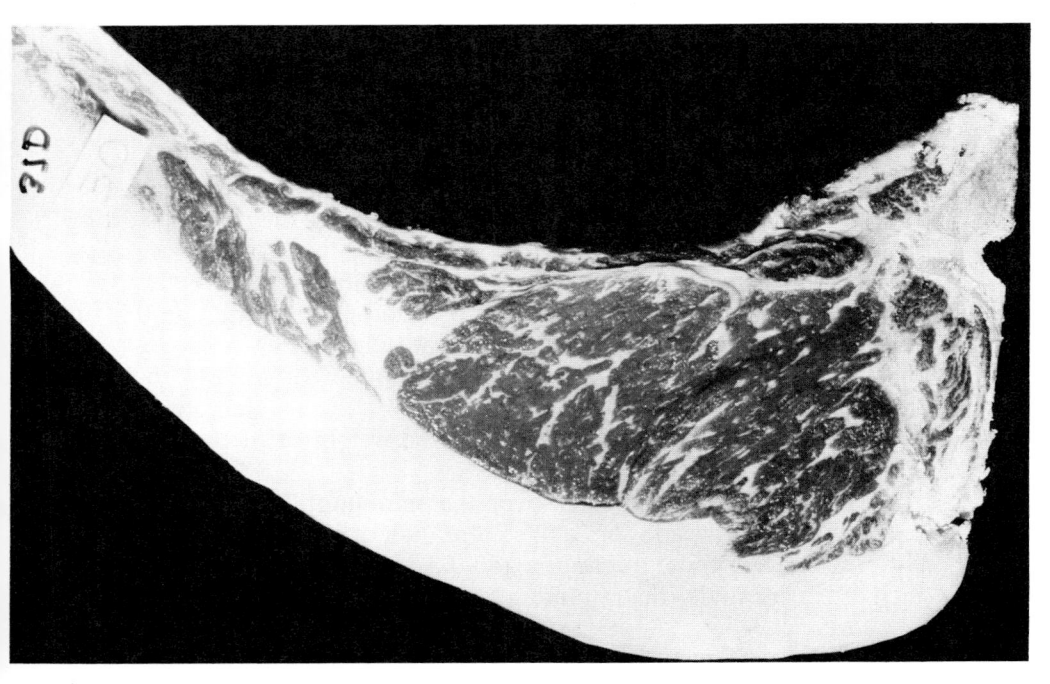

Yield grade No. 5, rib-eye section. Cutability 43.4–45.4. Photograph courtesy National Live Stock & Meat Board.

Cattle of North America

Yield Grade.—During the 1950's attention was directed to the wide variation in the quantity of salable meat obtained from different carcasses within the same quality grade. In the choice grade, for example, it was found that one carcass might produce 52 per cent of salable meat, while another yielded only 46 per cent. Such differences were largely due to variations in the amount of fat and the degree of muscling and could result in variations of as much as $80 in the value of the carcass or live animal.

In 1965 the USDA set up standards for yield grades which indicate the percentage of salable meat in carcass. The industry displayed little interest at first, but as performance records entered the picture and the factors determining yield grade were shown to be highly heritable, the economic aspects involved in this measure of salable meat forced its recognition.

Yield grades are applicable to any of the quality grades for carcasses but in practice are considered only in connection with prime and choice. They are determined by the measurement of the fat thickness and the rib-eye area at the twelfth rib, plus an estimate of the percentage of kidney fat in the hot carcass.

The detailed procedure for determining yield grade and the relationship between this and cutability follows:

Determining Yield Grade (1965 Standards)

1. Obtain preliminary yield grade by measuring fat thickness over twelfth rib at a point three-fourths of the way across the rib eye, starting from the vertebrae end (Table 1 opposite).

2. Estimate kidney fat as a percentage of hot-carcass weight.

3. Adjust preliminary yield grade by adding or subtracting 0.2 for each 1 per cent over or under 3.5 per cent kidney fat.

4. Determine rib-eye area with grid (number of squares included in rib eye divided by 10).

5. Obtain minimum rib-eye area for the weight of the carcass from Table 2.

6. If actual rib-eye area is over the minimum, subtract 0.3 from preliminary yield grade for each square inch of rib eye over minimum. If rib-eye area is under minimum, add 0.3 to yield grade for each square inch of rib eye under the minimum (0.5 square inch and over counts as 1 square inch).

7. Determine estimated percentage of closely trimmed, boneless, retail cuts from the round, loin, rib, and chuck from Table 3 by using adjusted yield grade.

TABLE 1.
PRELIMINARY YIELD GRADE

Fat Thickness	Preliminary Yield Grade
0.1	2.25
0.2	2.50
0.3	2.75
0.4	3.00
0.5	3.25
0.6	3.50
0.7	3.75
0.8	4.00
0.9	4.25
1.0	4.50
1.1	4.75
1.2	5.00
1.3	5.25
1.4	5.50
1.5	5.75

TABLE 2.
ADJUSTMENT FOR RIB-EYE AREA

Warm-Carcass Weight	Minimum Rib-Eye Area
350	8.0
375	8.3
400	8.6
425	8.9
450	9.2
475	9.5
500	9.8
525	10.1
550	10.4
575	10.7
600	11.0
625	11.3
650	11.6
675	11.9
700	12.2

TABLE 3. PERCENTAGE YIELD OF BONELESS MAJOR CUTS FOR CORRESPONDING YIELD GRADES*

Yield Grade	Yield of Cuts	Yield Grade	Yield of Cuts	Yield Grade	Yield of Cuts	Yield Grade	Yield of Cuts
1.0	54.6	2.4	51.4	3.8	48.2	5.0	45.4
1.2	54.2	2.6	51.0	4.0	47.7	5.2	45.0
1.4	53.7	2.8	50.5	4.2	47.3	5.4	44.5
1.6	53.3	3.0	50.0	4.4	46.8	5.6	44.1
1.8	52.8	3.2	49.6	4.6	46.1	5.8	43.6
2.0	52.3	3.4	49.1	4.8	45.7	5.9	43.4
2.2	51.9	3.6	48.7				

*For bone-in cuts add 11 per cent.
Source: Agricultural Marketing Service, USDA.

One of the basic reasons for the introduction of yield grades was the recognition of the fact that cutability, like quality, is to a large extent an expression of a genetic trait. It was hoped that this trait, once recognized, could be converted to market values, which would eventually be communicated all the way back to the cow-calf man, the ultimate producer. This work, which many experiment stations and the more progressive producers are doing on carcass evaluation of finished animals, should eventually crystallize in a more realistic price-yield relationship than that which exists today. It appears certain that in the years to come yield grade will become an important element in the pricing of feeders as well as slaughter cattle.

Feeder and Slaughter Cattle

Four main classes of cattle are recognized by the commercial trade: stocker and feeder cattle, fat cattle, cows, and bulls.

Stocker and feeder cattle are the calves and yearlings of either sex destined for the feed lot. In recent years this category has included an increasing number of steers of the dairy breeds, particularly Holstein-Friesian. Fat cattle are the finished product of the feed lot. Cows and bulls, in this connotation, are culled animals.

There are three basic marketing channels for the classes of cattle: direct selling, by the producer to the feed lot, or the feed lot to the packer; the country market, where sales are conducted by auction; and the stockyard or terminal market, where sales may be made by either private treaty or auction. The auction has been introduced at the terminal market in recent years in an attempt to increase receipts.

Feeder cattle are being sold directly to the feed lots in greater numbers every year.

Fat cattle from the larger feed lots go directly to the packer. The farmer who raises and feeds out a few steers usually sells them at the auction market or terminal, where the packer-buyer picks them up. Continuously increasing numbers of fat cattle are sold by grade and weight. Over 20 per cent of the cattle killed in federally inspected packing plants in 1971 were bought on this basis.

Culled cows and bulls of both dairy and beef herds generally find their way to the auction markets.

Breeding Stock

There is no particular pattern in the sale of breeding stock. Public-auction sales are a favorite method of moving animals, though many purebred cattle—bulls in particular—are sold by private treaty.

In recent years the bull-testing stations have also become an increasingly important source of superior bulls.

Futures Trading

Trading in live-cattle futures was inaugurated on the Chicago Mercantile Exchange late in 1964. Its basic functions are to provide a hedge for the cattle feeder and to provide speculative trade in choice-grade cattle. The unit traded is 40,000 pounds of choice-grade live steer. The live-cattle contracts have proved extremely popular, though there is some question how much they are used in an actual hedge.

Elaborate regulations are prescribed by the exchange for actual delivery on a contract. In practice transfer of live cattle rarely occurs, though there have been occasions when literally thousands of head have been delivered.

Futures trading in frozen boneless beef was introduced by the exchange in 1970, and feeder-cattle futures were traded for the first time on the Mercantile Exchange on November 30, 1971. Trading in these futures has so far been insignificant.

MILK

The control of quality and quantity and the procedures for the sale of milk by the producer are well organized in commercial-dairy regions throughout the United States. In the nineteenth century delivery of bulk milk by the farmer to the housewife in town was replaced by delivery of bottled milk by horse-drawn milk wagon. The milk can picked up at the dairyman's mailbox disappeared after World War II, and the tank truck now makes regular calls to take milk directly from a refrigerated stainless-steel tank placed in a spotless milkhouse.

In most dairying areas state and municipal regulations require a high level of sanitation in the production of milk. These regulations are often politically slanted and needlessly involved but are adequate to ensure pure milk. The processor to whom milk is delivered provides a further check on purity by microscopic examination for bacteria and brucellosis on every delivery from every producer.

The minimum price of milk paid the producer is controlled by federal and state laws. Federal control is exercised through the milk-marketing orders, which are promulgated for a regional market area. These orders are administered by an official of the USDA and issued over the signature of the secretary of agriculture. Milk-marketing orders establish the minimum price dairymen are to be paid for all milk sold to first buyers.

Most dairymen now belong to co-operative associations which perform the very important bargaining function for their members. Some extend their activities into processing and distribution. Milk prices are established by negotiations between the co-operative associations and other handlers and the USDA official. The order establishing a price must be approved by two-thirds of the dairymen coming under it. Public hearings must be held on any changes.

The formula employed in establishing the milk price usually takes into account the supply and demand for milk and its use, such as fluid milk for direct consumption, milk for the production of cream, and milk to be used for manufacturing the wide variety of dairy products, such as ice cream, cheese, and powdered milk. The minimum price is arrived at by adjusting for the average quantity of milk going into each of these different products. The price so established is set for a base butterfat content, often 3.5 per cent. All deliveries of each dairyman are tested for butterfat content, and an adjustment is made in his payments for variations from the base.

PACKING PLANTS

At the opening of the twentieth century cattle slaughter was centralized in large plants in the Middle West, largely the corn belt, extending south from Chicago and west to the Missouri River. The large stockyards of this area attracted cattle for feeding as well as for slaughter from the farms and ranches of most of the nation lying between the Appalachians and the Rockies, with the exception of the Deep South.

In those years many of the cattle moved from range to yards to packer without ever sampling a mouthful of grain; but as the practice of feeding increased, more and more steers, and heifers too, moved from the yards to the feed lot, back to the yards, and on to the packer. This pattern continued until the mid-1940's.

During the decade following World War II the large packers began an exodus from Chicago and the Middle West, re-establishing in the corn and milo regions of the West and Southwest. Freight savings resulted from slaughtering closer to the supply of cattle and grain. Further economy accrued when the new location was closer to the point of consumption. Another element involved was labor. As union monopoly increased, demands were made which could not be met. When obsolescence forced the closing of the older packing plants, logic dictated a move to more strategic locations.

Eastern Colorado, West Texas, Arizona, southwestern California, the Columbia River Basin in Oregon, and other points which became centers of cattle feeding also attracted packing plants. This pattern became stabilized during the 1960's, as the ten major concentrations of stockyards and packing plants which had dominated the industry before World War II were succeeded by twice this many large feeding and packing centers. By 1970 there were 726 large packinghouses and 6,446 small commercial plants. The large plants, units with a capacity of upwards of 500 head a day, accounted for 87 per cent of the total cattle slaughtered. The units classified as "other commercial plants" by the USDA killed 12 per cent of the total slaughter. These plants are small, largely unmechanized units which run an average kill of about 15 cattle and a few dozen hogs a week.

Following the enactment of the Wholesome Meat Act in 1967, nearly all commercial slaughterhouses were brought under USDA inspection or its equivalent. Previously intrastate movements of meat were exempt from federal inspection. The new law required that state inspection must meet standards equal to those of the USDA and must be observed by all commercial slaughterers. The 460,000 cattle killed annually in farm slaughter (1.3 per cent of the national total) are now practically the only beef in the United States not subject to rigid inspection.

The large plants process their cattle on an endless overhead moving conveyer known as the Wilcox Chain, which is geared to handle upwards of 60 head an hour. In the average operation 100 men can process 120 head in an hour from stunning to delivering the split carcass to the cooler. The large plants kill 100,000 to 300,000 head annually. A few are in the 500,000 head range.

The slaughtering operation begins when the cattle are driven to the stunning pen, a steel box which is open on the top and can be partly rotated to drop the unconscious animal to the floor. Stunning is accomplished with a captive-bolt gun. After the animal drops, one leg is shackled, the body is hoisted to a hook on the endless chain, and is stuck and bled. From this point on, the chain moves the animal continuously as the necessary operations are performed, and it is systematically reduced to two dressed sides ready for the cooler. On a line with a capacity of 60 head an hour the chain moves at a rate of 8 feet a minute; at a capacity of 120 head an hour the speed is 16 feet a minute.

The principal steps performed along the line are head removal, skinning, either by hand or by means of a mechanical hide puller, eviscerating, splitting (dividing the carcass in halves by sawing down the center of the backbone), shrouding (wrapping a tight cloth covering around each half of the carcass to prevent loss of moisture), inspection, weighing, and delivery to the cooler.

The various by-products and offal are removed as the body passes down the line and are delivered by short conveyers to the several processing points. Areas are provided for the inspection of the vital organs as they are removed and for the final inspection of the total carcass. The endless line of dressed carcasses then passes by the grading station where a USDA grader evaluates each half-side before it is weighed on automatic scales.

Conveniently located hoses are available at key locations for washing

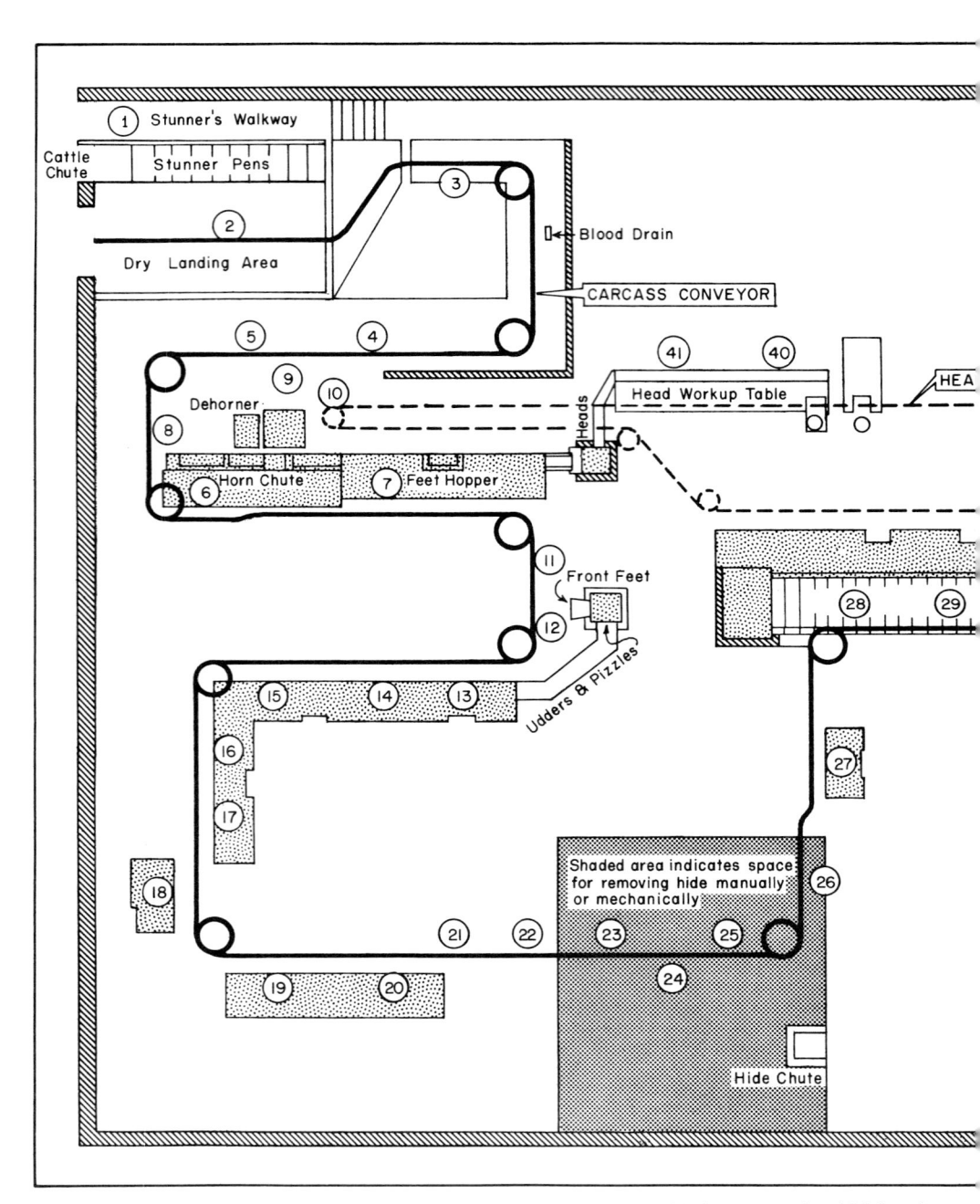

Diagram of a modern packing-plant kill floor, 115 by 80 feet, capacity 120 head

EXPLANATION

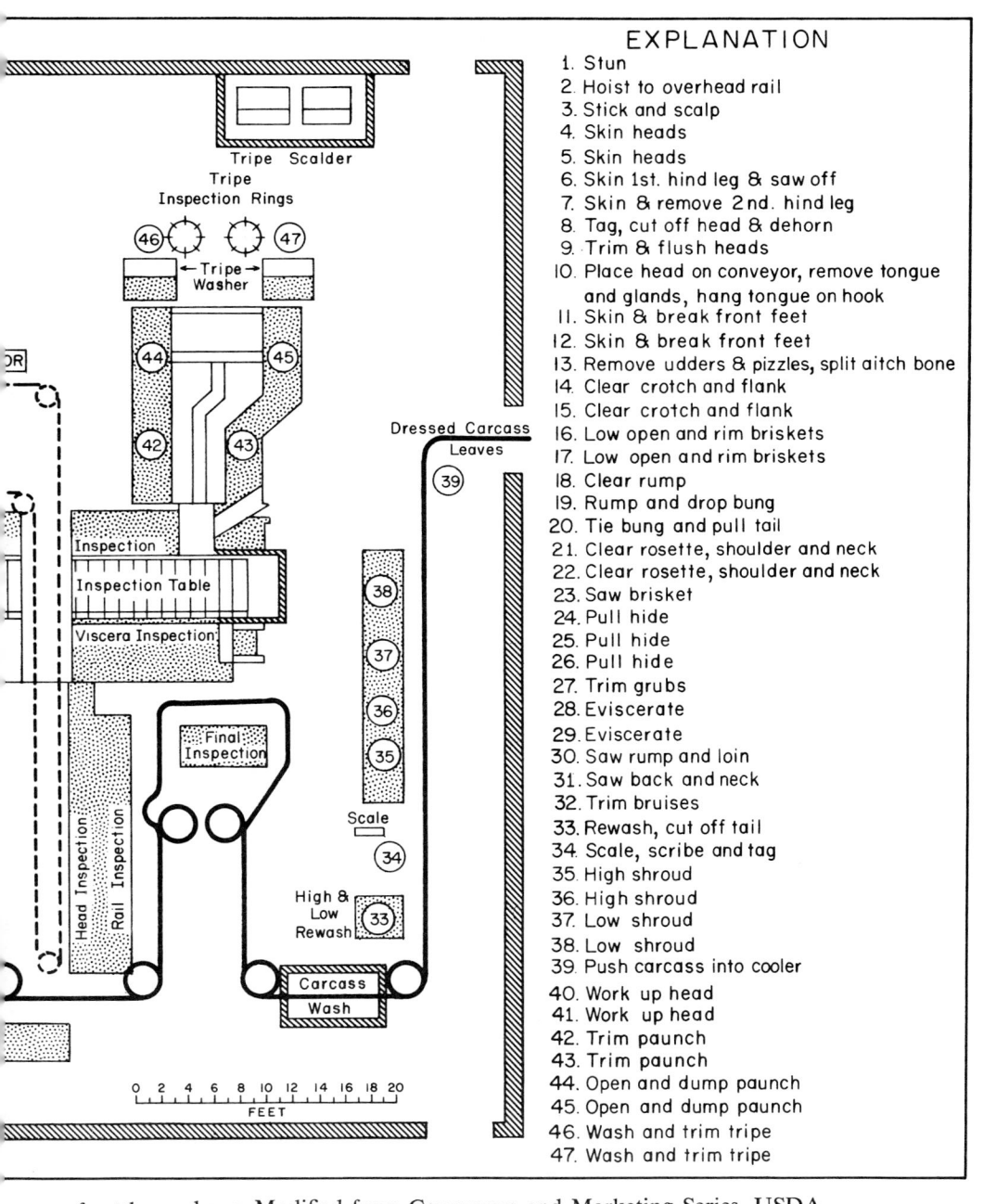

1. Stun
2. Hoist to overhead rail
3. Stick and scalp
4. Skin heads
5. Skin heads
6. Skin 1st. hind leg & saw off
7. Skin & remove 2nd. hind leg
8. Tag, cut off head & dehorn
9. Trim & flush heads
10. Place head on conveyor, remove tongue and glands, hang tongue on hook
11. Skin & break front feet
12. Skin & break front feet
13. Remove udders & pizzles, split aitch bone
14. Clear crotch and flank
15. Clear crotch and flank
16. Low open and rim briskets
17. Low open and rim briskets
18. Clear rump
19. Rump and drop bung
20. Tie bung and pull tail
21. Clear rosette, shoulder and neck
22. Clear rosette, shoulder and neck
23. Saw brisket
24. Pull hide
25. Pull hide
26. Pull hide
27. Trim grubs
28. Eviscerate
29. Eviscerate
30. Saw rump and loin
31. Saw back and neck
32. Trim bruises
33. Rewash, cut off tail
34. Scale, scribe and tag
35. High shroud
36. High shroud
37. Low shroud
38. Low shroud
39. Push carcass into cooler
40. Work up head
41. Work up head
42. Trim paunch
43. Trim paunch
44. Open and dump paunch
45. Open and dump paunch
46. Wash and trim tripe
47. Wash and trim tripe

of cattle per hour. Modified from Consumers and Marketing Series, USDA.

606

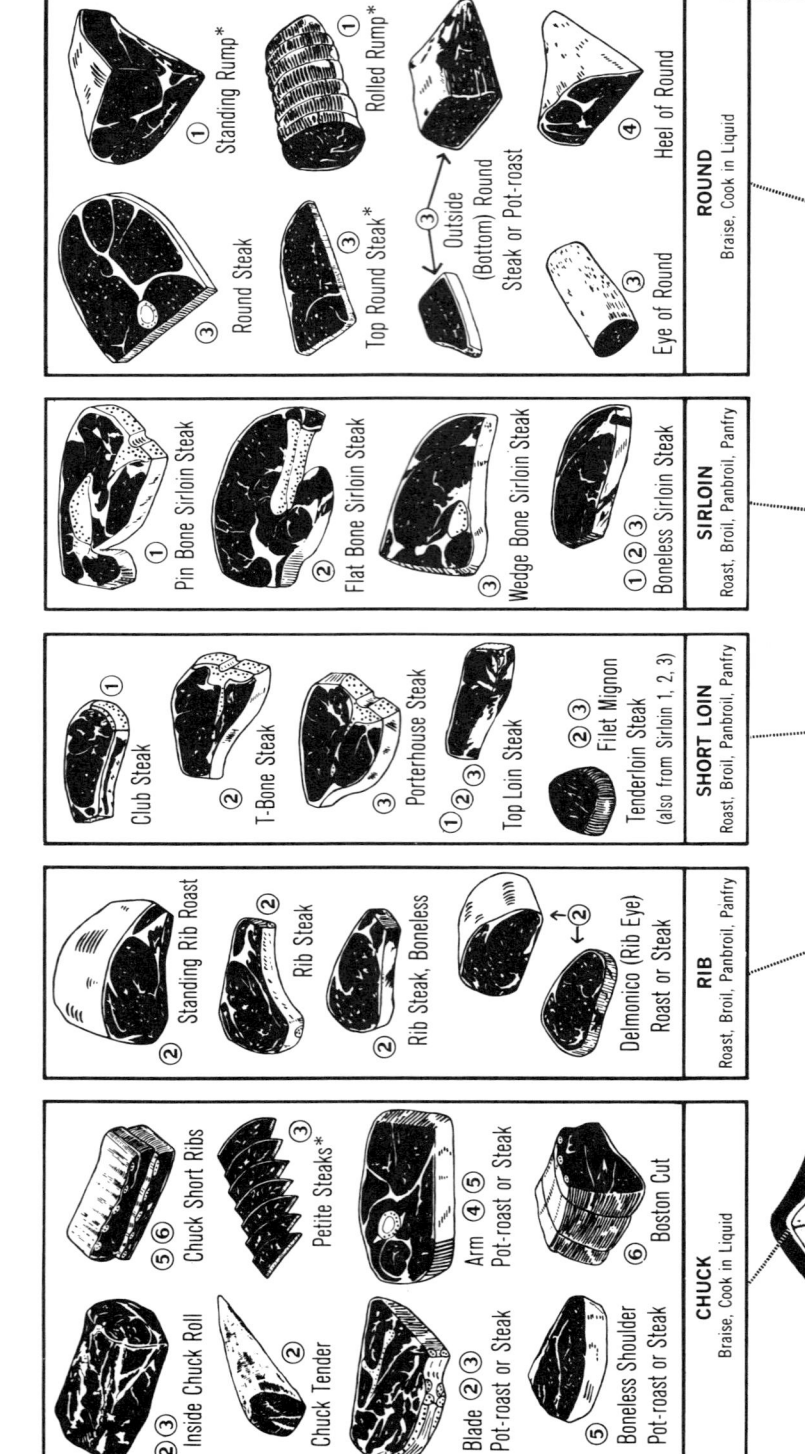

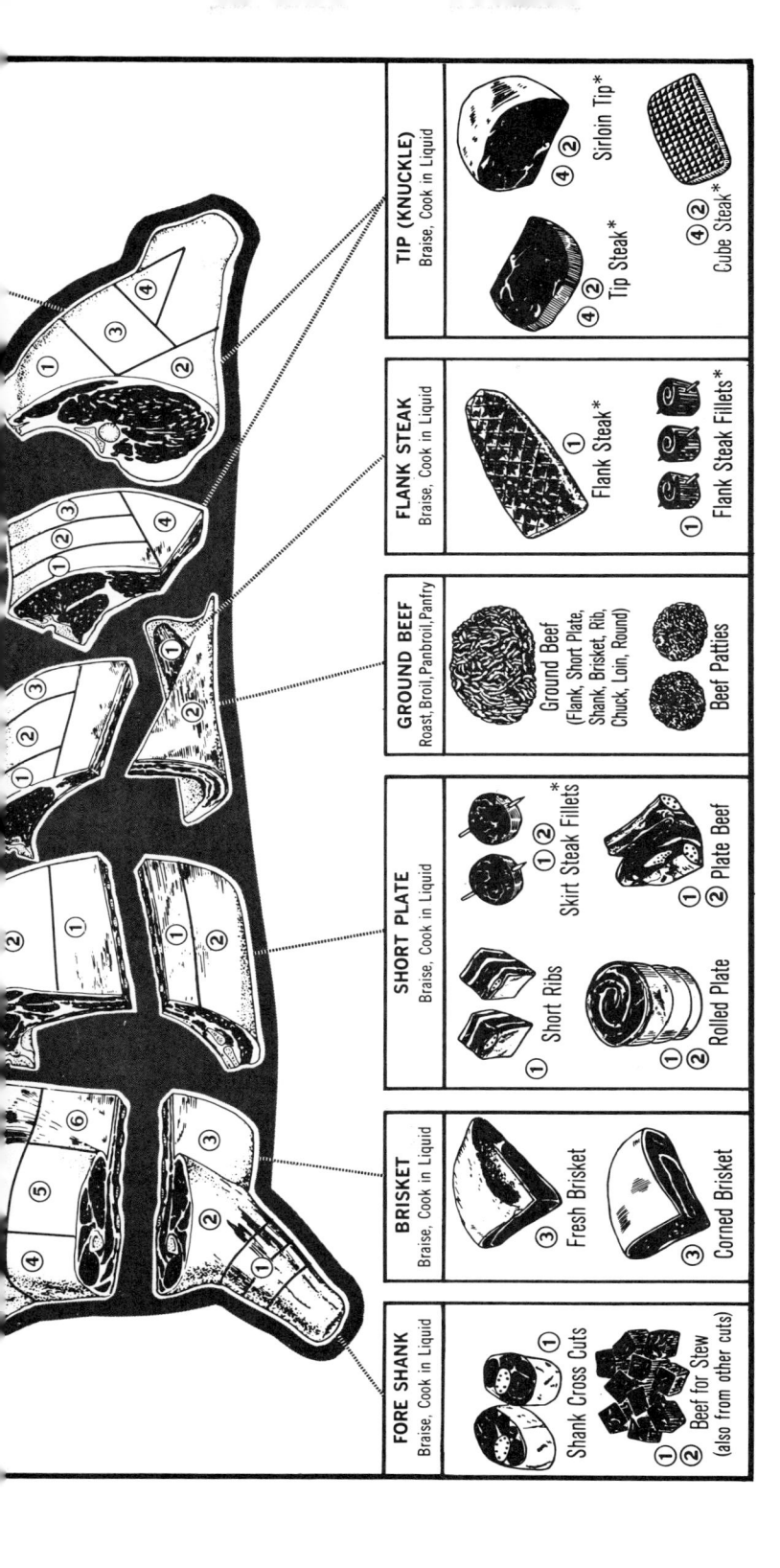

*May be roasted, broiled, panbroiled, or panfried from high-quality beef.

Retail cuts of beef and the parts of the carcass from which they are derived. Courtesy of National Live Stock and Meat Board.

the carcasses and for continual housekeeping chores. Power tools are generally used throughout the entire operation.

The smaller plants are less fully mechanized; fewer by-products are recovered; and less floor space is required. Plants with capacities below 30 to 40 head an hour seldom use the Wilcox Chain.

A typical minimum facility killing 10 to 50 head a week is usually housed in a small concrete building with a well-drained floor. A hand hoist is used to raise the body for bleeding. It is then lowered to a low metal stand, or cradle, for skinning, after which it is hoisted to an overhead rail and reduced to a finished carcass. Warm water for washing the carcass and floor is essential.

These small units meet a local demand from individuals who want to have a beef processed for their own use. Operators of such facilities also buy cattle and sell wholesale cuts, as well as packaged meal-size portions. While the volume of trade handled by these small plants is minor, the demand for their products is steady, and they will undoubtedly continue to exist.

For many years the basic product of the old-line packer was a side of beef, which moved to the wholesaler and eventually to the retailer. More recently large retailers, such as the chain supermarkets, established their own warehouses and bought carcasses for distribution to butchers in their stores. Next followed the "breaking" of the carcass by the packer, the wholesaler, or the warehouse, and now many large packers market a sizable part of their output as packaged cuts.

Packaged cuts are of two types, known in the trade as "retail breaks" and "HRI (hotel restaurant institution) breaks." The retail breaks are bone-in cuts, such as chucks, ribs, and rounds. The HRI breaks are ready cuts, consisting of large oven-size roasts and steaks.

In 1965 the Monfort Packing Company in Greeley, Colorado, started packing HRI breaks in plastic bags from which the air was removed by a vacuum pump and the bag then shrunk tightly to the cut by a quick pass through a warming area. This company was one of the pioneers in this process, known as Cryovac, which permits the packaged beef to age normally under refrigeration but without freezing. The date of packaging is stamped on the carton so that the user can determine his own period of aging.

Monfort also was one of the first packers to package in plastic bags which are then placed in the shipping carton. A small quantity of dry ice is added before the carton is sealed. On evaporation the dry ice displaces the air in the box with carbon dioxide. The carbon dioxide permits aging for such interval as the buyer may desire.

The breaking operation is increasing and has many variations. In addition to the consumer cuts, family-sized portions of like quality, such as steaks or roasts, are packaged ready to go on the retailer's counter. The role of the beef wholesaler and the meat-warehousing function of the packers and supermarkets appear certain to decline in the future.

The packer has long suffered the stigma among producers of retaining an undue portion of the final receipts from the dressed steer. Any justification for this opinion has long since been dissipated by the competition in the meat-packing business. Today the packer is doing well who can sell a dressed carcass for what he paid for the live animal. At 1971 prices a 1,250-pound steer costing the packer $400 would return to an efficient operator $400 for the dressed carcass. Operating cost and profit must be derived from the $24 the hide and offal bring after being reduced to a salable form.

CATTLE DISEASES

No area in the world (with the exception of the isolated Scandinavian peninsula) can match the United States for the health of its cattle population. This favorable situation, however, and the continuous vigilance necessary to maintain it are not always appreciated by the cattleman, who often finds onerous the regulations which have, in reality, protected the health of his herd throughout the years.

The USDA has been instrumental in effectively controlling or eradicating most of the serious diseases which affect the nation's livestock, and its programs have made major inroads against parasitic infestations. The department's Bureau of Animal Industry, established in 1884, was the first unit charged with the protection of animal health on a nationwide basis. In 1953 this Bureau was merged with the Agricultural Research Service, and in 1966 the Animal Health Division of the Agricultural Research Service was created and made responsible for health programs for all domestic animals. In 1971 the Animal Health Division was incorporated in the newly established Animal and Plant Health Service, whose name was subsequently changed to the Animal and Plant Health Inspection Service.

Major Health Hazards Satisfactorily Controlled

Many of the most serious bovine health hazards have been eradicated or brought under control. They include pleuropneumonia, tick fever, foot-and-mouth disease, tuberculosis, brucellosis, screwworm, and blackleg.

Contagious Bovine Pleuropneumonia.—This was the first cattle disease to be identified and subsequently eradicated in the United States. The disease was introduced in 1843 by a ship's-store cow unloaded at Brooklyn, New York. The disease spread rapidly from New York to Maryland and on to Virginia. When some years later England banned the entry of cattle from the United States because of the disease, the Bureau of Animal Industry was established within the Department of Agriculture and received as its initial assignment the task of eradicating pleuropneumonia.

By 1892 the eradication was completed. The importation of animals from countries where the disease is known to exist has been prohibited for many years. The blessing of eradication, however, has left the cattle of this nation particularly susceptible to reinfestation, and constant vigilance is essential.

Tick Fever (Bovine Piroplasmosis, or Texas Fever).—This disease, introduced by the fever tick, became the subject of various state laws restricting the movement of southern cattle in the years following the Civil War. These laws in turn led to the first tick-fever quarantine order of the Bureau of Animal Industry in 1889.

In 1905 the Department of Agriculture initiated a program of fever-tick eradication in co-operation with various states. Over the years the ticks were gradually eliminated, and since 1941 the United States has been essentially free of the disease. Only an occasional outbreak occurs outside the narrow quarantine buffer zone extending along the 500-mile international boundary with Mexico from Brownsville on the Gulf of Mexico to Del Rio, Texas. Maintenance of this quarantine zone is essential to prevent reinfestation. No animals may be brought into the rest of the United States from this zone without undergoing special disinfection procedures.

Foot-and-Mouth Disease.—This disease will probably continue to be the gravest threat to the bovine population of the United States and indeed to all of North America. There have been nine outbreaks in the United States since 1870, the last in 1929. Direct losses from the six outbreaks from 1902 to 1929 have been estimated at $253 million. The cost of an outbreak today, with the tremendous concentration of cattle in feed lots and dairies, would be astronomical. The 1968 outbreak in England nearly equaled the above amount, notwithstanding the fact that the total cattle population of England is only one-tenth that of the United States and the concentrations of cattle there do not compare with those of the larger dairies and feed lots in this country.

United States

The Emergency Animal Disease Eradication Organization (EADEO) of the Animal and Plant Health Inspection Service, USDA, has the responsibility of dealing with foot-and-mouth and other outbreaks or threatened outbreaks of disease which occur in the United States and its possessions, as well as in Mexico, Panama, and Central American countries with which treaty arrangements have been made for protective measures. This organization consists of a group of qualified veterinarians and a supporting staff who are trained to go into immediate action when a report of a suspected outbreak of any exotic animal disease is received.

Facilities are available for flying appropriate specimens from the suspect animal to the USDA diagnostic laboratory on Plum Island, New York. This unit is under maximum security and is the only site in the United States equipped to diagnose exotic diseases.

In the event of a suspected outbreak all animals which could possibly have had any contact with the suspect would be traced in minute detail by an emergency force organized by the EADEO. In the case of a major central market or feed lot this procedure could involve thousands of cattle.

If the diagnosis at Plum Island should be negative, the investigation would cease. If the diagnosis should be positive, immediate slaughter and disposal procedures would be activated for every animal that could have come into contact with the diseased one. In addition, the premises and all equipment, such as vehicles, touched by the suspected animals would be cleaned and disinfected. Arrangements for the use and transportation of essential equipment stationed at strategic locations are on a 24-hour-alert basis.

As a precautionary measure test exercises are carried out periodically by the EADEO. This is indeed modest insurance, considering that a major outbreak of foot-and-mouth disease could cost the nation hundreds of millions of dollars. Even more serious, a chronic outbreak could lead to the necessity of a vaccination program. Such a measure could cost the nation a billion dollars a year. This figure includes the estimated cost of vaccination, loss in condition of infected animals, and death loss attributable to the weakened condition.

No really satisfactory vaccination has been developed for foot-and-mouth disease. The best biologicals now available require a minimum of two, and preferably three, innoculations annually to give reasonable protection. Moreover, live-virus vaccines create carrier animals, and the disease becomes endemic.

Tuberculosis.—The federal-state program for the eradication of tuberculosis was approved in 1917 by the Bureau of Animal Industry. This program authorized the test-and-slaughter method of control and the indemnification of the owner of infected animals. Surveys had previously indicated that 5 per cent of the cattle population of the United States were tubercular.

By 1940 the prevalence of tuberculosis in the national herd had been reduced to less than 0.5 per cent. Since 1935 the reduction in the number of infected animals has been maintained at a practically constant rate of 17 per cent a year. It is estimated that in fiscal 1970 there was on the average only 1 case in 20,000 head, or a prevalence of 0.005 per cent.

The main effort against the disease now consists of tracing the sources of infected animals by the back-tag identification procedure described under "Brucellosis" below. Complete eradication of tuberculosis in the national herd is now considered in sight. The target date is 1980.

Brucellosis.—The state-federal co-operative program for the control and eradication of brucellosis was initiated in 1934, at the time of a government campaign to reduce cattle numbers during a severe drought in the western states. On-the-farm testing was started in 1935, when the number of animals found to be infected averaged 11 per cent for the country as a whole. The test-and-slaughter procedure was the only control measure employed until 1940.

A satisfactory live-bacteria vaccine, now internationally known as Strain 19, had been developed and used experimentally, but control programs based on vaccination encountered widespread opposition. The outcome was a plan for calfhood vaccination which was begun in 1940. The test-and-slaughter program was continued, however.

By 1949 a milk-ring test was considered advisable since the reduction in the number of infected herds made a broader diagnostic tool useful. The milk-ring test (Brucellosis Ring Test—BRT) indicates the presence of brucella-infected cows by examination of a small quantity of milk. When a positive reaction is obtained, the suspect herd is readily identified, after which it is an easy matter to locate and eliminate the infected animals.

The Market Cattle Testing Program was inaugurated in 1961. Under this program cattle to be slaughtered are back-tagged with an identifying number furnished by the Animal and Plant Health Inspection Service. USDA inspectors at the packing plant take a blood sample, which is examined in a co-operative state-federal brucellosis laboratory. When a reactor is discovered, the origin of the animal is traced by means of the back tag, and other reactors discovered in the herd are eliminated. Five million cows were blood-tested under the Market Cattle Testing Program in 1970.

Of the 3,153 counties in the United States, Puerto Rico, and the Virgin Islands, 52.5 per cent had qualified by July 1, 1971, as Certified Brucellosis Free, and 47.1 per cent as Modified Certified. The goal of complete eradication of the disease is now considered attainable by 1978.

Screwworm.—The screwworm fly is native to all the southern United States and southward to the frost line in South America. Screwworm wounds result in a high death loss, and young calves are particularly vulnerable. Until recently the only effective treatment was the disinfectant known as Smear 62. Its application requires complete immobilization of the animal by roping or confinement in a chute, and many animals went untreated. Victory over the screwworm finally came about through a program of mass sterilization of the male flies. It was found that if a preponderance of sterile male flies could be introduced into a normal population of the insects the females, which mate only once, would deposit sterile egg masses, and the population could be eliminated rapidly. The program was tested on the island of Curaçao in the Dutch West Indies. By 1954 the screwworm was eradicated on that island.

In 1958 an eradication program was begun in Florida, and portions of the nearby states of Alabama, Georgia, and South Carolina. By the next year the screwworm had been eradicated from these areas. A similar program was undertaken in 1962 in the Southwest, including the states of Arkansas, Louisiana, New Mexico, Oklahoma, Texas, and eventually Arizona and California. The program was so successful that self-sustaining screwworm populations were declared nonexistent in the continental United States in 1966.

There have been several serious screwworm outbreaks in recent years, but they have been successfully combated by the sterile-fly technique. A particularly severe outbreak in 1972 led to a joint Mexico–United States program to create a sterile-fly barrier at the Isthmus of Tehuantepec, the narrow neck of land in southern Mexico. Eradication of the screwworm fly from all of Mexico north of the isthmus will ultimately be effected.

Blackleg (Formerly Known as Charbone).—Once a serious cause of loss in beef-cattle herds, blackleg has been practically eliminated by the nearly universal practice of calfhood vaccination.

Health Hazards Not Controlled

The Agricultural Research Service estimated the annual loss to the

Cattle of North America

national herd from diseases other than those noted above and to parasites and insects at $1,768,000,000 annually over the five-year period 1961 to 1965. This amounts to an average loss of approximately $17 a head for every bovine in the United States. Details of these losses are given in the tables below.

Estimated Average Annual Losses
Infectious and Noninfectious Diseases

Cause of Loss	Value of Loss, Dollars	Cause of Loss	Value of Loss, Dollars
Mastitis	$411,090,000	Urinary calculi	$ 4,052,000
Calf losses, beef	113,843,000	Chemical poisoning	3,783,000
Bloat	104,940,000	Foot rot	2,969,000
Vibriosis, beef	104,194,000	Ketosis	1,899,000
Brucellosis	46,353,000	Bovine hyperkeratosis	1,791,000
Anaplasmosis	36,001,000	Grass tetany	662,000
Plant poisoning	17,303,000	Tuberculosis	622,000
Johne's Disease	13,800,000	Encephalitis	54,000
Leptospirosis	12,189,000	Rabies	27,000
Milk fever	10,649,000	Total	$886,221,000

Source: *Agriculture Handbook No. 291*, Agricultural Research Service, USDA.

Internal Parasites

Cause of Loss	Value of Loss, Dollars
Internal parasites (gastroenteritis)	$100,046,000
Coccidiosis	14,569,000
Trichomoniasis, genital	8,040,000
Liver fluke	3,022,000
Other parasitic diseases	323,000
Total	$126,000,000

Source: *Agriculture Handbook No. 291*, Agricultural Research Service, USDA.

Insects

Cause of Loss	Value of Loss, Dollars	Cause of Loss	Value of Loss, Dollars
Grub (heel fly)	$192,000,000	Lice	$ 47,000,000
Hornfly	179,000,000	Horsefly and deer fly	40,000,000
Stable fly	142,000,000	Mosquito	25,000,000
Face fly	68,000,000	Scabies mite	3,000,000
Tick	60,000,000	Total	$756,000,000

Source: *Agriculture Handbook No. 291*, Agricultural Research Service, USDA.

Five diseases (mastitis, bloat, vibriosis, brucellosis, and anaplasmosis), plus calf loss, internal parasites, and six insects (heel fly, hornfly, stable fly, face fly, tick, and louse) account for 87 per cent of the total death loss.

Mastitis.—This disease is an inflammation of the udder resulting from bacterial infection. Dairy cattle are most seriously affected. The greatest damage is in loss of milk production. Nothing has come of efforts of the dairy industry to organize a national program to abate mastitis.

Calf Loss.—As compiled by the Agricultural Research Service calf loss applies only to beef animals. Death of young calves between birth and weaning is commonly accepted among producers large and small as a hazard of the business. It is probable that the actual death loss among young calves is considerably higher than the service estimate, a country-wide average of around 2 per cent. Cattlemen are notoriously reticent about death loss among calves, and such numbers as are given are often understated. There are also substantial losses during calfhood in the national dairy herd which are not included in the number shown for calf losses.

Scours, or diarrhea, is the most deadly of calfhood diseases. Calves from a few days to two months of age are most seriously affected. It is now generally recognized that death is usually due to dehydration caused by the loss of body fluids. Antibiotics were used for years to combat calfhood scours with little success, the calf succumbing from dehydration even though the causitive organism was eliminated. In recent years intravenous injections to replace the body fluids in addition to treatment with antibiotics have been successfully used to combat scours.

Shipping fever is another killer. The term is applied loosely to almost any affliction a calf suffers at the time it is shipped. Two factors have now materially reduced losses attributable to shipping fever. First, most cattle leaving the producer are now shipped by truck. This has considerably lessened the time they are under stress from movement. Second, preconditioning has come into practice. This consists in weaning the calves and then getting them accustomed to a ration of dry feed before they are exposed to the stress of shipping.

Bloat.—This is essentially a feed-lot disease which occurs when cattle are on a high-concentrate ration. Legume pastures, particularly alfalfa, also induce bloat if animals are not properly conditioned before being allowed to graze.

Feed-lot bloat seems to have no common remedy. Most of the trouble is with chronic "bloaters," and the best recourse is to market such animals quickly.

Vibriosis.—This is a venereal disease transmitted by a carrier cow to a clean bull in the act of coitus. It is then carried by the infected bull to any clean cows he breeds. The newly infected cow aborts or absorbs her calf during that pregnancy, but will usually calve regularly thereafter though she remains a carrier.

The insidious nature of vibriosis arises from the difficulty of diagnosis. Positive identification can be made only from examination of an aborted fetus. Other methods of detection, such as vaginal smears, have not been developed to the point where they can be considered 100 per cent accurate. Calving losses can run high in a herd without the owner's being aware of the cause, even with good veterinary advice.

An effective vaccine for vibriosis is available, but annual booster vaccinations are necessary to obtain continuous immunity.

Anaplasmosis.—This disease was at one time often confused with tick fever (piroplasmosis). The first signs are anemia, listlessness, and fever. Anaplasmosis is caused by microscopic parasites in the red-blood cells. The parasites are usually transferred from an infected animal to a clean one by either an insect bite or a hypodermic needle. For a number of years, anaplasmosis has been spreading northward from the southern states.

An effective vaccine has been developed which gives protection for about 12 months. Annual innoculations are necessary. When the disease is in the active stage, treatment involves transfusions of blood or plasma and is expensive. Infection in a carrier animal which has survived the active stage can be cleared up by heavy antibiotic treatment. This procedure eliminates the danger of insects transmitting the anaplasma parasites to other animals.

Internal Parasites.—Over 80 per cent of the losses caused by internal parasites are produced by the large variety of stomach and intestinal worms. Effective drugs for the treatment of animals suffering from these internal parasites are available at moderate cost. In some areas of the southern states, when infestation is known to be heavy, calves are routinely drenched. Throughout the rest of the country, however, cattle are usually treated only after there has been considerable loss in condition.

Insects.—The major fly pests—the heel, horn, stable, and face flies—account for 77 per cent of the loss occasioned by insects and external

UNITED STATES DEPARTMENT OF AGRICULTURE

USDA organization chart. Based on chart in *United States Government Organization Manual, 1970–71*.

Cattle of North America

parasites. Effective insecticides are available for the elimination of all the flies and lice. Spraying or dipping with these products is now commonplace on many ranches. Used routinely, either method will eliminate these pests.

GOVERNMENT AND CATTLE

It has long been the pride of the cattleman that he has operated independently of government subsidies. Yet inevitably the federal government has played an important role in the affairs of both the beef and the dairy industries. The states also have their agricultural departments. They are involved in such intrastate activities as brand inspection, inspection of public scales, some phases of animal health, co-operative marketing, and beef promotion.

Outside the regulatory control of milk producers and processors and, in some instances, of slaughtering facilities, the municipalities are little involved in the cattle industry.

Various aspects of the federal government's participation in the cattle industry have been referred to in pertinent sections of this book, such as those on cattle diseases and marketing.

The Forest Service of the USDA and the Bureau of Land Management of the Department of the Interior control the leasing of federal lands to cattlemen. The Forest Service leases the National Forest lands, and the Bureau of Land Management leases the public lands available for grazing under the Taylor Act. This is the only phase of the cattle industry in which these services are involved.

The Food and Drug Administration of the Department of Health, Education, and Welfare issues regulations governing the use of food additives, chemicals, and drugs by all phases of the cattle industry.

The principal functions of the nine USDA entities involved in the cattle business are briefly described below.

Commodity Credit Corporation

Price Support.—The beef-cattle industry is the only major segment of United States agriculture which has never received a direct subsidy from government in any form of price support. The dairy industry has had a floor put under the price of some dairy products—butter, for example—by the Commodity Credit Corporation.

Agricultural Marketing Service

Meat and Poultry Inspection Programs.—All cattle slaughtered for sale are required by law to undergo either state or federal inspection. If the meat enters into interstate commerce, it must be federally inspected.

United States

Federal officers also inspect all plants in every country that export meat to the United States. Approval of such plants is necessary before their meat can pass through United States ports.

Livestock Division, Meat Grading Branch.—The grading systems described in the section "Marketing" above are voluntary, and the service is rendered at the request of the packer. The official grader stamps each carcass with the quality grade. The carcass is then "rolled" by the packer if it grades choice or better ("rolled" is the term used for ink-branding the grade on the exterior of a carcass so that most of the retail cuts will show the mark). The packer pays the charge for the service, which in 1970 was 4.6 cents per 100 pounds. Nearly all beef slaughtered in large commercial plants in the United States is graded for quality, and a growing volume, now about 27 per cent, is graded for yield.

Packer and Stockyards Administration

This organization is concerned with regulatory functions in the marketing and interstate movement of cattle. Fair-trade practices, scale inspection in stockyards and packing plants, financial protection to the seller on public markets, and supervision of marketing charges are its major activities.

Economic Research Service

Many phases of the cattle industry are covered under the programs involved in the study of agricultural production and marketing. The reports published by this service are widely distributed among the many interests concerned in one way or another with the cattle business.

Statistical Reporting Service

Numerous market reports on all classes of cattle are published by this service on a daily, weekly, monthly, and annual basis. The releases include information on the numbers sold in the principal marketing sections of the country, the prices for various grades and classes of both live animals and carcasses, and forecasts of future demand and supply.

Agricultural Research Service

Animal Science Research Division.—Research is conducted on all phases of beef- and dairy-cattle husbandry. This work is carried on at the USDA Agricultural Research Center at Beltsville, Maryland, and at five other stations: at Miles City, Montana; Clay Center, Nebraska; Jeanerette, Louisiana; Fort Reno, Oklahoma; and Brooksville, Florida.

Dairy Cattle Research Branch.—This section compiles and analyzes the data obtained by the Dairy Herd Improvement Associations, a nationwide affiliation of milk-testing co-operatives. The national sire-evaluation summary for the five recognized dairy breeds is published annually by this branch.

Veterinary Science Research Division.—This division conducts investigational work on animal diseases, parasites, and insects harmful to animals.

Animal and Plant Health Inspection Service

This service was created in 1972, initially as the Animal and Plant Health Service, to establish in an independent agency the regulatory and control activities formerly under the Agricultural Research Service.

Animal Health Division.—This division formulates and administers programs for the control and eradication of serious threats to animal health, enforces quarantine regulations, and carries out test and slaughter procedures on outbreaks of serious diseases where authorized by law to do so.

Cooperative State Research Service

The 53 state experiment stations and their many branches are continuously pursuing research and experiment projects involving both beef and dairy cattle. The stations are dependent on federal grants for about one-third of their total budgets. The service participates in the allocation of the funds to the experiment stations and co-ordinates the research programs of the stations with federal research work.

Federal Extension Service

The federal, state, and county governments jointly employ the agents in charge of the extension work involving agriculture, home economics, 4-H clubs, and rural development. Foremost among these workers is the county agent, one of whom is located in nearly every county of the nation. Many programs are concerned with beef and dairy animals. The Federal Extension Service co-ordinates the work of the states involved in these programs and allocates the federal funds to the states.

The Federal Extension Service is involved in the Beef Improvement Federation, an organization composed of state Beef Cattle Improvement Associations, Performance Registry International, the breed association, and private individuals working with performance-testing programs, sire evaluation, and various other aspects of modern beef production.

OUTLOOK FOR CATTLE

Passing reference has been made throughout the section on the United

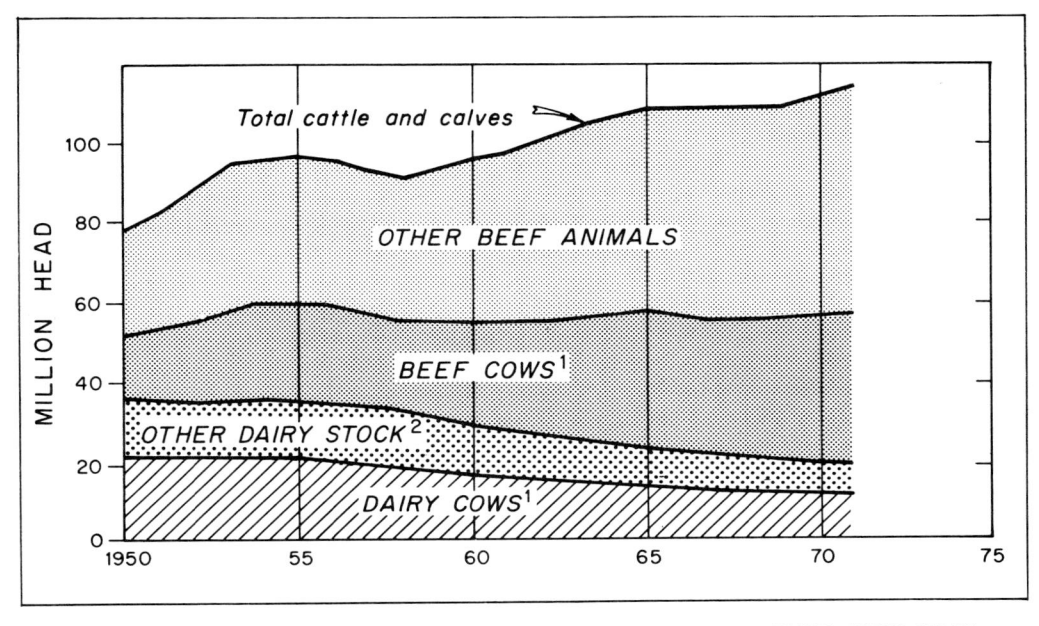

Breakdown of total cattle population. Based on data from USDA.

States to the future of many facets of the cattle business. A summary of these various aspects is presented here.

The trend in the total cattle population continues upward as the consumption of beef per capita increases. The decline in the number of dairy cattle is more than offset by the increase in beef-cattle numbers. The rate of increase in total cattle will probably be lower in the future.

The price of milk in recent years has followed a steady upward trend that has corresponded to the rate of inflation. Beef prices, however, have shown an erratic pattern owing to the lack of co-ordination in production and demand. The historic 8-to-10-year cycle in beef production and prices is the result.

Dairy Cattle

The dairy industry has reached the stage of maturity where rapid change is ordinarily avoided. The per capita consumption of milk has been in a definite decline since 1964, a decline which has not been

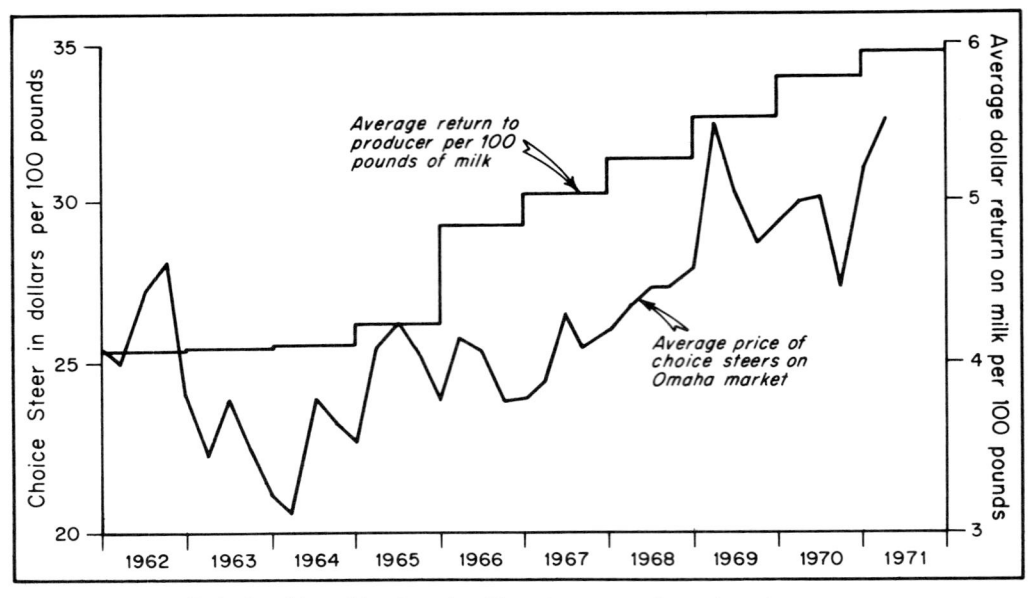

Relationship of beef and milk prices. Based on data from USDA.

offset by population expansion, though there was some slight recovery in 1970 and 1971.

Following the long-range trend, milk production per cow will continue to increase while total milk production will probably fall. The price of milk will continue to advance in line with the inflationary trend. As milk demand varies, production automatically adjusts to a new level almost overnight. When prices go down, the dairyman sells a bit earlier the cows he has marked for culling. If milk is in short supply, he keeps his marginal producers a little longer.

Milk and butter substitutes are continually cutting deeper into the milk market. Oleomargarine has replaced butter on many tables. Non-dairy coffee mixes are being produced in greater volume every year, some even by major marketers of dairy products. Such competitive products seem certain to make further inroads on future milk consumption.

The Holstein-Friesian became the American milk cow years ago and is probably now near the peak of her success, numbering 80 per cent or more of the national dairy herd. This proportion of Holsteins will increase more slowly in the future. The hard core of the other dairy breeds is now being approached. Many a small dairyman has a love

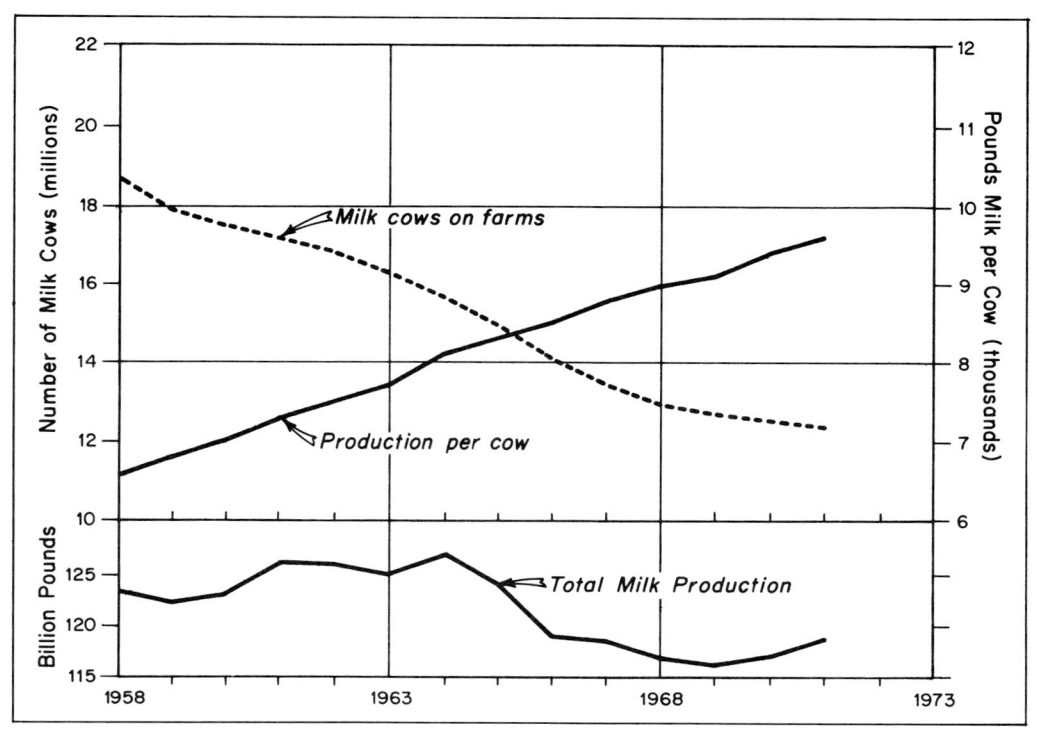

Milk production, number of milk cows, and milk per cow.

for his Jersey herd that is far greater than the desire for larger financial returns. And the same is true for owners of each of the other dairy breeds. On some farms these breed loyalties will pass to the next generation. While the proportion of Holstein-Friesians in the national herd will continue to gain, most of the other dairy breeds will be present in the year 2000.

There is increasing participation by the dairyman in the production of beef. Holstein steer calves in some areas are going into the feed lots in increasing numbers every year, and the cattle feeders like them. The cow-calf man is finding that a dairy bull can get more milk into his cow herd. Such trends as these, which are only beginning to take form, are indicative of a leaning toward the dual-purpose milk-beef type of cow.

Beef Cattle

The production of beef for the retail market has become segregated

Cattle of North America

into three levels, represented respectively by the producer, the feed lot, and the packer. The purebred breeder, once recognized as a separate entity in beef production, today finds himself at the crossroads. One view is that he is being integrated with the producer in the same manner that the purebred breeder of dairy cattle has become largely a producer of milk. The other view is that crossbreeding of beef cattle will become as systemized as it is now in hog raising and that the purebred breeder will eventually become more important to the commercial producer than ever. Only time will determine which path is followed. Wider use of artificial insemination, the application of performance records, and the introduction of the new breeds are the forces responsible for this era of change.

The function of the stocker, who roughed a calf through the winter to sell as a feeder the following summer, is being absorbed by either the producer or the feed lot.

There is some tendency toward integration in the major activities of the beef industry. With the general acceptance of custom-feeding, larger producers have adopted the practice of holding onto their cattle throughout the feeding period and selling directly to the packer. The custom feed lot provides this service.

There are also examples of integration of the feed lot and the packer. Monfort Packing Company, which operates the two largest feed lots in the country, was a pioneer in this practice. There are no outstanding examples of complete integration encompassing all three levels from producer to feeder to packer. It appears, however, that the producer-feeder will increase in number as long as favorable fat-cattle prices prevail. His feeding ventures will probably decline when prices drop. The feed-lot–packer trend will probably increase in the years to come if for no other reason than to ensure the packer a steady supply of uniform raw material.

For several decades cattle producers have been an economic enigma. They can be placed in three categories: the rancher, the diversified farmer, and the incidental producer with only a few head of cattle. The average return for all of them has been below an economic level, probably less than 2 per cent of the going value of their investment. The question can well be asked, Why do they continue in business?

The rancher, as the term is used here, derives his income solely from the sale of the commercial cattle he raises. Historically he has lived in regions IV, V, and VI, indicated on the map under "Twentieth Century" in the section "Management Practices." During the past few years he has been moving into region II, and even into region III.

United States

In the old, established cattle areas he is likely to be either a second- or third-generation descendant of the founder of the operation. There are also a growing number of newcomers, men of outside means, who have bought land and raised cattle. The inheritors do not have to show a profit on the current value of their ranches, since they made no investment. The newcomers can continue in business without making a reasonable return because they have other means. The generation now inheriting, however, is gradually selling out. It seems likely that this pattern will continue in the future. By one means or another beef-cattle production should continue at the current level in regions IV, V, and VI, and sizable increases are possible with improvement in cattle prices. It also appears possible that a better return on investment may be in the wind, but the extent to which prices drop after the present surge in the cattle cycle will tell the story. During the 1960's ranchers became established in increasing numbers in region II, where many diversified farmers also added cattle to their operations. Under current conditions this is the most profitable region for beef production in the United States. Here a well-planned ranch can be made to show a reasonable return on investment. The diversified farmer by converting his marginal cropland to improved pasture can also show a reasonable profit in cattle. Here, also, the incidental producer is increasing in number. Urban-employed owners of small parcels of land have found that a 10-to-20-head cow herd is more profitable than any other use to which they could put their acreage. Beef production is certain to increase in this region at all levels.

Region III is still the largest producer of beef cattle. Except for the drier plains area in the west of this region, beef are largely in the hands of the diversified farmer and the incidental producer. Both could be classed as small producers with few, if any, herds exceeding 200 head. Throughout the present century this region has produced one-third of the nation's beef cattle with numbers doubling during the past 20 years. The rate of increase should continue for some time to come.

A number of factors are involved in this increase. The intensified utilization of crop residues, particularly cornstalks, for fall and winter feeding has been one. Pasturing small plots of land that had lain idle and putting marginal cropland in grass are others. Since none of these opportunities have been fully exploited, there is still much latitude for expansion.

In recent years the incidental producer has raised most of the beef cattle in region I. He will probably continue at about the current level of production and will be joined by dairymen, particularly in the western states of region I—Michigan, Wisconsin, and Minnesota—who are moving into the picture with Holstein steers.

Dairy cows outnumber beef cows four to one in these three states, though the beef-cow herd has been increasing in recent years. This regional increase in production will have little impact on national figures however, since the area accounts for only 5 per cent of the total beef cattle.

The new breeds and the revival of interest in crossbreeding which they engendered are the current dilemma of the cattle producer. The new breeds are here, and more are on the way. They will undoubtedly stay and run their course. The compelling fact about the new breeds is that they all are arriving in one morning's mail. Many new breeds have been introduced into the United States in the past, but they arrived in small lots and did not fight for the spotlight at the same time.

The story of the "old new breeds" should be remembered when one is speculating on the future of the new ones. There are wide differences between modern breeding practices and those of the nineteenth century, but a grower's love of his own breed, and the rise and decline of fads in the kinds of cattle that are sought after today and forgotten tomorrow, are elements that will continue to shape the industry. The rise and decline of Shorthorn popularity is a common story. There were early arrivals of the Shorthorn breed after its development in England, but the first importation of Shorthorns for which continuity can be established dates from 1817. In the same year a shipment of Devon cattle was received which likewise established continuity for that breed on American soil. The history of the Devon is not as spectacular as that of the Shorthorn, but the breed has continued to enjoy a meager though consistent level of popularity.

Both the Angus and the Red Poll arrived in the United States for the first time in 1873. These were really new breeds of the day, both following half a century after the Shorthorns and the Devons. The Angus went on to a realm of popularity that is still growing, while the Red Poll has continued holding its own as a "farmer's cow" although at a much lower level. The Simmental, in its entrée as a new breed, arrived in 1886 and, after being held in different parts of the country for a number of years, disappeared. The promotional efforts of their backers probably played a major role in the degree of popularity the nineteenth century imports achieved. The skillful publicity given the new breeds today might also determine the "winners."

Regardless of their role in future crossbreeding schemes, many of the new breeds are here to stay. The Simmental; the Limousin; the

Gelbvieh; the Fleckvieh, or German Simmental; the Chianina; the Marchigiana; the Romagnola—all are beautiful breeds of cattle with excellent attributes for beef production. Some of them will find a permanent home in the United States; others may disappear. The breeds most successful will probably be those sponsored by the most dynamic breed societies.

Reasonable guidelines by which to forecast the future of crossbreeding are lacking. Some opponents of the practice predict a debacle similar to that which resulted from the feverish selection for the compressed type of cattle a few decades ago. That is hardly a reasonable analogy, for the performance-testing techniques which are coming into wider use every day should detect any major retrogression before it becomes widespread.

Proponents of crossing expound on the results achieved by hybridization of corn, chickens, and hogs and claim that it will do the same thing for cattle. This position overlooks the fact that, because of the longer generation interval, cattle cannot be handled and bred like corn or chickens or hogs.

Most disturbing of all in attempting to evaluate the future of crossbreeding is the divergence of opinion among authorities. One highly regarded specialist stated that he thinks "the commercial cow of the future will be a crossbred." Another authority of equal standing, addressing a group of cattlemen, defined crossbreeding as "a very effective method of making your herd a living, breathing demonstration of all the bovine colors, patterns, shapes, and sizes," and then proceeded to define upgrading as a technique that in 25 years will be advocated to overcome the accomplishments of today's crossbreeding. Only time can evaluate the results of the many hybridization efforts that are in the planning stages today.

Synthetics and other substitutes for meat are becoming more of a threat to the cattle producer every day. "All-beef" frankfurters in many supermarkets are now outnumbered 6 to 1 by the "all-meat" variety. The pork and chicken the all-meat product contains is just as much a substitute for beef as a vegetable product would be. Soybean steaks are not yet on the market, but food processors are ready to bring out such products whenever their price surveys show that the time is right. In the current rising and inflationary market, beef is particularly vulnerable because of the high price level it has recently attained. It appears certain that the use of beef, as well as milk, substitutes will increase in the future. The extent to which this development will affect the livestock industries can only be determined by time.

Cattle of North America

The feed-lot sector seems to be stabilized for the present, with sufficient capacity to meet the growing demand for beef. Custom feeding will probably increase.

On-the-farm feeding will likely continue at about the present level. Farmer-feeders cannot compare in efficiency with the large commercial plants, but the smaller feeder has no labor charges. His larger neighbor also utilizes family labor, and at every level this gives his operation much more supervision than is possible in the large commercial plant. Many are in a position to increase the size of their operations without adding to overhead.

Repugnant as the concept may be to the old-time cattlemen, there is no denying the tremendous impact that government involvement has had on all segments of the cattle industry in the past 50 years. Disease eradication and control, sanitation, beef grading, milk-marketing orders, dairy-sire evaluation, and, more recently, beef-sire evaluation are among the accomplishments that have worked wonders for the cattle industry. Government participation in these activities seems certain to increase. The question facing the beef industry is how much longer the resolve of its founders can continue to resist some form of government subsidy.

BEYOND THE HORIZON

New concepts continue to attract attention in the cattle industry. Some of these deserve mention here.

Confining Beef Cows for Calving and Breeding

For a number of years experiments have been run on maintaining brood cows under dry-lot conditions. In some instances the confinement period has been continuous. In others it has extended only from calving to weaning or from calving to the end of the breeding season. Some of these operations have been quite successful.

The advantages claimed for such confinement handling are greater control over nutrition, better supervision during calving, and more accurate heat detection for artificial insemination. A major disadvantage is that all feed must be cut and hauled to the cow instead of allowing her to do her own harvesting. Whether the advantages offset the increased labor cost, as well as the increased risk of disease which comes with close confinement, is still an open question largely dependent on location.

Estrus Synchronization.—For the past decade livestock scientists have been working on methods to control estrus in the cow so that she can be bred at a predetermined time. Of the various drugs and treatments used, the most promising so far is synthetic progesterone, the female sex hormone which delays estrus. The compound is administered for one heat period either as a feed additive or by injection or implant. After treatment is stopped, the cows should come in heat within one to four days.

On a practical level controlled estrus would be a boon in some herds in which artificial insemination is used and in certain operations where the cow herd is handled under confined conditions. For large herds under range conditions estrus control would be of doubtful advantage. For practical purposes it must still be considered in the experimental stage.

Superovulation, Multiple Births, and Ovum Transplant
Superovulation refers to the formation of an unnaturally large number of Graafian follicles in the ovaries of a female to produce a corresponding number of ova. This is induced by the injection of hormones, usually in connection with procedures for estrus control.

Twinning in beef cows occurs about once in every 200 births, probably less frequently in the average commercial operation. Twin calves are of no great advantage in most cow-calf operations today. One of the two calves will usually be subjected to various handicaps. If the mother does not lose one of the calves, she will probably not have enough milk to feed both of them properly. In spite of the problems, however, considerable attention is being devoted to techniques for producing superovulation and multiple births.

Multiple births in experimental groups of cows at the United States Range Live Stock Experiment Station, Miles City, Montana, have produced 129 per cent calf crops under carefully controlled conditions. After injection for superovulation on 81 cows the Fort Reno Experiment Station in Oklahoma produced 52 pregnancies by artificial insemination. These resulted in 29 single births, 13 sets of twins, 8 sets of triplets, 2 sets of quadruplets, and 1 set of quintuplets, a total of 92 calves.

Experimental "broods" of calves have been taken from a cow at birth and raised under laboratory conditions on prescription feeds.

In other cases fertilized ova have been transplanted from a donor cow and carried to birth in host cows. This procedure, like those above still in the early experimental stage, enables the outstanding cow to produce several calves a year.

Sex Control
Sex in cattle is determined when the single male spermatozoon unites

with the female ovum. The sperm may contain either an X or a Y chromosome on what amounts to an equal-chance basis. The ovum contains only the X chromosome. If the ovum is fertilized by an X-carrying sperm, the sex is fixed as female; if the fertilization is by a Y-carrying sperm, the sex is male. Thus the probability is practically even for a male or a female calf.

Research work in West Germany and Sweden as well as in the United States has indicated some possibility of separating the X and Y chromosomes in the sperm. If this were accomplished, the X sperm would produce females, and the Y sperm would produce males. Even if separation of the X and Y chromosomes were not completely accomplished, the ratio of X to Y chromosomes might be decreased, and the ratio of females to males would then follow a similar pattern.

The X sperm is thought to be heavier than the Y sperm. Gravitational or centrifugal means of separation have been tried, as has also electrical-field and chemical methods, but procedure for separation of the X and Y sperm has been accomplished at only an early experimental level.

Research and Experimentation.—Unlike other major industries, no private sector of the cattle industry is large enough to finance basic research projects. If such research is to be done, under the present economy, it must be undertaken by government. The extensive facilities and highly trained personnel of the federal and state experiment stations are continually involved in investigation work in which there is vast duplication of effort. Although there is organizational mechanism for co-ordination of the activities of this huge complex of scientific institutions, effective co-ordination is not achieved. This inefficiency is particularly evident in the work that the experiment stations do with cattle. Innumerable projects have been devoted to proving that stilbestrol increases feed efficiency 10 per cent and that the nutrition level is of prime importance in calving heifers. The common flies, however, still continue to take an annual toll of over half a billion dollars annually, from United States cattle, according to USDA estimates, and the death loss of newborn calves in the average range operation continues at the rate of 5 to 15 per cent. The maze of governmental red tape and the aspiration of the individual experiment stations to obtain recognition in all new developments are the main factors involved in the inefficient operation of the national experimental plant.

Appendix

Human and Cattle Populations of North America, 1972

Country	Area, Sq. Mi.	Human Population	Cattle Population
Caribbean and Guianas:			
Barbados	166	254,000	18,000
Cuba	44,218	8,600,000	7,600,000
Dominican Republic	18,800	4,200,000	1,100,000
Haiti	10,700	5,300,000	940,000
Jamaica	4,200	2,000,000	250,000
Martinique	424	340,000	40,000
Trinidad & Tobago	1,980	1,100,000	63,000
Total Caribbean	80,488	21,794,000	10,011,000
French Guiana	35,000	42,000	2,000
Guyana	83,000	764,000	257,000
Surinam	63,000	403,000	50,000
Total Guianas	181,000	1,209,000	309,000
Central America:			
Costa Rica	19,600	1,800,000	1,574,000
El Salvador	8,300	3,500,000	1,350,000
Guatemala	42,100	5,500,000	1,376,000
Honduras	43,400	2,700,000	1,830,000
Nicaragua	50,200	1,900,000	1,700,000
Panama	29,200	1,500,000	1,297,000
Total Central America	192,800	16,900,000	9,127,000
Mexico	761,000	50,600,000	24,876,000
Canada	3,851,800	21,681,000	13,660,000
United States	3,615,200	203,185,000	118,283,000
Grand total	8,682,288	315,369,000	176,266,000

Cattle of North America

U.S. CATTLE POPULATION, 1972
(Thousands of Head)

State	Milk Cows	Beef Cows	Total Cattle
Alabama	134	951	2,050
Arizona	53	345	1,295
Arkansas	96	982	1,912
California	816	888	4,775
Colorado	101	1,154	3,716
Connecticut	62	5	119
Delaware	14	5	32
Florida	197	971	1,939
Georgia	146	887	2,042
Idaho	151	626	1,882
Illinois	286	802	3,400
Indiana	237	456	1,956
Iowa	465	1,737	7,773
Kansas	192	1,919	6,757
Kentucky	334	1,136	2,916
Louisiana	167	942	1,807
Maine	66	9	142
Maryland	164	57	430
Massachusetts	62	8	118
Michigan	473	138	1,542
Minnesota	971	547	3,998
Mississippi	181	1,341	2,613
Missouri	337	2,081	5,238
Montana	39	1,644	3,165
Nebraska	185	2,056	6,780
Nevada	14	344	658
New Hampshire	36	2	71
New Jersey	64	11	125
New Mexico	33	654	1,441
New York	973	67	1,812
North Carolina	182	381	1,103
North Dakota	132	1,013	2,278
Ohio	444	390	2,244
Oklahoma	144	2,237	5,441
Oregon	103	664	1,593
Pennsylvania	696	97	1,763
Rhode Island	7	1	12
South Carolina	65	294	687
South Dakota	174	1,829	4,543

Appendix

Tennessee	284	1,029	2,472
Texas	355	5,452	12,829
Utah	81	357	874
Vermont	205	7	351
Virginia	216	515	1,460
Washington	192	379	1,362
West Virginia	52	210	475
Wisconsin	1,866	279	4,421
Wyoming	17	735	1,490
Total, 48 states	12,264	38,634	117,662
Alaska	1.7	2.5	9
Hawaii	13	89	249
Puerto Rico	113	64	367
Grand total	12,392	38,789	118,287

Source: Crop Reporting Service, USDA.

633

Cattle of North America

BREED ASSOCIATIONS IN CANADA

Beef Breeds

Canadian Aberdeen-Angus Association
Box 663
Guelph, Ontario

Canadian Charolais Association
4816 Macleod Trail
Calgary 6, Alberta

Canadian Galloway Association
Box 204
Gull Lake, Saskatchewan

Canadian Hereford Association
1706 First Street, S.E.
Calgary 21, Alberta

Canadian Highland Cattle Society
Box 777
Duncan, British Columbia

Canadian Shorthorn Association
5 Douglas Street
Guelph, Ontario

Canadian Santa Gertrudis Association
2349 Yonge Street
Toronto 12, Ontario

New Breeds

Italian White Cattle Association
Edmonton, Alberta

Canadian Limousin Association
Box 144
Midnapore, Alberta

Canadian Lincoln Red Association
Box 644
Rocky Mountain House, Alberta

Canadian Maine-Anjou Association
10603 100th Avenue
Edmonton, Alberta

Canadian Murray Grey Association
R.R. 1
Bentley, Alberta

Canadian Simmental Association
P.O. Box 8550, Station F
Calgary 13, Alberta

Canadian Welsh Black Cattle Society
Box 52
Hillspring, Alberta

Dairy Breeds

Canadian Cattle Breeders Society
Box 547
Granby, Quebec

Ayrshire Breeders Association of Canada
1160 Carling Avenue
Ottawa 3, Ontario

Canadian Brown Swiss Association
Box 593
Brighton, Ontario

Canadian Guernsey Breeders' Association
368 Woolwich Street
Guelph, Ontario

Holstein-Friesian Association
Brantford, Ontario

Canadian Jersey Cattle Club
290 Lawrence Avenue West
Toronto 20, Ontario

Canadian Red Poll Cattle Association
Salwell, Alberta

Appendix

BREED ASSOCIATIONS IN THE UNITED STATES

Beef Breeds

American Angus Association
3201 Frederick Boulevard
St. Joseph, Missouri 64501

Red Angus Association of America
Box 776
Denton, Texas 76201

American-International Charolais Association
1610 Old Spanish Trail
Houston, Texas 77025

Devon Cattle Association, Inc.
Goldendale, Washington 98620

American Galloway Breeders Association
1020 Rapid Street
Rapid City, South Dakota 57701

Galloway Cattle Society of America, Inc.
Hennepen, Illinois 61327

Galloway Performance International
P.O. Box 620
Eureka, Kansas 67045

Belted Galloway Society
Summetville, Ohio 43962

American Hereford Association
Hereford Drive
Kansas City, Missouri 64130

American Scotch Highland Breeders Association
Edgemont, South Dakota 57735

Pan American Zebu Association
P.O. Box 1946
Cotulla, Texas

Red Poll Cattle Club of America
3275 Holdrege Street
Lincoln, Nebraska 68503

Red Poll Beef Breeders, International
P.O. Box 176
Ross, Ohio 45061

American Shorthorn Association
8288 Haskill Street
Omaha, Nebraska 68124

Sussex Cattle Association of America
Drawer AA
Refugio, Texas 78377

American Developed Breeds

American Brahman Breeders Association
4815 Gulf Freeway
Houston, Texas 77023

Barzona Breed Association of America
1107 Copper Base Road
Prescott, Arizona 86301

Beefmaster Breeders Universal
Gunter Hotel
San Antonio, Texas 78206

Foundation Beefmaster Association
201 Wyandot Street
Denver, Colorado 80223

International Braford Association
P.O. Box 1030
Fort Pierce, Florida 33450

International Brangus Breeders Association, Inc.
908 Livestock Exchange Building
Kansas City, Missouri

American Red Brangus Association
620 Colorado Building
Austin, Texas

Santa Gertrudis Breeders International
P.O. Box 1257
Kingsville, Texas 78363

Cattle of North America

New Breeds

American Chianina Association, Inc.
Box 11537
Kansas City, Missouri 64138

American Gelbvieh Association
Route 1
Newkirk, Oklahoma 74647

North American Limousin
  Foundation
309 Livestock Exchange Building
Denver, Colorado 80216

International Maine-Anjou
  Association
P.O. Box 5636
Livestock Exchange Building
Kansas City, Missouri 64102

American Murray Grey Association,
  Inc.
Rt. 4, Box 179-A
Shelbyville, Kentucky 40065

American Simmental Association
P.O. Box 24
Bozeman, Montana 59715

South Devon Breed Society
Box S.D.
Albert Lea, Minnesota

Dairy Breeds

Ayrshire Breeders Association
Brandon, Vermont 95733

Brown Swiss Cattle Breeders'
  Association of America
P.O. Box 1038
Beloit, Wisconsin 53511

Brown Swiss Beef International, Inc.
Box 10381
Beloit, Wisconsin 53511

American Dexter Cattle Association
Decorah, Iowa 52101

Dutch Belted Cattle Association of
  America
6000 N.W. 32 Avenue
Miami, Florida

American Guernsey Cattle Club
Peterborough, New Hampshire
  03458

Holstein-Friesian Association of
  America
P.O. Box 808
Brattleboro, Vermont 05301

American Jersey Cattle Club
1521 East Broad Street
Columbus, Ohio 43205

American Milking Shorthorn Society
313 South Glenstone Avenue
Springfield, Missouri 65802

Red and White Dairy Cattle
  Association
P.O. Box 771
Elgin, Illinois 60120

Bibliography

Bibliography

"Aberdeen-Angus: The Story of Our Breed," Webster City, Iowa, Aberdeen-Angus Journal Publishing Co., Inc., 1969.

Allen, Lewis F. *American Cattle: Their History, Breeding and Management.* New York, O. Judd Company, 1887.

Andrew, D. A. *The Hereford in Canada, 1860–1960,* Toronto, 1962.

Archer, Sellers G., and Clarence E. Bunch. *The American Grass Book: A Manual of Pasture and Range Practices.* Norman, University of Oklahoma Press, 1953.

Archibald, E. S. *Hybridization of Domestic Cattle and Buffalo.* Ottawa, Canada, Experimental Farms Service, Department of Agriculture, 1966.

Atlas of World History. New York, Rand McNally & Company, 1957.

Baker, A. H. *Livestock and Complete Stock Doctor.* Minneapolis, H. L. Baldwin Publishing Company, 1912.

Baker, Gladys L., et al. *Century of Service.* Report of U.S. Department of Agriculture. Washington, Government Printing Office, 1963.

Bidwell, Percy Wells, and John I. Falconer. *History of Agriculture in the Northern United States, 1620–1860.* Washington, Carnegie Institution, May, 1925.

Briggs, Hilton M. *Modern Breeds of Livestock.* New York, Macmillan Company, 1970.

Caldwell, William H. *The Guernsey.* Peterborough, N.H., American Guernsey Cattle Club, 1941.

Canadian Cattle Herd Book (Livre de Généalogie du Betail Canadien, Premier Volume). Ottawa, Société des Eleveurs de Betail Canadien, 1909.

"The Century Ahead." Seminar on Agricultural Administration in the Land-Grant System. Fort Collins, Colorado State University, 1963.

Cole, H. H., ed. *Introduction to Livestock Production Including Dairy and Poultry*. San Francisco and London, W. H. Freeman & Company, 1962.

Cunha, T. J., M. Koger, and A. C. Warnick. *Crossbreeding Beef Cattle*. Gainsville, University of Florida Press, 1963.

Cuba '67: Image of a Country. Havana, Book Institute of Havana, 1967.

Davidson, Gordon Charles. *The North West Company*. Berkeley, University of California Press, 1918.

Deakin, Alan, G. W. Mair, and A. G. Smith. *Publication 479, Technical Bulletin 2*. Ottawa, Department of Agriculture, 1933.

de Alba, Jorge. "The Charolais Breed in Mexico," *Cattleman*, Vol. XXXIII, No. 5 (1943).

Dobie, J. Frank. *The Longhorns*. Boston, Little, Brown and Company, 1941.

Edwards, Philip Leget. *Diary: The Great Cattle Drive to Oregon from California in 1837*. San Francisco, Grabhorn Press, 1932.

Ensminger, M. E. *Beef Cattle Science*. Danville, Ill., Interstate Printers & Publishers, Inc., 1968.

Feher, Joseph. *Hawaii: A Pictorial History*. Honolulu, Bishop-Prentiss Press, 1969.

Ferragut, Leon. *Principal Breeds of Cattle and Study of Various Aspects of Commercial Production of Cattle in Cuba*. Havana, Editorial Neptuno, 1948.

Ferris, Robert G., ed. *Prospector, Cowhand, and Sodbuster*. Washington, U.S. Department of the Interior, 1967.

―――. *Explorers and Settlers*, Vol. V. Washington, U.S. Department of the Interior, 1968.

Fowler, Stewart H. *Beef Production in the South*. Danville, Ill., Interstate Printers & Publishers, Inc., 1969.

Gard, Wayne. *The Chisholm Trail*. Norman, University of Oklahoma Press, 1954.

Gasser, G. W. *Livestock in Alaska*. Juneau, Alaska Development Board, 1946.

Georgeson, C. C. *Brief History of Cattle Breeding in Alaska*. Report of U.S. Department of Agriculture. Washington, Government Printing Office, 1929.

Gow, R. M. *The Jersey*. New York, American Jersey Cattle Club, 1936.

Gray, Lewis Cecil, et al. *History of Agriculture in the Southern United States to 1860*. Washington, Carnegie Institution, 1933.

Haystead, Ladd, and Gilbert C. Fite. *The Agricultural Regions of the United States*. Norman, University of Oklahoma Press, 1963.

Hinman, Claude H. *Dual-Purpose Cattle*. Springfield, Ill., Roberts Brothers, 1953.

Bibliography

Hodgson, Ralph E. *Germ Plasma Resources.* Publication No. 66 of the American Association for the Advancement of Science. Washington, 1961.
Jennings, Robert. *Cattle and Their Diseases.* Philadelphia, John Potter, 1863.
John, Roger McKinley. "A History of the Brahman Cattle Industry in Texas." Thesis, Texas Agricultural and Mechanical University, 1966.
Johnson, Willis Fletcher. *The History of Cuba.* New York, B. F. Buck & Company, 1920.
Juergenson, Elwood M. *Approved Practices in Beef Cattle Production.* Danville, Ill., Interstate Printers & Publishers, Inc., 1964.
Kellogg, Charles E., and David C. Knapp. *Science in the Public Service.* New York, McGraw-Hill Book Company, 1966.
King, Thomas E. *The Great White Cattle.* Chicago, Wolf and Krautter, Inc., 1967.
Kuyendall, Ralph S., and A. Grover Day. *Hawaii: A History.* New York, Prentice-Hall, Inc., 1948.
Logan, V. S., and P. E. Sylvestre. *Hybridization of Domestic Beef Cattle and Buffalo.* Ottawa, Experimental Farms Service, Department of Agriculture, 1950.
López, Ignacio Díaz. *Growth of Cattle Industry of Cuba.* Twentieth Annual Report, Bureau of Animal Husbandry, U.S. Department of Agriculture. Washington, Government Printing Office, 1904.
Luaces, Roberto L. *Native Cattle: The Zebu Cattle in the Improvement of the Cattle Industry.* Havana, 1916.
MacEwan, J. W. G. *The Breeds of Farm Live Stock in Canada.* London, Thomas Nelson & Sons, Ltd., 1941.
McKay, Douglas. *The Honourable Company: A History of the Hudson's Bay Company.* Freeport, N.Y., Books for Libraries Press, 1936.
Mansolo, D. R. "Cattle Raising in Cuba." *Cattleman,* Vol. XVII, No. 5 (1930), 42–43.
Mason, I. L. *A World Dictionary of Livestock Breeds, Types and Varieties.* Farnham Royal, England, Commonwealth Agricultural Bureau, 1969.
———. *Comparative Beef Performance of the Large Cattle Breeds of Western Europe.* Animal Breeding Abstract. Edinburgh, Commonwealth Bureau of Animal Breeding and Genetics, 1971.
Morse, E. W. *The Ancestry of Domesticated Cattle.* Twenty-seventh Report, Bureau of Animal Husbandry, U.S. Department of Agriculture. Washington, Government Printing Office, 1912.
Mothershead, Harmon Ross. *The Swan Land and Cattle Company, Ltd.* Norman, University of Oklahoma Press, 1971.
Neumann, A. L., and Roscoe R. Snapp. *Beef Cattle.* New York, John Wiley & Sons, Inc., 1969.
Oliphant, J. Orin. *On the Cattle Ranges of the Oregon Country.* Seattle, University of Washington Press, 1968.

Ornduff, Donald R. *The Hereford in America*. Kansas City, Hereford History Press, 1969.

Phillips, Ralph W. "World Distribution of the Major Types of Cattle," *Journal of Heredity*, 1961.

———. "Untapped Sources of Animal Germ Plasma," *Journal of Heredity*, 1961.

Plumb, C. S. *Beginnings in Animal Husbandry*. St. Paul, Webb Publishing Company, 1912.

Post, Lauren C. *The Old Cattle Industry of Southwest Louisiana*. Reprinted from *McNeese Review*, 1957.

Reaman, George Elmore. *History of the Holstein-Friesian Breed in Canada*. Toronto, Collins, 1946.

Ross, P. H. *Dairy Practice at Kenai Station, Alaska*. Report of U.S. Department of Agriculture. Washington, Government Printing Office, 1908.

Ruddick, J. A., et al. *The Dairy Industry in Canada*. Toronto, Ryerson Press, 1937.

Sanders, Alvin Howard. "The Taurine World," *National Geographic Magazine*, December, 1925.

Schmid, A. *Rassenkunde des Rindes*. Bern, Benteli A.G., 1942.

Scott, David R. *A History of the United States*. New York, Harper and Brothers, 1883.

Thomas, Gerald W. *Progress and Change in the Agricultural Industry*. Dubuque, Wm. C. Brown Company, 1969.

Thompson, James Westfall. *History of Livestock Raising in the United States, 1607–1860*. Report of U.S. Department of Agriculture. Washington, Government Printing Office, 1942.

Towne, Charles Wayland, and Edward Norris Wentworth. *Cattle and Men*. Norman, University of Oklahoma Press, 1955.

U.S. Department of Agriculture. *Agricultural Statistics*. Washington, U.S. Government Printing Office, 1970.

———. *Fact Book of U.S. Agriculture*. Washington, U.S. Government Printing Office, 1970.

———. *Handbook of Agricultural Charts*. Washington, U.S. Government Printing Office, 1971.

———. *Livestock and Meat Statistics*. 1970 Supplement to Statistical Bulletin No. 222. Washington, U.S. Government Printing Office, 1970.

———. *Losses in Agriculture*. Washington, U.S. Government Printing Office, 1965.

———. *Occupations and Trends in the Dairy Products Industry*. Washington, U.S. Government Printing Office, 1970.

———. *Report of the Commissioner of Agriculture for the Year 1866*. Washington, U.S. Government Printing Office, 1867.

———. *Report on Exploratory Investigations of Agricultural Problems of*

Alaska. Miscellaneous Publication No. 700. Washington, U.S. Government Printing Office, n.d. (ca. 1942).

Warwick, E. J. *Fifty Years of Progress in Breeding Cattle.*

White, W. T., Ralph W. Phillips, and E. C. Elting. "Yaks and Yak-Cattle Hybrids in Alaska," *Journal of Heredity*, 1946.

Wilcox, E. V. *Farm Animals.* New York, Doubleday, Page and Company, 1910.

Work, John. *The Snake Country Expedition of 1830–1831.* Norman, University of Oklahoma Press, 1971.

Index

Aberdeen-Angus (Angus): in Cuba, 27, 38, 51; in Dominican Republic, 57; in Jamaica, 83, 89, 93; in Middle America, 153–54; in Costa Rica, 159; in El Salvador, 172; in Guatemala, 184; in Nicaragua, 211; in Mexico, 236, 238; in Canada, 268, 282–84, 287, 300, 302–303, 305, 310, 312, 315, 320, 339; in United States, 268, 353, 362, 377, 379–82, 384, 390–92, 394–97, 443, 446, 454–56, 473, 483, 514, 518, 520–21, 525, 540–41, 626; in Alaska, 483, 561; in Hawaii, 564–65
Africander, in United States: 350, 443–46, 478
Alaska: Siberian cattle, 482–83, 555; Galloway, 483–85, 555–57; Holstein-Friesian, 483, 558–60; Milking Shorthorn, 483, 558; Guernsey, 483; Jersey, 483, 559; Hereford, 483, 561; Angus, 483, 561; Yak, 558–59; Red Danish, 558; Charolais, 561
Alberta, Canada: 259, 264; Longhorn, 268, 317; Shorthorn, 268, 317; Hereford, 268, 288–89; Brown Swiss, 277; Aberdeen-Angus, 283; Charolais, 284–87; Highland (West Highland), 289–91; Hays Converter, 293–95; Limousin, 299–300; Murray Grey, 302; Simmental (Pie Rouge de l'Est), 303; Welsh Black, 304; Cattalo, 305
American Brahman: 16, 86–87, 95; on Barbados, 18; in Cuba, 27, 32–33, 41, 43, 51; in Dominican Republic, 57, 60, 65; in Haiti, 75; on Martinique, 101–103; on Trinidad, 107–108; in Guyana, 124, 126, 128, 132; in Surinam, 138–39, 143; in Middle America, 153–54; in Costa Rica, 157, 159, 163; in El Salvador, 172, 179; in Guatemala, 183–84, 189–91; in Honduras, 197–98, 203–205; of Nicaragua, 210–11, 216–18; of Panama, 222, 227; in Mexico, 237, 239–40, 246–47; of Canada, 304–305; in United States, 362, 387, 401–402, 412, 439–43, 448, 451–57, 460–63, 507, 516, 519–20, 532, 540; of Puerto Rico, 574
Angel cattle, of United States: 360–61; *see also* Red Danish
Armoricaine, of France: 300–301, 473; *see also* Maine-Anjou
Ayrshire: in Cuba, 38–39; in Jamaica, 83; in Trinidad, 106; in El Salvador, 172; in Canada, 274–77, 279, 309, 311–12, 320, 339; in United States, 353, 372, 390, 412–14; in Puerto Rico, 573

Baja California: 229; Criollo (pure) (Chinampo), 197, 234–35; *see also* Mexico
Barbados: 11, 14, 136; Spanish cattle, 18, 356; Zebu-Creole cross, 18, 27; Holstein-Friesian, 18; Friesian, 18; Jersey, 18; American Brahman type, 18; Santa Gertrudis, 18; Senepol, 18; Charolais, 18; cattle diseases, 21
Barrow cattle: 440; *see also* Zebu
Barzona, in United States: 350, 443–47, 540
Beefmaster: in Costa Rica, 159; in United States, 447–51
Belted Galloway: in Canada, 288; in United States, 392, 420
Bison: in United States, 348, 357–59; in Canada, 271, 305–306; *see also* Cattalo
Black and White Lowland cattle, in United States: 359; *see also* Dutch Friesian
Black Angus: *see* Aberdeen-Angus
"Black Canadian": *see* Canadian (French Canadian)
"Black Jersey": *see* Canadian (French Canadian)

643

Cattle of North America

Blanco Orejinegro, of Colombia: 197
Blonde d'Aquitaine: of Canada, 296; of United States, 470–71
Braford, of United States: 451–54, 516, 532
Brahman-Hereford cross: *see* Braford
Brangus: 89; in Dominican Republic, 58; in Panama, 222, 224, 227; in United States, 362, 454–56, 516, 532, 540
British Columbia, Canada: 259; Hereford, 288, 320; South Devon, 304; Aberdeen-Angus, 320; Shorthorn, 320; Charolais, 320; Jersey, 320; Guernsey, 320; Holstein-Friesian, 320
British White (Park) cattle, in United States: 350, 478–79
Brittany, cattle of: 266, 271, 300, 505
Brown Swiss: in Cuba, 16, 38–39, 43; on Barbados, 18; in Dominican Republic, 57, 63; in Haiti, 75–76; on Martinique, 101; in Costa Rica, 160, 162–63; in El Salvador, 172, 179; of Honduras, 198, 201, 205; of Nicaragua, 210–11, 213–14, 218; of Panama, 222, 227; of Mexico, 239, 247; in Canada, 63, 277–78, 293–94; in United States, 353, 414–17, 470–71, 482; of Puerto Rico, 573
Buffalo (water, swamp): in Trinidad, 15–16, 107–12, 114, 116; of Surinam, 137, 139, 141; in Hawaii, 565
Buffalo, American: *see* bison
Buffalo, Murrah, on Trinidad: 107–108
Buffalo, Nihli, on Trinidad: 108

Calf clock auction, in Canada: 328–29
Canada: 15, 18, 234, 259–66; Holstein-Friesian, 27, 268, 270, 276–80, 293, 306, 309, 311–15, 320, 339; Hereford, 38, 268, 283, 287–89, 292–94, 303, 305–306, 310, 320; Canadian (French Canadian), 266, 270–75, 312, 339; Spanish cattle, 266–67; Devon, 266, 296–99; Native cattle, 266–67; Jersey, 266, 271–73, 278–81, 320, 339; Guernsey, 266; 271–72, 278, 309–10, 320, 339; Longhorn, 266, 292, 317; Dutch Black and White, 268; Aberdeen-Angus, 268, 282–84, 287, 300, 302–303, 305, 310, 312, 315, 320, 339; Shorthorn, 268, 288–89, 291–93, 299–300, 302–303, 306, 310, 312, 320, 339; Charolais, 268–69, 284–87, 297, 310, 315, 320, 322, 339; Pie Rouge de l'Est (Simmental), 269, 296–97, 303; Simmental, 269, 295–99, 310; Limousin, 269, 296–99, 310; Maine-Anjou, 269, 296, 299–302, 310; Fleckvieh (German Simmental), 269, 296; Chianina (Italian), 269, 296–97; Gelbvieh (German Yellow), 269, 296; Ayrshire, 274–77, 279, 309, 311–12, 320, 339; Brown Swiss, 277–78, 293–94; Holstein, 278–80, 294; Red and White Holstein-Friesian, 280; Red Poll, 281–82; Galloway, 287–88, 391; Belted Galloway, 288; Polled Hereford, 289, 305, 315; Highland (West Highland), 289–91; Santa Gertrudis, 291; Polled Shorthorn, 292; Hays Converter, 293–95; Parthenay, 296; Blonde d'Aquitaine, 296; Lincoln Red, 296; 299–300; South Devon, 296, 303–304; Welsh Black, 296, 304; Murray Grey, 296, 302–304; Tarantaise, 296; Pinzgauer, 296; Red and White Lowland (German Red and White), 301–302; American Brahman, 303–305; Cattalo, 305–306; Yak, 306–307; beef grading in, 330–32; cattle diseases, 268, 295–96, 302, 333–34; *see also* Maritime Provinces, Prairie Provinces, diseases
Canadian (French Canadian) cattle: of Canada, 266, 270–75, 312, 339; of United States, 350, 352–53, 479–80, 504–505
Cattalo: in Canada, 305–306; in United States, 348, 357–60
Charbray: in Dominican Republic, 58, 62; on Trinidad, 107–108; in Costa Rica, 159; in United States, 387–88, 460–61
Charolais: on Barbados, 18; in Cuba, 27, 33–35, 43, 48, 51; in Dominican Republic, 62, 65; in Jamaica, 90, 93; on Martinique, 101–103; on Trinidad, 107–108; of Surinam, 138; in Costa Rica, 159; in El Salvador, 172; in Guatemala, 184; of Nicaragua, 211; of Panama, 222; of Mexico, 234, 236, 240–41; of Canada, 268–69, 284–87, 297, 310, 315, 320, 322, 339; of United States, 34, 185, 268, 350–51, 354, 385–87, 473, 514, 516, 518, 520–21, 525; of Alaska, 561; of Puerto Rico, 574–75
Chianina (Italian): in Canada, 269, 296–97; in United States, 470–71, 473, 627
Chiapas, Mexico: Zebu, 246; American Brahman, 246; Indo-Brazil, 246; Criollo (mixed), 246; *see also* Mexico
Chihuahua, Mexico: Hereford, 238; Aberdeen-Angus, 238; Criollo, 238, 245; beef cattle of, 242–46; *see also* Mexico
Chinampo: *see* Baja California, Criollo
Coastal Plains, Mexico: beef cattle of, 246–47; *see also* Chiapas, Oaxaca, Veracruz
Corriente: *see* Sonora, Mexico, Criollo
Costa Rica: 147, 149–50, 154–56, 209; Jersey, 152, 158, 160; Guernsey, 152, 158–63; Criollo, 153, 157–59, 162–63, 165, 183, 210; American Brahman, 157, 159, 163; Guzerat, 157; Nellore, 157; Holstein-Friesian, 159–60, 162–63; Red Brahman, 159; Charolais, 159; Charbray, 159; Aberdeen-Angus, 159; Beefmaster, 159; Brown Swiss, 160, 162–63; cattle diseases, 162, 166–67; *see also* diseases.
Cream Pot: *see* Old Rock Stock
Creole (Criollo): *see* Spanish cattle
Criollo (Creole): 15–16; in Cuba, 16, 26–31, 40, 43, 51; in Dominican Republic, 57–59, 65; in Haiti, 74–75; in Jamaica, 82–83; on Martinique, 100–102, 104; on Trinidad,

Index

106–108, 114; in French Guiana, 119–20; in Guyana, 124, 127–29; of Surinam, 137–39, 143; of Middle America, 152–54; in Costa Rica, 153, 157–59, 162–63, 165; in El Salvador, 153, 170–72, 174, 178–79; in Guatemala, 153, 182–92; of Honduras, 153, 194–97, 201, 204; of Nicaragua, 153, 210; of Panama, 157, 221; of Mexico, 234–35, 238; *see also* Zebu-Creole cross; Baja California; Sonora, Mexico

Criollo (mixed): *see* Zebu-Creole cross

Cuba: 4–11, 14, 23–26; Criollo, 16, 26–31, 40, 43, 51; Holstein-Friesian, 16, 27, 36, 41, 43, 51; Brown Swiss, 16, 38–39, 43; Zebu, 16, 26–29, 31–32, 38, 40; Spanish cattle, 26, 28; Charolais, 27, 33–35, 43, 48, 51; Angus, 27, 38, 51; Hereford, 27, 35, 38–39, 51; Jersey, 27, 38–39, 51; Guernsey, 27, 38–39, 51; Red Poll, 27, 38–39; American Brahman, 27, 32–33, 41, 43, 51; Santa Gertrudis, 27, 35, 43, 51; Criollo-Zebu types, 27; Tinima, 31; Shorthorn, 35, 38, 43; Milking Shorthorn, 38; Devon, 38; Ayrshire, 38–39; South Devon, 38; Illawara Shorthorn, 38; Limousin, 38; Indo-Brazil, 38; Gyr, 38; Guzerat, 38; Nellore, 38; Holstein, 39, 43; cattle diseases, 40, 51–52; *see also* diseases

De Lidia: in Mexico, 152, 237–38, 248–51; in Middle America, 152

Devon cattle: in Cuba, 38; in Jamaica, 83; in Canada, 266, 296–99; in United States, 359, 372, 388–91, 461, 626; in Hawaii, 564

Devonshire cattle, of United States: 359–61, 388–89; *see also* Devon, South Devon

Dexter, in United States: 417–19

Diseases: tick-borne, 15, 21, 51, 95, 142, 153, 177, 190, 216, 226, 247, 253, 268, 295–96, 334; foot-and-mouth, 21, 51, 68, 78–79, 95, 103–104, 114, 131, 142, 190, 218, 226–27, 238–39, 354–55, 385, 470, 576; tuberculosis, 21, 51, 79, 96, 104, 131, 142, 167, 203, 213, 216, 226, 334, 568, 576; blackleg, 21, 40, 51, 96, 114, 131, 166, 177, 190, 203, 216, 226, 249, 254, 334, 555, 568; internal parasites, 21, 51, 68, 79, 131, 142, 203, 216; screwworm, 40, 69, 79, 131, 142–43, 162, 166, 177, 190, 203, 216, 226, 253–54, 371–72, 519; external parasites, 51, 68, 166; brucellosis, 51, 68–69, 79, 96, 104, 114, 131, 167, 177, 191, 203, 213, 216, 218, 226–27, 254, 334, 568, 576, 601; vibrosis, 52, 114, 177, 334; rabies, 52, 69, 114, 128, 132, 142, 167, 177, 190, 216, 226, 254; measles, 52; rinderpest, 68; piroplasmosis, 68, 79, 95, 114, 178, 203, 576; anaplasmosis, 68, 79, 95, 114, 178, 191, 203, 226, 568; liver fluke, 69, 79, 96; gastro-intestinal worms, 69, 115; leptospirosis, 79; anthrax, 79, 96, 167, 177, 190, 216, 254; malignant edema, 166; 203, 216; hemorrhagic septicemia, 167, 177, 203, 216, 226, 334; weed (escobilla) poison, 177, 216; pleuropneumonia, 302; acute anemia, 374; salt sick, 374; trypanosomiasis, 467; *see also* diseases, health hazards, United States; diseases, United States—controlled; *see also under* countries and provinces

Diseases, health hazards, United States: losses from, 614; mastitis, 615; bloat, 615–16; vibrosis, 615; brucellosis, 615; anaplasmosis, 615–16; internal parasites, 615–16; insects, 615–16; scours, 615; shipping fever, 615

Diseases, United States—controlled: pleuropneumonia, 609–10; tick fever (piroplasmosis, or Texas fever), 609–10; foot-and-mouth, 609–11; tuberculosis, 609, 611–12; brucellosis, 609, 612–13; screwworm, 609, 613; blackleg, 613

Dominican Republic: 7, 10, 14, 55–57; Romana Red, 16, 57, 60–63, 65; Spanish cattle, 57–58; Criollo, 57, 60–63, 65–66, 75; Zebu, 57, 61–63; Holstein-Friesian, 57, 63; Brown Swiss, 57, 63; American Brahman, 57, 60, 65; Aberdeen-Angus, 57; Zebu types, 58–59, 66; Brangus, 58; Jamaica Black, 58; Santa Gertrudis, 58; Charbray, 58, 62; Mysore, 61; Nellore, 61; Charolais, 62, 65; cattle diseases, 68–69

Durango, Mexico: Hereford, 238; Aberdeen-Angus, 238; Criollo, 238, 245; beef cattle of, 242–46; dairy cattle of, 247; *see also* Mexico

Durham: in United States, 377, 405, 500, 524; in Hawaii, 564–65; *see also* Shorthorn

Dutch Belted, in United States: 419–20

Dutch Black and White: in United States, 268, 353, 426; in Canada, 268; *see also* Holstein-Friesian

Dutch Friesian (Holland Friesian): of Surinam, 138; in United States, 407, 427, 437–39; *see also* Holstein-Friesian

El Salvador: 147, 150, 154, 169–70, 181, 193; Criollo (pure), 153, 170–71, 174, 178–79, 210; Spanish cattle, 170; Criollo (mixed), 171–72; Zebu, 170–72, 178–79; American Brahman, 172, 179; Indo-Brazil, 172; Angus, 172; Charolais, 172; Holstein-Friesian, 172, 179; Ayrshire, 172; Brown Swiss, 172, 179; cattle diseases, 177–79, 216; *see also* diseases

English Friesian: in Barbados, 18; in Trinidad, 107

Fleckvieh (German Simmental): in Canada, 269, 296; in United States, 470, 474, 627; *see also* United States, Simmental

Florida Scrub, of United States: 359, 363–65, 369–74, 376, 452, 504, 507

645

Cattle of North America

Food and Agricultural Organization (FAO) of United Nations: 51
Fort Niobrara National Wild Life Refuge, Nebraska: 368
French Canadian cattle: *see* Canadian (French Canadian) cattle
French Guiana: 10, 117–18, 135; Zebu, 119; Criollo, 119–20; Red Sindhi, 120

Galloway: in Canada, 287–88; in United States, 391–92, 483–85; in Alaska, 483–85, 555–57
Gelbvieh (German Yellow): in Canada, 269, 296; in United States, 470–71, 627
German Red and White: *see* Red and White Lowland
Gir, of United States: 401, 442, 448
Grading, beef, in Canada: 330–32
Grading, beef, in United States: USDA quality grades, 579–91; USDA yield grades, 592–600
Guatemala: 147, 150, 154, 181–82, 193, 229; Criollo (pure), 153, 182–83, 188; Zebu, 170, 182–85, 188, 190–91; Criollo (mixed), 182–84, 189–92; Spanish cattle, 182–83; American Brahman, 183–84, 189–91; Aberdeen-Angus, 184; Charolais, 184; Santa Gertrudis, 183–84, 188–91; Holstein-Friesian, 185, 188, 191; diseases, 190–91; *see also* diseases
Guernsey: on Barbados, 18; in Cuba, 27, 38–39, 51; in Jamaica, 83; in Guyana, 124; in Middle America, 152–53; in Costa Rica, 152, 158–63; of Nicaragua, 152, 211, 218; of Honduras, 198; of Panama, 222; of Canada, 266, 271–72, 278, 309–10, 320, 339; of United States, 353, 390, 412, 421–25, 430, 483; of Alaska, 483
Guyana: 10, 121–23, 125; Spanish cattle, 124; Zebu-Criollo cross, 124; Zebu, 124, 128; Criollo, 124, 127–28; American Brahman, 124, 126, 128, 132; Santa Gertrudis, 124, 126, 128, 132; Holstein-Friesian, 124–25, 129; Guernsey, 124; Jersey, 124; Hereford, 124, 128; Romana Red, 126; Criollo-Holstein, 129; cattle diseases, 128; 131; *see also* diseases
Guzerat: in Cuba, 38; in Jamaica, 85; on Trinidad, 107; of Honduras, 152, 197; in Costa Rica, 157; of Mexico, 240; of United States, 401, 442, 448, 463
Gyr: *see* Gir

Haiti: 7, 11, 14, 55, 71–73; Spanish cattle, 73–74; Criollo, 74–75; Creole, 74–76, 80; American Brahman, 75; Brown Swiss, 75–76; Jersey, 75; Holstein-Friesian, 75–76; Zebu-Creole, 77; cattle diseases, 79; *see also* diseases

Hawaii: Mexican cattle (Hawaiian wild), 562, 565; Shorthorn (Durham), 564–65; Milking Shorthorn, 564; Hereford, 564–65, 570; Angus, 564–65; Devon, 564; Holstein-Friesian, 570
Hawaiian wild cattle: 359, 374–78, 562, 565
Hays Converter: in Canada, 293–95; in United States, 470–71
Hereford: in Cuba, 27, 35, 38–39, 51; in Jamaica, 83; in Trinidad, 106; in Guyana, 124, 128; of Surinam, 138, 143; in Middle America, 153; in Mexico, 236, 238, 255; in Canada, 38, 268, 283, 287–89, 292–94, 303, 305–306, 310, 320; in United States, 268, 353, 359, 362, 365, 377–78, 381, 390, 392–96, 398, 400, 406–407, 442, 444, 446, 448–49, 451–53, 458–59, 463, 471, 483, 506, 514, 516, 518–21, 524–25, 540; of Alaska, 483, 561; of Hawaii, 564–65, 570
Highland (Scotch Highland, West Highland): in United States: 399–400; in Canada, 289–91
Hispaniola: *see* Haiti, Dominican Republic
Hissar, in Jamaica: 85
Holstein: in Cuba, 39; in Dominican Republic, 63; in Haiti, 75–76; in Jamaica, 89–90; in Canada, 278–80, 294; in United States, 426–28, 554, 623, 626; *see also* Holstein-Friesian, Dutch Friesian
Holstein-Friesian: on Barbados, 18; in Cuba, 16, 27, 36, 41, 43, 51; in Dominican Republic, 57; in Haiti, 75–76; in Jamaica, 83–84, 89–90; on Martinique, 101, 103; in Trinidad, 107, 110, 114–15; in Guyana, 124–25, 129; in Middle America, 153; in Costa Rica, 159–60, 162–63; in El Salvador, 172, 179; in Guatemala, 185, 188, 191; of Honduras, 195, 201, 204–205; of Nicaragua, 211, 218; of Panama, 222, 227; of Mexico, 185, 236, 239, 247; in Canada, 27, 268, 270, 276–80, 293, 306, 309, 311–15, 330, 339; in United States, 63, 185, 268, 270, 278–80, 353, 390, 414, 424–30, 436–39, 471, 482–83, 526, 600, 622–23; in Alaska, 483, 558–60; in Hawaii, 570; in Puerto Rico, 573; *see also* Dutch Black and White, Holstein, Dutch Friesian
Honduras: 147, 150, 154, 207, 229; Red Poll, 152, 195–98; Nellore, 152, 197; Guzerat, 152, 197; Criollo (pure), 153, 194–97, 201, 204; Spanish cattle, 194–95, 197; Zebu, 195, 201, 204; Criollo (mixed), 197–98, 203–205; American Brahman, 197–98, 203, 205; Santa Gertrudis, 198; Guernsey, 198; Jersey, 198; Brown Swiss, 198, 201, 205; Holstein-Friesian, 198, 201, 204–205; cattle diseases, 203–204; *see also* diseases
Hungarian cattle, in United States: 360

Illawara Shorthorn, of Cuba: 38

646

Index

Indo-Brazil; in Cuba, 38; in El Salvador, 172; in Mexico, 237, 239–40, 246; in United States, 400–402, 442

Jacques Cream Pot, of United States: 361
Jamaica: 7, 10, 14–18, 81–82; Jamaica Hope, 16, 83–85, 89–91, 93, 95–96; Jamaica Red, 16, 87–91, 93, 95; Jamaica Black, 16, 89; Jamaica Brahman, 16, 83, 85–87, 89–90, 93, 95; Zebu, 27, 82–83, 85–89; Criollo, 82–83; Spanish cattle, 82; Zebu-Creole, 83, 85, 89, 91; Hereford, 83; Shorthorn, 83; Aberdeen-Angus, 83, 89, 93; Devon, 83; Red Poll, 83, 87, 91, 93, 197; Ayrshire, 83; Jersey, 83–84; Holstein-Friesian, 83–84, 89–90; Sahiwal, 83–84; Mysore, 85; Nellore, 85, 197; Hissar, 85; Guzerat, 55, 197; South Devon, 87–89, 93; Charolais, 90–93; Santa Gertrudis, 90, 93; Brown Swiss, 91; cattle diseases, 95–96; *see also* diseases
Jamaica Black: in Jamaica, 16, 89; in Dominican Republic, 58
Jamaica Brahman, of Jamaica: 16, 83, 85–87, 89–90, 93, 95
Jamaica Hope, of Jamaica: 16, 83–85, 90–91, 93, 95–96
Jamaica Red Poll (Jamaica Red): in Jamaica, 16, 87–89; in Trinidad, 107–108
Jersey: on Barbados, 18; in Cuba, 27, 38–39, 51; in Haiti, 75; in Jamaica, 83–84; in Guyana, 124; of Surinam, 139; in Middle America, 152–53; in Costa Rica, 152, 158, 160; in Honduras, 198; in Nicaragua, 152, 211; in Panama, 222; in Mexico, 239, 247; in Canada, 266, 271–73, 278–81, 320, 339; in United States, 273, 280–81, 353, 362, 372, 390, 412–13, 424, 430–34, 436–39, 471, 482–83, 623; in Alaska, 483, 559; in Puerto Rico, 573

King Ranch, Texas: Santa Gertrudis, 35, 462–66; British White (Park) cattle, 350, 478–79; Charolais, 385; Africander, 478

Limousin: in Cuba, 38; in Canada, 269, 296–99, 310; in United States, 470, 473–73, 516, 626
Lincoln Red: of Canada, 296, 299–300; of United States, 470, 472; *see also* Shorthorn
Longhorn: in Canada, 268, 292, 317; in United States, 353, 357–58, 363–69, 371, 373–74, 376, 379, 381, 394, 406–407, 457, 499, 504, 519, 521; *see also* Shorthorn, Hereford
Longhorn (English), of United States: 383, 480–81

Maine-Anjou: in Canada, 269, 296, 299–302, 310; in United States, 470, 472–73
Makaweli, of Hawaii: 461
Manitoba, Canada: 264; Brown Swiss, 277;

Holstein-Friesian, 278; Hereford, 288; Highland (West Highland), 289–91; Polled Shorthorn, 292; American Brahman, 304–305
Marchigiana, of United States: 470, 473, 627
Maritime Provinces, Canada: Guernsey, 278, 309; Jersey, 278; Holstein-Friesian, 309; Ayrshire, 309; Hereford, 310; Angus, 310; Shorthorn, 310; Charolais, 310; Chianina, 310; Limousin, 310; Maine-Anjou, 310; Simmental, 310; *see also* Nova Scotia, New Brunswick, Prince Edward Island
Martinique: 11, 14, 33, 99–100; Spanish cattle, 100; Creole-Zebu, 100, 103–104; Creole, 100–102, 104; American Brahman, 101–103; Charolais, 101–103; Red Sindhi, 101; Holstein-Friesian, 101, 103; Brown Swiss, 101; Friesian (French Pie Noir), 103; cattle diseases, 100, 103–104; *see also* diseases
Meuse-Rhine-Yssel: 437; *see also* German Red and White, Red and White Lowland
Mexico: 3–10, 15, 23, 26, 55, 147, 149–50, 170, 181, 194, 229–34; Zebu, 27, 152, 234, 237, 239, 241–42, 246–47, 252, 254–55; De Lidia, 152, 237–38, 248–51; Holstein-Friesian, 185, 236, 239, 247; Charolais, 234, 236, 240–41; Spanish cattle, 234, 237; Criollo (pure), 234–35, 238, 356, 363; Criollo (mixed), 236–37, 241–42, 245–48, 251–52, 254–55; Hereford, 236, 238, 255; Aberdeen-Angus, 236, 238; Indo-Brazil, 237, 239–40, 246; American Brahman, 237, 239–40, 246–47; Brown Swiss, 239, 247; Jersey, 239, 247; Guzerat, 240; Nellore, 240; Gyr, 240; Red Sindhi, 240; Santa Gertrudis, 240; beef cattle of Sonora, Chihuahua, Durango, Zacatecas, 242; beef cattle of Coastal Plains, 246–47; cattle diseases, 238–39, 247, 249, 253–54; *see also* diseases
Milking Shorthorn: in Cuba, 38; in United States, 378–79, 434–36, 439, 483; in Alaska, 483, 558; in Hawaii, 564
Murray Grey: in Canada, 296, 303–304; in United States, 470, 473–74
Mysore: in Dominican Republic, 61; in Jamaica, 85; in Trinidad, 107

Native cattle of United States: 266–67, 352–53, 358–63, 390, 393, 440, 481, 490, 495, 505–506, 516, 524; imported to Canada, 266–67
N'Dama: in Virgin Islands, 15, 466–69, 481
Nellore: in Cuba, 38; in Dominican Republic, 61; in Jamaica, 85; in Trinidad, 107; in Costa Rica, 157; of Honduras, 152, 197; in Mexico, 240; in United States, 401, 442, 448, 463
Nelthropp: *see* Senepol

647

Cattle of North America

New Brunswick, Canada: 264; Canadian (mixed), 266; Jersey, 280–81; Red Poll, 281; Shorthorn, 291; Devon, 297–98; *see also* Maritime Provinces

Nicaragua: 147, 150–51, 154, 193–94, 207–10; Jersey, 152, 211; Guernsey, 152, 211, 218; Criollo (pure), 153, 210; Zebu, 210, 218; Brown Swiss, 210–11, 213–14, 218; American Brahman, 210, 216, 218; Criollo (mixed), 210–11, 213, 216–17; Charolais, 211; Aberdeen-Angus, 211; Holstein-Friesian, 211, 218; cattle diseases, 213, 216, 218; *see also* diseases

Normandy, cattle of, in Canada: 266, 271

Normandy (breed), in United States: 350, 353, 481

Nova Scotia, Canada: 262, 264; Canadian (mixed), 266; Guernsey, 278; *see also* Maritime Provinces

Oaxaca, Mexico: Zebu, 246; American Brahman, 246; Indo-Brazil, 246; Criollo (mixed), 246; *see also* Mexico

Old Rock Stock, of United States: 361

Ontario, Canada: 264; Ayrshire, 276–77, 312; Jersey, 276, 280–81; Holstein-Friesian, 276–78, 312–15; Brown Swiss, 277; Guernsey, 278; Friesian, 278; Red Poll, 281; Aberdeen-Angus, 283, 315; Galloway, 287–88; Shorthorn, 288, 291; Polled Hereford, 289, 315; Santa Gertrudis, 291; Devon, 297–98; Canadian, 312; Charolais, 315

Panama: 10, 147, 149–50, 154, 207, 219–22; Red Poll, 152, 222; Criollo, 157, 221; Criollo (mixed), 221–22; American Brahman, 222, 227; Santa Gertrudis, 222, 227; Charolais, 222; Brangus, 222, 224, 227; Guernsey, 222; Jersey, 222; Brown Swiss, 222, 227; Holstein-Friesian, 222, 227; Zebu, 227; cattle diseases, 226–27; *see also* diseases

Parthenay, of Canada: 296

Pie Noir, French: on Martinique, 103

Pie Rouge de l'Est (Simmental), in Canada: 269, 296–97

Pinzgauer, of Canada: 296

Polled Durham: *see* Polled Shorthorn

Polled Hereford: in Canada, 289, 305, 215; in United States, 393–99

Polled Shorthorn: in Canada, 292, 396; in United States, 408–10, 434

Prairie Provinces, Canada: Longhorn, 268, 316–17; Shorthorn, 268, 291–93, 316–17; Hereford, 268, 288, 316–18; Aberdeen-Angus, 283, 319; Highland (West Highland), 289–91; Polled Shorthorn, 292; *see also* Manitoba, Alberta, Saskatchewan

Prince cattle: 422; *see also* Guernsey

Prince Edward Island, Canada: 264; Hereford, 310; Angus, 310; Shorthorn, 310; Charolais, 310; Chianina, 310; Limousin, 310; Maine-Anjou, 310; Simmental, 310; *see also* Maritime Provinces

Puerto Rico: 7–11, 14, 16, 23–24, 32, 55, 66–67, 571; Criollo, 27, 31, 572–73; Romana Red, 61; Spanish cattle, 572; Zebu-Criollo cross, 572–73; Zebu, 572, 575; Holstein-Friesian, 573; Jersey, 573; Brown Swiss, 573; Ayrshire, 573; American Brahman, 574; Charolais, 574–75

Quebec, Canada: 262, 264; Canadian (French Canadian), 266, 271, 274–76, 311; Jersey, 271, 276, 280–81; Guernsey, 271; Ayrshire, 274–77, 311; Holstein-Friesian, 276–77, 311; Aberdeen-Angus, 282, 312; Shorthorn, 312

Red and White cattle, Dutch, of United States: 359

Red and White Dairy cattle (Red and White Holstein-Friesian), of United States: 436–39

Red and White Holstein-Friesian, of Canada: 280; *see also* Red and White Dairy cattle

Red and White Lowland: 301–302; *see also* Meuse-Rhine-Yssel

Red Angus: in Canada, 300; in United States, 381–85, 412, 437; *see also* Aberdeen-Angus

Red Brahman, in Costa Rica: 159; *see also* American Brahman

Red Brangus, of United States: 455–57; *see also* American Brahman, Aberdeen-Angus, Brangus

Red Danish: in United States, 350, 481; in Alaska, 558

Red Poll: in Cuba, 27; in Jamaica, 83, 87, 91, 93; in Trinidad, 106; of Honduras, 195–98; in Panama, 152, 222; in United States, 372, 378, 390, 396–97, 402–405, 467–69, 626

Red Rubies: *see* Devon

Red Sindhi: in Martinique, 101; in French Guiana, 120; in Mexico, 240; in United States, 350, 353, 482

Romagnola, of United States: 470, 473–74, 627

Romana Red: in Dominican Republic, 16, 57, 60–63, 65; in Guyana, 126

Rouge de l'Ouest (Red of the West): 301; *see also* Armoricaine, of France

Sabre, of United States: 461–62

Sahiwal: in Jamaica, 83–84; in Trinidad, 107, 115

Santa Gertrudis: on Barbados, 18; in Cuba, 27, 35, 43, 51; in Dominican Republic, 58; in Jamaica, 90, 93; in Guyana, 124, 126, 128, 132; of Surinam, 138; of Honduras, 198; in Mexico, 240; in Canada, 291; in United States, 362, 443, 446, 462–66, 470, 478, 516, 520, 532, 540

648

Index

Saskatchewan, Canada: 264; Native, 267; Hereford, 288; Polled Shorthorn, 292; South Devon, 304; Cattalo, 305; *see also* Prairie Provinces
Scotch Highland: *see* Highland
Senegal cattle: *see* N'Dama
Senepol (Nelthropp): on St. Croix, 16, 350, 466–70, 481; on Barbados, 18
Shorthorn: in Cuba, 35, 38; Jamaica, 83; in Canada, 268, 288–89, 291–93, 299–300, 302–303, 306, 310, 312, 320, 339; in United States, 353, 359, 361, 378, 381, 387, 390, 394–97, 405–10, 424, 434, 442, 446, 448–49, 461, 463, 473, 478, 480, 493, 500, 514, 520, 524, 626; *see also* Durham
Siberian cattle: in Alaska, 350, 482–84, 555–56
Simmental: in Canada, 269, 295–99, 310; in United States, 353, 470, 473–75, 626; *see also* Fleckvieh (German Simmental), Pie Rouge de l'Est (French Simmental)
Slesvig cattle, of United States: 360–61; *see also* Red Danish
Sonora, Mexico: Criollo (pure) (*corriente*), 234, 238; Hereford, 238; Aberdeen-Angus, 238; beef cattle, 242–46; Criollo (mixed), 245; *see also* Mexico
South Devon: in Cuba, 38; in Jamaica, 87–89, 93; in Canada, 296, 303–304; in United States, 359, 388–89, 470, 475–78
Spanish cattle: in Caribbean area, 14–15; on Barbados, 18; in Cuba, 26, 28; in Dominican Republic, 57; in Haiti, 73–74; in Jamaica, 82; on Martinique, 100; in Trinidad, 106, 108; in Guyana, 124, 127; in Surinam, 137; of Middle America, 151–52; in El Salvador, 170; in Guatemala, 182–83; of Honduras, 194–95, 197; in Mexico, 234, 237; in Canada, 266–67; in United States, 350, 355–58, 363–64, 371, 524; in Puerto Rico, 572; *see also* Criollo
Surinam: 10, 117, 121, 135–37, 144; water buffalo, 137, 139, 141; Spanish cattle, 137; Creole, 137–39, 143; Zebu, 138; Dutch Friesian, 138–39, 143; Santa Gertrudis, 138; American Brahman, 138–39, 143; Hereford, 38, 143; Charolais, 138; Jersey, 139; cattle diseases, 142–43; *see also* diseases
Sussex cattle, in United States: 410–12
Swiss Brown, of Switzerland: *see* Brown Swiss

Tarantaise, of Canada: 296
Tasmania Grey: *see* Murray Grey
Tinima, of Cuba: 31
Tobago: *see* Trinidad
Trinidad 7, 10, 14–15, 105–106; Spanish cattle, 106, 108; Zebu, 106–107, 114–15; Creole, 106–108, 114, 356; Red Poll, 106, 222; Hereford, 106; Ayrshire, 106; Nellore, 107; Mysore, 107; Guzerat, 107; Sahiwal, 107, 115; Friesian, 107; Holstein-Friesian, 107, 110, 114–15; Jamaica Red, 107–108; American Brahman, 107–108; Charolais, 107–108; Charbray, 107–108; water buffalo, 107, 111, 114, 116; Murrah buffalo, 107–108; Nihli buffalo, 108; swamp buffalo, 111–12; cattle diseases, 114–15; *see also* diseases
United States: bison (American buffalo), 348, 357–59; Zebu, 350, 358, 374, 387, 400–402, 439–43, 446, 448–49, 454–55, 461, 463–65, 469, 478, 482, 506, 515–16, 519–20, 531–32; N'Dama, 350, 466–69, 481; Africander, 350, 443–46; Senepol (Nelthropp), 350, 466–70, 481; Canadian (French Canadian), 350, 352–53, 479–80; Red Sindhi, 350, 353, 482; Red Danish, 350, 481; Yak, 350, 484–85; Siberian cattle, 350, 482–84; British White (Park) cattle, 350, 478–79; Charolais, 34, 185, 268, 270, 278–80, 350–57, 385–88, 405, 460–61, 463, 473, 514, 516, 518, 520–21, 525, 543; Hereford, 268, 353, 359, 362, 365, 377–78, 381, 390, 392–96, 398, 400, 406–407, 442, 444, 446, 448–49, 451–53, 458–59, 463, 471, 483, 500, 506, 514, 516, 518–21, 524–25, 540; Aberdeen-Angus, 268, 353, 362, 377, 379–82, 384, 390–92, 394–97, 443, 446, 454–56, 473, 483, 514, 518, 520–21, 525, 540–41, 626; Ayrshire, 353, 372, 390, 412–14; Brown Swiss, 353, 414–17, 470–71, 482; Guernsey, 353, 390, 412, 421–25, 430, 483; Dutch Black and White, 268, 353, 426; Holstein-Friesian, 63, 185, 268, 270, 278–80, 353, 390, 414, 424–30, 436–39, 471, 482–83, 526, 622–23; Jersey, 273, 280–81, 353, 362, 372, 390, 412–13, 424, 430–34, 474–75, 482–83, 623; Simmental, 353, 470, 473–75, 626–27; Normandy, 266, 271, 350, 353; Devon, 359, 372, 388–91, 461, 626; South Devon, 359, 388–89, 470, 475–78; Red Poll, 378, 390, 396–97, 402–405, 467–69, 626; Durham, 377, 405, 524; Milking Shorthorn, 378–79, 434–36, 439, 483; Red Angus, 381–85, 412, 437; Longhorn (English), 383, 480–81; Galloway, 391–92, 483–85; Belted Galloway, 392, 420; Polled Hereford, 393–99; Highland (Scotch Highland), 399–400; Indo-Brazil, 400–402, 442; Guzerat, 401, 442, 448, 463; Gir, 401, 442, 448; Nellore, 401, 442, 448, 463; Polled Shorthorn, 408–10, 434; Sussex, 410–12, 461–62; Dexter, 417–19; Dutch Belted, 419–20; Holstein, 426–28, 554, 623, 626; Dutch Friesian, 407, 427, 437–39; Red and White Dairy cattle (Red and White Holstein), 436–39; Fleckvieh (German Simmental), 470, 474, 627; Limousin, 470–73, 516, 626; Maine-Anjou, 470, 472–73; Murray Grey, 470, 473–74; Lincoln Red, 470, 472; Welsh Black, 470, 478; Hays Converter, 470–71; Gelbvieh (German Yellow), 470–71, 627; Blonde

649

Cattle of North America

d'Aquitaine, 470–71; Chianina, 470–71, 473, 627; Marchigiana, 470, 473, 627; Romagnola, 470, 473–74, 627; Normandy, 481; *see also* United States, early cattle of; United States, breeds developed in

United States, breeds developed in: Barzona, 350, 443–47, 550; American Brahman, 362, 387, 401–402, 412, 439–43, 448, 451–57, 460–63, 506, 516, 519–20, 532, 540; Brangus, 362, 454–56, 516, 532, 540; Santa Gertrudis, 362, 443, 446, 462–66, 470, 478, 516, 520, 532, 540; Charbray, 387–88, 460–61; Beefmaster, 447–51; Braford (Victoria), 451–54, 516, 532; Red Brangus, 455–57; Cattalo, 457–60; Makaweli, 461; Sabre, 461–62; Senepol, 466–70; *see also* United States; United States, early cattle of

United States, early cattle of: Native, 266–67, 352–53, 358–63, 390, 393, 440, 481, 490, 495, 505–506, 516, 524; Spanish cattle, 350, 355–58, 363–64, 371, 524; Longhorn, 353, 357–58, 363–69, 371, 373–74, 376, 379, 381, 394, 406–407, 442, 459, 499, 504, 519, 521; Shorthorn, 353, 359, 361, 378, 381, 387, 390, 394–97, 405–10, 424, 434, 442, 446, 448–49, 461, 463, 473, 478, 480, 493, 514, 524, 626; Florida Scrub, 359, 363–65, 369–74, 376, 452, 500, 504, 507, 520; Hawaiian wild, 359, 374–78; Devonshire, 359–61, 388–89; Black and White Lowland, 359; Red and White (Dutch), 359; Angel, 360–61; Slesvig, 360–61; Hungarian, 360; Old Rock Stock, 361; Cream Pot, 361; Jacques Cream Pot, 361; *see also* United States; United States, cattle developed in

Veracruz, Mexico: Zebu, 246–47; American Brahman, 246–47; Indo-Brazil, 246–47; Criollo (mixed), 246–47; *see also* Mexico

Victoria cattle: 452; *see also* Braford

Virgin Islands: 7; N'Dama, 15; Senepol cattle, 16, 466–69

Welsh Black: of Canada, 296, 304; of United States, 470–78

West Highland: *see* Highland

West Indies, cattle of: 15–16

Wichita Mountains Wild Life Refuge (Cache, Okla.): 367–68, 374

Yak: in Canada, 306–307; in Alaska, 350, 484–85

Yellow cattle: *see* Angel, Slesvig

Zacatecas, Mexico: Hereford, 239; Holstein-Friesian, 239; Brown Swiss, 239; Jersey, 239; beef cattle of, 242–46; Criollo, 245; *see also* Mexico

Zebu: 15–16; in Cuba, 16, 26–29, 31–32, 38, 40; in Dominican Republic, 57–59, 61–63; in Jamaica, 27, 82–89; on Trinidad, 106–107, 114–15; on French Guiana, 119; in Guyana, 124, 128; in Surinam, 138; of Middle America, 152–54; in El Salvador, 170–72, 178–79; in Guatemala, 170, 182–85, 188, 190–91; of Honduras, 195, 201, 204; of Nicaragua, 210, 218; of Panama, 227; of Mexico, 27, 152, 234, 237, 239, 241–42, 246–47, 252, 254–55; in United States, 350, 358, 374, 387, 400–402, 439–43, 446, 448–49, 454–55, 461, 463–65, 469, 478, 482, 506, 515–16, 519–20, 531–32; in Puerto Rico, 572, 575

Zebu-Angus cross: *see* Jamaica Black, Brangus

Zebu-Creole cross: on Barbados, 18, 27; Cuba, 27; in Dominican Republic, 60, 65–66; in Haiti, 77; in Jamaica, 83, 85, 89, 91; on Martinique, 100, 103–104; in Guyana, 124; in El Salvador, 171–72; of Honduras, 197–98, 203–205; of Nicaragua, 210–11, 213, 216–17; of Panama, 221–22; of Mexico, 236–37, 241–42, 245–48, 251–52, 254–55; of Puerto Rico, 572–73

650

A000019606507